Calcium Signaling: From Physiology to Diseases

Calcium Signaling: From Physiology to Diseases

Editor: Roman Brown

MURPHY & MOORE
www.murphy-moorepublishing.com

www.murphy-moorepublishing.com

MURPHY & MOORE

Cataloging-in-Publication Data

Calcium signaling : from physiology to diseases / edited by Roman Brown.
p. cm.
Includes bibliographical references and index.
ISBN 978-1-63987-091-2
1. Calcium--Physiological effect. 2. Cellular signal transduction. 3. Intracellular calcium.
4. Calcium in the body. 5. Calcium channels. I. Brown, Roman.
QP535.C2 C35 2022
572.516--dc23

Murphy & Moore Publishing,
1 Rockefeller Plaza,
New York City, NY 10020, USA

ISBN 978-1-63987-091-2 (Hardback)

Contents

Preface

This book was inspired by the evolution of our times; to answer the curiosity of inquisitive minds. Many developments have occurred across the globe in the recent past which has transformed the progress in the field.

The usage of calcium ions to drive intracellular processes and for communication is called calcium signaling. It is generally a step in signal transduction. Calcium ions can also function as second messengers wherein they fulfill tasks such as neuronal transmission, cellular motility, muscle contraction and cell growth. Within neurons, concomitant rise in mitochondrial and cytosolic calcium ions are integral to the synchronization of mitochondrial energy metabolism and neuronal electrical activity. However, high level of cytoplasmic calcium ions can cause the cell apostosis. The influx of calcium ions also triggers the development of the oocyte in numerous species. In the case of fish and echinoderms, this influx takes place as a single rise in concentration, while in mammals, the concentration oscillates. This book provides comprehensive insights into the research on calcium signaling. It presents researches and studies performed by experts across the globe. This book will serve as a reference to a broad spectrum of readers.

This book was developed from a mere concept to drafts to chapters and finally compiled together as a complete text to benefit the readers across all nations. To ensure the quality of the content we instilled two significant steps in our procedure. The first was to appoint an editorial team that would verify the data and statistics provided in the book and also select the most appropriate and valuable contributions from the plentiful contributions we received from authors worldwide. The next step was to appoint an expert of the topic as the Editor-in-Chief, who would head the project and finally make the necessary amendments and modifications to make the text reader-friendly. I was then commissioned to examine all the material to present the topics in the most comprehensible and productive format.

I would like to take this opportunity to thank all the contributing authors who were supportive enough to contribute their time and knowledge to this project. I also wish to convey my regards to my family who have been extremely supportive during the entire project.

Editor

1

Endothelial Ca^{2+} Signaling and the Resistance to Anticancer Treatments: Partners in Crime

Francesco Moccia

Laboratory of General Physiology, Department of Biology and Biotechnology "L. Spallanzani", University of Pavia, I-27100 Pavia, Italy; francesco.moccia@unipv.it

Abstract: Intracellular Ca^{2+} signaling drives angiogenesis and vasculogenesis by stimulating proliferation, migration, and tube formation in both vascular endothelial cells and endothelial colony forming cells (ECFCs), which represent the only endothelial precursor truly belonging to the endothelial phenotype. In addition, local Ca^{2+} signals at the endoplasmic reticulum (ER)–mitochondria interface regulate endothelial cell fate by stimulating survival or apoptosis depending on the extent of the mitochondrial Ca^{2+} increase. The present article aims at describing how remodeling of the endothelial Ca^{2+} toolkit contributes to establish intrinsic or acquired resistance to standard anti-cancer therapies. The endothelial Ca^{2+} toolkit undergoes a major alteration in tumor endothelial cells and tumor-associated ECFCs. These include changes in TRPV4 expression and increase in the expression of P2X7 receptors, Piezo2, Stim1, Orai1, TRPC1, TRPC5, Connexin 40 and dysregulation of the ER Ca^{2+} handling machinery. Additionally, remodeling of the endothelial Ca^{2+} toolkit could involve nicotinic acetylcholine receptors, gasotransmitters-gated channels, two-pore channels and Na^+/H^+ exchanger. Targeting the endothelial Ca^{2+} toolkit could represent an alternative adjuvant therapy to circumvent patients' resistance to current anti-cancer treatments.

Keywords: Ca^{2+} signaling; tumor; endothelial cells; endothelial progenitor cells; endothelial colony forming cells; anticancer therapies; VEGF; resistance to apoptosis

---

1. Introduction

An increase in intracellular Ca^{2+} concentration ($[Ca^{2+}]_i$) has long been known to play a crucial role in angiogenesis and arterial remodeling [1–5]. Accordingly, growth factors and cytokines, such as vascular endothelial growth factor (VEGF), epidermal growth factor (EGF), basic fibroblast growth factor (bFGF), insulin-like growth factor-1 (IGF-1), angiopoietin and stromal derived factor-1α (SDF-1α), trigger robust Ca^{2+} signals in vascular endothelial cells [6–12], which recruit a number of downstream Ca^{2+}-dependent pro-angiogenic decoders. These include, but are not limited to, the transcription factors, Nuclear factor of activated T-cells (NFAT), Nuclear factor-kappaB (NF-κB) and cAMP responsive element binding protein (CREB) [8,13,14], myosin light chain kinase (MLCK) and myosin 2 [8,15], endothelial nitric oxide synthase (eNOS) [16,17], extracellular signal–regulated kinases $\frac{1}{2}$ (ERK 1/2) [18,19] and Akt [19,20]. Not surprisingly, therefore, subsequent studies clearly revealed that endothelial Ca^{2+} signals may also drive tumor angiogenesis, growth and metastasis [3,21–24]. However, the process of tumor vascularization is far more complex than originally envisaged [25]. Accordingly, the angiogenic switch, which is the initial step in the multistep process that ensures cancer cells with an adequate supply of oxygen and nutrients and provides them with an escape route to enter peripheral circulation, is triggered by the recruitment of bone marrow-derived endothelial progenitor cells (EPCs), according to a process termed vasculogenesis [26–28]. Similar to mature endothelial cells, EPCs require an increase in $[Ca^{2+}]_i$ to proliferate, assembly into capillary-like tubular networks in vitro and form patent neovessels in vivo [29–31]. Of note, intracellular Ca^{2+} signals finely regulate proliferation and in vitro tubulogenesis also in tumor-derived EPCs (T-EPCs) [23,32,33]. An established tenet of neoplastic transformation is the remodeling of the Ca^{2+} machinery in malignant cells, which contributes to the distinct hallmarks of cancer described by Hanahan and Weinberg [34–36]. Tumor endothelial cells (T-ECs) and T-EPCs do not derive from the malignant clone, but they display a dramatic dysregulation of their Ca^{2+} signaling toolkit [29,32,37]. The present article surveys the most recent updates on the remodeling of endothelial Ca^{2+} signals during tumor vascularization. In particular, it has been outlined which Ca^{2+}-permeable channels and Ca^{2+}-transporting systems are up- or down-regulated in T-ECs and T-EPCs and how they impact on neovessel formation and/or apoptosis resistance in the presence of anti-cancer drugs. Finally, the hypothesis that the remodeling of endothelial Ca^{2+} signals may be deeply involved in tumor resistance to standard therapeutic treatments, including chemotherapy, radiotherapy and anti-angiogenic therapy is widely discussed.

2. Ca^{2+} Signaling in Normal Endothelial Cells: A Brief Introduction

The resting $[Ca^{2+}]_i$ in vascular endothelial cells is set at around 100–200 nM by the concerted interaction of three Ca^{2+}-transporting systems, which extrude Ca^{2+} across the plasma membrane, such as the Plasma-Membrane Ca^{2+}-ATPase and the Na^+/Ca^{2+} exchanger (NCX), or sequester cytosolic Ca^{2+} into the endoplasmic reticulum (ER), the largest intracellular Ca^{2+} reservoir [2,38–40], such as the SarcoEndoplasmic Reticulum Ca^{2+}-ATPase (SERCA). Endothelial cells lie at the interface between the vascular wall and the underlying tissue; therefore, they are continuously exposed to a myriad of low levels soluble factors, including growth factors, hormones and transmitters, which may induce highly localized events of inositol-1,4,5-trisphosphate ($InsP_3$)-dependent Ca^{2+} release from the ER even in the absence of global cytosolic elevations in $[Ca^{2+}]_i$ [41–45]. These spontaneous $InsP_3$-dependent Ca^{2+} microdomains are redirected towards the mitochondrial matrix through the direct physical association specific components of the outer mitochondrial membrane (OMM) with specialized ER regions, which are known as mitochondrial-associated membranes (MAMs) [46]. This constitutive ER-to-mitochondria Ca^{2+} shuttle drives cellular bioenergetics by activating intramitochondrial Ca^{2+}-dependent dehydrogenases, such as pyruvate dehydrogenase, NAD-isocitrate dehydrogenase and oxoglutarate dehydrogenase [47–49]. This pro-survival Ca^{2+} transfer may be switched into a pro-death Ca^{2+} signal by various apoptotic stimuli [46,47,50]. For instance, hydrogen peroxide (H_2O_2), menadione, resveratrol, ceramide, and etoposide boost the $InsP_3$-dependent ER-to-mitochondria Ca^{2+} communication, thereby causing a massive increase in mitochondrial Ca^{2+} concentration ($[Ca^{2+}]_{mit}$), which ultimately results in the opening of mitochondrial permeability transition pore and in the release of pro-apoptotic factors into the cytosol [46,51–53]. The hypoxic microenvironment of a growing tumor may then trigger an oxygen (O_2)-sensitive transcriptional program in tumor cells by activating two basic helix-loop-helix transcription factors, i.e., the hypoxia-inducible factors HIF-1 and HIF-2, which drive the expression of a myriad of growth factors and cytokines [54]. These include, but are not limited to, VEGF, EGF, bFGF, IGF-1, angiopoietin and SDF-1α [27,54], which are liberated into peripheral circulation according to a concentration gradient, which delivers a strong pro-angiogenic signal to vascular endothelial cells residing in close proximity to the primary tumor site [3,33]. Growth factors bind to their specific tyrosine kinase receptors (TKRs), such as VEGFR-2 (KDR/Flk-1), EGFR (ErbB-1), and IGF-1R, thereby stimulating phospholipase Cγ (PLCγ) to cleave phosphatidylinositol 4,5-bisphosphate (PIP_2) into the two intracellular second messengers, $InsP_3$ and diacylglycerol (DAG) [1,2,55]. The following increase in cytosolic $InsP_3$ levels further stimulates ER-dependent Ca^{2+} release through $InsP_3$ receptors ($InsP_3Rs$), which can be amplified by the recruitment of adjoining ryanodine receptors (RyRs) through the process of Ca^{2+}-induced Ca^{2+} release (CICR) [1,2]. The following drop in ER Ca^{2+} concentration ($[Ca^{2+}]_{ER}$) is detected by Stromal Interaction Molecule 1 (Stim1), a sensor of ER Ca^{2+} levels, which is prompted to aggregate into oligomers and relocate towards ER-plasma membrane junctions, known as puncta and positioned in close vicinity to the plasma membrane (10–20 nm). Herein, Stim1 interacts with and gates the Ca^{2+}-permeable channel, Orai1, thereby triggering the so-called store-operated Ca^{2+} entry (SOCE), the most important Ca^{2+} entry route in endothelial cells [42,56–59]. In addition, Stim1 may recruit additional Ca^{2+}-permeable channels, which belong to the Canonical Transient Receptor Potential (TRPC) sub-family [2,59]. Accordingly, the TRP superfamily of cation channels comprises 28 members, subdivided into six sub-families: TRPC, TRPV (Vanilloid), TRPM (Melastatin), TRPP (Polycystin), TRPML (Mucolipin) and TRPA (Ankyrin) based on the homology of their amino acid sequences [60]. More specifically, endothelial SOCE could involve TRPC1 and TRPC4, which are recruited by Stim1 into a supermolecular heteromeric complex [61], whose Ca^{2+} selectivity is determined by Orai1 [62,63]. Moreover, TRPC3 and TRPC6 may mediate DAG-induced Ca^{2+} entry in several types of endothelial cells [64,65]. This toolkit of Ca^{2+} release/entry channels may be differently exploited by growth factors to stimulate angiogenesis by eliciting diverse patterns of Ca^{2+} signals depending on the vascular bed. For instance, VEGF triggers a biphasic increase in $[Ca^{2+}]_i$ in human umbilical vein endothelial cells (HUVECs), which consists in an initial $InsP_3$-dependent Ca^{2+} peak followed by a plateau phase of intermediate amplitude due to SOCE activation [57,58]. Likewise, $InsP_3$ and SOCE shape VEGF- and EGF-induced intracellular Ca^{2+} oscillations in sheep uterine artery endothelial cells [66] and in rat microvascular endothelial cells (CMECs) [7], respectively. VEGF-induced Ca^{2+} influx in HUVECs may, however, be sustained by TRPC3, which causes Na^+ accumulation beneath the plasma membrane and stimulates the forward (i.e., Ca^{2+} entry) mode of NCX [18]. Moreover, the DAG-gated channel, TRPC6, underlies the monotonic increase in $[Ca^{2+}]_i$ induced by VEGF in human dermal microvascular endothelial cells (HDMECs) [67]. Finally, TRPC1 is engaged by bFGF to mediate Ca^{2+} entry in HDMECs [68]. These data have been recently confirmed by directly monitoring angiogenesis in developing zebrafish; this model showed that VEGF stimulated biphasic Ca^{2+} signals to drive migration in stalk cells and intracellular Ca^{2+} oscillations to promote proliferation in tip cells [8]. Besides growth factors-activated channels, vascular endothelial cells dispose of a larger toolkit of plasmalemmal Ca^{2+}-permeable channels that can be recruited by a multitude of chemical and physical stimuli [2]. For instance, endothelial Ca^{2+} entry may be mediated by additional intracellular second messengers, such as arachidonic acid (AA) and AA metabolites, i.e., epoxyeicosatrienoic acids (EETs) and 2-arachidonoylglycerol, which activate TRPV4 [69,70]; NO, which gates TRPC5 [71]; adenosine 5′-diphosphoribose (ADPR) and low micromolorar doses of H_2O_2, which converge on TRPM2 activation [72]; and cyclic nucleotides [73]. Moreover, vascular endothelial cells are endowed with several Ca^{2+}-permeable ionotropic receptors, including ATP-sensitive P_{2X} receptors [74], acetylcholine-sensitive nicotinic receptors [75], and *N*-methyl-D-aspartate (NMDA) receptors [76]. Finally, mechanical stimuli (e.g., laminar shear stress, pulsatile stretch, and changes in the local osmotic pressure) elicit Ca^{2+} influx by recruiting a variety of mechano-sensitive channels, such as TRPP2 [77], heteromeric TRPC1-TRPP2 [78], TRPV4 [79], TRPC1-TRPP2-TRPV4 [80], and Piezo1 [81]. Recently, the Ca^{2+} toolkit has also been explored in human EPCs [29]; most of the work has been carried out in endothelial colony forming cells (ECFCs), which represent the only EPC subset truly belonging to the endothelial, rather than the myeloid, lineage [82]. VEGF triggers pro-angiogenic intracellular Ca^{2+} oscillations in ECFCs by triggering the interaction between $InsP_3$-dependent Ca^{2+} release and SOCE, which is mediated by Stim1, Orai1 and TRPC1 [83,84]. Conversely, RyRs and the DAG-sensitive channels, TRPC3 and TRPC6, are absent and do not contribute to Ca^{2+} signaling [84–86]. Of note, AA may promote proliferation by directly activating TRPV4 and inducing NO release in the presence of extracellular growth factors and cytokines [86]. Finally, the $InsP_3$-dependent ER-to-mitochondria Ca^{2+} shuttle is at work and finely regulates the sensitivity to apoptotic stimuli in ECFCs, too [87].

Herein, the mechanisms whereby the remodeling of the endothelial transportome, i.e., the specific arsenal of ion channels and transporters expressed by vascular endothelial cells and ECFCs, confers resistance to anti-cancer therapies have been subdivided into two main categories: (1) enhanced neovascularization, which attenuates the therapeutic outcome of anticancer treatments by nourishing cancer cells with O_2 and nutrients and removing their catabolic waste, and further provides them with a direct access to peripheral circulation, thereby favoring metastasis (Figure 1 and Table 1); and (2) resistance to apoptosis, which hampers the cellular stress induced by chemo- and radiotherapy on tumor endothelial cells and interferes with the dismantling of cancer vasculature (Figure 2 and Table 2).

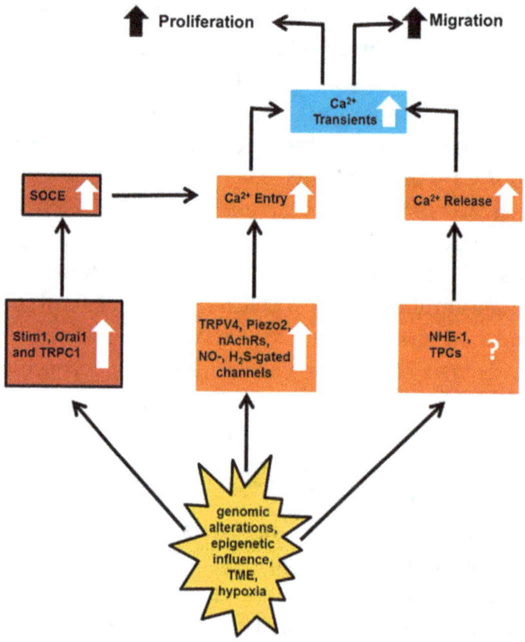

Figure 1. The endothelial Ca^{2+} transportome is remodeled to sustain tumor vascularization. The equence of events is illustrated by the black arrows. Upward arrows indicate the over-expression of a specific Ca^{2+}-permeable channel or transporter and the stimulation of a precise cellular process. See the text for further details.

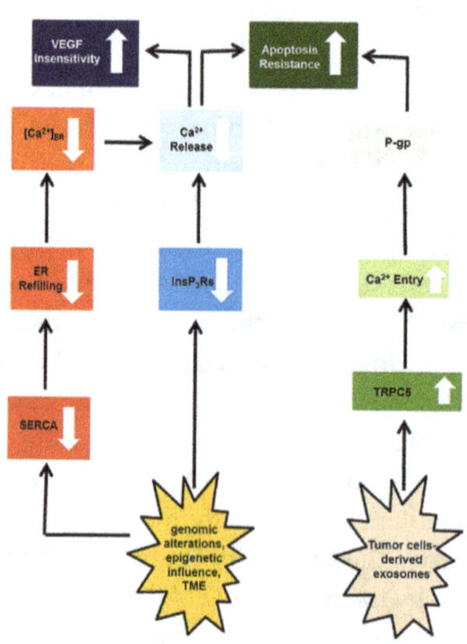

Figure 2. The endothelial Ca^{2+} transportome is remodeled to promote tumor endothelial cell resistance to apoptosis. The sequence of events is illustrated by the black arrows. Downward arrows indicate the down-regulation of a specific Ca^{2+}-permeable channel/transporter or of a precise cellular process. Upward arrows indicate the over-expression of a specific Ca^{2+}-permeable channel or the stimulation of a precise cellular process.

Table 1. Channels and transporters directly supporting tumor vascularization.

Channel/ Transporter	Tumor and Cell Type (T-EC, T-ECFC T-EPC)	Expression Levels (Transcripts and/or Proteins)	Effect on Tumor Vascularization	Strategy to Target Tumor Vascularization	Ref.
TRPV4	Breast Cancer: T-ECs	↑	Stimulates B-TEC proliferation, migration and in vitro tubulogenesis	Channel blockade with shTRPV4 or with CAI (0.1–10 μM)	[88,89]
TRPV4	Lewis Lung Carcinoma: T-ECs (isolated from prostate adenocarcinoma)	↓	Inhibits T-EC mechanosensation, proliferation and migration in vitro and promotes the formation of a malfunctioning, leaky and exceedingly expanded vascular network in vivo	Injection of TRPV4 agonist GSK (10 μg/kg) to normalize tumor vasculature and favor cisplatin-induced tumor regression	[90–92]
Piezo2 proteins	Glioma: T-ECs	↑	Regulates tumor angiogenesis, vascular leakage and permeability	Blockade with siPiezo2	[93]
P2X7Rs	Breast cancer: T-ECs	↑	Inhibits B-TEC migration and normalizes B-TECs-derived vessels in vitro	Activated by BzATP (50 μM)	[94]
Stim1, Orai1, TRPC1	Renal cellular carcinoma: T-ECFCs	↑	Stimulate T-EPC proliferation and in vitro tubulogenesis	Blockade with siStim1 and siOrai1 and with YM-58483/BTP2 (20 μM), La^{3+} (10 μM), Gd^{3+} (10 μM), CAI (2–10 μM), 2-APB (50 μM), and genistein (50 μM)	[95]
Stim1, Orai1, TRPC1	Breast cancer: T-ECFCs	=	Control T-ECFC proliferation and in vitro tubulogenesis	Blockade with YM-58483/BTP2 (20 μM), La^{3+} (10 μM), and CAI (10 μM)	[96]
Stim1, Orai1, TRPC1	Infantile hemangioma: T-ECFCs	↑	Control T-ECFCs proliferation in vitro	Blockade with with YM-58483/BTP2 (20 μM), La^{3+} (10 μM), and Pyr6 (10 μM)	[97]
α7-nAchRs	Lewis lung carcinoma: T-ECs and T-EPCs	Not determined	Controls tumor growth and angiogenesis in vivo	Blockade with mecamylamine (1.0 μg/kg) or hexamethonium (1.0 μg/kg)	[98,99]
			Stimulates EPC proliferation, migration and tubulogenesis in vitro and EPC recruitment in vivo	Blockade in vitro with mecamylamine (1 μM) and α-bungarotoxin (10 nM) and in vivo with mecamylamine (0.24 mg/kg per day)	[100,101]
Connexin40	Melanoma and urogenital cancers: T-EC	↑	Stimulates tumor angiogenesis and growth in vivo	Blockade in vivo with $^{40}Gap^{27}$ peptide (100 μg)	[102]
NHE-1	Breast cancer: TECs	Not determined	Stimulates B-TEC migration in vitro	Blocked with siNHE-1 and with cariporide (50 μM)	[103]

The generic term EPC, in this context, refers to circulating pro-angiogenic cells which cannot be grouped into the ECFC sub-family and are likely to belong to the myeloid lineage.

Table 2. Components of the endothelial Ca^{2+} toolkit that determine endothelial cell resistance to chemotherapeutic drugs.

Channel/ Transporter	Tumor and Cell Type (T-EC and T-EPC)	Expression Levels	Effect on Tumor Vascularization	Strategy to Target Tumor Vascularization	Ref.
TRPC5	Breast Cancer: T-ECs	↑	Stimulates endothelial resistance to adriamycin	Channel blockade with the specific blocking antibody T5E3 (concentration not reported)	[104]
$InsP_3Rs$	RCC: T-ECFCs	↓	Favor T-ECFC resistance to rapamycin	Preventing $InsP_3$-dependent ER–mitochondria Ca^{2+} shuttle with selective $InsP_3R$ inhibitors or cytosolic Ca^{2+} buffers (e.g., BAPTA)	[87]

3. Enhanced Neovascularization

3.1. Vanilloid Transient Receptor Potential 4 (TRPV4)

TRPV4 has been the first endothelial Ca^{2+}-permeable channel to be clearly involved in tumor vascularization [88]. TRPV4 is gated by an array of chemical and physical cues and represents, therefore, the archetypal of polymodal TRP channels [60]. For instance, TRPV4 may be activated by physiological stimuli, including AA and its cytochrome P450-derived metabolites mediators, i.e., EETs, acidic pH, hypotonic swelling, mechanical deformation, heat (>17–24 °C), and dimethylallyl pyrophosphate (DMAPP) [105,106]. Furthermore, TRPV4-mediated Ca^{2+} entry is elicited by manifold synthetic compounds, including the α-phorbol esters, phorbol 12-myristate 13-acetate (PMA) and 4α -phorbol 12,13-didecanoate (4α PDD), and the small molecule drugs, GSK1016790A (GSK) and JNc-440 [60,107]. TRPV4 has long been known to stimulate angiogenesis and arteriogenesis [4,5,108] by stimulating endothelial cell proliferation [5,109] and migration [110]. TRPV4-mediated Ca^{2+} entry is translated into a pro-angiogenic signal by several decoders, such as the Ca^{2+}-dependent transcription factors NFAT cytoplasmic 1 (NFATc1), myocyte enhancer factor 2C (MEF2C), and Kv channel interacting protein 3, calsenilin (KCNIP3/CSEN/DREAM), which drive endothelial cell proliferation, [4], β1-integrin and phosphatidylinositol 3-kinase (PI3-K), which promote endothelial cell motility [111]. The opening of only few TRPV4 channels, that tend to assemble into a four-channel cluster, results in spatially-restricted cytosolic Ca^{2+} microdomains, known as Ca^{2+} sparklets, which selectively recruit the downstream Ca^{2+}-dependent effectors [112,113]. A recent study revealed that TRPV4 was dramatically up-regulated in breast tumor-derived endothelial cells (B-TECs) and that TRPV4-mediated Ca^{2+} entry significantly increased the rate of cell migration as compared to control cells [88]. TRPV4 promoted B-TEC motility by eliciting local Ca^{2+} pulses at the leading edge of migrating cells [88], which were reminiscent of TRPV4-dependent Ca^{2+} sparklets [112]. TRPV4 was physiologically gated by AA [89], which is quite abundant in breast cancer microenvironment [114]. Likewise, cytosolic phospholipase A2 (PLA2), which cleaves AA from membrane phospholipids in response to physiological stimuli [115] is up-regulated and promotes cancer development by stimulating angiogenesis in several types of tumors, including breast cancer [116]. Therefore, TRPV4 might represent a novel and specific target to treat breast cancer as it is only barely expressed and does not drive migration in healthy endothelial cells [88].

Subsequently, the role of TRPV4 was investigated in prostate adenocarcinoma-derived endothelial cells (A-TECs). Unlike B-TECs, TRPV4 was down-regulated in A-TECs, which increased their sensitivity towards extracellular matrix stiffness, boosted their migration rate and favored the development of an aberrant (i.e., non-uniform, abnormally dilated and leaky) tumor vascular network [90]. This feature gains therapeutic relevance as the resultant hostile (i.e., low extracellular pH, hypoxia, and high interstitial pressure) microenvironment fuels tumor progression and hampers the efficacy of chemotherapy, radiation therapy, anti-angiogenic therapy immunotherapy [117,118]. Accordingly, overexpression or pharmacological activation of TRPV4 with GSK restored A-TEC mechanosensitivity and normalized their abnormal tube formation in vitro by inhibiting enhanced basal Rho activity [91]. Moreover, the daily intraperitoneal injection of GSK was able to normalize tumor vasculature in a xenograft mouse model of Lewis Lung Carcinoma (LLC), thereby improving cisplatin delivery and causing significant tumor shrinkage [91]. In addition, TRPV4-mediated Ca^{2+} entry reduced A-TEC proliferation in vitro by inhibiting the extracellular signal-regulated kinases 1/2 [92]. This mechanism further contributes to GSK-induced dismantling of LLC vasculature in vivo [92]. Therefore, remodeling of TRPV4-mediated Ca^{2+} entry may be used to effectively target tumor vascularization, although the most effective approach may depend on the tumor type. Accordingly, TRPV4 should be inhibited to halt tumor vascularization in breast cancer, while it must be stimulated to normalize tumor vasculature in LLC [23].

3.2. Piezo Proteins

Piezol and Piezo2 proteins are two recently identified non-selective cation channels that mediate mechanosensory transduction in mammalian cells [119,120]. Piezo proteins are gigantic homotetrameric complexes endowed with one or four ion-conducting pores: each subunit comprises over 2500 amino acids and presents 24–40 predicted transmembrane domains [119]. Piezo channels are Ca^{2+}-permeable and, therefore, lead to robust Ca^{2+} entry in response to mechanical deformation of the plasma membrane; unlike TRPV4 channels [105], Piezo-mediated Ca^{2+} entry is directly activated by tension within the lipid bilayer of the plasma membrane rather than by physical coupling to the sub-membranal cytoskeleton or intracellular second messengers [120]. A recent study demonstrated that the endothelial Piezol was activated by laminal shear stress to drive embryonic vascular development [81]. Piezol promoted vascular endothelial cell migration, alignment and re-alignment along the direction of blood flow by engaging the Ca^{2+}-dependent decoders, eNOS and calpain [81,121]. More recently, Piezo2 was found to be up-regulated in T-ECs from mouse xenografted with GL261 glioma cells [93]. Knocking down Piezo2 with a selective small interfering RNA (siRNA) reduced glioma angiogenesis and normalized tumor neovessels [93]. Moreover, suppressing Piezo2 expression decreased VEGF- and interleukin-1β-induced angiogenesis in the mouse corneal neovascularization model [93]. Finally, Piezo2-mediated Ca^{2+} entry elicited the Ca^{2+}-dependent transcription of Wnt11 and, consequently, the nuclear translocation of β-catenin in HUVECs, thereby promoting their angiogenic activity in vitro [93]. Although this mechanism remains to be confirmed in T-ECs, Piezo2 stands out as a crucial regulator of tumor angiogenesis and should be probed as a novel target for more effective anti-cancer treatments.

3.3. P2X7 Receptors

ATP and its metabolite, adenosine, are major constituents of tumor microenvironment and may differently affect tumor growth, immune cells and tumor-host interaction by activating a wealth of metabotropic (i.e., P2Y1, P2Y2, P2Y4, P2Y6, P2Y11, P2Y12, P2Y13, and P2Y14) and ionotropic (P2X1–2X8,) receptors [122]. Of note, ATP has long been known to stimulate angiogenesis though metabotropic P2y receptors [123]. Nevertheless, a recent investigation demonstrated that P2X7 stimulates tumor angiogenesis in vivo. Two different tumor cell lines, i.e., HEK293 and CT26 colon carcinoma cells, were transfected with P2X7 receptors and subsequently xenografted into immunodeficient or immunocompetent BALB/c mice, respectively. Tumor growth and angiogenesis were significantly enhanced by P_{2X7} expression; consequently, pharmacological inhibition (with AZ10606120) or genetic silencing of P_{2X7} decreased tumor growth and dramatically reduced vascular density [124]. This study further confirmed that P2X7 receptors were significantly up-regulated in several types of cancer cells, including those from breast cancer, and stimulated angiogenesis by promoting VEGF release [124]. More recently, it was found that P2X7 receptors were over-expressed also in B-TECs [94]. This study revealed that the activation of these purinergic receptors with high doses of ATP (>20 μM) and BzATP, a selective P_{2X7} agonist, inhibited B-TEC, but not HDMEC, migration in vitro. The anti-angiogenic effect of P_{2X7} was mediated by the Ca^{2+}-sensitive adenylate cyclase 10 (AC10), which increased cyclic adenosine monophosphate (cAMP) and recruited EPAC-1 to dampen cell migration by inducing cytoskeletal remodeling [94]. Moreover, P2X7 receptors-induced cAMP production stabilized bidimensional tumor vessels by favoring pericyte attraction towards B-TECs and reducing endothelial permeability [94]. Intriguingly, hypoxia prevented the anti-angiogenic ability of P_{2X7} receptors by likely reducing their expression [94,125]. These data, therefore, strongly suggest that stimulating P2X7 receptors could provide an efficient strategy to normalize tumor vasculature, thereby enhancing the delivery of cytotoxic drugs and of O_2 for radiotherapy. In this context, it should be pointed out that P2X7 receptors target hematopoietic EPCs to glioblastoma [126]. Although this investigation was conducted on healthy cells, and remains therefore to be validated in T-EPCs, it suggests that ATP may differentially affect tumor endothelial cells and T-EPCs. Alternatively, the effect exerted by P2X7 con T-EPC fate could be cancer-dependent and needs to be further investigated.

3.4. *Stim1, Orai1 and Canonical Transient Receptor Potential 1 (TRPC1)*

SOCE represents the most important Ca^{2+} entry pathway supporting the pro-angiogenic activity of human ECFCs [29,33,56]. Accordingly, TRPV4 boosted ECFC proliferation rate only when accompanied by the administration of a robust dose of growth factors [86,127], whereas TRPV1 stimulated ECFC proliferation and tubulogenesis by mediating the intracellular intake of anandamide in a Ca^{2+}-independent manner [128]. SOCE is activated by the pharmacological (by blocking SERCA-mediated Ca^{2+} sequestration) or physiological (by stimulating $InsP_3Rs$) depletion of the ER Ca^{2+} stores and is mediated by the dynamic interplay between Stim1, Orai1 and TRPC1 [29,56,58,84]. It is, however, still unknown whether Orai1 and TRPC1 form two independent Stim1-gated Ca^{2+}-permeable routes [129] or assemble into a unique heteromeric supermolecular complex in ECFCs [130]. A recent series of studies revealed that SOCE maintained VEGF-induced intracellular Ca^{2+} oscillations and promoted ECFC proliferation and in vitro tubulogenesis by recruiting the Ca^{2+}-dependent transcription factor, NF-κB [58,83]. Of note, SOCE was significantly enhanced in metastatic renal cellular carcinoma (RCC)-derived ECFCs (RCC-ECFCs) due to the up-regulation of Stim1, Orai1 and TRPC1 [95]. Similar to normal cells, the pharmacological blockade of SOCE with YM-58483/BTP2 or with low micromolar doses of lanthanides prevented proliferation and tube formation in RCC-ECFCs [95]. This finding strongly suggests that SOCE could provide an alternative target for the treatment of metastatic RCC [32,131], which develops either intrinsic or acquired refractoriness towards conventional treatments, such as anti-VEGF inhibitors and anti-mammalian target of rapamycin (mTOR) blockers [132]. As more extensively discussed below, the overall remodeling of the intracellular Ca^{2+} toolkit in T-ECFCs could indeed be responsible for the relative or complete failure of standard therapies in RCC patients. Conversely, SOCE was not significantly up-regulated in breast cancer-derived ECFCs (BC-ECFCs) [96]. Accordingly, Orai1 and TRPC1 expression were not significantly altered, while Stim1 was significantly more abundant as compared to control cells. Nevertheless, a tight stoichiometric ratio between Stim1, Orai1 and TRPC1 is required for SOCE to be activated [133]. If all Orai1 and TRPC1 channel proteins are gated by the physiological levels of Stim1, any increase in Stim1 expression will not be sufficient to enhance SOCE as there will be no further channels available on the plasma membrane. Similar to RCC-ECFCs, however, the pharmacological inhibition of SOCE abrogated BC-ECFC proliferation and tube formation, thereby confirming that Orai1 and TRPC1 could serve as reliable targets to interfere with tumor vascularization, although this hypothesis remains to be validated in vivo [134,135]. The strict requirement of Stim1 for tumor vascularization is further suggested by the recent finding that Stim1 transcription in hypoxic tumors is finely regulated by HIF-1 [136]. SOCE, in turn, was found to stimulate HIF-1 accumulation in hypoxic cancer cells by engaging Ca^{2+}/calmodulin-dependent protein kinase II and p300 [136]. Therefore, targeting SOCE could also affect the expression of the primary transcription factor responsible for RCC and breast cancer growth and metastasis [137,138]. Of note, HIF-1 has been shown to control also TRPC1 expression [139], although it is still unclear whether this regulation also occurs in tumor microenvironment and, if so, why TRPC1 is up-regulated in RCC-ECFCs, but not in RCC-ECFCs.

The role played by SOCE in tumor vascularization has, finally, been uncovered also in infantile hemangioma (IH), the most common childhood malignancy which may cause disfigurement, ulceration and obstruction and, if not treated, ultimately leads to patients' death [140]. IH is a vascular tumor that arises as a consequence of dysregulation of angiogenesis and vasculogenesis [140]. The clonal expansion of an endothelial progenitor/stem cell population, which is closely reminiscent of ECFCs, is deeply involved in IH vascularization [141,142]. A recent investigation provided the evidence that Stim1, Orai1 and TRPC1 drive the higher rate of IH-derived ECFC (IH-ECFC) growth as compared to control cells [97]. Stim1, Orai1 and TRPC1 were not up-regulated in IH-ECFCs; however, the ER Ca^{2+} store was depleted to such an extent that Stim1 was basally activated and gated the constitutive activation of Orai1 and TRPC1 [97]. Stim2 displays a lower Ca^{2+} affinity as respect to Stim1 and supports basal Ca^{2+} entry in HUVECs [143]. Nevertheless, the pharmacological abrogation of Stim2 silencing did not affect constitutive SOCE in IH-ECFCs [97]. Constitutive SOCE boosted IH-ECFC proliferation by enhancing NO release [97], thereby emerging as an alternative target to treat IH in propranolol-resistant patients [144].

3.5. Neuronal Nicotinic Receptors (nAchRs)

nAchRs belong to a super-family of Cys-loop ligand-gated non-selective cation channels that are physiologically activated by acetylcholine, mediate fast synaptic transmission in neurons and, by virtue of their resolvable Ca^{2+}-permeability, control a number of Ca^{2+}-dependent processes, including neurotransmitter release and synaptic plasticity [145,146]. However, nAchRs are also largely expressed in non-neuronal brain cells, such as astrocytes, in epithelial cells and in several types of vascular cells, including smooth muscle cells and endothelial cells [75,147–149]. It has been established that α7 homomeric nAchRs (α7-nAchRs) promote endothelial cell proliferation, migration and tube formation both in vitro and in vivo by recruiting an array of Ca^{2+}-dependent effectors [98,99]. These include eNOS, mitogen-activated protein kinase, phosphoinositide 3-kinase (PI3K), NF-κB, matrix metalloproteinase-2 and -9 [98,99,150]. In addition, α7-nAChRs were shown to induce the JAK2/STAT3 signaling cascade to promote endothelial cell survival [151]. Intriguingly, α7-nAchRs possess the highest Ca^{2+}-permeability among the known nAchR subtypes [152]. These pieces of evidence ignited the hypothesis that nicotine accelerated tumor growth by stimulating endothelial α7-nAchRs, thereby promoting angiogenesis and tumor vascularization [147,149]. In support of this model, nicotine induced tumor growth in a mouse model of LLC by stimulating endothelial cell proliferation and tube formation. Nicotine-induced tumor vascularization was significantly reduced by pharmacological blockade (with mecamylamine or hexamethonium) as well as by genetic silencing of α7-nAChRs. The signaling pathways recruited by α7-nAChRs to sustain tumor angiogenesis were not deeply investigated, but nicotine stimulated endothelial cells to release NO, prostacyclin and VEGF [98,99]. It should, however, be pointed out that the expression and role of α7-nAChRs in T-ECs has not been investigated, yet. Nevertheless, hypoxia has been shown to increase α7-nAChRs expression in a mouse model of hindlimb ischemia [98], whereas α7-nAChRs may stimulate HIF-1αtranscription [153]. These observations support the hypothesis that α7-nAChRs are actually over-expressed in T-ECs.

In addition to promoting angiogenesis, nicotine could recruit α7-nAChRs to boost vasculogenesis. A recent study revealed that nicotine induced proliferation, migration and tube formation also in ECFCs and that this effect was inhibited by mecamylamine or α-bungarotoxin [100]. Moreover, nicotine triggered EPC mobilization from bone marrow in a cohort of mice xenografted with colorectal cancer cells, thereby fostering tumor growth and vascularization [101]. Lastly, exposure to second hand smoke stimulated tumor angiogenesis and increased the number of circulating EPCs in a mouse model of LLC by enhancing VEGF release: mecamylamine, however, halted VEGF release, thereby reducing tumor size and capillary density. The pro-angiogenic effect of nicotine was, therefore, likely to be mediated by nAchRs [154]. We are yet to know whether and how α7-nAChRs are altered in T-ECs and T-EPCs. Nevertheless, these ionotropic receptors could be regarded as a promising target for alternative anti-angiogenic therapies.

3.6. Gasotransmitters-Activated Ca^{2+}-Permeable Channels

Gaseous mediators or gasotransmitters are endogenous signaling messengers that, although being toxic at high concentrations, regulate a multitude of physiological processes, ranging from the regulation of vascular tone to synaptic plasticity and mitochondrial bioenergetics [155–158]. The gasotransmitters NO and hydrogen sulphide (H_2S) have recently been shown to stimulate endothelial cells through an increase in $[Ca^{2+}]_i$ [6,156,159], while the role of CO in angiogenesis is less clear [160]. NO promotes angiogenesis and disease progression in several types of malignancies [161,162], including breast cancer [163]. The administration of two structurally unrelated NO donors, i.e., S-nitroso-*N*-acetylpenicillamine (SNAP) or sodium nitroprusside (SNP), was recently found to trigger Ca^{2+} influx and migration in B-TECs [164]. These effects were mimicked by elevating endogenous NO release with L-arginine [164], which is the physiological substrate for eNOS [156]. Of note, AA-induced TRPV4 activation in B-TECs was inhibited by preventing NO production with N^{G}-nitro-L-arginine methyl ester (L-NAME) [88,164]; moreover, AA- and NO-induced Ca^{2+} entry were both sensitive to protein kinase A (PKA) inhibition [164]. It is, therefore, likely that NO elicits Ca^{2+} entry in B-TECs by gating TRPV4. In agreement with this hypothesis, TRPV4 may be activated by NO through direct S-nitrosylation [71] and is phosphorylated by PKA upon AA stimulation in vascular endothelial cells [70]. Finally, NO-induced Ca^{2+} entry and migration were dramatically reduced in HDMECs [164], in which TRPV4 expression was significantly down-regulated [88]. Besides TRPV4, however, NO is able to recruit multiple TRP channels, such as TRPC1, TRPC4, TRPC5, TRPV1, and TRPV3 [71], some of which are up-regulated in T-ECFCs [95,165]. Unfortunately, it is still unclear whether NO elicits intracellular Ca^{2+} entry in these cells. Although future work is mandatory to understand whether NO stimulates TRP channels, as well as other Ca^{2+}-permeable channels, to promote tumor vascularization, endothelial Ca^{2+} signaling is emerging as an attractive target to prevent its pro-tumorigenic effect.

H_2S has also been shown to promote angiogenesis in a Ca^{2+}-dependent manner. For instance, H_2S mediated VEGF-induced Ea.hy926 cell proliferation and migration by inducing $InsP_3$-dependent ER Ca^{2+} release without the contribution of extracellular Ca^{2+} entry [6]. The components of the endothelial Ca^{2+} toolkit recruited by H_2S may, however, vary depending on the vascular bed [166]. H_2S induced ER-dependent Ca^{2+} release through $InsP_3Rs$ and RyRs followed by a sustained SOCE in primary cultures of human saphenous vein endothelial cells [167], whereas it recruited the reverse mode of NCX by gating a Na^+- and Ca^{2+}-permeable pathway in rat aortic endothelial cells [168] and HDMECs [21]. Conversely, NaHS did not elicit any resolvable elevation in $[Ca^{2+}]_i$ in ECFCs [6] and its role in neovasculogenesis in vivo operated by truly endothelial precursors remains to be elucidated [169]. H_2S-induced Ca^{2+} signals may be translated into a pro-angiogenic signal by multiple Ca^{2+}-dependent decoders, including the PI3K/Akt and the ERK/p38 signaling pathways [155,156]. Growing evidence demonstrated that H_2S drove disease progression and angiogenesis in several types of tumor, such as RCC and colorectal cancer [169]. Intriguingly, sodium hydrosulfide (NaHS), a widely employed H_2S donor, induced intracellular Ca^{2+} signals in both B-TECs and HDMECs; however, NaHS-elicited Ca^{2+} signals were enhanced and arose within a significantly lower range (nanomolar vs. micromolar) in B-TECs [21]. Consequently, NaHS promoted proliferation and migration in B-TECs, but not in control endothelial cells [21]. The Ca^{2+} response to H_2S was mediated by a Ca^{2+}-permeable non-selective cation channel and was sustained by membrane hyperpolarization through the activation of a K^+ conductance [21], likely an ATP-sensitive K^+ channel [157]. The molecular nature of this Ca^{2+}-permeable route is yet to be identified [166]. Nevertheless, H_2S is able to stimulate TRPV3 and TRPV6 in bone marrow-derived mesenchymal cells by direct sulfhydration of some of Cys residues within their protein structure [170]. Moreover, H_2S activated TRPA1 in RIN14B cells [171]. Deciphering the molecular target of H_2S in tumor endothelium is, therefore, mandatory to devise alternative anti-cancer treatments. In addition, both eNOS and cys-tathionine gamma lyase (CSE), the enzyme which catalyzes H_2S production in vascular endothelial cells, are Ca^{2+}-sensitive [16,42,172]. Therefore, targeting a Ca^{2+} entry/release pathway tightly coupled to either eNOS (i.e., Orai1, [173]) or CSE (yet to be identified) has the potential to interfere with multiple pro-angiogenic pathways and, therefore, exert a more profound anti-tumor effect.

3.7. *Connexin 40 (Cx40)*

Connexin (Cx) hemichannels, also termed connexons, have long been known to provide the building blocks of gap junctions, thereby enabling the transfer of small solutes, ions and signaling molecules, such as Ca^{2+} and $InsP_3$, between adjacent cells [174]. Three diverse Cx isoform exist in vascular endothelial cells, i.e., Cx37, Cx40, and Cx43, and synchronize robust NO release induced by extracellular autacoids by mediating intercellular Ca^{2+} communication [175,176]. In addition, unopposed Cx hemichannels were found to mediate extracellular Ca^{2+} entry and NO release in endothelial cells from different vascular beds [16,74,177–179]. Earlier work suggested that Cxs served as tumor suppressors and were down-regulated in cancer, thereby affecting vascular integrity and reducing vascular leakage [180–182]. However, a recent study challenged this dogma by showing that Cx40 was over-expressed in T-ECs and promoted disease progression and angiogenesis in melanoma and urogenital cancers [102]. Cx40 stimulated tumor growth by inducing eNOS recruitment, which strongly suggest that intracellular Ca^{2+} levels increased during the angiogenic process [102]. Intriguingly, targeting Cx40 function with ^{40}Gap27, a peptide that has long been use to inhibit Cx40-mediated intercellular communication and extracellular Ca^{2+} entry [1,16,178], normalized tumor vasculature and enhanced the efficacy of the chemotherapeutic drug, cyclophosphamide [102]. Therefore, although these findings remain to be confirmed in other tumor types, and the role served by Ca^{2+} is still unclear, Cx40 deserves careful consideration for the design of new anticancer drugs.

3.8. *Na^+/H^+ Exchanger-1 (NHE-1)*

The Na^+/H^+ exchanger NHE-1 is a reversible electroneutral antiporter that maintains cytosolic pH by expelling H^+ at expense of the inwardly directed Na^+ electrochemical gradient with a 1:1 stoichiometric ratio [183]. NHE-1 induces endothelial cell proliferation, migration and tube formation by means of several Ca^{2+}-dependent effectors, such as calpain [184], eNOS [185], and ERK 1/2 [186]. Accordingly, thrombin-induced NHE-1 activation was able to increase sub-membranal Na^+ levels, thereby switching NCX into the reverse mode and mediating extracellular Ca^{2+} entry in HUVECs [187,188]. Moreover, NHE-1-induced cytosolic alkalinization triggered ER-dependent Ca^{2+} release through $InsP_3Rs$ in bovine aortic endothelial cells and human pulmonary artery endothelial cells [189,190]. NHE1 is constitutively activated in cancer cells to favor extracellular acidification and stimulate metastasis and invasion by facilitating protease-mediated degradation of the extracellular matrix [191]. In addition, NHE-1 is transcriptionally regulated by HIF-1 and is up-regulated in a multitude of carcinomas [191,192]. A recent series of studies demonstrated that NHE-1 was over-expressed in endothelial cells exposed to tumor microenvironment [193] and was able to boost vascularization, invasion and metastasis in several types of tumors, including breast cancer [191,194]. Accordingly, aldosterone-induced NHE-1 activation promoted B-TEC proliferation, migration and cytosolic alkalinization [103]. Further work is required to assess whether NHE-1 activation stimulates tumor vascularization through an increase in $[Ca^{2+}]_i$. However, NHE-1 blockers, including cariporide and the more specific 3-methyl-4-flouro analog of 5-aryl-4(4-(5-methyl-14-imidazol-4-yl) piperidin-1-yl)pyrimidine (Compound 9t), have been put forward as alternative anti-cancer drugs [195].

3.9. *Two-Pore Channels (TPCs)*

The ER is the largest endogenous Ca^{2+} store in vascular endothelial cells by accounting for ≈75% of the total Ca^{2+} storage capacity [2]. The remainder 25% of the total stored Ca^{2+} is located within the mitochondria and the acidic Ca^{2+} stores of the endolysosomal (EL) system [2]. As more widely illustrated in [196], the EL Ca^{2+} store releases Ca^{2+} through many Ca^{2+}-permeable channels, including Mucolipin TRP 1 (TRPML1), Melastatin TRPM 2 (TRPM2) and two-pore channels 1 and 2 (TPC1–2) [197,198]. The newly discovered second messenger, nicotinic acid adenine dinucleotide phosphate (NAADP), is the physiological stimulus that gates TPC1–2 in response to extracellular stimulation [196,199,200]. NAADP-induced EL Ca^{2+} release is, in turn, amplified by juxtaposed ER-embedded $InsP_3Rs$ and RyRs through the CICR process, thereby initiating a regenerative Ca^{2+} wave [196,200,201]. NAADP-gated TPC2 channels are also expressed in vascular ECs [202,203], whereas *N*-ECFCs display larger amounts of TPC1 [86,204]. A recent study demonstrated that NAADP-induced Ca^{2+} signals promoted tumor vascularization and metastasis in murine models xenografted with B16 melanoma cells [205]. Of note, the pharmacological blockade of NAADP-induced Ca^{2+} release with Ned-19 dampened melanoma growth, vascularization and lung metastasis [205]. Future work will have to assess whether TPC2 channels are up-regulated in T-ECs and whether

NAADP-induced Ca^{2+} signaling also drive T-ECFC incorporation into tumor neovessels. However, TPCs stand out as promising targets to develop alternative anti-angiogenic treatments.

4. Resistance to Apoptosis

4.1. Canonical Transient Receptor Potential 5 (TRPC)

TRPC5 forms a homotetrameric Ca^{2+}-permeable channel that is gated upon PLCβ activation by Gq/11-coupled membrane receptors through a yet to be identified signaling cascade [2,206,207]. Accordingly, although some studies reported that TRPC5 is recruited in a store-dependent manner by Stim1 [208], it has been proposed that TRPC5 activation by PLCβ does not involve ER store depletion [209]. In addition, TRPC5-mediated Ca^{2+} entry is elicited by several physiological messengers, including reduced thioredoxin, protons, sphingosine-1-phosphate, lysophospholipids, NO and Ca^{2+} itself [207]. Finally, TRPC5 presents a spontaneous activity that is increased by lanthanides, cold temperatures (47 °C to 25 °C) and membrane stretch; consequently, TRPC5 serves as a cold sensor in the peripheral nervous system [210]. Of note, TRPC5 may establish physical associations with a multitude of molecular partners, including TRPC1, TRPC4 and TRPC6, which regulate its membrane localization and biophysical properties [207]. TRPC5 differently tunes angiogenesis depending on the vascular bed. For instance, TRPC5 promoted proliferation and tube formation by inducing intracellular Ca^{2+} oscillations in EA.hy926 cells [211]. Conversely, a TRPC6-TRPC5 channel interaction inhibited angiogenesis by decreasing the rate of migration in bovine aortic ECs (BOECs). In this context, TRPC6-mediated Ca^{2+} entry triggered an ERK-mediated phosphorylation cascade that leads to MLCK activation and TRPC5 externalization on the plasma membrane [212,213]. It has recently been shown that endothelial TRPC5 could underlie the development of chemoresistance to anticancer drugs in both breast cancer [104,214,215] and colorectal carcinoma [216]. P-glycoprotein (P-gp), also termed multidrug resistance protein 1 (MDR1), is a multidrug efflux transporter that expels xenobiotics out from the cytoplasm into the extracellular milieu [217]. P-gp overexpression, therefore, confers resistance to malignant cells, which become insensitive to a wide range of cancer chemotherapeutics, including adriamycin, vincristine, taxol, and anthracyclines [217]. Earlier evidence demonstrated that TRPC5 was up-regulated and induced P-gp overexpression by hyper-stimulating the Ca^{2+}-dependent transcription factor, NFATc3 in chemoresistant MCF-7 breast cancer cells [214]. In agreement with this observation, microRNA 320a (miR-320a), which is able to associate with and degrade TRPC5 and NFATc3 transcripts in normal cells, was down-regulated in chemoresistant breast cancer cells due to the hypermethylation of its promoter sequence [218]. TRPC5 up-regulation induced resistance to adriamycin, paclixatel, epirubicin, mitoxantrone and vincristine [214]. Additionally, TRPC5-mediated Ca^{2+} entry promoted transcription of HIF-1α gene, thereby boosting VEGF release and enhancing tumor angiogenesis [219]. Remarkably, TRPC5-based chemoresistance could be shuttled to tumor endothelial via intercellular communication. Adriamycin-resistant MCF-7 cells could pack the up-regulated TRPC5 channels into mobile extracellular vesicles (EVs), which are released in tumor microenvironment and transferred their signaling content to surrounding endothelial cells. This scenario is supported by the observations that HDMECs exposed to TRPC5-contanining EVs, which were collected from adriamycin MCF-7 breast cancer cells, over-expressed the TRPC3-NFATc3-P-gp signaling pathway and developed resistance to adriamycin-induced apoptosis [104]. Moreover, TRPC5-containing vesicles were identified in peripheral blood of breast cancer patients receiving chemotherapy and of nude mice bearing adriamycin-resistant MCF-7 tumor xenografts [215]. Furthermore, P-gp production was enriched in tumor endothelium of adriamycin-resistant MCF-7 xenografts than in other sites and was sensitive to TRPC5 inhibition with a specific siRNA (siTRPC5) [104]. These data, therefore, suggest that TRPC5 provide a promising target to design alternative adjuvant anticancer treatments [220]. Accordingly, a blocking TRPC5 antibody reduced P-gp expression, retarded cancer growth and boosted paclitaxel-induced tumor regression in chemoresistant breast cancer in vivo [104,214,215]. The endothelial effects of TRPC5 in breast cancer are seemingly limited to T-ECs, as BC-ECFCs do not express this channel [96]. Future work will have to assess whether, besides conferring B-TECs with the resistance to chemotherapeutic drugs, TRPC5 up-regulation accelerates breast cancer angiogenesis.

4.2. Inositol-1,4,5-Trisphosphate ($InsP_3$) Receptors ($InsP_3Rs$)

$InsP_3Rs$ are non-selective cation channels which mediate ER-dependent Ca^{2+} release, thereby controlling multiple endothelial cell functions, including bioenergetics, apoptosis, angiogenesis and vasculogenesis (see Paragraph 2. Ca^{2+} signaling in normal endothelial cells: a brief introduction). Three distinct $InsP_3R$ isoforms exist in both vascular endothelial cells and ECFCs [2,29], i.e., $InsP_3R1$, $InsP_3R2$ and $InsP_3R3$, which may associate into homo- or hetero-tetrameric ER-embedded channels [2]. It has recently been shown that $InsP_3Rs$ were dramatically down-regulated in RCC-ECFCs, thereby preventing the onset of VEGF-induced intracellular Ca^{2+} oscillations, proliferation and in vitro tubulogenesis [95] (Figure 3). More specifically, RCC-ECFCs only expressed $InsP_3R1$, while $InsP_3R2$ and $InsP_3R3$ were absent [95]. This result was surprising as $InsP_3R1$ was transcriptionally regulated by HIF-2 in human RCC cancer cell lines [221]. The failure of the pro-angiogenic Ca^{2+} response to VEGF also involves the chronic reduction in the ER Ca^{2+} concentration ($[Ca^{2+}]_{ER}$) in RCC-ECFCs, as monitored by using an ER-targeted aequorin Ca^{2+} indicator [87,222]. Therefore, in contrast with the widely accepted belief that VEGF sustains the angiogenic switch [223], VEGF does not stimulate ECFC-dependent neovessel formation in RCC patients [27,32]. This observation shed novel light on the refractoriness to anti-VEGF therapies in individuals suffering from RCC [224–226]. It has long been known that humanized monoclonal anti-VEGF antibodies, such as bevacizumab, or small molecule tyrosine kinase inhibitors, such as sorafenib and sunitinib, did not increase the overall survival of RCC patients, who ultimately developed secondary (acquired) resistance and succumbed because of tumor relapse and metastasis. In addition, targeting VEGF-dependent pathway proved to be ineffective in a large cohort of subjects, who displayed intrinsic refractoriness to these anti-VEGF drugs and did not show any improvement in their progression free survival [227,228]. ECFCs play a key role during the early phases of the angiogenic switch that supports tumor vascularization and metastasis [27,28,229]. If tumor vasculature is dismantled by anti-VEGF drugs, the following drop in P_{O2} will release in circulation a cytokine storm that attracts ECFCs from their bone marrow and/or vascular niches. ECFCs will home to the shrunk tumor, but, being insensitive to VEGF, will not be affected by the presence of anti-VEGF drugs. Consequently, they will proliferate in response to the mixture of growth factors liberated in tumor microenvironment and will restore blood supply to cancer cells [32]. Remodeling of the Ca^{2+} toolkit in RCC-ECFCs could, therefore, underlie the resistance to anti-angiogenic therapies in RCC patients. Similar data were obtained in BC-ECFCs, in which the significant reduction in $[Ca^{2+}]_{ER}$ prevented VEGF from triggering robust intracellular Ca^{2+} oscillations, proliferation and tube formation, although the pattern of $InsP_3R$ expression remained unchanged [37,96]. Again, this result is consistent with notion that also breast cancer patients present intrinsic or secondary refractoriness to anti-VEGF therapies [54,230]. The reduction in $[Ca^{2+}]_{ER}$ observed in several types of tumor-associated ECFCs, including IH-ECFCs [95–97], was likely to reflect the down-regulation of SERCA2B activity [87,95]. Accordingly, ATP-induced $InsP_3$-dependent ER Ca^{2+} release in RCC-ECFCs decayed to resting Ca^{2+} levels with slower kinetics as compared to normal ECFCs [95]. Intriguingly, the gene expression profile of RCC- and BC-ECFCs resulted to be dramatically different with respect to normal cells [37]: BC-ECFCs and RCC-ECFCs presented, respectively, 382 and 71 differently expressed genes (DEGs) as compared to healthy cells, including TMTC1 [37]. TMTC1 is a tetratricopeptide repeat-containing adapter protein, which binds to and inhibits SERCA2B, thereby reducing ER Ca^{2+} levels and dampening agonist-induced intracellular Ca^{2+} release [231]. It is conceivable that TMTC1 up-regulation in T-ECFCs contributes to the chronic underfilling of their ER Ca^{2+} reservoir. In further agreement with this observation, electron microscopy revealed that both RCC- and BC-ECFCs presented dramatic ultrastructural differences as compared to control cells [87,96]. In particular, T-ECFCs presented a remarkable expansion of ER volume, whereas mitochondria were more abundant and very often elongated as compared to *N*-ECFCs [87,96]. This ultrastructural remodeling is consistent with the ER stress caused by the chronic reduction in $[Ca^{2+}]_{ER}$ [232,233].

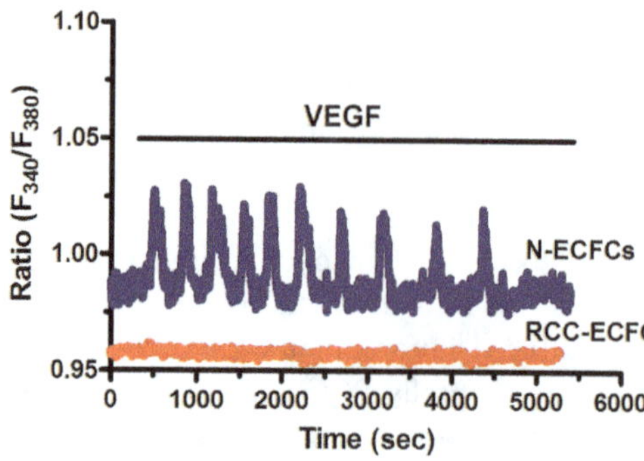

Figure 3. VEGF does not trigger pro-angiogenic Ca^{2+} oscillations in tumor-derived endothelial colony forming cells. VEGF (10 ng/mL) triggers intracellular Ca^{2+} oscillations in *N*-ECFCs, but not in RCC-ECFCs. Adapted from [95].

5. Targeting the Endothelial Ca^{2+} Toolkit to Circumvent the Resistance to Anticancer Treatments

Remodeling of the Ca^{2+} toolkit in tumor cells led many authors to search for alternative strategies to treat cancer. Intracellular Ca^{2+} signaling controls most, if not all, the so-called cancer hallmarks and could, therefore, be targeted to inhibit or, at least, retard tumor growth and metastasis [35,36,234–240]. As shown above, the Ca^{2+} transportome is also altered in stromal cancer cells [32,241,242], including endothelial cells and ECFCs. Remodeling of the endothelial Ca^{2+} toolkit could play a crucial role in the refractoriness to anticancer treatments, by supporting tumor vascularization and decreasing the susceptibility to pro-apoptotic stimuli. Therefore, the endothelial Ca^{2+} transportome might provide an efficient target for adjuvant therapies to conventional anti-cancer treatments. Three strategies could be pursued to improve the therapeutic outcome of standard therapies by interfering with the endothelial Ca^{2+} machinery: (1) blocking Ca^{2+} signaling to dampen angiogenesis and/or vasculogenesis; (2) stimulating Ca^{2+} entry to normalize tumor vessels, thereby improving the delivery and efficacy of chemo-, radio- and immunotherapy; and (3) manipulating Ca^{2+} signaling to endothelial cell apoptosis and dismantle tumor vasculature.

SOCE is, perhaps, the most suitable target to affect tumor vasculature by inhibiting both angiogenesis and vasculogenesis. Although there is no report of SOCE expression in T-ECs, Stim1 and Orai1 control proliferation and tube formation in normal endothelial cells, such as rat CMECs [7], bovine brain capillary endothelial cells [243], mouse lymphatic endothelial cells [244], and HUVEC [57,58]. Moreover, the pharmacological blockade of SOCE attenuates the rate of cell growth and abrogates in vitro tubulogenesis in RCC-, IH- and BC-derived ECFCs [95–97]. In addition, SOCE controls proliferation, migration and metastasis in a multitude of different cancer cell lines [134,135], which expands the cellular targets of SOCE inhibitors to the whole tumor microenvironment. We [131,135] and others [133,245] have recently described the Orai1 and TRPC1 inhibitors, some of which have been listed in Table 1, that could serve as a molecular template to design novel anticancer drugs. Unfortunately, none of these drugs have reached the milestone of being approved by US Food and Drug Administration (FDA) due to their scarce selectivity and high toxicity. For instance, carboxyamidotriazole (CAI), a non-selective blocker of Ca^{2+} signaling, was originally used to inhibit angiogenesis in vitro and tumor vascularization in vivo [131,246–248]. Depending on the cell type, CAI was able to block SOCE by occluding the mitochondrial Ca^{2+} uniporter [249,250] or reducing $InsP_3$ synthesis, which in turn prevents $InsP_3$-dependent Ca^{2+} release and Stim activation [56,251,252]. Intriguingly, CAI inhibited proliferation and tube formation also in RCC- and BC-ECFCs by preventing

$InsP_3$-dependent ER Ca^{2+} release [95,96]. Additionally, CAI was found to block growth and motility in several types of cancer cell lines [252,253]. Therefore, phase I-III clinical trials were launched to assess CAI toxicity and tolerability in patients suffering from several types of malignancies, including RCC, breast cancer, ovarian cancer, melanoma, non-small cell lung carcinoma, and gastrointestinal (stomach and pancreas) adenocarcinomas [56,248,254,255]. As discussed elsewhere [131], this drug caused disease stabilization when administrated alone or as adjuvant of chemo- or radio-therapy, and induced well tolerable side effects in most patients, such diarrhea, nausea and/or vomiting, fatigue and constipation. The therapy was discontinued only in RCC patients, who underwent disease progression and suffered from unacceptable toxicities, such as neuropsychiatric difficulties and asthenia [256]. As mentioned earlier, however, the effect of CAI is not directed towards the SOCE machinery, but is indirect. In addition to SOCE, CAI may also block TRPV4 and ER leakage channels [89,95,96,131]. A recent investigation, however, screened a library of >1800 FDA-approved drugs to search for specific SOCE blockers and identified five novel compounds, i.e., leflunomide, teriflunomide, lansoprazole, tolvaptan and roflumilast, that could be successfully used in therapy (leflunomide and terifluonomide) or provide the template to design more selective Orai1 inhibitors (i.e., lansoprazole, tolvaptan and roflumilast) [257].

An alternative strategy consists in stimulating endothelial Ca^{2+} signaling to induce tumor normalization by activating distinct Ca^{2+} entry routes depending on the tumor type. For instance, TRPV4-mediated Ca^{2+} entry drives tumor normalization in LLC [90,91], whereas P_{2X7} receptors could be targeted to normalize tumor vasculature in breast cancer [94]. Tumor normalization, in turn, represents a promising adjuvant approach to facilitate cancer therapy by increasing the diffusion of chemotherapeutic drugs, improving radiotherapy efficiency and favoring the recruitment of tumor-killing immune cells [118,258]. Several synthetic agonists may selectively induce TRPV4 opening, such as 4αPDD derivatives, RN-1747, and JNc-440. Moreover, GlaxoSmithKline commercialized several patent applications of small molecule TRPV4 activators, the most famous of which is GSK [105,259]. Likewise, BzATP is regarded as the most potent P_{2X7} receptor agonist, while 2-meSATP and ATPγS are only partial agonists and $\alpha\beta$-meATP and $\beta\gamma$-meATP exert a rather weaker on activation [260]. Clearly, further studies are required to uncover additional components of the endothelial Ca^{2+} toolkit potentially implicated in tumor normalization. Nevertheless, a recent investigation reported that angiopoietins, which induce vessel maturation by regulating the interaction between luminal endothelial cells and mural cells, such as vascular smooth muscle cells and pericytes, stimulate HUVEC migration by promoting ER-dependent Ca^{2+} release through $InsP_3Rs$ and RyRs [261]. These findings lend further support to the hypothesis that targeting the endothelial Ca^{2+} signaling provides a suitable means to accelerate the dismantling of tumor vasculature by standard anticancer therapies.

Finally, the endothelial Ca^{2+} machinery could be properly manipulated to enhance the pro-apoptotic outcome of chemo- and radiation-therapy. For instance, TRPC5-mediated Ca^{2+} entry could be inhibited in B-TECs by taking advantage of a battery of novel small molecule inhibitors, such as Pico145 [262], 3,5,7-trihydroxy-2-(2-bromophenyl)-4H-chromen-4-one (AM12) [263], 2-aminobenzimidazole derivatives [264], ML204 [265], and neuroactive steroids [266]. Alternatively, the $[Ca^{2+}]_{ER}$ could be augmented to such an extent to induce the pro-apoptotic $InsP_3$-driven ER-to-mitochondria Ca^{2+} transfer. Pinton's group demonstrated that phototherapy induces a p53-dependent increase in $[Ca^{2+}]_{mit}$, which leads to tumor disruption in vivo [239,267]. Moreover, cytotoxic ER-dependent Ca^{2+} mobilization could be promoted by conjugating thapsigargin, a selective SERCA inhibitor, with a protease-specific peptide carrier, which is cleaved by the prostate-specific membrane antigen (PMSA) [268]. PMSA is widely expressed in the endothelium of many solid tumors [269,270], including RCC, thereby selectively favoring thapsigargin release in TME and inducing cancer and stromal cell apoptosis [268,271]. This prodrug has been termed mipsagargin or prodrug G202 and has recently been probed in a phase I clinical trials in patients suffering from refractory, advanced or metastatic solid tumors [272]. We do not know yet whether $[Ca^{2+}]_{ER}$ is also decreased in the endothelium of tumor neovessels, as ECFCs are likely to be replaced/diluted by local endothelial cells after the angiogenic switch [27]. Nevertheless, mipsagargin is likely to cause pro-apoptotic Ca^{2+} release in all stromal cells, including T-ECs.

As outlined elsewhere [23,32,135], caution is warranted when targeting a ubiquitous intracellular second messenger, such as Ca^{2+}. It should, however, be pointed out that several inhibitors of voltage-gated Ca^{2+} channels, such as verapamil, nifedipine and nitrendipin, are routinely employed in clinical practice to treat severe cardiovascular disorders, including hypertension, arrhythmia, acute myocardial infarction-induced heart failure and chronic stable angina [135,273]. In agreement with this observation, a phase I clinical trial is currently assessing the therapeutic outcome of Ca^{2+} electroporation on cutaneous metastases of solid tumors as compared to standard electrochemotherapy with bleomycin (https://clinicaltrials.gov/ct2/show/NCT01941901). Ca^{2+} electroporation is predicted to enhance the rate of cancer cell death by resulting in cytotoxic Ca^{2+} accumulation in the cytosol and in mitochondria [36]. Finally, the pharmacological inhibition of intracellular Ca^{2+} signaling did not elicit any intolerable side effects, such as immune depression, bleeding or neuropathic disorders, in least three distinct models of human cancer xenografts [24,205,274].

6. Conclusions

The present article discussed how remodeling of the endothelial Ca^{2+} toolkit (or transportome) could contribute to the resistance to anti-cancer treatments, which hampers from the very beginning their therapeutic outcome (intrinsic resistance) or leads to tumor relapse (acquired resistance) and patients' death. The intimate relationship between endothelial Ca^{2+} signaling and refractoriness to anti-cancer treatments cannot be fully appreciated by studying normal/healthy endothelial cells [32]. For instance, the role of VEGF in promoting tumor neovascularization has been extensively acknowledged based upon the observation that VEGF triggers pro-angiogenic Ca^{2+} signals in normal endothelial cells [6,58,251] and ECFCs. Nevertheless, VEGF does not stimulate proliferation and tube formation in T-ECFCs, which play a crucial role in sustaining the angiogenic switch and are likely to restore tumor vasculature prior to recurrence of disease progression. Therefore, to be effective in the patients, a strategy aiming at targeting the Ca^{2+} toolkit must be first probed on tumor-associated endothelial cells and ECFCs in vitro, as the their Ca^{2+} machinery could be different from that of naïve cells. The protocol to isolate ECFCs from peripheral blood does not require an unreasonable volume of blood (≈40 mL), but it takes no less than three weeks [31] due to lack of ECFCs-specific membrane antigens. It will be imperative to speed up this procedure to accelerate the therapeutic translation of the findings generated by basic research. Isolating T-ECs represents a more technically demanding challenge, but several strategies were designed to collect and expand T-ECs from several types of solid cancers [275]. Further work on patients-derived T-ECs or T-ECFCs is mandatory to identify novel components of the endothelial Ca^{2+} toolkit involved in the refractoriness to anti-cancer therapies. Most of the attention is, of course, currently paid to Stim1 and Orai1 [33,131] and to the multiple TRP channel subfamilies that drive physiological angiogenesis [2,131,276,277]. Additional components of the endothelial Ca^{2+} transportome deserve careful investigation. For instance, Orai3 was found to up-regulated in several types of T-ECFCs [95,165], but its role in tumor vascularization is currently unknown. Of note, Orai3 may replace Orai1 as the pore-forming subunit of store-operated channels in cancer cells and could, therefore, emerge as a promising target for anti-cancer therapies [278]. NCX provides another unconventional Ca^{2+} entry that regulates proliferation and tube formation in healthy endothelial cells [18,279], but has been scarcely investigated in tumor neovessels. Finally, nAchRs are not the only ionotropic receptors expressed in vascular endothelial cells. *N*-methyl-D-aspartate receptors (NMDARs) are widely expressed in brain microvascular endothelial cells [280], in which they recruit eNOS and stimulate NO release in response to synaptic activity [281]. Aberrant glutamate signaling has been associated to glioma growth [282] and NMDARs-mediated Ca^{2+} entry could engage Ca^{2+}-dependent decoders other than eNOS in brain endothelium. Therefore, the expression and role of endothelial NMDARs in glioblastoma should be carefully evaluated.

Acknowledgments: The author gratefully acknowledges the contribution and support of all the colleagues and students who participated in the studies described in the present article. In particular, I would like to thank Vittorio Rosti and Germano Guerra. I am also grateful to Teresa Soda for her continuous support and comprehension.

References

1. Moccia, F.; Tanzi, F.; Munaron, L. Endothelial remodelling and intracellular calcium machinery. *Curr. Mol. Med.* **2014**, *14*, 457–480. [CrossRef] [PubMed]
2. Moccia, F.; Berra-Romani, R.; Tanzi, F. Update on vascular endothelial Ca^{2+} signalling: A tale of ion channels, pumps and transporters. *World J. Biol. Chem.* **2012**, *3*, 127–158. [CrossRef] [PubMed]
3. Munaron, L.; Pla, A.F. Endothelial Calcium Machinery and Angiogenesis: Understanding Physiology to Interfere with Pathology. *Curr. Med. Chem.* **2009**, *16*, 4691–4703. [CrossRef] [PubMed]
4. Troidl, C.; Nef, H.; Voss, S.; Schilp, A.; Kostin, S.; Troidl, K.; Szardien, S.; Rolf, A.; Schmitz-Rixen, T.; Schaper, W.; et al. Calcium-dependent signalling is essential during collateral growth in the pig hind limb-ischemia model. *J. Mol. Cell. Cardiol.* **2010**, *49*, 142–151. [CrossRef] [PubMed]
5. Troidl, C.; Troidl, K.; Schierling, W.; Cai, W.J.; Nef, H.; Mollmann, H.; Kostin, S.; Schimanski, S.; Hammer, L.; Elsasser, A.; et al. Trpv4 induces collateral vessel growth during regeneration of the arterial circulation. *J. Cell. Mol. Med.* **2009**, *13*, 2613–2621. [CrossRef] [PubMed]
6. Potenza, D.M.; Guerra, G.; Avanzato, D.; Poletto, V.; Pareek, S.; Guido, D.; Gallanti, A.; Rosti, V.; Munaron, L.; Tanzi, F.; et al. Hydrogen sulphide triggers VEGF-induced intracellular Ca^{2+} signals in human endothelial cells but not in their immature progenitors. *Cell Calcium* **2014**, *56*, 225–236. [CrossRef] [PubMed]
7. Moccia, F.; Berra-Romani, R.; Tritto, S.; Signorelli, S.; Taglietti, V.; Tanzi, F. Epidermal growth factor induces intracellular Ca^{2+} oscillations in microvascular endothelial cells. *J. Cell. Physiol.* **2003**, *194*, 139–150. [CrossRef] [PubMed]
8. Noren, D.P.; Chou, W.H.; Lee, S.H.; Qutub, A.A.; Warmflash, A.; Wagner, D.S.; Popel, A.S.; Levchenko, A. Endothelial cells decode VEGF-mediated Ca^{2+} signaling patterns to produce distinct functional responses. *Sci. Signal.* **2016**, *9*, ra20. [CrossRef] [PubMed]
9. Munaron, L.; Fiorio Pla, A. Calcium influx induced by activation of tyrosine kinase receptors in cultured bovine aortic endothelial cells. *J. Cell. Physiol.* **2000**, *185*, 454–463. [CrossRef]
10. Gupta, S.K.; Lysko, P.G.; Pillarisetti, K.; Ohlstein, E.; Stadel, J.M. Chemokine receptors in human endothelial cells. Functional expression of CXCR4 and its transcriptional regulation by inflammatory cytokines. *J. Biol. Chem.* **1998**, *273*, 4282–4287. [CrossRef] [PubMed]
11. Moccia, F.; Bonetti, E.; Dragoni, S.; Fontana, J.; Lodola, F.; Romani, R.B.; Laforenza, U.; Rosti, V.; Tanzi, F. Hematopoietic progenitor and stem cells circulate by surfing on intracellular Ca^{2+} waves: A novel target for cell-based therapy and anti-cancer treatment? *Curr. Signal Trans. Ther.* **2012**, *7*, 161–176. [CrossRef]
12. Yang, C.; Ohk, J.; Lee, J.Y.; Kim, E.J.; Kim, J.; Han, S.; Park, D.; Jung, H.; Kim, C. Calmodulin Mediates Ca^{2+}-Dependent Inhibition of Tie2 Signaling and Acts as a Developmental Brake During Embryonic Angiogenesis. *Arterioscler. Thromb. Vasc. Biol.* **2016**, *36*, 1406–1416. [CrossRef] [PubMed]
13. Zhu, L.P.; Luo, Y.G.; Chen, T.X.; Chen, F.R.; Wang, T.; Hu, Q. Ca^{2+} oscillation frequency regulates agonist-stimulated gene expression in vascular endothelial cells. *J. Cell Sci.* **2008**, *121*, 2511–2518. [CrossRef] [PubMed]
14. Chen, F.; Zhu, L.; Cai, L.; Zhang, J.; Zeng, X.; Li, J.; Su, Y.; Hu, Q. A stromal interaction molecule 1 variant up-regulates matrix metalloproteinase-2 expression by strengthening nucleoplasmic Ca^{2+} signaling. *Biochim. Biophys. Acta* **2016**, *1863*, 617–629. [CrossRef] [PubMed]
15. Tsai, F.C.; Seki, A.; Yang, H.W.; Hayer, A.; Carrasco, S.; Malmersjo, S.; Meyer, T. A polarized Ca^{2+}, diacylglycerol and STIM1 signalling system regulates directed cell migration. *Nat. Cell Biol.* **2014**, *16*, 133–144. [CrossRef] [PubMed]
16. Berra-Romani, R.; Avelino-Cruz, J.E.; Raqeeb, A.; Della Corte, A.; Cinelli, M.; Montagnani, S.; Guerra, G.; Moccia, F.; Tanzi, F. Ca^{2+}-dependent nitric oxide release in the injured endothelium of excised rat aorta: A promising mechanism applying in vascular prosthetic devices in aging patients. *BMC Surg.* **2013**, *13* (Suppl. 2), S40. [CrossRef] [PubMed]

17. Charoensin, S.; Eroglu, E.; Opelt, M.; Bischof, H.; Madreiter-Sokolowski, C.T.; Kirsch, A.; Depaoli, M.R.; Frank, S.; Schrammel, A.; Mayer, B.; et al. Intact mitochondrial Ca^{2+} uniport is essential for agonist-induced activation of endothelial nitric oxide synthase (eNOS). *Free Radic. Biol. Med.* **2017**, *102*, 248–259. [CrossRef] [PubMed]
18. Andrikopoulos, P.; Eccles, S.A.; Yaqoob, M.M. Coupling between the TRPC3 ion channel and the NCX1 transporter contributed to VEGF-induced ERK1/2 activation and angiogenesis in human primary endothelial cells. *Cell. Signal.* **2017**, *37*, 12–30. [CrossRef] [PubMed]
19. Lyubchenko, T.; Woodward, H.; Veo, K.D.; Burns, N.; Nijmeh, H.; Liubchenko, G.A.; Stenmark, K.R.; Gerasimovskaya, E.V. P2Y1 and P2Y13 purinergic receptors mediate Ca^{2+} signaling and proliferative responses in pulmonary artery vasa vasorum endothelial cells. *Am. J. Physiol. Cell Physiol.* **2011**, *300*, C266–C275. [CrossRef] [PubMed]
20. Sameermahmood, Z.; Balasubramanyam, M.; Saravanan, T.; Rema, M. Curcumin modulates SDF-1alpha/CXCR4-induced migration of human retinal endothelial cells (HRECs). *Investig. Ophthalmol. Vis. Sci.* **2008**, *49*, 3305–3311. [CrossRef] [PubMed]
21. Pupo, E.; Pla, A.F.; Avanzato, D.; Moccia, F.; Cruz, J.E.; Tanzi, F.; Merlino, A.; Mancardi, D.; Munaron, L. Hydrogen sulfide promotes calcium signals and migration in tumor-derived endothelial cells. *Free Radic. Biol. Med.* **2011**, *51*, 1765–1773. [CrossRef] [PubMed]
22. Fiorio Pla, A.; Munaron, L. Functional properties of ion channels and transporters in tumour vascularization. *Philos. Trans. R. Soc. Lond. B Biol. Sci.* **2014**, *369*, 20130103. [CrossRef] [PubMed]
23. Moccia, F. Remodelling of the Ca^{2+} Toolkit in Tumor Endothelium as a Crucial Responsible for the Resistance to Anticancer Therapies. *Curr. Signal Trans. Ther.* **2017**, *12*. [CrossRef]
24. Chen, Y.F.; Chiu, W.T.; Chen, Y.T.; Lin, P.Y.; Huang, H.J.; Chou, C.Y.; Chang, H.C.; Tang, M.J.; Shen, M.R. Calcium store sensor stromal-interaction molecule 1-dependent signaling plays an important role in cervical cancer growth, migration, and angiogenesis. *Proc. Natl. Acad. Sci. USA* **2011**, *108*, 15225–15230. [CrossRef] [PubMed]
25. Jain, R.K.; Carmeliet, P. SnapShot: Tumor angiogenesis. *Cell* **2012**, *149*, 1408.e1. [CrossRef] [PubMed]
26. Gao, D.C.; Nolan, D.; McDonnell, K.; Vahdat, L.; Benezra, R.; Altorki, N.; Mittal, V. Bone marrow-derived endothelial progenitor cells contribute to the angiogenic switch in tumor growth and metastatic progression. *Biochim. Biophys. Acta* **2009**, *1796*, 33–40. [CrossRef] [PubMed]
27. Moccia, F.; Zuccolo, E.; Poletto, V.; Cinelli, M.; Bonetti, E.; Guerra, G.; Rosti, V. Endothelial progenitor cells support tumour growth and metastatisation: Implications for the resistance to anti-angiogenic therapy. *Tumour Biol.* **2015**, *36*, 6603–6614. [CrossRef] [PubMed]
28. Yoder, M.C.; Ingram, D.A. The definition of EPCs and other bone marrow cells contributing to neoangiogenesis and tumor growth: Is there common ground for understanding the roles of numerous marrow-derived cells in the neoangiogenic process? *Biochim. Biophys. Acta* **2009**, *1796*, 50–54. [CrossRef] [PubMed]
29. Moccia, F.; Guerra, G. Ca^{2+} Signalling in Endothelial Progenitor Cells: Friend or Foe? *J. Cell. Physiol.* **2016**, *231*, 314–327. [CrossRef] [PubMed]
30. Maeng, Y.S.; Choi, H.J.; Kwon, J.Y.; Park, Y.W.; Choi, K.S.; Min, J.K.; Kim, Y.H.; Suh, P.G.; Kang, K.S.; Won, M.H.; et al. Endothelial progenitor cell homing: Prominent role of the IGF2-IGF2R-PLCbeta2 axis. *Blood* **2009**, *113*, 233–243. [CrossRef] [PubMed]
31. Moccia, F.; Ruffinatti, F.A.; Zuccolo, E. Intracellular Ca^{2+} Signals to Reconstruct A Broken Heart: Still A Theoretical Approach? *Curr. Drug Targets* **2015**, *16*, 793–815. [CrossRef] [PubMed]
32. Moccia, F.; Poletto, V. May the remodeling of the Ca^{2+} toolkit in endothelial progenitor cells derived from cancer patients suggest alternative targets for anti-angiogenic treatment? *Biochim. Biophys. Acta* **2015**, *1853*, 1958–1973. [CrossRef] [PubMed]
33. Moccia, F.; Lodola, F.; Dragoni, S.; Bonetti, E.; Bottino, C.; Guerra, G.; Laforenza, U.; Rosti, V.; Tanzi, F. Ca^{2+} signalling in endothelial progenitor cells: A novel means to improve cell-based therapy and impair tumour vascularisation. *Curr. Vasc. Pharmacol.* **2014**, *12*, 87–105. [CrossRef] [PubMed]
34. Bergers, G.; Hanahan, D. Modes of resistance to anti-angiogenic therapy. *Nat. Rev. Cancer* **2008**, *8*, 592–603. [CrossRef] [PubMed]

35. Prevarskaya, N.; Ouadid-Ahidouch, H.; Skryma, R.; Shuba, Y. Remodelling of Ca^{2+} transport in cancer: How it contributes to cancer hallmarks? *Philos. Trans. R. Soc. Lond. B Biol. Sci.* **2014**, *369*, 20130097. [CrossRef] [PubMed]
36. Monteith, G.R.; Prevarskaya, N.; Roberts-Thomson, S.J. The calcium-cancer signalling nexus. *Nat. Rev. Cancer* **2017**, *17*, 367–380. [CrossRef] [PubMed]
37. Moccia, F.; Fotia, V.; Tancredi, R.; Della Porta, M.G.; Rosti, V.; Bonetti, E.; Poletto, V.; Marchini, S.; Beltrame, L.; Gallizzi, G.; et al. Breast and renal cancer-Derived endothelial colony forming cells share a common gene signature. *Eur. J. Cancer* **2017**, *77*, 155–164. [CrossRef] [PubMed]
38. Moccia, F.; Berra-Romani, R.; Baruffi, S.; Spaggiari, S.; Signorelli, S.; Castelli, L.; Magistretti, J.; Taglietti, V.; Tanzi, F. Ca^{2+} uptake by the endoplasmic reticulum Ca^{2+}-ATPase in rat microvascular endothelial cells. *Biochem. J.* **2002**, *364 Pt 1*, 235–244. [CrossRef] [PubMed]
39. Berra-Romani, R.; Raqeeb, A.; Guzman-Silva, A.; Torres-Jacome, J.; Tanzi, F.; Moccia, F. Na^{+}-Ca^{2+} exchanger contributes to Ca^{2+} extrusion in ATP-stimulated endothelium of intact rat aorta. *Biochem. Biophys. Res. Commun.* **2010**, *395*, 126–130. [CrossRef] [PubMed]
40. Paszty, K.; Caride, A.J.; Bajzer, Z.; Offord, C.P.; Padanyi, R.; Hegedus, L.; Varga, K.; Strehler, E.E.; Enyedi, A. Plasma membrane Ca^{2+}-ATPases can shape the pattern of Ca^{2+} transients induced by store-operated Ca^{2+} entry. *Sci. Signal.* **2015**, *8*, ra19. [CrossRef] [PubMed]
41. Cardenas, C.; Foskett, J.K. Mitochondrial Ca^{2+} signals in autophagy. *Cell Calcium* **2012**, *52*, 44–51. [CrossRef] [PubMed]
42. Zuccolo, E.; Lim, D.; Kheder, D.A.; Perna, A.; Catarsi, P.; Botta, L.; Rosti, V.; Riboni, L.; Sancini, G.; Tanzi, F.; et al. Acetylcholine induces intracellular Ca^{2+} oscillations and nitric oxide release in mouse brain endothelial cells. *Cell Calcium* **2017**, *66*, 33–47. [CrossRef] [PubMed]
43. Huang, T.Y.; Chu, T.F.; Chen, H.I.; Jen, C.J. Heterogeneity of $[Ca^{2+}](i)$ signaling in intact rat aortic endothelium. *FASEB J.* **2000**, *14*, 797–804. [PubMed]
44. Duza, T.; Sarelius, I.H. Localized transient increases in endothelial cell Ca^{2+} in arterioles in situ: Implications for coordination of vascular function. *Am. J. Physiol. Heart Circ. Physiol.* **2004**, *286*, H2322–H2331. [CrossRef] [PubMed]
45. Kansui, Y.; Garland, C.J.; Dora, K.A. Enhanced spontaneous Ca^{2+} events in endothelial cells reflect signalling through myoendothelial gap junctions in pressurized mesenteric arteries. *Cell Calcium* **2008**, *44*, 135–146. [CrossRef] [PubMed]
46. Pedriali, G.; Rimessi, A.; Sbano, L.; Giorgi, C.; Wieckowski, M.R.; Previati, M.; Pinton, P. Regulation of Endoplasmic Reticulum-Mitochondria Ca^{2+} Transfer and Its Importance for Anti-Cancer Therapies. *Front. Oncol.* **2017**, *7*, 180. [CrossRef] [PubMed]
47. De Stefani, D.; Rizzuto, R.; Pozzan, T. Enjoy the Trip: Calcium in Mitochondria Back and Forth. *Annu. Rev. Biochem.* **2016**, *85*, 161–192. [CrossRef] [PubMed]
48. Marcu, R.; Wiczer, B.M.; Neeley, C.K.; Hawkins, B.J. Mitochondrial matrix Ca^{2+} accumulation regulates cytosolic NAD(+)/NADH metabolism, protein acetylation, and sirtuin expression. *Mol. Cell. Biol.* **2014**, *34*, 2890–2902. [CrossRef] [PubMed]
49. Dong, Z.; Shanmughapriya, S.; Tomar, D.; Siddiqui, N.; Lynch, S.; Nemani, N.; Breves, S.L.; Zhang, X.; Tripathi, A.; Palaniappan, P.; et al. Mitochondrial Ca^{2+} Uniporter Is a Mitochondrial Luminal Redox Sensor that Augments MCU Channel Activity. *Mol. Cell* **2017**, *65*, 1014.e7–1028.e7. [CrossRef] [PubMed]
50. Bittremieux, M.; Parys, J.B.; Pinton, P.; Bultynck, G. ER functions of oncogenes and tumor suppressors: Modulators of intracellular Ca^{2+} signaling. *Biochim. Biophys. Acta* **2016**, *1863 Pt B*, 1364–1378. [CrossRef] [PubMed]
51. Pinton, P.; Giorgi, C.; Pandolfi, P.P. The role of PML in the control of apoptotic cell fate: A new key player at ER-mitochondria sites. *Cell Death Differ.* **2011**, *18*, 1450–1456. [CrossRef] [PubMed]
52. Zhu, H.; Jin, Q.; Li, Y.; Ma, Q.; Wang, J.; Li, D.; Zhou, H.; Chen, Y. Melatonin protected cardiac microvascular endothelial cells against oxidative stress injury via suppression of IP3R-$[Ca^{2+}]$c/VDAC-$[Ca^{2+}]$m axis by activation of MAPK/ERK signaling pathway. *Cell Stress Chaperones* **2018**, *23*, 101–113. [CrossRef] [PubMed]
53. Madreiter-Sokolowski, C.T.; Gottschalk, B.; Parichatikanond, W.; Eroglu, E.; Klec, C.; Waldeck-Weiermair, M.; Malli, R.; Graier, W.F. Resveratrol Specifically Kills Cancer Cells by a Devastating Increase in the Ca^{2+} Coupling Between the Greatly Tethered Endoplasmic Reticulum and Mitochondria. *Cell. Physiol. Biochem.* **2016**, *39*, 1404–1420. [CrossRef] [PubMed]

54. Carmeliet, P.; Jain, R.K. Molecular mechanisms and clinical applications of angiogenesis. *Nature* **2011**, *473*, 298–307. [CrossRef] [PubMed]
55. Berridge, M.J.; Bootman, M.D.; Roderick, H.L. Calcium signalling: Dynamics, homeostasis and remodelling. *Nat. Rev. Mol. Cell Biol.* **2003**, *4*, 517–529. [CrossRef] [PubMed]
56. Moccia, F.; Dragoni, S.; Lodola, F.; Bonetti, E.; Bottino, C.; Guerra, G.; Laforenza, U.; Rosti, V.; Tanzi, F. Store-dependent Ca^{2+} entry in endothelial progenitor cells as a perspective tool to enhance cell-based therapy and adverse tumour vascularization. *Curr. Med. Chem.* **2012**, *19*, 5802–5818. [CrossRef] [PubMed]
57. Abdullaev, I.F.; Bisaillon, J.M.; Potier, M.; Gonzalez, J.C.; Motiani, R.K.; Trebak, M. Stim1 and Orai1 mediate CRAC currents and store-operated calcium entry important for endothelial cell proliferation. *Circ. Res.* **2008**, *103*, 1289–1299. [CrossRef] [PubMed]
58. Li, J.; Cubbon, R.M.; Wilson, L.A.; Amer, M.S.; McKeown, L.; Hou, B.; Majeed, Y.; Tumova, S.; Seymour, V.A.L.; Taylor, H.; et al. Orai1 and CRAC channel dependence of VEGF-activated Ca^{2+} entry and endothelial tube formation. *Circ. Res.* **2011**, *108*, 1190–1198. [CrossRef] [PubMed]
59. Blatter, L.A. Tissue Specificity: SOCE: Implications for Ca^{2+} Handling in Endothelial Cells. *Adv. Exp. Med. Biol.* **2017**, *993*, 343–361. [PubMed]
60. Gees, M.; Colsoul, B.; Nilius, B. The role of transient receptor potential cation channels in Ca^{2+} signaling. *Cold Spring Harb. Perspect. Biol.* **2010**, *2*, a003962. [CrossRef] [PubMed]
61. Sundivakkam, P.C.; Freichel, M.; Singh, V.; Yuan, J.P.; Vogel, S.M.; Flockerzi, V.; Malik, A.B.; Tiruppathi, C. The Ca^{2+} sensor stromal interaction molecule 1 (STIM1) is necessary and sufficient for the store-operated Ca^{2+} entry function of transient receptor potential canonical (TRPC) 1 and 4 channels in endothelial cells. *Mol. Pharmacol.* **2012**, *81*, 510–526. [CrossRef] [PubMed]
62. Cioffi, D.L.; Wu, S.; Chen, H.; Alexeyev, M.; St Croix, C.M.; Pitt, B.R.; Uhlig, S.; Stevens, T. Orai1 determines calcium selectivity of an endogenous TRPC heterotetramer channel. *Circ. Res.* **2012**, *110*, 1435–1444. [CrossRef] [PubMed]
63. Xu, N.; Cioffi, D.L.; Alexeyev, M.; Rich, T.C.; Stevens, T. Sodium entry through endothelial store-operated calcium entry channels: Regulation by Orai1. *Am. J. Physiol. Cell Physiol.* **2015**, *308*, C277–C288. [CrossRef] [PubMed]
64. Antigny, F.; Jousset, H.; Konig, S.; Frieden, M. Thapsigargin activates Ca^{2+} entry both by store-dependent, STIM1/Orai1-mediated, and store-independent, TRPC3/PLC/PKC-mediated pathways in human endothelial cells. *Cell Calcium* **2011**, *49*, 115–127. [CrossRef] [PubMed]
65. Weissmann, N.; Sydykov, A.; Kalwa, H.; Storch, U.; Fuchs, B.; Mederos y Schnitzler, M.; Brandes, R.P.; Grimminger, F.; Meissner, M.; Freichel, M.; et al. Activation of TRPC6 channels is essential for lung ischaemia-reperfusion induced oedema in mice. *Nat. Commun.* **2012**, *3*, 649. [CrossRef] [PubMed]
66. Boeldt, D.S.; Grummer, M.A.; Magness, R.R.; Bird, I.M. Altered VEGF-stimulated Ca^{2+} signaling in part underlies pregnancy-adapted eNOS activity in UAEC. *J. Endocrinol.* **2014**, *223*, 1–11. [CrossRef] [PubMed]
67. Hamdollah Zadeh, M.A.; Glass, C.A.; Magnussen, A.; Hancox, J.C.; Bates, D.O. VEGF-mediated elevated intracellular calcium and angiogenesis in human microvascular endothelial cells in vitro are inhibited by dominant negative TRPC6. *Microcirculation* **2008**, *15*, 605–614. [CrossRef] [PubMed]
68. Antoniotti, S.; Lovisolo, D.; Fiorio Pla, A.; Munaron, L. Expression and functional role of bTRPC1 channels in native endothelial cells. *FEBS Lett.* **2002**, *510*, 189–195. [CrossRef]
69. Ho, W.S.; Zheng, X.; Zhang, D.X. Role of endothelial TRPV4 channels in vascular actions of the endocannabinoid, 2-arachidonoylglycerol. *Br. J. Pharmacol.* **2015**, *172*, 5251–5264. [CrossRef] [PubMed]
70. Zheng, X.; Zinkevich, N.S.; Gebremedhin, D.; Gauthier, K.M.; Nishijima, Y.; Fang, J.; Wilcox, D.A.; Campbell, W.B.; Gutterman, D.D.; Zhang, D.X. Arachidonic acid-induced dilation in human coronary arterioles: Convergence of signaling mechanisms on endothelial TRPV4-mediated Ca^{2+} entry. *J. Am. Heart Assoc.* **2013**, *2*, e000080. [CrossRef] [PubMed]
71. Yoshida, T.; Inoue, R.; Morii, T.; Takahashi, N.; Yamamoto, S.; Hara, Y.; Tominaga, M.; Shimizu, S.; Sato, Y.; Mori, Y. Nitric oxide activates TRP channels by cysteine S-nitrosylation. *Nat. Chem. Biol.* **2006**, *2*, 596–607. [CrossRef] [PubMed]
72. Dietrich, A.; Gudermann, T. Another TRP to endothelial dysfunction: TRPM2 and endothelial permeability. *Circ. Res.* **2008**, *102*, 275–277. [CrossRef] [PubMed]
73. Kwan, H.Y.; Cheng, K.T.; Ma, Y.; Huang, Y.; Tang, N.L.; Yu, S.; Yao, X. CNGA2 contributes to ATP-induced noncapacitative Ca^{2+} influx in vascular endothelial cells. *J. Vasc. Res.* **2010**, *47*, 148–156. [CrossRef] [PubMed]

74. Berra-Romani, R.; Raqeeb, A.; Avelino-Cruz, J.E.; Moccia, F.; Oldani, A.; Speroni, F.; Taglietti, V.; Tanzi, F. Ca^{2+} signaling in injured in situ endothelium of rat aorta. *Cell Calcium* **2008**, *44*, 298–309. [CrossRef] [PubMed]
75. Moccia, F.; Frost, C.; Berra-Romani, R.; Tanzi, F.; Adams, D.J. Expression and function of neuronal nicotinic ACh receptors in rat microvascular endothelial cells. *Am. J. Physiol. Heart Circ. Physiol.* **2004**, *286*, H486–H491. [CrossRef] [PubMed]
76. LeMaistre, J.L.; Sanders, S.A.; Stobart, M.J.; Lu, L.; Knox, J.D.; Anderson, H.D.; Anderson, C.M. Coactivation of NMDA receptors by glutamate and D-serine induces dilation of isolated middle cerebral arteries. *J. Cereb. Blood Flow Metab.* **2012**, *32*, 537–547. [CrossRef] [PubMed]
77. AbouAlaiwi, W.A.; Takahashi, M.; Mell, B.R.; Jones, T.J.; Ratnam, S.; Kolb, R.J.; Nauli, S.M. Ciliary polycystin-2 is a mechanosensitive calcium channel involved in nitric oxide signaling cascades. *Circ. Res.* **2009**, *104*, 860–869. [CrossRef] [PubMed]
78. Berrout, J.; Jin, M.; O'Neil, R.G. Critical role of TRPP2 and TRPC1 channels in stretch-induced injury of blood-brain barrier endothelial cells. *Brain Res.* **2012**, *1436*, 1–12. [CrossRef] [PubMed]
79. Filosa, J.A.; Yao, X.; Rath, G. TRPV4 and the regulation of vascular tone. *J. Cardiovasc. Pharmacol.* **2013**, *61*, 113–119. [CrossRef] [PubMed]
80. Du, J.; Ma, X.; Shen, B.; Huang, Y.; Birnbaumer, L.; Yao, X. TRPV4, TRPC1, and TRPP2 assemble to form a flow-sensitive heteromeric channel. *FASEB J.* **2014**, *28*, 4677–4685. [CrossRef] [PubMed]
81. Li, J.; Hou, B.; Tumova, S.; Muraki, K.; Bruns, A.; Ludlow, M.J.; Sedo, A.; Hyman, A.J.; McKeown, L.; Young, R.S.; et al. Piezo1 integration of vascular architecture with physiological force. *Nature* **2014**, *515*, 279–282. [CrossRef] [PubMed]
82. Medina, R.J.; Barber, C.L.; Sabatier, F.; Dignat-George, F.; Melero-Martin, J.M.; Khosrotehrani, K.; Ohneda, O.; Randi, A.M.; Chan, J.K.Y.; Yamaguchi, T.; et al. Endothelial Progenitors: A Consensus Statement on Nomenclature. *Stem Cells Transl. Med.* **2017**, *6*, 1316–1320. [CrossRef] [PubMed]
83. Dragoni, S.; Laforenza, U.; Bonetti, E.; Lodola, F.; Bottino, C.; Berra-Romani, R.; Carlo Bongio, G.; Cinelli, M.P.; Guerra, G.; Pedrazzoli, P.; et al. Vascular endothelial growth factor stimulates endothelial colony forming cells proliferation and tubulogenesis by inducing oscillations in intracellular Ca^{2+} concentration. *Stem Cells* **2011**, *29*, 1898–1907. [CrossRef] [PubMed]
84. Sanchez-Hernandez, Y.; Laforenza, U.; Bonetti, E.; Fontana, J.; Dragoni, S.; Russo, M.; Avelino-Cruz, J.E.; Schinelli, S.; Testa, D.; Guerra, G.; et al. Store-operated Ca^{2+} entry is expressed in human endothelial progenitor cells. *Stem Cells Dev.* **2010**, *19*, 1967–1981. [CrossRef] [PubMed]
85. Dragoni, S.; Laforenza, U.; Bonetti, E.; Lodola, F.; Bottino, C.; Guerra, G.; Borghesi, A.; Stronati, M.; Rosti, V.; Tanzi, F.; et al. Canonical transient receptor potential 3 channel triggers vascular endothelial growth factor-induced intracellular Ca^{2+} oscillations in endothelial progenitor cells isolated from umbilical cord blood. *Stem Cells Dev.* **2013**, *22*, 2561–2580. [CrossRef] [PubMed]
86. Zuccolo, E.; Dragoni, S.; Poletto, V.; Catarsi, P.; Guido, D.; Rappa, A.; Reforgiato, M.; Lodola, F.; Lim, D.; Rosti, V.; et al. Arachidonic acid-evoked Ca^{2+} signals promote nitric oxide release and proliferation in human endothelial colony forming cells. *Vascul. Pharmacol.* **2016**, *87*, 159–171. [CrossRef] [PubMed]
87. Poletto, V.; Dragoni, S.; Lim, D.; Biggiogera, M.; Aronica, A.; Cinelli, M.; De Luca, A.; Rosti, V.; Porta, C.; Guerra, G.; et al. Endoplasmic Reticulum Ca^{2+} Handling and Apoptotic Resistance in Tumor-Derived Endothelial Colony Forming Cells. *J. Cell. Biochem.* **2016**, *117*, 2260–2271. [CrossRef] [PubMed]
88. Pla, A.F.; Ong, H.L.; Cheng, K.T.; Brossa, A.; Bussolati, B.; Lockwich, T.; Paria, B.; Munaron, L.; Ambudkar, I.S. TRPV4 mediates tumor-derived endothelial cell migration via arachidonic acid-activated actin remodeling. *Oncogene* **2012**, *31*, 200–212.
89. Fiorio Pla, A.; Grange, C.; Antoniotti, S.; Tomatis, C.; Merlino, A.; Bussolati, B.; Munaron, L. Arachidonic acid-induced Ca^{2+} entry is involved in early steps of tumor angiogenesis. *Mol. Cancer Res.* **2008**, *6*, 535–545. [PubMed]
90. Adapala, R.K.; Thoppil, R.J.; Ghosh, K.; Cappelli, H.C.; Dudley, A.C.; Paruchuri, S.; Keshamouni, V.; Klagsbrun, M.; Meszaros, J.G.; Chilian, W.M.; et al. Activation of mechanosensitive ion channel TRPV4 normalizes tumor vasculature and improves cancer therapy. *Oncogene* **2016**, *35*, 314–322. [CrossRef] [PubMed]
91. Thoppil, R.J.; Cappelli, H.C.; Adapala, R.K.; Kanugula, A.K.; Paruchuri, S.; Thodeti, C.K. TRPV4 channels regulate tumor angiogenesis via modulation of Rho/Rho kinase pathway. *Oncotarget* **2016**, *7*, 25849–25861. [CrossRef] [PubMed]

92. Thoppil, R.J.; Adapala, R.K.; Cappelli, H.C.; Kondeti, V.; Dudley, A.C.; Gary Meszaros, J.; Paruchuri, S.; Thodeti, C.K. TRPV4 channel activation selectively inhibits tumor endothelial cell proliferation. *Sci. Rep.* **2015**, *5*, 14257. [CrossRef] [PubMed]
93. Yang, H.; Liu, C.; Zhou, R.M.; Yao, J.; Li, X.M.; Shen, Y.; Cheng, H.; Yuan, J.; Yan, B.; Jiang, Q. Piezo2 protein: A novel regulator of tumor angiogenesis and hyperpermeability. *Oncotarget* **2016**, *7*, 44630–44643. [CrossRef] [PubMed]
94. Avanzato, D.; Genova, T.; Fiorio Pla, A.; Bernardini, M.; Bianco, S.; Bussolati, B.; Mancardi, D.; Giraudo, E.; Maione, F.; Cassoni, P.; et al. Activation of P2X7 and P2Y11 purinergic receptors inhibits migration and normalizes tumor-derived endothelial cells via cAMP signaling. *Sci. Rep.* **2016**, *6*, 32602. [CrossRef] [PubMed]
95. Lodola, F.; Laforenza, U.; Bonetti, E.; Lim, D.; Dragoni, S.; Bottino, C.; Ong, H.L.; Guerra, G.; Ganini, C.; Massa, M.; et al. Store-operated Ca^{2+} entry is remodelled and controls in vitro angiogenesis in endothelial progenitor cells isolated from tumoral patients. *PLoS ONE* **2012**, *7*, e42541. [CrossRef] [PubMed]
96. Lodola, F.; Laforenza, U.; Cattaneo, F.; Ruffinatti, F.A.; Poletto, V.; Massa, M.; Tancredi, R.; Zuccolo, E.; Khdar, A.D.; Riccardi, A.; et al. VEGF-induced intracellular Ca^{2+} oscillations are down-regulated and do not stimulate angiogenesis in breast cancer-derived endothelial colony forming cells. *Oncotarget* **2017**, *8*, 95223–95246. [PubMed]
97. Zuccolo, E.; Bottino, C.; Diofano, F.; Poletto, V.; Codazzi, A.C.; Mannarino, S.; Campanelli, R.; Fois, G.; Marseglia, G.L.; Guerra, G.; et al. Constitutive Store-Operated Ca^{2+} Entry Leads to Enhanced Nitric Oxide Production and Proliferation in Infantile Hemangioma-Derived Endothelial Colony-Forming Cells. *Stem Cells Dev.* **2016**, *25*, 301–319. [CrossRef] [PubMed]
98. Heeschen, C.; Weis, M.; Aicher, A.; Dimmeler, S.; Cooke, J.P. A novel angiogenic pathway mediated by non-neuronal nicotinic acetylcholine receptors. *J. Clin. Investig.* **2002**, *110*, 527–536. [CrossRef] [PubMed]
99. Heeschen, C.; Jang, J.J.; Weis, M.; Pathak, A.; Kaji, S.; Hu, R.S.; Tsao, P.S.; Johnson, F.L.; Cooke, J.P. Nicotine stimulates angiogenesis and promotes tumor growth and atherosclerosis. *Nat. Med.* **2001**, *7*, 833–839. [CrossRef] [PubMed]
100. Yu, M.; Liu, Q.; Sun, J.; Yi, K.; Wu, L.; Tan, X. Nicotine improves the functional activity of late endothelial progenitor cells via nicotinic acetylcholine receptors. *Biochem. Cell Biol.* **2011**, *89*, 405–410. [CrossRef] [PubMed]
101. Natori, T.; Sata, M.; Washida, M.; Hirata, Y.; Nagai, R.; Makuuchi, M. Nicotine enhances neovascularization and promotes tumor growth. *Mol. Cells* **2003**, *16*, 143–146. [PubMed]
102. Alonso, F.; Domingos-Pereira, S.; Le Gal, L.; Derre, L.; Meda, P.; Jichlinski, P.; Nardelli-Haefliger, D.; Haefliger, J.A. Targeting endothelial connexin40 inhibits tumor growth by reducing angiogenesis and improving vessel perfusion. *Oncotarget* **2016**, *7*, 14015–14028. [CrossRef] [PubMed]
103. Rigiracciolo, D.C.; Scarpelli, A.; Lappano, R.; Pisano, A.; Santolla, M.F.; Avino, S.; De Marco, P.; Bussolati, B.; Maggiolini, M.; De Francesco, E.M. GPER is involved in the stimulatory effects of aldosterone in breast cancer cells and breast tumor-derived endothelial cells. *Oncotarget* **2016**, *7*, 94–111. [CrossRef] [PubMed]
104. Dong, Y.; Pan, Q.; Jiang, L.; Chen, Z.; Zhang, F.; Liu, Y.; Xing, H.; Shi, M.; Li, J.; Li, X.; et al. Tumor endothelial expression of P-glycoprotein upon microvesicular transfer of TrpC5 derived from adriamycin-resistant breast cancer cells. *Biochem. Biophys. Res. Commun.* **2014**, *446*, 85–90. [CrossRef] [PubMed]
105. White, J.P.; Cibelli, M.; Urban, L.; Nilius, B.; McGeown, J.G.; Nagy, I. TRPV4: Molecular Conductor of a Diverse Orchestra. *Physiol. Rev.* **2016**, *96*, 911–973. [CrossRef] [PubMed]
106. Everaerts, W.; Nilius, B.; Owsianik, G. The vanilloid transient receptor potential channel TRPV4: From structure to disease. *Prog. Biophys. Mol. Biol.* **2010**, *103*, 2–17. [CrossRef] [PubMed]
107. He, D.; Pan, Q.; Chen, Z.; Sun, C.; Zhang, P.; Mao, A.; Zhu, Y.; Li, H.; Lu, C.; Xie, M.; et al. Treatment of hypertension by increasing impaired endothelial TRPV4-KCa2.3 interaction. *EMBO Mol. Med.* **2017**, *9*, 1491–1503. [CrossRef] [PubMed]
108. Chen, C.K.; Hsu, P.Y.; Wang, T.M.; Miao, Z.F.; Lin, R.T.; Juo, S.H. TRPV4 Activation Contributes Functional Recovery from Ischemic Stroke via Angiogenesis and Neurogenesis. *Mol. Neurobiol.* **2017**. [CrossRef] [PubMed]
109. Hatano, N.; Suzuki, H.; Itoh, Y.; Muraki, K. TRPV4 partially participates in proliferation of human brain capillary endothelial cells. *Life Sci.* **2013**, *92*, 317–324. [CrossRef] [PubMed]

110. Matthews, B.D.; Thodeti, C.K.; Tytell, J.D.; Mammoto, A.; Overby, D.R.; Ingber, D.E. Ultra-rapid activation of TRPV4 ion channels by mechanical forces applied to cell surface beta1 integrins. *Integr. Biol. (Camb.)* **2010**, *2*, 435–442. [CrossRef] [PubMed]
111. Thodeti, C.K.; Matthews, B.; Ravi, A.; Mammoto, A.; Ghosh, K.; Bracha, A.L.; Ingber, D.E. TRPV4 Channels Mediate Cyclic Strain-Induced Endothelial Cell Reorientation Through Integrin-to-Integrin Signaling. *Circ. Res.* **2009**, *104*, 1123–1130. [CrossRef] [PubMed]
112. Sonkusare, S.K.; Bonev, A.D.; Ledoux, J.; Liedtke, W.; Kotlikoff, M.I.; Heppner, T.J.; Hill-Eubanks, D.C.; Nelson, M.T. Elementary Ca^{2+} signals through endothelial TRPV4 channels regulate vascular function. *Science* **2012**, *336*, 597–601. [CrossRef] [PubMed]
113. Zhao, L.; Sullivan, M.N.; Chase, M.; Gonzales, A.L.; Earley, S. Calcineurin/nuclear factor of activated T cells-coupled vanilliod transient receptor potential channel 4 Ca^{2+} sparklets stimulate airway smooth muscle cell proliferation. *Am. J. Respir. Cell Mol. Biol.* **2014**, *50*, 1064–1075. [CrossRef] [PubMed]
114. Wen, Z.H.; Su, Y.C.; Lai, P.L.; Zhang, Y.; Xu, Y.F.; Zhao, A.; Yao, G.Y.; Jia, C.H.; Lin, J.; Xu, S.; et al. Critical role of arachidonic acid-activated mTOR signaling in breast carcinogenesis and angiogenesis. *Oncogene* **2013**, *32*, 160–170. [CrossRef] [PubMed]
115. Munaron, L. Shuffling the cards in signal transduction: Calcium, arachidonic acid and mechanosensitivity. *World J. Biol. Chem.* **2011**, *2*, 59–66. [CrossRef] [PubMed]
116. Kim, E.; Tunset, H.M.; Cebulla, J.; Vettukattil, R.; Helgesen, H.; Feuerherm, A.J.; Engebraten, O.; Maelandsmo, G.M.; Johansen, B.; Moestue, S.A. Anti-vascular effects of the cytosolic phospholipase A2 inhibitor AVX235 in a patient-derived basal-like breast cancer model. *BMC Cancer* **2016**, *16*, 191. [CrossRef] [PubMed]
117. Carmeliet, P.; Jain, R.K. Principles and mechanisms of vessel normalization for cancer and other angiogenic diseases. *Nat. Rev. Drug Discov.* **2011**, *10*, 417–427. [CrossRef] [PubMed]
118. Goel, S.; Duda, D.G.; Xu, L.; Munn, L.L.; Boucher, Y.; Fukumura, D.; Jain, R.K. Normalization of the vasculature for treatment of cancer and other diseases. *Physiol. Rev.* **2011**, *91*, 1071–1121. [CrossRef] [PubMed]
119. Bagriantsev, S.N.; Gracheva, E.O.; Gallagher, P.G. Piezo proteins: Regulators of mechanosensation and other cellular processes. *J. Biol. Chem.* **2014**, *289*, 31673–31681. [CrossRef] [PubMed]
120. Honore, E.; Martins, J.R.; Penton, D.; Patel, A.; Demolombe, S. The Piezo Mechanosensitive Ion Channels: May the Force Be with You! *Rev. Physiol. Biochem. Pharmacol.* **2015**, *169*, 25–41. [PubMed]
121. Ranade, S.S.; Qiu, Z.; Woo, S.H.; Hur, S.S.; Murthy, S.E.; Cahalan, S.M.; Xu, J.; Mathur, J.; Bandell, M.; Coste, B.; et al. Piezo1, a mechanically activated ion channel, is required for vascular development in mice. *Proc. Natl. Acad. Sci. USA* **2014**, *111*, 10347–10352. [CrossRef] [PubMed]
122. Di Virgilio, F.; Adinolfi, E. Extracellular purines, purinergic receptors and tumor growth. *Oncogene* **2017**, *36*, 293–303. [CrossRef] [PubMed]
123. Burnstock, G. Purinergic Signaling in the Cardiovascular System. *Circ. Res.* **2017**, *120*, 207–228. [CrossRef] [PubMed]
124. Adinolfi, E.; Raffaghello, L.; Giuliani, A.L.; Cavazzini, L.; Capece, M.; Chiozzi, P.; Bianchi, G.; Kroemer, G.; Pistoia, V.; Di Virgilio, F. Expression of P2X7 receptor increases in vivo tumor growth. *Cancer Res.* **2012**, *72*, 2957–2969. [CrossRef] [PubMed]
125. Azimi, I.; Beilby, H.; Davis, F.M.; Marcial, D.L.; Kenny, P.A.; Thompson, E.W.; Roberts-Thomson, S.J.; Monteith, G.R. Altered purinergic receptor-Ca^{2+} signaling associated with hypoxia-induced epithelial-mesenchymal transition in breast cancer cells. *Mol. Oncol.* **2016**, *10*, 166–178. [CrossRef] [PubMed]
126. Fang, J.; Chen, X.; Wang, S.; Xie, T.; Du, X.; Liu, H.; Li, X.; Chen, J.; Zhang, B.; Liang, H.; et al. The expression of P2X(7) receptors in EPCs and their potential role in the targeting of EPCs to brain gliomas. *Cancer Biol. Ther.* **2015**, *16*, 498–510. [CrossRef] [PubMed]
127. Dragoni, S.; Guerra, G.; Fiorio Pla, A.; Bertoni, G.; Rappa, A.; Poletto, V.; Bottino, C.; Aronica, A.; Lodola, F.; Cinelli, M.P.; et al. A functional Transient Receptor Potential Vanilloid 4 (TRPV4) channel is expressed in human endothelial progenitor cells. *J. Cell. Physiol.* **2015**, *230*, 95–104. [CrossRef] [PubMed]
128. Hofmann, N.A.; Barth, S.; Waldeck-Weiermair, M.; Klec, C.; Strunk, D.; Malli, R.; Graier, W.F. TRPV1 mediates cellular uptake of anandamide and thus promotes endothelial cell proliferation and network-formation. *Biol. Open* **2014**, *3*, 1164–1172. [CrossRef] [PubMed]

129. Ong, H.L.; Jang, S.I.; Ambudkar, I.S. Distinct contributions of Orai1 and TRPC1 to agonist-induced [Ca^{2+}](i) signals determine specificity of Ca^{2+}-dependent gene expression. *PLoS ONE* **2012**, *7*, e47146. [CrossRef] [PubMed]
130. Gueguinou, M.; Harnois, T.; Crottes, D.; Uguen, A.; Deliot, N.; Gambade, A.; Chantome, A.; Haelters, J.P.; Jaffres, P.A.; Jourdan, M.L.; et al. SK3/TRPC1/Orai1 complex regulates SOCE-dependent colon cancer cell migration: A novel opportunity to modulate anti-EGFR mAb action by the alkyl-lipid Ohmline. *Oncotarget* **2016**, *7*, 36168–36184. [CrossRef] [PubMed]
131. Moccia, F.; Dragoni, S.; Poletto, V.; Rosti, V.; Tanzi, F.; Ganini, C.; Porta, C. Orai1 and Transient Receptor Potential Channels as novel molecular targets to impair tumor neovascularisation in renal cell carcinoma and other malignancies. *Anticancer Agents Med. Chem.* **2014**, *14*, 296–312. [CrossRef] [PubMed]
132. Porta, C.; Giglione, P.; Paglino, C. Targeted therapy for renal cell carcinoma: Focus on 2nd and 3rd line. *Expert Opin. Pharmacother.* **2016**, *17*, 643–655. [CrossRef] [PubMed]
133. Prakriya, M.; Lewis, R.S. Store-Operated Calcium Channels. *Physiol. Rev.* **2015**, *95*, 1383–1436. [CrossRef] [PubMed]
134. Vashisht, A.; Trebak, M.; Motiani, R.K. STIM and Orai proteins as novel targets for cancer therapy. A Review in the Theme: Cell and Molecular Processes in Cancer Metastasis. *Am. J. Physiol. Cell Physiol.* **2015**, *309*, C457–C469. [CrossRef] [PubMed]
135. Moccia, F.; Zuccolo, E.; Poletto, V.; Turin, I.; Guerra, G.; Pedrazzoli, P.; Rosti, V.; Porta, C.; Montagna, D. Targeting Stim and Orai Proteins as an Alternative Approach in Anticancer Therapy. *Curr. Med. Chem.* **2016**, *23*, 3450–3480. [CrossRef] [PubMed]
136. Li, Y.; Guo, B.; Xie, Q.; Ye, D.; Zhang, D.; Zhu, Y.; Chen, H.; Zhu, B. STIM1 Mediates Hypoxia-Driven Hepatocarcinogenesis via Interaction with HIF-1. *Cell Rep.* **2015**, *12*, 388–395. [CrossRef] [PubMed]
137. Gudas, L.J.; Fu, L.; Minton, D.R.; Mongan, N.P.; Nanus, D.M. The role of HIF1alpha in renal cell carcinoma tumorigenesis. *J. Mol. Med. (Berl.)* **2014**, *92*, 825–836. [CrossRef] [PubMed]
138. Semenza, G.L. Regulation of the breast cancer stem cell phenotype by hypoxia-inducible factors. *Clin. Sci. (Lond.)* **2015**, *129*, 1037–1045. [CrossRef] [PubMed]
139. Wang, J.; Weigand, L.; Lu, W.; Sylvester, J.T.; Semenza, G.L.; Shimoda, L.A. Hypoxia inducible factor 1 mediates hypoxia-induced TRPC expression and elevated intracellular Ca^{2+} in pulmonary arterial smooth muscle cells. *Circ. Res.* **2006**, *98*, 1528–1537. [CrossRef] [PubMed]
140. Leaute-Labreze, C.; Harper, J.I.; Hoeger, P.H. Infantile haemangioma. *Lancet* **2017**, *390*, 85–94. [CrossRef]
141. Bischoff, J. Progenitor cells in infantile hemangioma. *J. Craniofac. Surg.* **2009**, *20* (Suppl. 1), 695–697. [CrossRef] [PubMed]
142. Khan, Z.A.; Melero-Martin, J.M.; Wu, X.; Paruchuri, S.; Boscolo, E.; Mulliken, J.B.; Bischoff, J. Endothelial progenitor cells from infantile hemangioma and umbilical cord blood display unique cellular responses to endostatin. *Blood* **2006**, *108*, 915–921. [CrossRef] [PubMed]
143. Brandman, O.; Liou, J.; Park, W.S.; Meyer, T. STIM2 is a feedback regulator that stabilizes basal cytosolic and endoplasmic reticulum Ca^{2+} levels. *Cell* **2007**, *131*, 1327–1339. [CrossRef] [PubMed]
144. Greenberger, S.; Bischoff, J. Infantile hemangioma-mechanism(s) of drug action on a vascular tumor. *Cold Spring Harb. Perspect. Med.* **2011**, *1*, a006460. [CrossRef] [PubMed]
145. Zoli, M.; Pucci, S.; Vilella, A.; Gotti, C. Neuronal and extraneuronal nicotinic acetylcholine receptors. *Curr. Neuropharmacol.* **2017**. [CrossRef]
146. Yakel, J.L. Nicotinic ACh receptors in the hippocampal circuit; functional expression and role in synaptic plasticity. *J. Physiol.* **2014**, *592*, 4147–4153. [CrossRef] [PubMed]
147. Egleton, R.D.; Brown, K.C.; Dasgupta, P. Nicotinic acetylcholine receptors in cancer: Multiple roles in proliferation and inhibition of apoptosis. *Trends Pharmacol. Sci.* **2008**, *29*, 151–158. [CrossRef] [PubMed]
148. Cooke, J.P.; Ghebremariam, Y.T. Endothelial nicotinic acetylcholine receptors and angiogenesis. *Trends Cardiovasc. Med.* **2008**, *18*, 247–253. [CrossRef] [PubMed]
149. Egleton, R.D.; Brown, K.C.; Dasgupta, P. Angiogenic activity of nicotinic acetylcholine receptors: Implications in tobacco-related vascular diseases. *Pharmacol. Ther.* **2009**, *121*, 205–223. [CrossRef] [PubMed]
150. Dom, A.M.; Buckley, A.W.; Brown, K.C.; Egleton, R.D.; Marcelo, A.J.; Proper, N.A.; Weller, D.E.; Shah, Y.H.; Lau, J.K.; Dasgupta, P. The alpha7-nicotinic acetylcholine receptor and MMP-2/-9 pathway mediate the proangiogenic effect of nicotine in human retinal endothelial cells. *Investig. Ophthalmol. Vis. Sci.* **2011**, *52*, 4428–4438. [CrossRef] [PubMed]

151. Smedlund, K.; Tano, J.Y.; Margiotta, J.; Vazquez, G. Evidence for operation of nicotinic and muscarinic acetylcholine receptor-dependent survival pathways in human coronary artery endothelial cells. *J. Cell Biochem.* **2011**, *112*, 1978–1984. [CrossRef] [PubMed]
152. Fucile, S. Ca^{2+} permeability of nicotinic acetylcholine receptors. *Cell Calcium* **2004**, *35*, 1–8. [CrossRef] [PubMed]
153. Shi, D.; Guo, W.; Chen, W.; Fu, L.; Wang, J.; Tian, Y.; Xiao, X.; Kang, T.; Huang, W.; Deng, W. Nicotine promotes proliferation of human nasopharyngeal carcinoma cells by regulating alpha7AChR, ERK, HIF-1alpha and VEGF/PEDF signaling. *PLoS ONE* **2012**, *7*, e43898.
154. Zhu, B.Q.; Heeschen, C.; Sievers, R.E.; Karliner, J.S.; Parmley, W.W.; Glantz, S.A.; Cooke, J.P. Second hand smoke stimulates tumor angiogenesis and growth. *Cancer Cell* **2003**, *4*, 191–196. [CrossRef]
155. Altaany, Z.; Moccia, F.; Munaron, L.; Mancardi, D.; Wang, R. Hydrogen sulfide and endothelial dysfunction: Relationship with nitric oxide. *Curr. Med. Chem.* **2014**, *21*, 3646–3661. [CrossRef] [PubMed]
156. Mancardi, D.; Pla, A.F.; Moccia, F.; Tanzi, F.; Munaron, L. Old and new gasotransmitters in the cardiovascular system: Focus on the role of nitric oxide and hydrogen sulfide in endothelial cells and cardiomyocytes. *Curr. Pharm. Biotechnol.* **2011**, *12*, 1406–1415. [CrossRef] [PubMed]
157. Wang, R. Physiological implications of hydrogen sulfide: A whiff exploration that blossomed. *Physiol. Rev.* **2012**, *92*, 791–896. [CrossRef] [PubMed]
158. Hartmann, C.; Nussbaum, B.; Calzia, E.; Radermacher, P.; Wepler, M. Gaseous Mediators and Mitochondrial Function: The Future of Pharmacologically Induced Suspended Animation? *Front. Physiol.* **2017**, *8*, 691. [CrossRef] [PubMed]
159. Munaron, L. Intracellular calcium, endothelial cells and angiogenesis. *Recent Pat. Anticancer Drug Discov.* **2006**, *1*, 105–119. [CrossRef] [PubMed]
160. Loboda, A.; Jozkowicz, A.; Dulak, J. Carbon monoxide: Pro- or anti-angiogenic agent? Comment on Ahmad et al. (Thromb Haemost 2015; 113: 329–337). *Thromb. Haemost.* **2015**, *114*, 432–433. [CrossRef] [PubMed]
161. Tran, A.N.; Boyd, N.H.; Walker, K.; Hjelmeland, A.B. NOS Expression and NO Function in Glioma and Implications for Patient Therapies. *Antioxid. Redox Signal.* **2017**, *26*, 986–999. [CrossRef] [PubMed]
162. Mocellin, S. Nitric oxide: Cancer target or anticancer agent? *Curr. Cancer Drug Targets* **2009**, *9*, 214–236. [CrossRef] [PubMed]
163. Basudhar, D.; Somasundaram, V.; de Oliveira, G.A.; Kesarwala, A.; Heinecke, J.L.; Cheng, R.Y.; Glynn, S.A.; Ambs, S.; Wink, D.A.; Ridnour, L.A. Nitric Oxide Synthase-2-Derived Nitric Oxide Drives Multiple Pathways of Breast Cancer Progression. *Antioxid. Redox Signal.* **2017**, *26*, 1044–1058. [CrossRef] [PubMed]
164. Fiorio Pla, A.; Genova, T.; Pupo, E.; Tomatis, C.; Genazzani, A.; Zaninetti, R.; Munaron, L. Multiple roles of protein kinase a in arachidonic acid-mediated Ca^{2+} entry and tumor-derived human endothelial cell migration. *Mol. Cancer Res.* **2010**, *8*, 1466–1476. [CrossRef] [PubMed]
165. Dragoni, S.; Laforenza, U.; Bonetti, E.; Reforgiato, M.; Poletto, V.; Lodola, F.; Bottino, C.; Guido, D.; Rappa, A.; Pareek, S.; et al. Enhanced expression of Stim, Orai, and TRPC transcripts and proteins in endothelial progenitor cells isolated from patients with primary myelofibrosis. *PLoS ONE* **2014**, *9*, e91099. [CrossRef] [PubMed]
166. Munaron, L.; Avanzato, D.; Moccia, F.; Mancardi, D. Hydrogen sulfide as a regulator of calcium channels. *Cell Calcium* **2013**, *53*, 77–84. [CrossRef] [PubMed]
167. Bauer, C.C.; Boyle, J.P.; Porter, K.E.; Peers, C. Modulation of Ca^{2+} signalling in human vascular endothelial cells by hydrogen sulfide. *Atherosclerosis* **2010**, *209*, 374–380. [CrossRef] [PubMed]
168. Moccia, F.; Bertoni, G.; Pla, A.F.; Dragoni, S.; Pupo, E.; Merlino, A.; Mancardi, D.; Munaron, L.; Tanzi, F. Hydrogen sulfide regulates intracellular Ca^{2+} concentration in endothelial cells from excised rat aorta. *Curr. Pharm. Biotechnol.* **2011**, *12*, 1416–1426. [CrossRef] [PubMed]
169. Katsouda, A.; Bibli, S.I.; Pyriochou, A.; Szabo, C.; Papapetropoulos, A. Regulation and role of endogenously produced hydrogen sulfide in angiogenesis. *Pharmacol. Res.* **2016**, *113 Pt A*, 175–185. [CrossRef] [PubMed]
170. Liu, Y.; Yang, R.; Liu, X.; Zhou, Y.; Qu, C.; Kikuiri, T.; Wang, S.; Zandi, E.; Du, J.; Ambudkar, I.S.; et al. Hydrogen Sulfide Maintains Mesenchymal Stem Cell Function and Bone Homeostasis via Regulation of Ca^{2+} Channel Sulfhydration. *Cell Stem Cell* **2014**, *15*, 66–78. [CrossRef] [PubMed]
171. Ujike, A.; Otsuguro, K.; Miyamoto, R.; Yamaguchi, S.; Ito, S. Bidirectional effects of hydrogen sulfide via ATP-sensitive K(+) channels and transient receptor potential A1 channels in RIN14B cells. *Eur. J. Pharmacol.* **2015**, *764*, 463–470. [CrossRef] [PubMed]

172. Yang, G.; Wu, L.; Jiang, B.; Yang, W.; Qi, J.; Cao, K.; Meng, Q.; Mustafa, A.K.; Mu, W.; Zhang, S.; et al. H2S as a physiologic vasorelaxant: Hypertension in mice with deletion of cystathionine gamma-lyase. *Science* **2008**, *322*, 587–590. [CrossRef] [PubMed]
173. Dedkova, E.N.; Blatter, L.A. Nitric oxide inhibits capacitative Ca^{2+} entry and enhances endoplasmic reticulum Ca^{2+} uptake in bovine vascular endothelial cells. *J. Physiol.* **2002**, *539 Pt 1*, 77–91. [CrossRef] [PubMed]
174. Saez, J.C.; Leybaert, L. Hunting for connexin hemichannels. *FEBS Lett.* **2014**, *588*, 1205–1211. [CrossRef] [PubMed]
175. Boeldt, D.S.; Krupp, J.; Yi, F.X.; Khurshid, N.; Shah, D.M.; Bird, I.M. Positive versus negative effects of VEGF165 on Ca^{2+} signaling and NO production in human endothelial cells. *Am. J. Physiol. Heart Circ. Physiol.* **2017**, *312*, H173–H181. [CrossRef] [PubMed]
176. Boittin, F.X.; Alonso, F.; Le Gal, L.; Allagnat, F.; Beny, J.L.; Haefliger, J.A. Connexins and M3 muscarinic receptors contribute to heterogeneous Ca^{2+} signaling in mouse aortic endothelium. *Cell. Physiol. Biochem.* **2013**, *31*, 166–178. [CrossRef] [PubMed]
177. Berra-Romani, R.; Raqeeb, A.; Torres-Jácome, J.; Guzman-Silva, A.; Guerra, G.; Tanzi, F.; Moccia, F. The mechanism of injury-induced intracellular calcium concentration oscillations in the endothelium of excised rat aorta. *J. Vasc. Res.* **2012**, *49*, 65–76. [CrossRef] [PubMed]
178. De Bock, M.; Culot, M.; Wang, N.; Bol, M.; Decrock, E.; De Vuyst, E.; da Costa, A.; Dauwe, I.; Vinken, M.; Simon, A.M.; et al. Connexin channels provide a target to manipulate brain endothelial calcium dynamics and blood-brain barrier permeability. *J. Cereb. Blood Flow Metab.* **2011**, *31*, 1942–1957. [CrossRef] [PubMed]
179. De Bock, M.; Wang, N.; Decrock, E.; Bol, M.; Gadicherla, A.K.; Culot, M.; Cecchelli, R.; Bultynck, G.; Leybaert, L. Endothelial calcium dynamics, connexin channels and blood-brain barrier function. *Prog. Neurobiol.* **2013**, *108*, 1–20. [CrossRef] [PubMed]
180. McLachlan, E.; Shao, Q.; Wang, H.L.; Langlois, S.; Laird, D.W. Connexins act as tumor suppressors in three-dimensional mammary cell organoids by regulating differentiation and angiogenesis. *Cancer Res.* **2006**, *66*, 9886–9894. [CrossRef] [PubMed]
181. Trosko, J.E.; Ruch, R.J. Gap junctions as targets for cancer chemoprevention and chemotherapy. *Curr. Drug Targets* **2002**, *3*, 465–482. [CrossRef] [PubMed]
182. Zhang, J.; O'Carroll, S.J.; Henare, K.; Ching, L.M.; Ormonde, S.; Nicholson, L.F.; Danesh-Meyer, H.V.; Green, C.R. Connexin hemichannel induced vascular leak suggests a new paradigm for cancer therapy. *FEBS Lett.* **2014**, *588*, 1365–1371. [CrossRef] [PubMed]
183. Counillon, L.; Bouret, Y.; Marchiq, I.; Pouyssegur, J. Na(+)/H(+) antiporter (NHE1) and lactate/H(+) symporters (MCTs) in pH homeostasis and cancer metabolism. *Biochim. Biophys. Acta* **2016**, *1863*, 2465–2480. [CrossRef] [PubMed]
184. Mo, X.G.; Chen, Q.W.; Li, X.S.; Zheng, M.M.; Ke, D.Z.; Deng, W.; Li, G.Q.; Jiang, J.; Wu, Z.Q.; Wang, L.; et al. Suppression of NHE1 by small interfering RNA inhibits HIF-1alpha-induced angiogenesis in vitro via modulation of calpain activity. *Microvasc. Res.* **2011**, *81*, 160–168. [CrossRef] [PubMed]
185. Ayajiki, K.; Kindermann, M.; Hecker, M.; Fleming, I.; Busse, R. Intracellular pH and tyrosine phosphorylation but not calcium determine shear stress-induced nitric oxide production in native endothelial cells. *Circ. Res.* **1996**, *78*, 750–758. [CrossRef] [PubMed]
186. Yuen, N.; Lam, T.I.; Wallace, B.K.; Klug, N.R.; Anderson, S.E.; O'Donnell, M.E. Ischemic factor-induced increases in cerebral microvascular endothelial cell Na/H exchange activity and abundance: Evidence for involvement of ERK1/2 MAP kinase. *Am. J. Physiol. Cell Physiol.* **2014**, *306*, C931–C942. [CrossRef] [PubMed]
187. Ghigo, D.; Bussolino, F.; Garbarino, G.; Heller, R.; Turrini, F.; Pescarmona, G.; Cragoe, E.J., Jr.; Pegoraro, L.; Bosia, A. Role of Na^+/H^+ exchange in thrombin-induced platelet-activating factor production by human endothelial cells. *J. Biol. Chem.* **1988**, *263*, 19437–19446. [PubMed]
188. Siffert, W.; Akkerman, J.W. Na^+/H^+ exchange and Ca^{2+} influx. *FEBS Lett.* **1989**, *259*, 1–4. [CrossRef]
189. Danthuluri, N.R.; Kim, D.; Brock, T.A. Intracellular alkalinization leads to Ca^{2+} mobilization from agonist-sensitive pools in bovine aortic endothelial cells. *J. Biol. Chem.* **1990**, *265*, 19071–19076. [PubMed]
190. Nishio, K.; Suzuki, Y.; Takeshita, K.; Aoki, T.; Kudo, H.; Sato, N.; Naoki, K.; Miyao, N.; Ishii, M.; Yamaguchi, K. Effects of hypercapnia and hypocapnia on $[Ca^{2+}]i$ mobilization in human pulmonary artery endothelial cells. *J. Appl. Physiol. (1985)* **2001**, *90*, 2094–2100. [CrossRef] [PubMed]
191. Amith, S.R.; Fliegel, L. Regulation of the Na^+/H^+ Exchanger (NHE1) in Breast Cancer Metastasis. *Cancer Res.* **2013**, *73*, 1259–1264. [CrossRef] [PubMed]

192. Reshkin, S.J.; Greco, M.R.; Cardone, R.A. Role of pHi, and proton transporters in oncogene-driven neoplastic transformation. *Philos. Trans. R. Soc. Lond. B Biol. Sci.* **2014**, *369*, 20130100. [CrossRef] [PubMed]
193. Pedersen, A.K.; Mendes Lopes de Melo, J.; Morup, N.; Tritsaris, K.; Pedersen, S.F. Tumor microenvironment conditions alter Akt and Na^+/H^+ exchanger NHE1 expression in endothelial cells more than hypoxia alone: Implications for endothelial cell function in cancer. *BMC Cancer* **2017**, *17*, 542. [CrossRef] [PubMed]
194. Orive, G.; Reshkin, S.J.; Harguindey, S.; Pedraz, J.L. Hydrogen ion dynamics and the Na^+/H^+ exchanger in cancer angiogenesis and antiangiogenesis. *Br. J. Cancer* **2003**, *89*, 1395–1399. [CrossRef] [PubMed]
195. Spugnini, E.P.; Sonveaux, P.; Stock, C.; Perez-Sayans, M.; De Milito, A.; Avnet, S.; Garcia, A.G.; Harguindey, S.; Fais, S. Proton channels and exchangers in cancer. *Biochim. Biophys. Acta* **2015**, *1848 Pt B*, 2715–2726. [CrossRef] [PubMed]
196. Morgan, A.J.; Platt, F.M.; Lloyd-Evans, E.; Galione, A. Molecular mechanisms of endolysosomal Ca^{2+} signalling in health and disease. *Biochem. J.* **2011**, *439*, 349–374. [CrossRef] [PubMed]
197. Jha, A.; Brailoiu, E.; Muallem, S. How does NAADP release lysosomal Ca^{2+}? *Channels* **2014**, *8*, 174–175. [CrossRef] [PubMed]
198. Patel, S.; Docampo, R. Acidic calcium stores open for business: Expanding the potential for intracellular Ca^{2+} signaling. *Trends Cell Biol.* **2010**, *20*, 277–286. [CrossRef] [PubMed]
199. Cosker, F.; Cheviron, N.; Yamasaki, M.; Menteyne, A.; Lund, F.E.; Moutin, M.J.; Galione, A.; Cancela, J.M. The ecto-enzyme CD38 is a nicotinic acid adenine dinucleotide phosphate (NAADP) synthase that couples receptor activation to Ca^{2+} mobilization from lysosomes in pancreatic acinar cells. *J. Biol. Chem.* **2010**, *285*, 38251–38259. [CrossRef] [PubMed]
200. Galione, A. A primer of NAADP-mediated Ca^{2+} signalling: From sea urchin eggs to mammalian cells. *Cell Calcium* **2015**, *58*, 27–47. [CrossRef] [PubMed]
201. Moccia, F.; Nusco, G.A.; Lim, D.; Kyozuka, K.; Santella, L. NAADP and InsP3 play distinct roles at fertilization in starfish oocytes. *Dev. Biol.* **2006**, *294*, 24–38. [CrossRef] [PubMed]
202. Favia, A.; Desideri, M.; Gambara, G.; D'Alessio, A.; Ruas, M.; Esposito, B.; Del Bufalo, D.; Parrington, J.; Ziparo, E.; Palombi, F.; et al. VEGF-induced neoangiogenesis is mediated by NAADP and two-pore channel-2-dependent Ca^{2+} signaling. *Proc. Natl. Acad. Sci. USA* **2014**, *111*, E4706–E4715. [CrossRef] [PubMed]
203. Brailoiu, G.C.; Gurzu, B.; Gao, X.; Parkesh, R.; Aley, P.K.; Trifa, D.I.; Galione, A.; Dun, N.J.; Madesh, M.; Patel, S.; et al. Acidic NAADP-sensitive calcium stores in the endothelium: Agonist-specific recruitment and role in regulating blood pressure. *J. Biol. Chem.* **2010**, *285*, 37133–37137. [CrossRef] [PubMed]
204. Di Nezza, F.; Zuccolo, E.; Poletto, V.; Rosti, V.; De Luca, A.; Moccia, F.; Guerra, G.; Ambrosone, L. Liposomes as a Putative Tool to Investigate NAADP Signaling in Vasculogenesis. *J. Cell. Biochem.* **2017**, *118*, 3722–3729. [CrossRef] [PubMed]
205. Favia, A.; Pafumi, I.; Desideri, M.; Padula, F.; Montesano, C.; Passeri, D.; Nicoletti, C.; Orlandi, A.; Del Bufalo, D.; Sergi, M.; et al. NAADP-Dependent Ca^{2+} Signaling Controls Melanoma Progression, Metastatic Dissemination and Neoangiogenesis. *Sci. Rep.* **2016**, *6*, 18925. [CrossRef] [PubMed]
206. Schaefer, M.; Plant, T.D.; Obukhov, A.G.; Hofmann, T.; Gudermann, T.; Schultz, G. Receptor-mediated regulation of the nonselective cation channels TRPC4 and TRPC5. *J. Biol. Chem.* **2000**, *275*, 17517–17526. [CrossRef] [PubMed]
207. Zholos, A.V. TRPC5. *Handb. Exp. Pharmacol.* **2014**, *222*, 129–156. [PubMed]
208. Yuan, J.P.; Zeng, W.; Huang, G.N.; Worley, P.F.; Muallem, S. STIM1 heteromultimerizes TRPC channels to determine their function as store-operated channels. *Nat. Cell Biol.* **2007**, *9*, 636–645. [CrossRef] [PubMed]
209. DeHaven, W.I.; Jones, B.F.; Petranka, J.G.; Smyth, J.T.; Tomita, T.; Bird, G.S.; Putney, J.W., Jr. TRPC channels function independently of STIM1 and Orai1. *J. Physiol.* **2009**, *587 Pt 10*, 2275–2298. [CrossRef] [PubMed]
210. Zimmermann, K.; Lennerz, J.K.; Hein, A.; Link, A.S.; Kaczmarek, J.S.; Delling, M.; Uysal, S.; Pfeifer, J.D.; Riccio, A.; Clapham, D.E. Transient receptor potential cation channel, subfamily C, member 5 (TRPC5) is a cold-transducer in the peripheral nervous system. *Proc. Natl. Acad. Sci. USA* **2011**, *108*, 18114–18119. [CrossRef] [PubMed]
211. Antigny, F.; Girardin, N.; Frieden, M. Transient Receptor Potential Canonical Channels Are Required for in Vitro Endothelial Tube Formation. *J. Biol. Chem.* **2012**, *287*, 5917–5927. [CrossRef] [PubMed]

212. Chaudhuri, P.; Colles, S.M.; Bhat, M.; Van Wagoner, D.R.; Birnbaumer, L.; Graham, L.M. Elucidation of a TRPC6-TRPC5 channel cascade that restricts endothelial cell movement. *Mol. Biol. Cell* **2008**, *19*, 3203–3211. [CrossRef] [PubMed]
213. Chaudhuri, P.; Rosenbaum, M.A.; Birnbaumer, L.; Graham, L.M. Integration of TRPC6 and NADPH oxidase activation in lysophosphatidylcholine-induced TRPC5 externalization. *Am. J. Physiol. Cell Physiol.* **2017**, *313*, C541–C555. [CrossRef] [PubMed]
214. Ma, X.; Cai, Y.; He, D.; Zou, C.; Zhang, P.; Lo, C.Y.; Xu, Z.; Chan, F.L.; Yu, S.; Chen, Y.; et al. Transient receptor potential channel TRPC5 is essential for P-glycoprotein induction in drug-resistant cancer cells. *Proc. Natl. Acad. Sci. USA* **2012**, *109*, 16282–16287. [CrossRef] [PubMed]
215. Ma, X.; Chen, Z.; Hua, D.; He, D.; Wang, L.; Zhang, P.; Wang, J.; Cai, Y.; Gao, C.; Zhang, X.; et al. Essential role for TrpC5-containing extracellular vesicles in breast cancer with chemotherapeutic resistance. *Proc. Natl. Acad. Sci. USA* **2014**, *111*, 6389–6394. [CrossRef] [PubMed]
216. Wang, T.; Chen, Z.; Zhu, Y.; Pan, Q.; Liu, Y.; Qi, X.; Jin, L.; Jin, J.; Ma, X.; Hua, D. Inhibition of transient receptor potential channel 5 reverses 5-Fluorouracil resistance in human colorectal cancer cells. *J. Biol. Chem.* **2015**, *290*, 448–456. [CrossRef] [PubMed]
217. Pokharel, D.; Roseblade, A.; Oenarto, V.; Lu, J.F.; Bebawy, M. Proteins regulating the intercellular transfer and function of P-glycoprotein in multidrug-resistant cancer. *Ecancermedicalscience* **2017**, *11*, 768. [CrossRef] [PubMed]
218. He, D.X.; Gu, X.T.; Jiang, L.; Jin, J.; Ma, X. A methylation-based regulatory network for microRNA 320a in chemoresistant breast cancer. *Mol. Pharmacol.* **2014**, *86*, 536–547. [CrossRef] [PubMed]
219. Zhu, Y.; Pan, Q.; Meng, H.; Jiang, Y.; Mao, A.; Wang, T.; Hua, D.; Yao, X.; Jin, J.; Ma, X. Enhancement of vascular endothelial growth factor release in long-term drug-treated breast cancer via transient receptor potential channel 5-Ca^{2+}-hypoxia-inducible factor 1alpha pathway. *Pharmacol. Res.* **2015**, *93*, 36–42. [CrossRef] [PubMed]
220. He, D.X.; Ma, X. Transient receptor potential channel C5 in cancer chemoresistance. *Acta Pharmacol. Sin.* **2016**, *37*, 19–24. [CrossRef] [PubMed]
221. Messai, Y.; Noman, M.Z.; Hasmim, M.; Janji, B.; Tittarelli, A.; Boutet, M.; Baud, V.; Viry, E.; Billot, K.; Nanbakhsh, A.; et al. ITPR1 protects renal cancer cells against natural killer cells by inducing autophagy. *Cancer Res.* **2014**, *74*, 6820–6832. [CrossRef] [PubMed]
222. Lim, D.; Bertoli, A.; Sorgato, M.C.; Moccia, F. Generation and usage of aequorin lentiviral vectors for Ca^{2+} measurement in sub-cellular compartments of hard-to-transfect cells. *Cell Calcium* **2016**, *59*, 228–239. [CrossRef] [PubMed]
223. De la Puente, P.; Muz, B.; Azab, F.; Azab, A.K. Cell trafficking of endothelial progenitor cells in tumor progression. *Clin. Cancer. Res.* **2013**, *19*, 3360–3368. [CrossRef] [PubMed]
224. Escudier, B.; Szczylik, C.; Porta, C.; Gore, M. Treatment selection in metastatic renal cell carcinoma: Expert consensus. *Nat. Rev. Clin. Oncol.* **2012**, *9*, 327–337. [CrossRef] [PubMed]
225. Porta, C.; Paglino, C.; Imarisio, I.; Canipari, C.; Chen, K.; Neary, M.; Duh, M.S. Safety and treatment patterns of multikinase inhibitors in patients with metastatic renal cell carcinoma at a tertiary oncology center in Italy. *BMC Cancer* **2011**, *11*. [CrossRef] [PubMed]
226. Porta, C.; Paglino, C.; Imarisio, I.; Ganini, C.; Sacchi, L.; Quaglini, S.; Giunta, V.; De Amici, M. Changes in circulating pro-angiogenic cytokines, other than VEGF, before progression to sunitinib therapy in advanced renal cell carcinoma patients. *Oncology* **2013**, *84*, 115–122. [CrossRef] [PubMed]
227. Ribatti, D. Tumor refractoriness to anti-VEGF therapy. *Oncotarget* **2016**, *7*, 46668–46677. [CrossRef] [PubMed]
228. Hiles, J.J.; Kolesar, J.M. Role of sunitinib and sorafenib in the treatment of metastatic renal cell carcinoma. *Am. J. Health Syst. Pharm.* **2008**, *65*, 123–131. [CrossRef] [PubMed]
229. Naito, H.; Wakabayashi, T.; Kidoya, H.; Muramatsu, F.; Takara, K.; Eino, D.; Yamane, K.; Iba, T.; Takakura, N. Endothelial Side Population Cells Contribute to Tumor Angiogenesis and Antiangiogenic Drug Resistance. *Cancer Res.* **2016**, *76*, 3200–3210. [CrossRef] [PubMed]
230. Loges, S.; Schmidt, T.; Carmeliet, P. Mechanisms of resistance to anti-angiogenic therapy and development of third-generation anti-angiogenic drug candidates. *Genes Cancer* **2010**, *1*, 12–25. [CrossRef] [PubMed]
231. Sunryd, J.C.; Cheon, B.; Graham, J.B.; Giorda, K.M.; Fissore, R.A.; Hebert, D.N. TMTC1 and TMTC2 are novel endoplasmic reticulum tetratricopeptide repeat-containing adapter proteins involved in calcium homeostasis. *J. Biol. Chem.* **2014**, *289*, 16085–16099. [CrossRef] [PubMed]

232. Sammels, E.; Parys, J.B.; Missiaen, L.; De Smedt, H.; Bultynck, G. Intracellular Ca^{2+} storage in health and disease: A dynamic equilibrium. *Cell Calcium* **2010**, *47*, 297–314. [CrossRef] [PubMed]
233. Schuck, S.; Prinz, W.A.; Thorn, K.S.; Voss, C.; Walter, P. Membrane expansion alleviates endoplasmic reticulum stress independently of the unfolded protein response. *J. Cell Biol.* **2009**, *187*, 525–536. [CrossRef] [PubMed]
234. Dubois, C.; Vanden Abeele, F.; Prevarskaya, N. Targeting apoptosis by the remodelling of calcium-transporting proteins in cancerogenesis. *FEBS J.* **2013**, *280*, 5500–5510. [CrossRef] [PubMed]
235. Prevarskaya, N.; Skryma, R.; Shuba, Y. Ion channels and the hallmarks of cancer. *Trends Mol. Med.* **2010**, *16*, 107–121. [CrossRef] [PubMed]
236. Prevarskaya, N.; Skryma, R.; Shuba, Y. Targeting Ca^{2+} transport in cancer: Close reality or long perspective? *Expert Opin. Ther. Targets* **2013**, *17*, 225–241. [CrossRef] [PubMed]
237. Vanoverberghe, K.; Vanden Abeele, F.; Mariot, P.; Lepage, G.; Roudbaraki, M.; Bonnal, J.L.; Mauroy, B.; Shuba, Y.; Skryma, R.; Prevarskaya, N. Ca^{2+} homeostasis and apoptotic resistance of neuroendocrine-differentiated prostate cancer cells. *Cell Death Differ.* **2004**, *11*, 321–330. [CrossRef] [PubMed]
238. Cui, C.; Merritt, R.; Fu, L.; Pan, Z. Targeting calcium signaling in cancer therapy. *Acta Pharm. Sin. B* **2017**, *7*, 3–17. [CrossRef] [PubMed]
239. Bonora, M.; Giorgi, C.; Pinton, P. Novel frontiers in calcium signaling: A possible target for chemotherapy. *Pharmacol. Res.* **2015**, *99*, 82–85. [CrossRef] [PubMed]
240. Leanza, L.; Manago, A.; Zoratti, M.; Gulbins, E.; Szabo, I. Pharmacological targeting of ion channels for cancer therapy: In vivo evidences. *Biochim. Biophys. Acta* **2016**, *1863 Pt B*, 1385–1397. [CrossRef] [PubMed]
241. Nielsen, N.; Lindemann, O.; Schwab, A. TRP channels and STIM/ORAI proteins: Sensors and effectors of cancer and stroma cell migration. *Br. J. Pharmacol.* **2014**, *171*, 5524–5540. [CrossRef] [PubMed]
242. Munaron, L. Systems biology of ion channels and transporters in tumor angiogenesis: An omics view. *Biochim. Biophys. Acta* **2015**, *1848 Pt B*, 2647–2656. [CrossRef] [PubMed]
243. Kito, H.; Yamamura, H.; Suzuki, Y.; Yamamura, H.; Ohya, S.; Asai, K.; Imaizumi, Y. Regulation of store-operated Ca^{2+} entry activity by cell cycle dependent up-regulation of Orai2 in brain capillary endothelial cells. *Biochem. Biophys. Res. Commun.* **2015**, *459*, 457–462. [CrossRef] [PubMed]
244. Choi, D.; Park, E.; Jung, E.; Seong, Y.J.; Hong, M.; Lee, S.; Burford, J.; Gyarmati, G.; Peti-Peterdi, J.; Srikanth, S.; et al. ORAI1 Activates Proliferation of Lymphatic Endothelial Cells in Response to Laminar Flow Through Kruppel-Like Factors 2 and 4. *Circ. Res.* **2017**, *120*, 1426–1439. [CrossRef] [PubMed]
245. Tian, C.; Du, L.; Zhou, Y.; Li, M. Store-operated CRAC channel inhibitors: Opportunities and challenges. *Future Med. Chem.* **2016**, *8*, 817–832. [CrossRef] [PubMed]
246. Luzzi, K.J.; Varghese, H.J.; MacDonald, I.C.; Schmidt, E.E.; Kohn, E.C.; Morris, V.L.; Marshall, K.E.; Chambers, A.F.; Groom, A.C. Inhibition of angiogenesis in liver metastases by carboxyamidotriazole (CAI). *Angiogenesis* **1998**, *2*, 373–379. [CrossRef] [PubMed]
247. Oliver, V.K.; Patton, A.M.; Desai, S.; Lorang, D.; Libutti, S.K.; Kohn, E.C. Regulation of the pro-angiogenic microenvironment by carboxyamido-triazole. *J. Cell. Physiol.* **2003**, *197*, 139–148. [CrossRef] [PubMed]
248. Patton, A.M.; Kassis, J.; Doong, H.; Kohn, E.C. Calcium as a molecular target in angiogenesis. *Curr. Pharm. Des.* **2003**, *9*, 543–551. [CrossRef] [PubMed]
249. Mignen, O.; Brink, C.; Enfissi, A.; Nadkarni, A.; Shuttleworth, T.J.; Giovannucci, D.R.; Capiod, T. Carboxyamidotriazole-induced inhibition of mitochondrial calcium import blocks capacitative calcium entry and cell proliferation in HEK-293 cells. *J. Cell Sci.* **2005**, *118 Pt 23*, 5615–5623. [CrossRef] [PubMed]
250. Enfissi, A.; Prigent, S.; Colosetti, P.; Capiod, T. The blocking of capacitative calcium entry by 2-aminoethyl diphenylborate (2-APB) and carboxyamidotriazole (CAI) inhibits proliferation in Hep G2 and Huh-7 human hepatoma cells. *Cell Calcium* **2004**, *36*, 459–467. [CrossRef] [PubMed]
251. Faehling, M.; Kroll, J.; Fohr, K.J.; Fellbrich, G.; Mayr, U.; Trischler, G.; Waltenberger, J. Essential role of calcium in vascular endothelial growth factor A-induced signaling: Mechanism of the antiangiogenic effect of carboxyamidotriazole. *FASEB J.* **2002**, *16*, 1805–1807. [CrossRef] [PubMed]
252. Wu, Y.; Palad, A.J.; Wasilenko, W.J.; Blackmore, P.F.; Pincus, W.A.; Schechter, G.L.; Spoonster, J.R.; Kohn, E.C.; Somers, K.D. Inhibition of head and neck squamous cell carcinoma growth and invasion by the calcium influx inhibitor carboxyamido-triazole. *Clin. Cancer. Res.* **1997**, *3*, 1915–1921. [PubMed]
253. Moody, T.W.; Chiles, J.; Moody, E.; Sieczkiewicz, G.J.; Kohn, E.C. CAI inhibits the growth of small cell lung cancer cells. *Lung Cancer* **2003**, *39*, 279–288. [CrossRef]

254. Hussain, M.M.; Kotz, H.; Minasian, L.; Premkumar, A.; Sarosy, G.; Reed, E.; Zhai, S.; Steinberg, S.M.; Raggio, M.; Oliver, V.K.; et al. Phase II trial of carboxyamidotriazole in patients with relapsed epithelial ovarian cancer. *J. Clin. Oncol.* **2003**, *21*, 4356–4363. [CrossRef] [PubMed]
255. Griffioen, A.W.; Molema, G. Angiogenesis: Potentials for pharmacologic intervention in the treatment of cancer, cardiovascular diseases, and chronic inflammation. *Pharmacol. Rev.* **2000**, *52*, 237–268. [PubMed]
256. Stadler, W.M.; Rosner, G.; Small, E.; Hollis, D.; Rini, B.; Zaentz, S.D.; Mahoney, J.; Ratain, M.J. Successful implementation of the randomized discontinuation trial design: An application to the study of the putative antiangiogenic agent carboxyaminoimidazole in renal cell carcinoma—CALGB 69901. *J. Clin. Oncol.* **2005**, *23*, 3726–3732. [CrossRef] [PubMed]
257. Rahman, S.; Rahman, T. Unveiling some FDA-approved drugs as inhibitors of the store-operated Ca^{2+} entry pathway. *Sci. Rep.* **2017**, *7*, 12881. [CrossRef] [PubMed]
258. Viallard, C.; Larrivee, B. Tumor angiogenesis and vascular normalization: Alternative therapeutic targets. *Angiogenesis* **2017**, *20*, 409–426. [CrossRef] [PubMed]
259. Vincent, F.; Duncton, M.A. TRPV4 agonists and antagonists. *Curr. Top. Med. Chem.* **2011**, *11*, 2216–2226. [CrossRef] [PubMed]
260. Coddou, C.; Yan, Z.; Obsil, T.; Huidobro-Toro, J.P.; Stojilkovic, S.S. Activation and regulation of purinergic P2X receptor channels. *Pharmacol. Rev.* **2011**, *63*, 641–683. [CrossRef] [PubMed]
261. Pafumi, I.; Favia, A.; Gambara, G.; Papacci, F.; Ziparo, E.; Palombi, F.; Filippini, A. Regulation of Angiogenic Functions by Angiopoietins through Calcium-Dependent Signaling Pathways. *BioMed Res. Int.* **2015**, *2015*, 965271. [CrossRef] [PubMed]
262. Rubaiy, H.N.; Ludlow, M.J.; Bon, R.S.; Beech, D.J. Pico145—Powerful new tool for TRPC1/4/5 channels. *Channels* **2017**, *11*, 362–364. [CrossRef] [PubMed]
263. Naylor, J.; Minard, A.; Gaunt, H.J.; Amer, M.S.; Wilson, L.A.; Migliore, M.; Cheung, S.Y.; Rubaiy, H.N.; Blythe, N.M.; Musialowski, K.E.; et al. Natural and synthetic flavonoid modulation of TRPC5 channels. *Br. J. Pharmacol.* **2016**, *173*, 562–574. [CrossRef] [PubMed]
264. Zhu, Y.; Lu, Y.; Qu, C.; Miller, M.; Tian, J.; Thakur, D.P.; Zhu, J.; Deng, Z.; Hu, X.; Wu, M.; et al. Identification and optimization of 2-aminobenzimidazole derivatives as novel inhibitors of TRPC4 and TRPC5 channels. *Br. J. Pharmacol.* **2015**, *172*, 3495–3509. [CrossRef] [PubMed]
265. Miller, M.; Shi, J.; Zhu, Y.; Kustov, M.; Tian, J.B.; Stevens, A.; Wu, M.; Xu, J.; Long, S.; Yang, P.; et al. Identification of ML204, a novel potent antagonist that selectively modulates native TRPC4/C5 ion channels. *J. Biol. Chem.* **2011**, *286*, 33436–33446. [CrossRef] [PubMed]
266. Majeed, Y.; Amer, M.S.; Agarwal, A.K.; McKeown, L.; Porter, K.E.; O'Regan, D.J.; Naylor, J.; Fishwick, C.W.; Muraki, K.; Beech, D.J. Stereo-selective inhibition of transient receptor potential TRPC5 cation channels by neuroactive steroids. *Br. J. Pharmacol.* **2011**, *162*, 1509–1520. [CrossRef] [PubMed]
267. Giorgi, C.; Bonora, M.; Missiroli, S.; Poletti, F.; Ramirez, F.G.; Morciano, G.; Morganti, C.; Pandolfi, P.P.; Mammano, F.; Pinton, P. Intravital imaging reveals p53-dependent cancer cell death induced by phototherapy via calcium signaling. *Oncotarget* **2015**, *6*, 1435–1445. [CrossRef] [PubMed]
268. Doan, N.T.; Paulsen, E.S.; Sehgal, P.; Moller, J.V.; Nissen, P.; Denmeade, S.R.; Isaacs, J.T.; Dionne, C.A.; Christensen, S.B. Targeting thapsigargin towards tumors. *Steroids* **2015**, *97*, 2–7. [CrossRef] [PubMed]
269. Denmeade, S.R.; Mhaka, A.M.; Rosen, D.M.; Brennen, W.N.; Dalrymple, S.; Dach, I.; Olesen, C.; Gurel, B.; Demarzo, A.M.; Wilding, G.; et al. Engineering a prostate-specific membrane antigen-activated tumor endothelial cell prodrug for cancer therapy. *Sci. Transl. Med.* **2012**, *4*, 140ra86. [CrossRef] [PubMed]
270. Liu, H.; Moy, P.; Kim, S.; Xia, Y.; Rajasekaran, A.; Navarro, V.; Knudsen, B.; Bander, N.H. Monoclonal antibodies to the extracellular domain of prostate-specific membrane antigen also react with tumor vascular endothelium. *Cancer Res.* **1997**, *57*, 3629–3634. [PubMed]
271. Quynh Doan, N.T.; Christensen, S.B. Thapsigargin, Origin, Chemistry, Structure-Activity Relationships and Prodrug Development. *Curr. Pharm. Des.* **2015**, *21*, 5501–5517. [CrossRef] [PubMed]
272. Mahalingam, D.; Wilding, G.; Denmeade, S.; Sarantopoulas, J.; Cosgrove, D.; Cetnar, J.; Azad, N.; Bruce, J.; Kurman, M.; Allgood, V.E.; et al. Mipsagargin, a novel thapsigargin-based PSMA-activated prodrug: Results of a first-in-man phase I clinical trial in patients with refractory, advanced or metastatic solid tumours. *Br. J. Cancer* **2016**, *114*, 986–994. [CrossRef] [PubMed]
273. Monteith, G.R.; McAndrew, D.; Faddy, H.M.; Roberts-Thomson, S.J. Calcium and cancer: Targeting Ca^{2+} transport. *Nat. Rev. Cancer* **2007**, *7*, 519–530. [CrossRef] [PubMed]

274. Zhu, H.; Zhang, H.; Jin, F.; Fang, M.; Huang, M.; Yang, C.S.; Chen, T.; Fu, L.; Pan, Z. Elevated Orai1 expression mediates tumor-promoting intracellular Ca^{2+} oscillations in human esophageal squamous cell carcinoma. *Oncotarget* **2014**, *5*, 3455–3471. [CrossRef] [PubMed]
275. Dudley, A.C. Tumor endothelial cells. *Cold Spring Harb. Perspect. Med.* **2012**, *2*, a006536. [CrossRef] [PubMed]
276. Fiorio Pla, A.; Gkika, D. Emerging role of TRP channels in cell migration: From tumor vascularization to metastasis. *Front. Physiol.* **2013**, *4*, 311. [CrossRef] [PubMed]
277. Munaron, L.; Genova, T.; Avanzato, D.; Antoniotti, S.; Fiorio Pla, A. Targeting calcium channels to block tumor vascularization. *Recent Pat. Anticancer Drug Discov.* **2013**, *8*, 27–37. [CrossRef] [PubMed]
278. Motiani, R.K.; Abdullaev, I.F.; Trebak, M. A novel native store-operated calcium channel encoded by Orai3: Selective requirement of Orai3 versus Orai1 in estrogen receptor-positive versus estrogen receptor-negative breast cancer cells. *J. Biol. Chem.* **2010**, *285*, 19173–19183. [CrossRef] [PubMed]
279. Andrikopoulos, P.; Baba, A.; Matsuda, T.; Djamgoz, M.B.; Yaqoob, M.M.; Eccles, S.A. Ca^{2+} influx through reverse mode Na^+/Ca^{2+} exchange is critical for vascular endothelial growth factor-mediated extracellular signal-regulated kinase (ERK) 1/2 activation and angiogenic functions of human endothelial cells. *J. Biol. Chem.* **2011**, *286*, 37919–37931. [CrossRef] [PubMed]
280. Hogan-Cann, A.D.; Anderson, C.M. Physiological Roles of Non-Neuronal NMDA Receptors. *Trends Pharmacol. Sci.* **2016**, *37*, 750–767. [CrossRef] [PubMed]
281. Stobart, J.L.; Lu, L.; Anderson, H.D.; Mori, H.; Anderson, C.M. Astrocyte-induced cortical vasodilation is mediated by D-serine and endothelial nitric oxide synthase. *Proc. Natl. Acad. Sci. USA* **2013**, *110*, 3149–3154. [CrossRef] [PubMed]
282. Takano, T.; Lin, J.H.; Arcuino, G.; Gao, Q.; Yang, J.; Nedergaard, M. Glutamate release promotes growth of malignant gliomas. *Nat. Med.* **2001**, *7*, 1010–1015. [CrossRef] [PubMed]

2

Calcium Ion Channels: Roles in Infection and Sepsis Mechanisms of Calcium Channel Blocker Benefits in Immunocompromised Patients at Risk for Infection

John A. D'Elia and Larry A. Weinrauch *

E P Joslin Research Laboratory, Kidney and Hypertension Section, Joslin Diabetes Center, Department of Medicine, Mount Auburn Hospital, Harvard Medical School, Boston and Cambridge, 521 Mount Auburn Street Watertown, MA 02472, USA; jd'elia@joslin.harvard.edu

* Correspondence: lweinrauch@hms.harvard.edu

Abstract: Immunosuppression may occur for a number of reasons related to an individual's frailty, debility, disease or from therapeutic iatrogenic intervention or misadventure. A large percentage of morbidity and mortality in immunodeficient populations is related to an inadequate response to infectious agents with slow response to antibiotics, enhancements of antibiotic resistance in populations, and markedly increased prevalence of acute inflammatory response, septic and infection related death. Given known relationships between intracellular calcium ion concentrations and cytotoxicity and cellular death, we looked at currently available data linking blockade of calcium ion channels and potential decrease in expression of sepsis among immunosuppressed patients. Notable are relationships between calcium, calcium channel, vitamin D mechanisms associated with sepsis and demonstration of antibiotic-resistant pathogens that may utilize channels sensitive to calcium channel blocker. We note that sepsis shock syndrome represents loss of regulation of inflammatory response to infection and that vitamin D, parathyroid hormone, fibroblast growth factor, and klotho interact with sepsis defense mechanisms in which movement of calcium and phosphorus are part of the process. Given these observations we consider that further investigation of the effect of relatively inexpensive calcium channel blockade agents of infections in immunosuppressed populations might be worthwhile.

Keywords: calcium ion channels; calcium channel blockade; sepsis; infection; immunosuppression

---

1. Introduction

Immunosuppression may occur for a number of reasons related to an individual's frailty, debility, disease or from therapeutic iatrogenic intervention or misadventure. A large percentage of morbidity and mortality in immunodeficient populations is related to an inadequate response to infectious agents with slow response to antibiotics, enhancements of antibiotic resistance and markedly increased prevalence of acute inflammatory response, septic and infection related death. Given the known relationships between intracellular calcium ion concentrations, cytotoxicity and cellular death, we review currently available data linking blockade of calcium ion channels and potential decrease in expression of sepsis among immunosuppressed patients. We consider possibilities for therapeutic interference with calcium ion channels that may alter immune responses to invading organisms in immunocompromised populations (Table 1).

Table 1. Pathophysiologic Interactions of calcium ion channels with the production of pathogen mediated sepsis syndrome: Potential for therapeutic interference with calcium ion channels.

Pathogenic Organism	Calcium Ion Channel Effect
Organ specific toxicity	Envelope protein increases toxic level of cytosolic calcium in host cell
Cell membrane or wall	Antibiotic efflux may be diminished by calcium channel blockade
Organism replication	Interference with calcium dependent RNA transcription
Cytoplasm (mitochondria)	Potential to interrupt intracellular calcium shifts and interrupt calcium efflux
Host Defense	**Calcium Ion Channel Effect**
Immunocompetence	Calcium activated potassium channels regulate Lymphocyte activation Mitogenesis Cell volume
Cellular Immunity Permeability, necrosis, apoptosis	Release of intracellular calcium leads to mitochondrial permeability and influx of extracellular calcium and permeability, necrosis, apoptosis
Humoral Immunity	Calcium controls antibody formation
Inflammasome	Calcium has a role in the production of TNF alpha, IL-1 beta

2. Calcium, Calcium Channel, Vitamin D Mechanisms Associated with Sepsis

In a survey of the United Kingdom General Practice database (Table 2) over 500,000 patients were listed as hypertensive. In this survey, use of angiotensin converting enzyme inhibitors was associated with a significantly higher rate of hospitalization for sepsis as well as mortality at thirty days when compared to use of angiotensin receptor blockers or calcium channel blockers [1]. One retrospective cohort study revealed that when 387 patients hospitalized for pneumonia who had not been treated with calcium channel blockers, were compared to 387 similarly hospitalized patients with pneumonia who had received calcium channel blockers there was a significantly higher incidence of bacteremia, respiratory insufficiency, and transfer to intensive care unit compared among those not so treated [2]. One retrospective evaluation of immunosuppressed recipients of kidney allografts from the pre-angiotensin receptor blocker era noted significantly higher incidence of sepsis and shorter survivals of allograft function for 33 patients who had not received calcium channel blockers as opposed to 36 who did receive calcium channel blockers [3].

Table 2. Calcium Ion Channels and hypothetical role in expression of infections.

I. Relationship of clinical infections to calcium channels (retrospective studies)

- A. Pneumonia: acute infection without antibiotic resistance
  - 1. General population
  - 2. Recipients of Kidney Transplant allografts
  - a. Positive outcome results with calcium channel blockers
  - b. Negative outcome results with calcium channel blockers
- B. Sepsis with or without antibiotic-resistant pathogens
  - 1. Potential benefit of calcium channel blockers in combination with more expensive drugs in an attempt to reduce costs while minimizing side effects of higher doses
  - a. quinolone-resistant streptococcus
  - b. rifampicin-resistant mycobacterium
  - c. quinine-resistant plasmodium
  - d. praziquantel-resistant schistosome
  - e. amphotericin-resistant leishmania
  - f. eflornithine-resistant trypanosome

II. Role of calcium movement in white blood cell defense against pathogens

- A. Neutrophils and macrophages may benefit from calcium channel blockers
  - 1. Restoration of capacity to attack pathogen
  - 2. Limitation of capacity of pathogen to extrude antibiotic

Table 2. *Cont.*

III. Sepsis with shock following trauma

- A. Intracellular calcium movement from storage sites may be stabilized by calcium channel -blockers, limiting cell injury B. Capillary leak during sepsis1.
    1. Angiopoietin 2 as a factor in sepsis-related capillary leaking
    2. Flunarizine, which blocks both calcium influx and calcium movement from intracellular stores, in prevention of angiopoietin 2—related capillary leaking

IV. Pathogen colony growth mechanisms may or may not involve calcium

- A. Generation of anti-oxidants (catalase, dismutase)
    1. *Bortadella pertussis*
    2. *Pseudomonas aeruginosa*
- B. Efflux of calcium from host cells
    1. *Bacillus anthracis*
    2. *Clostridium perfringens*
    3. *Streptococcus pneumoniae*
- C. Colony growth mechanisms inhibited by calcium channel blockers
    1. *Aspergillus fumigatus*
    2. *Saccharomyces cerevisiae*
    3. *Candida albicans*
    4. *Cryptococcus neoformans*

V. Mechanisms of Calcium balance in kidney failure

- A. Vitamin D (and cathelicidin) deficiency association with infection risk
    1. Promotes availability of calcium through intestinal absorption
    2. Promotes bone calcification at the growing front of osteoid
- B. Parathyroid hormone excess promotes lysis of calcified bone with excess calcium/phosphorus that may deposit in soft tissue, including blood vessels with skin cellulitis (calciphylaxis).
    1. Inhibits resorption of phosphate in kidney proximal tubule.
    2. Promotes activity of 1-alpha hydroxylase in renal proximal tubule, which converts 25(OH) Vitamin D to its most active form $1,25(OH)_2$ vitamin D.
- C. Fibroblast Growth Factor 23 minimizes the accumulation of phosphate which is associated with an increased mortality rate.
    1. Cooperates with parathyroid hormone in the inhibition of resorption of phosphate in kidney proximal tubule.
    2. Competes with parathyroid hormone's action on 1-alpha hydroxylase in the proximal tubule to generate the most active form of Vitamin D.
- D. Klotho, the antiaging gene, cooperates with parathyroid hormone and fibroblast growth factor 23 by inhibiting phosphate resorption in the kidney proximal tubule.

Since the patients reviewed in studies of angiotensin-active medications and calcium channel blockers would have had hypertension, a precise mechanism for protection from sepsis would have to include protection from injury to blood vessels supplying skin, bronchus, urinary bladder unless there had been a blood vessel indwelling catheter-line, suggesting the vascular protection hypothesis might be too narrow. Consequently, researchers have developed other approaches. An experimental model for testing the impact of calcium channel blockers on sepsis involving ligation of the cecum with puncture of the wall of the intestine was one such example. This septic shock model was used to [4] demonstrate a lesser accumulation of oxygen radicals [5] as well as longer survival if diltiazem were injected prior to the onset of septic shock.

2.1. *Positive Results*

Several studies have addressed the use of dihydropyridine and non-dihydropyridine calcium channel blockers in the peri-operative period and in longer-term follow up (Table 2), demonstrating improved allograft function [6,7] that did not consistently appear to be secondary to blood pressure control [8,9]. One study did demonstrate stable serum creatinine associated with a fall in renal vascular resistance calculated from mean arterial pressure + renal blood flow [10]. A unique study found the occurrence of acute graft dysfunction (acute tubular necrosis) by biopsy to be significantly lower with verapamil vs. a non-calcium channel medication [11].

Rapid calcium influx into cytoplasm is associated with cell death [12,13] Clinical use for medications affecting calcium channels may involve control of aberrant cardiac rhythm while moderating vascular spasm [14]. As multiple types of calcium ion channels (N-type, L type and T type voltage-dependent calcium channels) became recognizable and multiple pharmaceutical agents with differing action profiles became available, further studies to find a role for these agents were performed.

The fact that some of these agents could directly influence biosynthesis of aldosterone in human adrenocortical cells [15] led to an early interest in their use in hypertension. Animal studies indicated an inhibition of the mineralocorticoid receptor as an additional property of dihydropyridine calcium channel blockers with felodipine being twice as powerful as amlodipine [16]. The other classes of calcium channel blockers (diltiazem, verapamil) were not effective here. Medium-sized blood vessels experience pulsatile flow as pressure is conveyed down to points of resistance and reflected backwards. Arterioles experience non-pulsatile constant flow, which can be increased with vasoconstriction or decreased with vasodilatation. Nitric oxide regulates arteriole flow while calcium-activated potassium channels regulate conduit artery flow [17]. Pathological concentrations of the vasoconstrictor angiotensin can cause damage to the muscular layer of larger blood vessels and to the endothelium of smaller blood vessels [18].

Uncontrolled hypertension may result in loss of vascular integrity with impaired endothelial resistance to sepsis [19,20] through activation of inflammation cascades which inhibit expression of nitric oxide synthase as well as the loss of control of glucose disposal such that oxidative stress disrupts calcium/potassium vascular physiology. These same mechanisms are operative in chronic loss of kidney function through injury to blood vessels and interstitial matrix, which diminishes production of $1,25(OH)_2$ Vitamin D (calcitriol) in renal proximal tubules. Since calcitriol has been shown to be additive to the inotropic effects of norepinephrine and vasopressin [21] and since receptors for $1,25(OH)_2$ D are found in increased amounts in hypertrophied heart muscle [22], the fact that calcitriol synthesis in the proximal tubule decreases as kidney function is lost may be seen as protective from accelerated hypertension. Since vitamin D is useful in controlling excessive secretion of parathyroid hormone in kidney failure, replacement therapy is routine in hemodialysis units. But a further benefit may be protection from sepsis as found in experimental model studies [23–26].

Initial enthusiasm for calcium channel blockers in renal transplantation related to their role in control of hypertension as well as the possibility that calcium channel blockers might be organ protective from intracellular calcium infusion related cell death (as seen in the necrotic myocardium during acute coronary occlusion). Then chronic myocardial stress in a rat hypertension model demonstrated protection by a calcium channel blocker (mibefradil) from myocardial scarring with an increase in interstitial collagen thought to be the result of increased fibroblast collagen production [27]. But since outcome studies had not shown long-term benefits of vasodilation following myocardial infarction, there was concern for perioperative hypotension with acute kidney allograft injury, which eventually became an impediment to the study of calcium channel blockers in renal transplant centers, particularly as the potential treatment population has aged from less than 55 to greater than 75 years of age under certain circumstances.

Thus, observations connecting calcium channel blockers with prolonged survival of kidney transplant allografts may have been due to both preservation of circulation [8–10] to the allograft at risk for ischemia due to the vasoconstrictive effect of the immunosuppressive agent, cyclosporine, but perhaps also to inhibition of mechanisms in the immunological rejection process [6,7]. An unanswered question in multiple centers from several countries is the method of preservation of donor kidneys at a point in time after nephrectomy when they are most susceptible to acute injury. A powerful vasodilator, the calcium channel blocker, lidoflazine, was shown to protect Lewis rat donor kidneys preserved in University of Wisconsin perfusate [28]. Additive effects of cyclosporine and verapamil on suppression of cytotoxic T lymphocyte cell proliferation [29] are considered to be operating via the calcium/calmodulin/calcineurin pathway to nuclear direction of the expression of antibodies through nuclear factor kappa Beta. Additive effects of cyclosporine and verapamil may also be found in suppression of the generation of cytokines by lymphocytes through inhibition of the incorporation of thymidine into DNA or uridine into RNA or leucine into protein.

Eventually the possibility that calcium channel blockers have targets other than on slow calcium channels at the cell borders has been extended to include effects on calcium movement into and out of the cytosol from storage areas within the cell (endoplasmic reticulum, sarcoplasmic reticulum, mitochondria) or on the calcium extrusion technique of pathogen survival in a host whose levels of calcium could be fatal to the pathogen [30]. Our experience with infection in diabetic patients with nephropathy has been limited to end-stage renal disease. Use of peritoneal dialysis for urgent removal of fluid to prevent pulmonary edema [31] or for chronic removal of uremic toxins had been associated with frequent instances of fever, abdominal pain and cloudy peritoneal effluent. When this triad occurred together, the diagnosis of peritonitis with a positive culture of effluent fluid would be expected. However, if only one of the diagnostic pointers were present, there might be some time before a second or third would appear. Our prospective study to determine whether frequent culture of peritoneal dialysis effluent fluid might yield an early warning prior to the clinical syndrome did not produce positive culture results worthy of the extensive expense [32]. At the height of an incursion of illicit distribution of cocaine into the New England sector of the USA, we found the availability of clean needles did not protect diabetic dialysis patients from life-threatening cellulitis, abscess formation, and sepsis [33] All the instances of visceral infection occurred in the setting of cocaine intake. Bacterial infection was detected four times as often in this target population while evidence of spread of hepatitis virus was ten-fold greater amongst users of cocaine. In our studies of cardiovascular end-points in the setting of diabetes associated with cardio-renal complications, there were serendipitous findings in the field of kidney transplantation ten years after the introduction of cyclosporine, which had marked a time of significant improvement in prevention of allograft rejection.

But when our retrospective review of 70 consecutive diabetic recipients of kidney allografts looking for cardiovascular end-points did confirm outcome benefits, it was purely by serendipity that the benefits were mainly derived through protection from attacks of sepsis [3]. Of the 70 study subjects we reviewed from this center, data on 69 was available for analysis. There were 36 individuals who had received calcium channel blockers and 33 who had not received calcium channel blockers. Of the 36, 21 were treated with diltiazem, 10 with nifedipine, and 5 with verapamil. One study subject who initially received nifedipine was subsequently switched to diltiazem. There were no significant outcome differences in cardiovascular events which were found to be less common than sepsis-related events. So, sepsis became the focus of outcome analysis in our retrospective review. Of the 19 individuals with a prior history of cardiovascular events, six who received calcium channel blockers had two sepsis-related events while 13 who had not received calcium channel blockers had 22 sepsis-related events. Of the 50 individuals without a prior history of cardiovascular events, 30 who had received calcium channel blockers had 13 sepsis-related events while 20 individuals who had not received calcium channel blockers experienced 39 sepsis-related events. While these unadjusted differences are highly significant, a weakness of the retrospective analysis would be unequal follow-up times. On the other hand, Kaplan-Meier survival curves were able to demonstrate 1.6 ($p < 0.02$) and 1.5 ($p < 0.05$)-fold greater survivals for patients and allografts, respectively for calcium channel blocker-treated vs non-treated subgroups at 24 months. When both calcium channel blocker and beta blocker medications were used, diabetic non-living allograft kidney recipients having been rigorously evaluated by experienced cardiologists [34] were documented to have durations of survival equivalent to those of non-diabetic recipients of living donor allografts using data from the UNOS Registry [35] and the US Renal Data System [36].

2.2. *Negative Results*

More recently we had the opportunity to perform a post-hoc evaluation of kidney transplantation outcomes in the International Folic Acid for Vascular Outcomes Reduction in Transplant (FAVORIT) trial, finding the incidence of infection risks to be much higher in diabetic recipients of allografts from living or non-living donors [37]. Since follow-up in this study began on average three years post-transplantation surgery, years four to ten were the observation period when most of the acute rejection and cardiovascular issues would probably have been resolved. Thus, the FAVORIT study was useful in documenting the importance of infection as the most serious outcome measure under the unique conditions involved. Of the 4110 individuals recruited into the FAVORIT trial, there were 2447 non-diabetic study subjects of whom 199 (8.2%) died during the study; 166 Type 1 diabetic study subjects of whom 44 (26.5%) died during the study; and 1497 Type 2 diabetic study subjects of whom 250 (16.7%) died during the study. Thus, there were 493 deaths with 191(38.7%) from cardiovascular causes and 286 (58.0%) from non-cardiovascular sources. Of the 286 non-cardiovascular deaths, 113 (59.2%) were due to infection while 76 (39.8%) were due to malignancy associated with immune-suppression.

Our results confirmed those from the UK National Health Service which had reported a 70% greater infection-related mortality for diabetic compared to non-diabetic kidney transplant recipients [38]. Between 2002 and 2007 when FAVORIT trial recruited 4110 renal transplant recipients, 4009 were followed to completion of the portion reviewed herein. Of the 4009, 2323 had received a non-living donor allograft while 1868 had received a living donor organ. Of the 2323 non-living allograft recipients, 871 (37.5%) were taking a calcium channel blocker while 1452(62.4%) were not taking a calcium channel blocker. Of the 1686 living donor organ recipients, 531 (31.5%) were taking a calcium channel blocker while 1155 (68.5%) were not taking a calcium channel blocker. Kaplan-Meier survival curves demonstrated no benefit for patient or allograft survivals over six years of follow-up. This is approximately nine years post-transplant surgery, a point at which both patient and allograft survival rates were not less than 75% for several sub-groups [39]. We have not analyzed this data according to the impact of statin medication upon groups with or without use of calcium channel blockers. Statins appear to inhibit a process by which excess growth of vascular smooth muscle cells pathologically associated with a subendothelial collection of matrix metalloprotein and LDL cholesterol. Studies also indicate that statins activate an L-type calcium that can be antagonized by calcium channel blockers, thus amlodipine cooperates in this statin end-point [40].

Rescue of neutrophilic phagocytic function through use of nifedipine among hemodialysis patients has been reported [41]. When calcium channel blockers were tested for protection from infections in the intra-venous tubing used as temporary or permanent access for hemodialysis, there was no clinical benefit reported [42]. Another instance of failure to find benefit from calcium channel blockers in the infection risk situation, occurred with use of verapamil for the first ten post-operative days following kidney transplantation [43]. Among 152 renal transplant recipients 16 of 77 (21%) who received 240 mg of verapamil for the first ten post-operative days developed a significant infection over the next 2–6 months compared with 4 of 75 (5%) who did not. 20 patients were hospitalized for infection in this brief follow up, 4 died (20%), including 3 of the 16 (19%) who had received verapamil vs. 1 of 4 (25%) who did not receive verapamil. However, episodes of infection occurred 6–20 weeks after verapamil, leave some biologic doubt as to causality. Despite levels of cyclosporine being higher during verapamil treatment, the occurrence of biopsy -proven allograft rejection was not significantly different between the two groups: 19/77 (25%) for the verapamil group vs. 22/75 (29%) for the no-verapamil group. The incidence of acute graft dysfunction, however, from a non-rejection cause (acute tubular necrosis) was lower in the verapamil group: 9/77 (12%) vs. 31/75 (41%) in the no-verapamil group. Since episodes of acute renal failure occurred in the immediate post-operative period, a statistical relationship with plausible biologic validity to verapamil is noted. And post-hoc evaluation of infection risk studies in populations with multiple risk factors are difficult to control for analysis. In a post-hoc evaluation of the FAVORIT study, which had identified infection as the single greatest long-term risk factor to survival of allograft recipients [37], there was no evidence for protection from serious infection for allograft recipients who had received calcium channel blockers vs. those who had not received calcium channel blockers [39].

3. Antibiotic-Resistant Pathogens May Utilize Channels Sensitive to Calcium Channel Blockers

Specific mechanisms for protection from sepsis by calcium channel blockers may include control of concentration of cytosolic calcium of mono- and poly-morphonuclear cells for efficient chemotaxis, migration, adhesion, phagocytosis [44,45] while limiting excessive cytokine response with massive capillary leaking found in the adult respiratory distress syndrome [46,47].

3.1. Treatment Resistance through Antibiotic Extrusion by the Pathogen or Host Cell

Ultimately, an effect on invading pathogens to limit natural selection of antibiotic-resistant strains would be a cost-effect efficiency goal (Table 2). Calcium channel blockers have been studied in quinolone-resistant pneumococcal pneumonia [48], rifampicin-resistant tuberculosis [49], quinine-resistant *Plasmodium falciparum* malaria [50,51], and praziquantel-resistant *Schistosoma mansoni* infestation [52,53], amphotericin-resistant *Leishmania donovani* [54], and eflornithine-resistant *Trypanosoma cruzi* [55]. Benidipine may be useful for treatment of *Toxoplasma gondii* infestation for individuals allergic to sulfonamides like sulfadiazine or sulfamethoxazole [56].

Pathogens able to efficiently extrude antibiotics quickly become resistant since they are being treated with progressively higher doses. Calcium channel blockers have been useful through closing the channel through which antibiotics are extruded. *Mycobacterium tuberculosis* has the capacity to extrude antibiotics which have entered through its tough outer membrane. The efflux mechanism [57] can be demonstrated following engulfing of the pathogen by macrophage. Verapamil inhibits the efflux pumping of antibiotics by closing the critical channel [58–60], which is not a typical calcium channel as shown by comparison with isomers and metabolites that are very effective at the extrusion channel, but ineffective at the calcium channel.

Since treatment of resistant mycobacteria with new medications is costly, studies have utilized verapamil to achieve in vitro and in vivo eradication by bedaquiline at lower effective dose [61,62]. One of the dangerous side effects of bedaquiline, an ATP synthase inhibitor, is prolongation of QTc interval, increasing the risk for ventricular arrhythmias. Verapamil suppresses this side effect. The calcium channel blocker, nimodipine used in slow, continuous infusion has been found effective in normalizing the electrocardiographic changes found in experimental cerebral malaria [54]. Resistance to rifampicin and other antibiotics like ethambutol is a major impediment to recovery from pulmonary tuberculosis for individuals who are immune-suppressed by malnutrition, uncontrolled glycemia, uremia, cancer chemotherapy. In mice, the use of verapamil has been shown to reduce drug resistance by decreasing the extrusion of rifampicin. A similar mechanism for prevention of uncontrolled infestation with Plasmodium falciparum due to drug resistance has been described with effective use of a calcium channel blocker.

Two biologic defensive mechanisms are responsible for the antibiotic resistance of *Mycobacterium tuberculosis* and *M. smegmatis*. The first is the antibiotic efflux process which has now been demonstrated on the cell membrane where a protein is described that forms the structure of a channel with a preference for cations [63]. The second is the finding that the same protein channel which extrudes cations and antibiotics also produces a toxin which causes sufficient necrosis to allow the pathogen to escape from the macrophage which had recently engulfed it. This same channel also allows the pathogen to take up nutrients before it escapes [64]. Several gram negative and gram-positive bacteria have been found to have the potential for antibiotic efflux. Among the more well-known gram-negative bacteria are *Bacteroides, Brucella, Campylobacter, Enterobacter, haemophilus, Neisseria, Pseudomonas, Vibrio*. Among the more well-known gram-positive bacteria are *Bacillus, Clostridium, Listeria, Mycobacterium, Staphylococcus*. A rare *Mycobacterium, M. abscessus* is an important pathogen in cystic fibrosis, an inherited disorder with an increased risk of pneumonia associated with airway obstruction. Although this pathogen is fast-growing, it is frequently antibiotic-resistant [65]. Data on calcium channel blocker impact in this rare form of tuberculosis may be forth-coming from current studies, particularly in India. It is now known that antibiotic-resistant *M. tuberculosis* not only works by extrusion of antibiotics, but also by preventing their entry in the first place.

3.2. Other Mechanisms of Pathogen Resistance Involving Calcium Ion Channels

The normal concentration of calcium in the plasma is in milli-molar range (2 mMol/L) while that of the cytosol of both host cells [66] and invading pathogens [67] is in micromoles (0.1 micromol/L), a thousand-fold difference. In addition to this initial environmental shock, the pathogen may then be attacked by a sudden burst of intracellular calcium at an even higher concentration as the host cell attempts eradication (Table 2). Two protective responses have been demonstrated in pathogens. The first adaptive mechanism seeks to protect from an impending state of oxidative stress through generation of anti-oxidant proteins, such as catalase and superoxide dismutase. Catalase may be responsible for survival of antibiotic-resistant *P. aeruginosa* [68], but for *Bordetella pertussis*, dismutase is more important than catalase for survival within polymorphonuclear leukocytes [69]. The second adaptive mechanism involves influx of cytosolic calcium across the plasma membrane of the host [30] by means of energy generated by ATPase, which mechanism has been observed in *Bacillus anthracis*, *Clostridium perfringens*, and *Streptococcus pneumoniae* (Table 2). It would appear the first of these mechanisms would be of relatively short-term benefit to the pathogen. The second of these two would allow for a more interference with host cellular defenses. Thus, there are attempts to develop drugs that inhibit the calcium efflux transport function of the pathogen [57].

Calcium-related mechanisms have been identified in certain fungi which may be invasive to humans. *Aspergillus fumigatus*, an allergen and an invasive pulmonary pathogen, utilizes the calcium/calmodulin/calcineurin pathway [70] in colony growth through extension and branching of hyphae. Cyclosporine is a growth inhibitor. Calcium pathways are actively-employed by *Saccharomyces cerevisiae*, *Candida allicins* and *Cryptococcus neoformans*. Inhibition of growth of *C. neoformans* by the calcineurin inhibitor, tacrolimus, has been detected [71]. Inhibition of development of hyphae of *Candida albicans* (Table 2) has been reported for verapamil [72].

Measurement of cytosolic calcium of the Sprague-Dawley rat [73] or human [74] hepatocyte by the fluorescent indicator, Fura 4f, indicated a level of ≈1 micro-molar could be achieved by ATP stimulation. Thus, the normal concentration of 0.1 micro-molar would be elevated ten-fold during hepatitis virus cell invasion through a mobilization of calcium. But in a counter move, hepatitis virus pathogens may direct cytosolic calcium levels above 1 micro-molar into temporary stores in mitochondria. Since the leading intra-cellular store of calcium is endoplasmic reticulum, the first phase calcium release would occur from this store. High concentration of this cation might lead to host cell injury. An example of unregulated calcium entry into cytosol has been studied in retinal cells which are destroyed in rodents exposed to HIV-1 virus [75].

Therefore, in a second phase while a short-term bolus injection of calcium might eradicate an invading pathogen, to protect the host cell from prolonged excess calcium this cation must be removed from cytosol to either intra-cellular stores by sarco-endoplasmic reticulum ATPase pumping or to extra-cellular environment by means of plasma membrane ATPase pumping. As extra-cellular calcium intake proceeds to replenish endoplasmic reticulum stores, a temporary high-plateau of calcium concentration may be followed by a lower resting concentration (≈0.1 micro-molar).

3.3. Store Operated Calcium Entry (SOCE)

With the advent of CCB it became apparent that cytotoxic levels of calcium entry across the plasma membrane could be blocked with beneficial results. Entry of calcium into the cytosol from intracellular storage zones in the endoplasmic or sarcoplasmic reticulum (Store operated calcium entry, SOCE) may have therapeutic implications [76]. The juxtaposition of the sarcoplasmic endoplasmic reticulum allows for calcium signaling to control entry for calcium into the cytosol through the plasma membrane. Studies of defense mechanisms in liver cells infected with hepatitis B virus identify SOCE as being therapeutic for a short period of time. SOCE mechanisms may be activated during times of cytogenetic, inflammatory and hemodynamic instability.

In the instance of the hepatitis B patient or the victim of an attack by poliovirus [77], calcium transfer most likely occurred in the setting of a single pathogen in an uncompromised host.

This pathogenesis is more clearly delineated than that found in a trauma victim with multi-organism sepsis and shock due to capillary leak, sometimes referred to as adult respiratory distress syndrome (ARDS) during which SOCE can be sudden and to a massive degree [46,47]. An instance of SOCE that can occur without invasion by a pathogen is that of cancer metastasis [78]. Inhibition of proliferation of human leiomyoma by means of calcium release from intracellular stores via voltage gated calcium channels has been shown to be enhanced by simvastatin [66].

3.4. Hyperglycemia and Calcium Ion Channels

A related series of observations involves activity of the cardiac ATPase enzyme that may demonstrate increased expression following exposure to the diabetes medication liraglutide, a member of the glucagon-like peptide class of anti-hyperglycemia agents [79]. Increased cardiomyocyte plasma membrane ATPase calcium pumping has been linked to initiation and acceleration of congestive heart failure symptoms thought to be due to elevated intra-cellular calcium concentration. On the other hand, several studies have shown improved heart function with use of the diabetes medications empagliflozin, canagliflozin, or dapagliflozin, members of the sodium/glucose transporter 2 (SGLT2) inhibition class of antihyperglycemic agents [80]. Reversal of elevated cardiomyocyte calcium concentration may explain the beneficial result in heart failure reported with these medications. This mechanism could apply as well to heart failure patients who do not have the problem of hyperglycemia at least on a short-term basis [81]. Infection outcome results with SGLT2 inhibition would certainly be of interest, particularly in the realm of antibiotic-resistant pathogens.

Restoration in the capacity for phagocytosis when hyperglycemia is reversed with anti-hyperglycemic agents is reproducible and confirmed. The mechanism of disarming of the neutrophil appears to have been an increase in calcium concentration in the cytosol. If the source of calcium were the external medium of the neutrophil, then the effect of nifedipine/amlodipine would have to be exerted at the plasma membrane. But if the source of cytosolic calcium were stores in the endoplasmic reticulum, then the mechanism would be the target of future research agents. Which-ever source of calcium were involved, it would appear the pathological increase in calcium concentration in the cytoplasm was brought under control by means of a calcium channel blocker. Dihydropyridine calcium channel blockers like amlodipine or nifedipine in hypertension or with sulfonylurea medications, like glyburide or glipizide, in hyperglycemia has been found in rescue of neutrophil phagocytic function [44,45]. While the untested hypothesis that use of statin for cholesterol control may lower cost of expensive antibiotics has some rational grounds due to calcium channel blocking activity it should be remembered that for the type 2 diabetes population it is this same mechanism which may be responsible for hyperglycemia due to inhibition of beta cell insulin secretion [82].

4. Loss of Regulation of Inflammatory Response to Infection: Mechanisms Associated with the Sepsis Shock Syndrome

An experimental model for testing the impact of calcium channel blockers on sepsis involves ligation of the cecum with puncture of the wall of the intestine. This model has demonstrated relatively longer survival if diltiazem were injected prior to the onset of septic shock [4] in association with decreased formation of oxygen radicals [5]. Specific mechanisms for protection from sepsis by calcium channel blockers include: a decrease in cytosolic calcium of inflammation mediating cells, thereby limiting excessive cytokine responses, such as occurs in the adult respiratory distress syndrome [46]; an improved capacity to combat pathogens (chemotaxis, movement, adhesion, phagocytosis) through an increase in cytosolic calcium of polymorphonuclear cells and macrophages by release from intracellular stores (endoplasmic reticulum, sarcoplasmic reticulum, mitochondria) through alternate channels while the slow L channels from the exterior are blocked [47]; and an effect on invading pathogens to limit their capacity to select strains capable of rapid development of resistance to antibiotics. More recent reports of protection from sepsis have emerged from studies of antibiotic-resistant *Myocobacteria tuberculosis.* Along these lines, one analog of a calcium channel blocker (verapamil) may be more effective than another in assisting rifampicin in its battle with *M. tuberculosis* for reasons not easily understood in terms of L-type calcium channels, but rather by efficiency of docking at a site critical to the rifampin efflux mechanism. It is entirely possible that no available calcium blocker can reverse capillary leak as a complication of infection. Newer analogs however may become available [83]. It is now known that antibiotic-resistant *M. tuberculosis* bacilli inhibit entry and accelerate extrusion of antibiotics. Control of cytosolic calcium entry or transport from internal sources has been found critical for macrophage and neutrophil defense functions. Channels from endo- and sarcoplasmic reticula may interact with trauma/infection stresses for release of calcium from relevant storage areas. The endoplasmic reticulum utilizes inositol-3-phosphate receptors while the sarcoplasmic reticulum utilizes ryanodine receptors.

Sepsis Leads to Capillary Leaking

Understanding of the of the sepsis syndrome has shifted focus from that of an overwhelming inflammation cascade to an emphasis upon leaking of capillary fluid, resulting in hemoconcentration with diminished blood flow to vital organs and eventually into a state of shock due to hypovolemic hypotension (Table 2). This capillary leak results from endothelial dysfunction secondary to changes in the angiopoietin system. Ordinarily, angiopoietin 1 maintains the vascular barrier, but the effect of bacterial invasion is to enhance the concentration of angiopoietin 2 through its release from storage sites in endothelial cells mediated by tumor necrosis factor alpha (TNF α). A higher concentration of angiopoietin 1 vs. angiopoietin 2 normally operates the vascular barrier through a growth factor receptor, Tie2, which is inhibited when angiopoietin 2 concentration increases. So, a marker for the severity of the state of sepsis would be the rising level of circulating angiopoietin 2, a for-warning of imminent hypovolemia due to interstitial space accumulation of electrolyte/protein-containing fluid with strong osmolar capacity. Although the local presence of angiopoietin 2 at the portal of entry may be beneficial as a way of washing out the invasive pathogen, the target of sepsis-related research would be counter-measures by which a prolonged elevated level of circulating angiopoietin 2 might be brought into balance [84,85]. Calcium entry into endothelial cells under stress appears to occur via T-type calcium channels rather than L-type. Thus, amlodipine might have no effect upon capillary leak while flunarizine might be highly effective. In addition, flunarizine has been shown to inhibit vasoconstriction by norepinephrine and to diminish the risk of small vessel thrombosis [86]. Angiotensin caused vasoconstriction can damage muscular layers of larger and endothelium of smaller blood vessels [18]. The result may be a loss of resistance to sepsis [19,20].

5. Vitamin D, Parathyroid Hormone, Fibroblast Growth Factor, and Klotho Interact with Sepsis Defense Mechanisms in Which Movement of Calcium and Phosphorus Are Part of the Process

Given various findings of calcium movement inside host defense cells (polymorphonuclears, monocytes, macrophages), the issue of interaction between Vitamin D treatment for protection is an obvious area of sepsis study since lack of vitamin D is associated with risk of pneumonia [87,88]. In this connection relevant roles of vitamin D consist of assisting parathyroid hormone in the absorption of calcium in the intestinal tract as well as an independent capacity in the absorption of the companion anion phosphate which does not appear to be a target for parathyroid hormone. This increased prevalence of upper respiratory tract infections has been noted in patients with low vitamin D levels without [87,88] or with [89,90] advanced kidney failure. Intensive Care Unit patients who are septic have been found to have significantly lower concentrations of Vitamin D binding protein than those who were not septic. And there was also a positive correlation between Vitamin D and the toll-like receptors needed for expression of cathelicidin LL-37 levels [25]. Cathelicidins and defensins are antimicrobial peptides which cooperate in protection from infection. Since these agents are found in bronchopulmonary sites under attack from pathogens, such as *Mycobacterium tuberculosis*, they can be expected to offer a measure of protection in the lung.

Genes are being identified which code for peptides from pathogens, setting in motion expression of receptors (called Toll-like) which assist in generation of cathelicidins [25]. Susceptibility to tuberculosis is associated with vitamin D deficiency as well as a lack of production of cathelicidins. Efficiently activated monocytes will be capable of producing cathelicidin protein plus advancing 25(OH) vitamin D to the intracellular active form $1,25(OH)_2$ D [25]. Downstream from the Vitamin D receptor is the peptide, cathelicidin, with bactericidal capacity against *Mycobacterium tuberculosis* (Table 2). This suggests that hydroxylation of the cholecalciferol by the 1-alpha hydroxylase activates cathelicidin through toll like receptor activity. Raising the question as to whether correction of Vitamin D deficiency directly confers bactericidal potential. A question to be answered is the relationship between intracellular levels of calcium, vitamin D, and calcium channel blockers. The relationship of calcium channel blockers to expression of bactericidal proteins, cathelicidin and Beta-defensin in association with intracellular elaboration of $1,25(OH)_2$ vitamin D3 remains to be fully elucidated.

Since the demonstration of lower levels of 25(OH) vitamin D, vitamin D binding protein, and cathelicidin (LL37) in the serum of critically patients in the ICU compared to healthy controls, details of these observations have become of interest in trauma centers (Table 2). Eleven healthy controls were compared to 25 non-septic ICU control subjects versus 24 septic ICU study subjects. The individuals with sepsis had significantly higher levels of BUN and creatinine with lower levels of serum albumin, consistent with acute kidney injury. There was a positive correlation between 25(OH) vitamin D and cathelicidin (LL-37) levels [89]. Filtered 25(OH) vitamin D in complex with vitamin D binding protein can be resorbed from the glomerular filtrate with the assistance of proximal tubular protein, megalin [91] while the binding protein is degraded, the 25(OH) D is converted into 1,25(OH)2 D, and both 25(OH) D + $1,25(OH)_2$ D are resorbed into the circulation. Retention of phosphorus with lowering of calcium (despite reflex increases in parathyroid hormone) has been shown to be related to mortality [92]. Kidney failure with vitamin D deficiency can generate fibroblast growth factor to assist in excretion of retained phosphorus if vitamin D is replaced [93]. Klotho, the anti-aging gene seems to have a role in lowering the elevated level of phosphorus in kidney failure in cooperation with Fibroblast Growth Factor 23. Some insight into renal failure as a risk factor for infection might be gained (Table 2) by studying response to phosphorus retention in the setting of combined effects of parathyroid hormone, fibroblast growth factor 23, and Klotho [94] cooperating to exclude excess phosphorus at the sodium/phosphate co-transporter in the proximal tubule. But while PTH and FGF 23 are capable of cooperating at the proximal tubular site for phosphate transport to decrease phosphorus resorption, they are simultaneously capable of competing with each other at a different site where PTH assists in activation of 1-alpha hydroxylase (Table 2) for completion of synthesis if $1,25(OH)_2$ cholecalciferol (calcitriol), while inhibition of 1-alpha hydroxylase by Fibroblast Growth Factor is occurring simultaneously by way of inhibiting excess intestinal absorption of calcium in the setting of risk for deposition of calcium phosphorus in vascular tissues (calciphylaxis). In terms of infection risk, if the function of white blood cells is inhibited at low levels of both phosphorus and vitamin D, then the suggestion might be that kidney dysfunction and infection are more closely associated with Fibroblast Growth Factor 23 than with parathyroid hormone.

6. Conclusions

It is doubtful that sponsors can be found to support placebo-controlled trials for long-term calcium channel blockade in immunosuppressed individuals (such as solid organ transplant recipients) for the purpose of determining whether there is a beneficial effect on non-cardiovascular health care outcomes. However, there are transplant registries and multicenter trial databases from which additional information might be uncovered. An effort should be made to elucidate whether benefits or risks associated with calcium channel blockers in transplant populations as an initial example of immunosuppressed study groups with equally well-recorded information. Since long-term risk of infection in recipients of kidney transplant allografts is more important than cardiovascular complications [37] and outcome results for protection from sepsis by calcium channel blockers are in conflict [3,39] further studies need to be explored in terms of the multiple mechanisms reviewed herein. With respect to other populations immunosuppressed either iatrogenically or by virtue of underlying disease, we consider that further observations suggesting that this inexpensive class of medications might enhance treatment of infectious diseases should direct future trials. Another untested possibility is the hypothesis that use of statins to lower doses and cost of expensive antibiotics [95] may be effective since these drugs also exhibit some calcium channel blocking activity [96].

Funding: This research received no external funding.

References

1. Dial, S.; Nessim, S.J.; Kezouh, A.; Benisty, J.; Suissa, S. Antihypertensive agents acting on the renin-angiotensin system, and the risk of sepsis. *Br. J. Clin. Pharmacol.* **2014**, *78*, 1151–1158. [CrossRef] [PubMed]
2. Zeng, L.; Hunter, K.; Gaughan, J.; Podder, S. Preadmission use of calcium channel blockers and outcomes after hospitalization with pneumonia: A retrospective propensity-matched cohort study. *Am. J. Ther.* **2017**, *24*, e30–e38. [CrossRef] [PubMed]
3. Weinrauch, L.A.; D'Elia, J.A.; Gleason, R.E.; Shaffer, D.; Monaco, A.P. Role of calcium channel blockers in diabetic renal transplant patients: Preliminary observations on protection from sepsis. *Clin. Nephrol.* **1995**, *44*, 185–192. [PubMed]
4. Meldrum, D.R.; Ayala, A.; Chaudry, I.H. Mechanism of diltiazem's immunomodulatory effects after hemorrhage and resuscitation. *Am. J. Physiol.* **1993**, *265*, C412–C421. [CrossRef] [PubMed]
5. Rose, S.; Baumann, H.; Tahques, G.P.; Sayeed, M.M. Diltiazem and superoxide dismutase modulate hepatic acute phase response in gram-negative sepsis. *Shock* **1994**, *1*, 87–93. [CrossRef] [PubMed]
6. Weir, M.R. Therapeutic benefits of calcium channel blockers in cyclosporine-treated organ transplant recipients: Blood pressure control and immunosuppression. *Am. J. Med.* **1991**, *90*, 32S–36S. [CrossRef]
7. Chrysostomou, A.; Walker, R.G.; Russ, G.R.; D'Apice, A.J.F.; Kincaid-Smith, I.; Mathew, T.H. Diltiazem in renal allograft recipients receiving cyclosporine. *Transplantation* **1993**, *55*, 300–304. [CrossRef] [PubMed]
8. Van Riemsdijk, I.C.; Mulder, P.; deFijter, J.W.; Bruijn, J.A.; van Hoof, J.P.; Hoitsma, A.J.; Tegzess, A.M.; Weimar, W. Addition of isradipine (Lomir) results in better renal function after renal transplantation: A double blind randomized placebo controlled multicenter study. *Transplantation* **2000**, *70*, 122–126. [PubMed]
9. Kuypers, D.R.; Neumayer, H.H.; Fritsche, K.; Rodicio, J.L.; Vanrenterghem, Y. Lacidipine Study Group. Calcium channel blockade and preservation of renal graft function in cyclosporine-treated recipients: A prospective randomized placebo-controlled 2-year study. *Transplantation* **2004**, *78*, 1204–1211. [CrossRef] [PubMed]
10. Ahmed, K.; Michael, B.; Burk, J.F. Effects of isradipine on renal hemodynamics in renal transplant patients treated with cyclosporine. *Clin. Nephrol.* **1997**, *48*, 307–310. [PubMed]
11. Neumayer, H.-H.; Kunzendorf, U.; Schreiber, M. Protective effects of calcium antagonists in human renal transplantation. *Kidney Int.* **1992**, *36*, S87–S93.
12. Varley, K.G.; Dhalla, N.S. Excitation-contraction coupling in heart. XII. Subcellular calcium transport in isoproterenol-induced myocardial necrosis. *Exp. Mol. Pathol.* **1973**, *19*, 94–105. [CrossRef]
13. Judah, J.D.; McLean, A.E.; McLean, E.K. Biochemical mechanisms of liver injury. *Am. J. Med.* **1970**, *49*, 609–616. [CrossRef]
14. Ellrodt, G.; Chew, C.Y.; Singh, B.N. Therapeutic implications of slow-channel blockade in cardio-circulatory disorders. *Circulation* **1980**, *62*, 669–679. [CrossRef] [PubMed]
15. Akizuki, O.; Inayoshi, A.; Kitayama, T.; Yao, K.; Shirakura, S.; Saski, K.; Kusaka, H.; Marshbara, M. Blockade of T-type voltage-dependent Ca^{2+} by benidipine, a dihydropyridine calcium channel blocker, inhibits aldosterone production in human adreno-cortical cell line NCI-H 295 R. *Eur. J. Pharmacol.* **2008**, *584*, 424–434. [CrossRef] [PubMed]
16. Dietz, J.D.; Du, S.; Bolten, C.W.; Payne, M.A.; Xia, C.; Blinn, J.R.; Funder, J.W.; Hu, X. A number of marketed dihydropyridine calcium channel blockers have mineralocorticoid receptor antagonist activity. *Hypertension* **2008**, *51*, 742–748. [CrossRef] [PubMed]

17. Bellien, J.; Joannides, R.; Iacob, M.; Anaud, P.; Thuillez, C. Calcium activated potassium channels and nitric oxide regulate human peripheral conduit artery mechanics. *Hypertension* **2005**, *46*, 210–216. [CrossRef] [PubMed]
18. Doerschug, K.C.; Delsing, A.S.; Schmidt, G.A.; Ashare, A. Renin-angiotensin system activation correlates with microvascular dysfunction in a prospective cohort study of clinical sepsis. *Crit. Care* **2010**, *14*, R24. [CrossRef] [PubMed]
19. Lund, D.D.; Brooks, R.M.; Faraci, F.M.; Heistad, D.D. Role of angiotensin II in endothelial dysfunction by lipopolysaccharide in mice. *Am. J. Physiol. Heart Circ. Physiol.* **2007**, *293*, H3726–H3731. [CrossRef] [PubMed]
20. Laesser, M.; Oi, Y.; Ewert, J.; Fandriks, L.; Aneman, A. The angiotensin II receptor blocker candesartan improves survival and mesenteric perfusion in an acute porcine endotoxic model. *Acta Anaethesiol. Scand.* **2004**, *48*, 198–204. [CrossRef]
21. Bukoski, R.D.; Xua, H. On the vascular inotropic action of 1,25(OH) vitamin D3. *Am. J. Hypertens.* **1993**, *6*, 388–396. [CrossRef] [PubMed]
22. Walthers, M.R.; Wicker, D.C.; Riggle, P.C. 1,25-dihydroxy D3 receptors identified in the rat heart. *J. Mol. Cell Cardiol.* **1986**, *18*, 67–72. [CrossRef]
23. Horiuchi, H.; Nagata, I.; Komorlya, K. Protective effect of vitamin D analogues on endotoxin shock in mice. *Agents Action* **1991**, *33*, 343–348. [CrossRef]
24. Asakura, H.; Aoshima, K.; Suga, Y.; Yamazaki, M.; Morishita, E.; Saito, M.; Miyamoto, K.-I.; Nakao, S. Beneficial effect of the active form of vitamin D3 against LPS-induced DIC, but not against tissue-factor-induced DIC in rats. *Thromb. Haemost.* **2001**, *85*, 287–290. [PubMed]
25. Liu, P.T.; Stenger, S.; Li, H.; Wenzel, L.; Tan, B.H.; Krutzik, S.R.; Ochoa, M.T.; Schauber, J.; Wu, K.; Meinken, C.; et al. Toll-like receptor triggering of a vitamin D-medicated human antimicrobial response. *Science* **2008**, *311*, 1770–1773. [CrossRef] [PubMed]
26. Moller, S.; Laigaard, F.; Olgaard, K.; Hemmingsen, C. Effect of 1,25-dihydroxy-vitaminD3 in experimental sepsis. *Int. J. Med. Sci.* **2007**, *4*, 190–195. [CrossRef] [PubMed]
27. Ramires, F.J.A.; Sun, Y.; Weber, K.T. Myocardial fibrosis associated with aldosterone or angiotensin II administration: Attenuation by calcium channel blockade. *J. Mol. Cell. Cardiol.* **1998**, *30*, 475–483. [CrossRef] [PubMed]
28. Jacobsson, J.; Odlind, B.; Tufveson, G.; Wahlberg, J. Improvement of renal preservation by adding lidoflazine to University of Wisconsin solution. An experimental study in the rat. *Cryobiology* **1992**, *29*, 305–309. [CrossRef]
29. Grgic, H.; Wulff, I.; Eichler, C.; Flothmann, R.; Kohler, R.; Hoyer, J. Blockade of T-lymphocyte Kca 3.1 and Kv1.3 channels as novel immunosuppression strategy to prevent allograft rejection. *Transplant. Proc.* **2009**, *41*, 2601–2606. [CrossRef] [PubMed]
30. Roach, J.W.; Sublett, J.; Gao, G.; Wang, Y.-D.; Tuomanen, E.I. Calcium efflux is essential for bacterial survival in the eukaryotic host. *Mol. Microbiol.* **2008**, *70*, 435–444.
31. Weinrauch, L.A.; Kaldany, A.; Miller, D.G.; Yoburn, D.C.; Belok, S.; Healy, R.W.; Leland, O.S.; D'Elia, J.A. Cardio-renal failure: Treatment of refractory biventricular failure by peritoneal dialysis. *Uremia Investig.* **1984**, *8*, 1–8. [CrossRef]
32. Cooper, G.; White, J.; D'Elia, J.; DeGirolami, P.; Arkin, C.; Kaldany, A.; Platt, R. Lack of utility of routine screening tests for early detection of peritonitis in patients requiring intermittent peritoneal dialysis. *Infect. Control* **1984**, *5*, 321–325. [CrossRef] [PubMed]
33. D'Elia, J.A.; Weinrauch, L.A.; Paine, D.F.; Domey, P.E.; Smith-Ossman, S.; Williams, M.E.; Kaldany, A. Increased infection rate in diabetic dialysis patients exposed to cocaine. *Am. J. Kidney Dis.* **1991**, *17*, 349–352. [CrossRef]
34. Lindley, E.M.; Hall, A.K.; Hess, J.; Abraham, J.; Smith, B.; Hopkins, P.N.; Shihab, F.; Welt, F.; Owan, T.; Fang, J.C. Cardiovascular risk assessment and management in pre-renal transplant candidates. *Am. J. Cardiol.* **2016**, *117*, 146–150. [CrossRef] [PubMed]
35. Cecka, J.M.; Terasaki, P.I. The UNOS Scientific Renal Transplant Registry. *Clin. Transpl.* **1993**, *1*, 1–18.
36. Almond, P.S.; Matas, A.; Gillingham, K.; Dunn, D.L.; Payne, W.D.; Gores, P.; Gruessner, R.; Nagarian, J.S. Risk factors for chronic rejection in renal allograft recipients. *Transplantation* **1993**, *55*, 752–756. [CrossRef] [PubMed]
37. Weinrauch, L.A.; D'Elia, J.A.; Weir, M.R.; Bunnapradist, S.; Finn, P.; Liu, J.; Claggett, B.; Monaco, A.P. Infection and malignancy outweigh cardiovascular mortality in kidney transplant recipients: Post-hoc analysis of the FAVORIT trial. *Am. J. Med.* **2018**, *131*, 165–172. [CrossRef] [PubMed]

38. Hayer, M.K.; Ferrugia, D.; Begaj, I.; Ray, D.; Sharif, A. Infection-related mortality for kidney allograft recipients with pre-transplant diabetes mellitus. *Diabetologia* **2014**, *57*, 554–561. [CrossRef] [PubMed]
39. Weinrauch, L.A.; Liu, J.; Claggett, B.; Finn, P.V.; Weir, M.R.; D'Elia, J.A. Calcium channel blockade and survival in recipients of successful renal transplant: An analysis of the FAVORIT trial results. *Int. J. Nephrol. Renov. Dis.* **2018**, *11*, 1–7. [CrossRef] [PubMed]
40. Clunn, G.F.; Sever, P.S.; Hughes, A.D. Calcium channel regulation in vascular smooth muscle cells: Synergistic effects of statins and calcium channel blockers. *Int. J. Cardiol.* **2010**, *139*, 2–6. [CrossRef] [PubMed]
41. Alexiewicz, J.M.; Smogorzewski, M.; Klin, M.; Akmal, M.; Massry, S.G. Effective treatment of hemodialysis patients with nifedipine on metabolism and function of polymorphonuclear leukocytes. *Am. J. Kidney Dis.* **1995**, *25*, 440–444. [CrossRef]
42. Obialo, C.I.; Conner, A.C.; Lebon, L.F. Calcium blocking agents do not ameliorate hemodialysis catheter bacteremia. *Dial. Transplant.* **2002**, *31*, 848–854.
43. Nanni, G.; Pannochia, N.; Tacchino, R.; Foco, M.; Piccioni, E.; Castagneto, M. Increased incidence of Infection in verapamil-treated kidney transplant recipients. *Transpl. Proc.* **2000**, *32*, 551–553. [CrossRef]
44. Seyrek, N.; Markinkowski, W.; Smorgorzewski, M.; Demerdash, T.M.; Massry, S.G. Amlodipine prevents and reverses the elevation in [Ca2+] and the impaired phagocytosis of PMNL of diabetic rats. *Nephrol. Dial. Transpl.* **1997**, *12*, 265–272. [CrossRef]
45. Krol, E.; Agueel, R.; Smorgorzewski, M.; Kumar, D.; Massry, S.G. Amlodipine reverses the elevation in [Ca^{2+}] and the impairment in PMNLs of NIDDM patients. *Kidney Int.* **2003**, *64*, 2188–2195. [CrossRef] [PubMed]
46. Lee, C.; Xu, D.Z.; Feketova, E.; Nemeth, Z.; Kannan, K.B.; Hasko, G.; Deitch, E.A.; Hauser, C.J. Calcium entry inhibition during resuscitation from shock attenuates inflammatory lung injury. *Shock* **2008**, *30*, 29–35. [CrossRef] [PubMed]
47. Hotchkiss, R.S.; Karl, I.E. Ca2+, a regulator of the inflammatory response—The good, the bad, and the possibilities. *Shock* **1997**, *7*, 308–310. [CrossRef] [PubMed]
48. Pletz, M.W.; Mikhaylov, N.; Schumacher, U.; van der Liden, M.; Duesberg, C.B.; Fuehner, T.; Klugman, C.P.; Welte, T.; Makarewicz, O. Antihypertensives suppress the emergence of fluoroquinolone-resistant mutants in pneumococci: An in vitro study. *Int. J. Med. Microbiol.* **2013**, *303*, 176–181. [CrossRef] [PubMed]
49. Song, L.; Cui, R.; Yang, Y.; Wu, X. Role of calcium channels in cellular anti-tuberculosis effects: Potential of voltage-gated calcium-channel blockers in tuberculosis therapy. *J. Microbiol. Immunol. Infect.* **2015**, *48*, 471–476. [CrossRef] [PubMed]
50. Scheibel, L.W.; Colambani, P.M.; Hess, A.D.; Aikawa, M.; Atkinson, T.; Milhous, W.K. Calcium and calmodulin antagonists inhibit human malaria parasites (*Plasmodium falciparum*): Implications for drug design. *Proc. Natl. Acad. Sci. USA* **1987**, *84*, 7311–7314. [CrossRef]
51. Martins, Y.C.; Clemmer, L.; Orjuela-Sanchez, P.; Zanini, G.M.; Ong, P.K.; Frangos, J.A.; Carvalho, L.J. Slow and continuous delivery of a low dose of nimodipine improves survival and electrocardiogram parameters in rescue therapy of mice with experimental cerebral malaria. *Malar. J.* **2013**, *12*, 138–154. [CrossRef] [PubMed]
52. Gryseels, B.; Mbaye, A.; DeVias, S.J.; Stelma, F.F.; Guisse, F.; Van Lieshout, L.; Faye, D.; Diop, M.; Ly, A.; Tchuem-Tchuente, L.A.; et al. Are poor responses to praziquantel for treatment of Schistosoma mansoni infections in Senegal due to resistance? An overview of the evidence. *Trop. Med. Int. Health* **2001**, *6*, 864–873. [CrossRef] [PubMed]
53. Silva-Mores, V.; Couto, F.F.; Vasconcelos, M.M.; Araujo, N.; Coelho, P.M.; Katyz, N.; Grnfell, R.F. Anti-Schistosomal activity of a calcium channel antagonist on Schistosoma and adult Schistosoma Mansoni worms. *Memorias do Instituo Oswaldo Cruz* **2013**, *108*, 600–604. [CrossRef]
54. Kashif, M.; Manna, P.P.; Akhter, Y.; Alaidarous, M.; Rub, A. Screening of novel inhibitors against Leishmania donovani calcium channel ion channel to fight Leishmaniasis. *Infect. Disord. Drug Targets* **2017**, *17*, 120–129. [CrossRef] [PubMed]
55. Pollo, L.A.E.; deMoraes, M.H.; Cisilotto, J.; Creczynski-Pasa, T.B.; Biavatti, M.W.; Steindel, M.; Sandjo, L.P. Synthesis and in vitro evaluation of Ca^{2+} channel blockers 1,4-dihydropyridines analogues against Trypanosoma cruzi and Leishmania amazonensis: SAR analysis. *Parasitol. Int.* **2017**, *66*, 789–797. [CrossRef] [PubMed]
56. Kanatani, S.; Fuks, J.M.; Olafsson, E.B.; Westermark, L.; Chambers, B.; Varas-Godoy, M.; Ulen, P.; Barrigan, A. Voltage-dependent calcium channel signaling mediates GABA a receptor-induced migratory activation of dendritic cells infected by Toxoplasma gondii. *PLoS Pathog.* **2017**, *13*, e1006739. [CrossRef] [PubMed]
57. Liz, X.-Z.; Nikaido, H. Efflux-mediated drug resistance in bacteria: An update. *Drugs* **2009**, *69*, 1555–1623.

58. Adams, K.N.; Szumowski, J.D.; Ramakrishnan, L. Verapamil and its metabolite nor-verapamil inhibit macrophage-induced, bacterial efflux pump-mediated tolerance to multiple anti-tubercular drugs. *J. Infect. Dis.* **2014**, *210*, 456–466. [CrossRef] [PubMed]
59. Adams, K.N.; Takaki, K.; Connolly, L.E. Drug tolerance in replicating mycobacteria mediated by a macrophage-induced efflux mechanism. *Cell* **2011**, *145*, 39–53. [CrossRef] [PubMed]
60. Gupta, S.; Cohen, K.A.; Winglee, K.; Maiga, M.; Diarra BBishai, W.R. Efflux inhibition with verapamil potentiates bedaquiline in Mycobacterium tuberculosis. *Antimicrob. Agents Chemother.* **2014**, *58*, 574–576. [CrossRef] [PubMed]
61. Srikrishna, G.; Gupta, S.; Dooley, K.E.; Bishai, W.R. Can the addition of verapamil to bedaquiline-containing regimens improve tuberculosis treatment outcomes? A novel approach to optimizing TB treatment. *Future Microbiol.* **2015**, *10*, 1257–1260. [CrossRef] [PubMed]
62. Diacon, A.H.; Donald, P.R.; Pym, A.; Grobusch, M.; Patientia, R.F.; Mahanyele, R.; Bantubani, N.; Narasimooloo, R.; DeMarez, R.; van Heeswijk, R.; et al. Randomized pilot trial of eight weeks of bedaquiline (tme207) treatment for multidrug-resistant tuberculosis: Long-term outcome, tolerability, and effect on emergence of drug resistance. *Antimicrob. Agents Chemother.* **2012**, *56*, 3271–3276. [CrossRef] [PubMed]
63. Siroy, A.; Mailaender, C.; Harder, D.; Koerber, S.; Wolschendorph, F.; Danilchanka, O.; Wang, Y.; Heinz, C.; Niederweis, M. RV 1698 of microtubular proteins, a new class of channel-forming outer membrane proteins. *J. Biol. Chem.* **2008**, *283*, 17827–17837. [CrossRef] [PubMed]
64. Danichanka, O.; Sun, J.; Pavlemok, M.; Maueroder, C.; Speer, A.; Siroy, A.; Mayhew, N.; Doomlos, K.S.; Munez, L.E.; Herrmann, M.; et al. An outer membrane channel protein of Mycobacterium tuberculosis with exotoxin activity. *Proc. Natl. Acad. Sci. USA* **2014**, *111*, 6750–6755. [CrossRef] [PubMed]
65. Bryant, J.M.; Grogono, D.M.; Greaves, D.; Foweraker, J.; Roddick, I.; Inns, T.; Reacher, M.; Haworth, C.S.; Curran, M.D.; Harris, S.R.; et al. Whole-genome sequencing to identify transmission of Mycobacterium abscessus between patients with cystic fibrosis: A prospective cohort study. *Lancet* **2013**, *9877*, 1551–1560. [CrossRef]
66. Borahay, M.A.; Killic, G.S.; Yallampalli, C.; Snyder, R.R.; Hankins, G.D.V.; Al-Hendry, A.; Boehning, D. Simvastatin potentially induces calcium-dependent apoptosis of human leiomyoma cells. *J. Biol. Chem.* **2014**, *289*, 35075–35086. [CrossRef] [PubMed]
67. Gangola, P.; Rosen, B.P. Maintenance of intracellular calcium in *Escherichia coli*. *J. Biol. Chem.* **1987**, *262*, 12570–12574. [PubMed]
68. Orlandi, V.T.; Martegani, E.; Bolognese, F. Catalase A is involved in the response to photo-oxidative stress in *Pseudomonas aeruginosa*. *Photodiagn. Photodyn. Ther.* **2018**, *22*, 233–240. [CrossRef] [PubMed]
69. Khelef, N.; DeShager, D.; Friedman, R.L. In vitro and in vivo analogue of Bordetella pertussis catalase and Fe-superoxide dismutase mutants. *FEMS Lett.* **1996**, *142*, 231–235. [CrossRef]
70. DiMarco, T.M.; Freitas, F.Z.; Almeida, R.S.; Brown, N.A.; des Reis, T.F.; Zambelli-Ramalho, L.N.; Savoldi, M.; Goldman, M.H.S. Functional characterization of an Aspergillus fumigatus calcium transporter (PmcA) that is essential for fungal infection. *PLoS ONE* **2012**, *7*, e37591. [CrossRef]
71. Krauss, P.R.; Nichols, C.B.; Heitman, J. Calcium- and calcineurin-independent roles for calmodulin in Cryptococcus neoformans morphogenesis and high-temperature growth. *Eukaryot. Cell* **2005**, *4*, 1079–1087. [CrossRef] [PubMed]
72. Yu, Q.; Ding, X.; Bing, Z.; Xu, N.; Jia, C.; Mao, J.; Zhang, B.; Xing, L.; Li, M. Inhibitory effect of verapamil on Candida albicans hyphal development, adhesion, and gastrointestinal colonization. *FEMS Yeast Res.* **2014**, *14*, 633–641. [CrossRef] [PubMed]
73. Casciano, J.C.; Duchemin, N.J.; Lamontagne, J.; Steel, L.F.; Bouchard, M.J. Hepatitis B virus modulates store operated calcium entry to enhance viral replication in primary hepatocytes. *PLoS ONE* **2017**, *12*, e0168328. [CrossRef] [PubMed]
74. Casciano, J.C.; Bouchard, M.J. Hepatitis virus X protein modulates cytosolic Ca+ signaling in primary human hepatocytes. *Virus Res.* **2018**, *246*, 23–27. [CrossRef] [PubMed]
75. Dreyer, E.B.; Kaiser, P.K.; Offermann, J.T.; Lipton, S.A. HIV-1 coat protein neurotoxicity prevented by calcium channel antagonists. *Science* **1990**, *248*, 364–367. [CrossRef] [PubMed]
76. Reddish, F.N.; Miller, C.L.; Gorkhali, R.; Yang, J.J. Calcium dynamics mediated by the endoplasmic/sarcoplasmic reticulum and related diseases. *Int. J. Mol. Sci.* **2017**, *18*, 1024. [CrossRef] [PubMed]

77. Liu, S.; Rodriguez, A.V.; Tosteson, M.T. Role of simvastatin and beta cyclodextrin on inhibition of poliovirus infection. *Biochem. Biophys. Res. Commun.* **2006**, *347*, 51–59. [CrossRef] [PubMed]
78. Xie, J.; Pan, H.; Yao, J.; Han, W. SOCE and cancer: Recent progress and new perspectives. *Int. J. Cancer* **2016**, *138*, 2067–2077. [CrossRef] [PubMed]
79. Packer, M. Should we be combining GLP-1 receptor agonists and SGLT2 inhibitors in treating diabetes? *Am. J. Med.* **2018**, *131*, 461–463. [CrossRef] [PubMed]
80. Packer, M.; Anker, S.D.; Butler, J.; Filippatos, G.; Zannad, F. Effects of sodium-glucose cotransporter 2 inhibitors for the treatment of patients with heart failure: Proposal of a novel mechanism of action. *J. Am. Med. Assoc.* **2017**, *2*, 1025–1029. [CrossRef] [PubMed]
81. Byrne, N.J.; Parajuli, N.; Levasseur, J.L.; Boisvenue, J.; Becker, D.I.; Masson, G.; Fedak, P.W.M.; Verma, S.; Dyck, J.R.B. Empagliflozin prevents worsening of cardiac function in an experimental model of pressure overload-induced heart failure. *J. Am. Coll. Cardiol. Basic Transl. Sci.* **2017**, *2*, 347–354. [CrossRef] [PubMed]
82. Brault, M.; Ray, J.; Gomez, Y.H.; Mantzoros, C.S.; Daskalopoulou, S.S. Statin treatment and new onset diabetes: A review of proposed mechanisms. *Metabolism* **2014**, *63*, 735–745. [CrossRef] [PubMed]
83. Singh, K.; Kumar, M.; Pavodai, E.; Naran, K.; Warner, D.F.; Rominski, P.G.; Chibale, K. Synthesis of new verapamil analogues and their evaluation in combination with rifampicin analogues against Mycobacterium tuberculosis and molecular docking studies in the binding of efflux protein Rv1258c. *Bioorgan. Med. Chem. Lect.* **2014**, *14*, 2985–2990. [CrossRef] [PubMed]
84. Parikh, S.M.; Mammoto, T.; Schultz, A.; Yuan, H.T.; Christiani, D.; Karamuchi, S.A.; Sukhatme, V.P. Excess Angiopoietin-2 may contribute to pulmonary vascular leak in sepsis in humans. *PLoS Med.* **2006**, *3*, e46. [CrossRef] [PubMed]
85. David, S.; Kumpers, P.; van Slyke, P.; Parikh, S.M. Mending leaky blood vessels: The angiopoietin-Tie2 pathway in sepsis. *J. Pharmacol. Exp. Ther.* **2013**, *345*, 2–6. [CrossRef] [PubMed]
86. Retzlaff, J.; Thamm, K.; Ghosh, C.C.; Ziegler, W.; Haller, H.; Parikh, S.M.; David, S. Flunarizine suppresses endothelial angiopoietin-2 in a calcium-dependent fashion. *Science* **2017**, *7*, 44113. [CrossRef] [PubMed]
87. Grinde, A.A.; Mansbach, J.M.; Carmargo, C.A., Jr. Association between serum 25-hydroxyvitamin D Level and upper respiratory infection in the Third National Health and Nutrition Examination Survey. *Arch. Intern. Med.* **2009**, *169*, 384–390. [CrossRef] [PubMed]
88. Laaksi, I.; Ruohola, J.P.; Tuohimaa, P.; Auviven, A.; Haataja, R.; Pihlajamaki, H.; Yikomi, T. An association of serum vitamin D concentrations <40 nmol/L, with acute respiratory tract infection in young Finnish men. *Am. J. Clin. Nutr.* **2007**, *86*, 714–717. [PubMed]
89. Jengh, L.; Yamshchikov, A.V.; Judd, S.E.; Blumberg, H.M.; Martin, G.S.; Ziegler, T.R.; Tangpricha, V. Alterations in vitamin D status and anti-microbial peptide levels in patients in the intensive care unit with sepsis. *J. Transl. Med.* **2009**, *7*, 28–36. [CrossRef] [PubMed]
90. Chonchol, M.; Greene, T.; Zang, Y.; Hoofnagle, A.N.; Cheung, A.K. Low vitamin D and high fibroblast growth factor 23 serum levels associate with infectious and cardiac deaths in the HEMO study. *J. Am. Soc. Nephrol.* **2016**, *27*, 227–237. [CrossRef] [PubMed]
91. Nykjaer, A.; Dragun, D.; Walther, D.; Vorum, H.; Jacobsen, C.; Herz, J.; Mersen, F.; Christensen, E.I.; Willnow, T.E. An endocytic pathway essential for renal uptake of the steroid 25-(OH) vitamin D. *Cell* **1999**, *96*, 507–515. [CrossRef]
92. Kestenbaum, B.; Sampson, J.N.; Rudser, K.D.; Patterson, D.J.; Seliger, S.L.; Young, B.; Sherrard, D.J.; Andress, D.L. Serum phosphate levels and mortality risk among people with chronic kidney disease. *J. Am. Soc. Nephrol.* **2005**, *16*, 520–528. [CrossRef] [PubMed]
93. Fliser, D.; Kolleritis, B.; Neyer, U.; Ankerst, D.P.; Lhotta, K.; Lingenhel, A.; Rita, E.; Kronenberg, F.; Kuen, E.; Konig, P.; et al. Fibroblast growth factor (FGF 23) predicts progression of chronic kidney disease: The mile to moderate kidney disease (MMKD) study. *J. Am. Soc. Nephrol.* **2007**, *18*, 2600–2608. [CrossRef] [PubMed]
94. Razzaque, M.S. The FGF 23-Klotho axis: Endocrine regulation of phosphate homeostasis. *Nat. Rev. Endocrinol.* **2009**, *5*, 611–619. [CrossRef] [PubMed]

95. Lee, M.-Y.; Lin, K.-D.; Hsu, W.-H.; Chang, H.-L.; Yang, Y.-H.; Hsiao, P.-J.; Shin, S.-J. Statin, calcium channel blocker and beta blocker therapy may decrease the incidence of tuberculosis infection in elderly Taiwanese patients with type 2 diabetes. *Int. J. Mol. Sci.* **2015**, *16*, 11369–11384. [CrossRef] [PubMed]
96. Ali, N.; Begum RFaisal, M.S.; Khan, A.; Mabi, M.; Shehzadi, G.; Ullah, S.; Ali, W. Current statin show calcium channel blocking activity through voltage gated channels. *BioMed. Cent. Pharmacol. Toxicol.* **2016**, *17*, 43–50. [CrossRef] [PubMed]

3

The Role of Endothelial Ca^{2+} Signaling in Neurovascular Coupling: A View from the Lumen

Germano Guerra [1], Angela Lucariello [2], Angelica Perna [1], Laura Botta [3], Antonio De Luca [2] and Francesco Moccia [3,*]

1 Department of Medicine and Health Sciences "Vincenzo Tiberio", University of Molise, via F. De Santis, 86100 Campobasso, Italy; germano.guerra@unimol.it (G.G.); angelicaperna@gmail.com (A.P.)
2 Department of Mental Health and Preventive Medicine, Section of Human Anatomy, University of Campania "L. Vanvitelli", 81100 Naples, Italy; angela.lucariello@gmail.com (A.L.); antonio.deluca@unicampania.it (A.D.L.)
3 Laboratory of General Physiology, Department of Biology and Biotechnology "L. Spallanzani", University of Pavia, via Forlanini 6, 27100 Pavia, Italy; laura.botta@unipv.it
* Correspondence: francesco.moccia@unipv.it

Abstract: Background: Neurovascular coupling (NVC) is the mechanism whereby an increase in neuronal activity (NA) leads to local elevation in cerebral blood flow (CBF) to match the metabolic requirements of firing neurons. Following synaptic activity, an increase in neuronal and/or astrocyte Ca^{2+} concentration leads to the synthesis of multiple vasoactive messengers. Curiously, the role of endothelial Ca^{2+} signaling in NVC has been rather neglected, although endothelial cells are known to control the vascular tone in a Ca^{2+}-dependent manner throughout peripheral vasculature. Methods: We analyzed the literature in search of the most recent updates on the potential role of endothelial Ca^{2+} signaling in NVC. Results: We found that several neurotransmitters (i.e., glutamate and acetylcholine) and neuromodulators (e.g., ATP) can induce dilation of cerebral vessels by inducing an increase in endothelial Ca^{2+} concentration. This, in turn, results in nitric oxide or prostaglandin E2 release or activate intermediate and small-conductance Ca^{2+}-activated K^+ channels, which are responsible for endothelial-dependent hyperpolarization (EDH). In addition, brain endothelial cells express multiple transient receptor potential (TRP) channels (i.e., TRPC3, TRPV3, TRPV4, TRPA1), which induce vasodilation by activating EDH. Conclusions: It is possible to conclude that endothelial Ca^{2+} signaling is an emerging pathway in the control of NVC.

Keywords: neurovascular coupling; neuronal activity; brain endothelial cells; Ca^{2+} signaling; glutamate; acetylcholine; ATP; nitric oxide; endothelial-dependent hyperpolarization; TRP channels

---

1. Introduction

The brain comprises only 2% of the total body mass, yet it accounts for 20% of the overall energy metabolism [1]. As the brain has a limited capacity to store energy and lacks survival mechanisms that render other organs, such as heart and liver, more tolerant to short periods of anoxia or ischemia, the continuous supply of oxygen (O_2) and nutrients and removal of catabolic waste are critical to maintain neuronal integrity and overall brain function. Accordingly, the brain receives up to 20% of cardiac output and consumes ≈20% and ≈25% of total body's O_2 and glucose [2–4]. Brain functions cease within seconds after the interruption of cerebral blood flow (CBF), while irreversible neuronal injury occurs within minutes [2,3]. Cerebral autoregulation is the mechanism whereby CBF remains relatively stable in spite of physiological fluctuations in arterial pressure, at least within a certain range. Thus, cerebral arteries constrict in response to an increase in arterial pressure and relax upon a decrease in blood pressure [4,5]. A subtler mechanism, known as functional hyperemia or neurovascular coupling (NVC), intervenes to locally increase the rate of CBF to active brain areas, thereby ensuring adequate matching between the enhanced metabolic needs of neural cells and blood supply [2,4,6]. Through NVC, a local elevation in neuronal activity (NA) causes a significant vasodilation of neighboring microvessels, which increases CBF and generates the blood oxygenation level-dependent (BOLD) signals that are used to monitor brain function through functional magnetic resonance imaging [7,8]. NVC is finely orchestrated by an intercellular signaling network comprised of neurons, astrocytes and vascular cells (endothelial cells, smooth muscle cells and pericytes), which altogether form the neurovascular unit (NVU) [4,6,9]. Synaptically-activated neurons may signal to adjacent vessels either directly or through the interposition of glial cells. Recent evidence has shown that astrocytes modulate NVC at arteriole levels, whereas they mediate neuronal-to-vascular communication at capillaries [2,4,10–14]. NA controls cerebrovascular tone through a number of Ca^{2+}-dependent vasoactive mediators, which regulate the contractile state of either vascular smooth muscle cells (VSMCs) (arteries and arterioles) or pericytes (capillaries). These include nitric oxide (NO) and the arachidonic acid (AA) derivatives, prostaglandin E2 (PGE2), epoxyeicosatrienoic acids (EETs) and (20-HETE) [2,4,9]. While the role played by astrocytes and mural cells, i.e., VSMCs and pericytes, in CBF regulation has been extensively investigated, the contribution of microvascular endothelial cells to NVC has been largely underestimated [7]. This is quite surprising as the endothelium regulates the vascular tone in both systemic and pulmonary circulation by lining the innermost layer of all blood vessels [15–17]. Endothelial cells decode a multitude of chemical (e.g., transmitters and autacoids) and physical (e.g., shear stress pulsatile stretch) signals by generating spatio-temporally-patterned intracellular Ca^{2+} signals, which control VSMC contractility by inducing the synthesis and release of the vasorelaxing factors, NO, prostacyclin (or PGI2), carbon monoxide and hydrogen sulfide, as well as the vasoconstricting prostaglandin H2 and thromboxane A2 [15,18–21]. Moreover, an increase in sub-membranal Ca^{2+} concentration stimulates intermediate- and small-conductance Ca^{2+}-activated K^+ channels (IK_{Ca} and SK_{Ca}, respectively), thereby causing an abrupt hyperpolarization in endothelial membrane potential, which is electronically transmitted to adjacent VSMCs and causes vessel dilation [16,22,23]. Herein, we aim at surveying the role of endothelial Ca^{2+} signaling in NVC by examining the physiological stimuli, e.g., neurotransmitters, neuromodulators and dietary molecules, that modulate CBF through an increase in intracellular Ca^{2+} concentration ($[Ca^{2+}]_i$) in brain microvascular endothelial cells.

2. Neurovascular Coupling

2.1. Cerebral Circulation and the Neurovascular Unit

The concept of NVU was first introduced during the first Stroke Progress Review Group meeting of the National Institute of Neurological Disorders and Stroke of the NIH (July 2001) to highlight the relevance to CBF regulation of the intimate association between neurons and cerebral vessels [4]. The NVU represents a functional unit where neurons, interneurons and astrocytes are in close proximity and are functionally coupled to smooth muscle cells, pericytes, endothelial cells and extracellular matrix (Figure 1). Within the NVU, neurons, glial and vascular cells establish a mutual influence on each other to ensure a highly efficient system to match blood and nutrient supply to the local needs of brain cells [2,4,6]. The interaction between neurons, glial cells and blood vessels may, however, vary along the cerebrovascular tree. Large cerebral arteries arise from Circle of Willis at the base of the brain and give rise to a heavily-interconnected network of pial arteries and arterioles, which run along the cortical surface. Cerebral vessels are lined by a single layer of endothelial cells (*tunica intima*), which is separated from an intermediate layer of smooth muscle cells (*tunica media*) by the internal elastic lamina. An outermost layer mainly comprised of collagen fibers and fibroblasts and enriched in perivascular nerves, known *as tunica adventitia*, is separated from the brain by the Virchow–Robin space, which is an extension of the sub-arachnoid space [4,6]. Penetrating arterioles branch off smaller pial arteries and dive into the brain tissue, thereby giving rise to parenchymal arterioles. The muscular component of the vascular wall is reduced to a single layer of VSMCs in intracerebral arterioles: of note, endothelial cells may project through the internal elastic lamina by establishing direct connections, known as myoendothelial projections, with the adjoining smooth muscle cells. Myoendothelial projections are enriched with gap junctions and allow the bidirectional transfer of information between endothelial cells and VSMCs [24]. While penetrating arterioles are still separated from the substance of the brain by the Virchow–Robin space, perivascular astrocytic processes (i.e., end feet) enter in contact and fuse together with the basal lamina of parenchymal arterioles, so that the perivascular space is obliterated [4,6]. Relevant to local CBF regulation, pyramidal cells and γ-aminobutyric acid (GABA) interneurons provide extensive innervation to parenchymal arterioles, often with the interposition of glial cells [4,10,25–28]. In addition, sub-cortical neurons from locus coeruleus, raphe nucleus and basal forebrain may also send projection fibers, containing respectively acetylcholine, norepinephrine and 5-hydroxytryptamine (5-HT), to intracortical microvessels and surrounding astrocytes [29]. Finally, parenchymal arterioles supply the cerebral circulation by giving rise to a dense network of intercommunicating capillary vessels, which are composed only of specialized endothelial cells and lack VSMCs. However, $\approx$30% of the brain capillary surface is covered by spatially-isolated contractile cells, i.e., pericytes, while the remaining endothelial capillary tubes are almost entirely wrapped by astrocytic end feet [30,31]. Astrocytes and pericytes located outside of brain capillaries may be innervated by local neurons, as described for astrocytes and VSMCs situated in the upper districts of the vascular tree [2]. The average capillary density of human brain amounts to $\approx$400 capillaries mm^{-2}, which is enough to ensure that each neuron is endowed with its own capillary and to reduce on average to less than 15 µM the diffusion distance for O_2, nutrients and catabolic waste [32,33]. Therefore, brain microvascular endothelial cells are ideally positioned to sense neurotransmitters released by axonal terminals during local synaptic activity and by sub-cortical projections provided that they express their specific membrane receptors. The following increase in $[Ca^{2+}]_i$ could, in turn, drive the synthesis of Ca^{2+}-dependent vasoactive mediators. At the same time, brain microvascular endothelial cells have the potential to regulate local CBF independently of NA by detecting changes in blood flow variations and in the concentration of blood-borne agonists through distinct patterns of intracellular Ca^{2+} elevation.

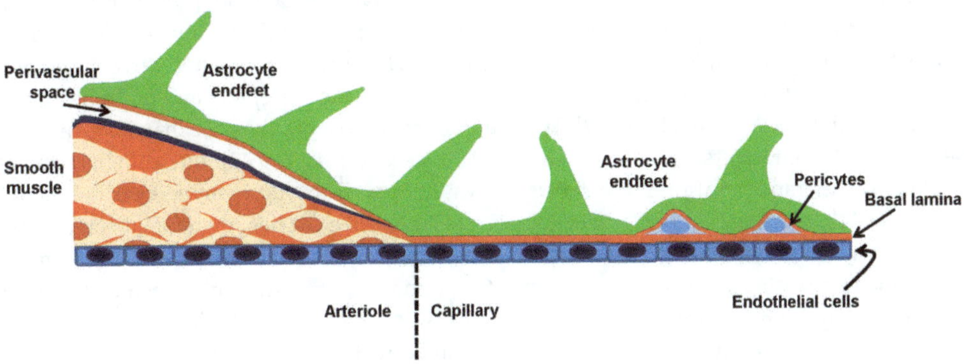

Figure 1. Cellular composition of the neurovascular unit. The vascular wall presents a different structure in arterioles and capillaries, which control the local supply of cerebral blood. Smooth muscle cells form one or more continuous layers around arterioles and change in their contractile state determine vessel diameter and regulate blood perfusion. Capillary diameter is regulated by contractile pericytes, which extend longitudinally and circumferentially along the capillary wall. Astrocyte end feet envelope arterioles and capillaries and are able to release vasoactive mediators, which regulate the contractile state of smooth muscle cells (arterioles) and pericytes (capillary) in response to neuronal activity.

2.2. Cellular and Biochemical Pathways of Neurovascular Coupling: The Role of Neurons and Astrocytes in Arterioles and Capillaries

The mechanisms whereby synaptic activity controls the microvascular tone vary depending on the brain structure and may differ along the vascular tree (Figure 2) [2,4,9]. An increase in intracellular Ca^{2+} concentration ($[Ca^{2+}]_i$) within the dendritic tree is the crucial signal that triggers the synthesis and release of vasoactive messengers in response to excitatory synaptic inputs [2,4,9,34]. For instance, glutamate stimulates post-synaptic *N*-methyl-D-aspartate (NMDA) and a-amino-3-hydroxy-5-methyl-4-isoxazol epropionic acid (AMPA) receptors to induce extracellular Ca^{2+} entry and recruit the Ca^{2+}/calmodulin (Ca^{2+}/CaM)-dependent neuronal nitric oxide (NO) synthase (nNOS) in hippocampal and cerebellar GABA interneurons [10,25,28,33,35]. The following NO release elicits arteriole vasodilation by inducing VSMC hyperpolarization and relaxation through a soluble guanylate cyclase/protein kinase G (PKG)-dependent mechanism (see below) [13,26,35–37]. Conversely, NMDA receptor (NMDAR)-mediated Ca^{2+} entry in synaptically-activated pyramidal neurons of the somatosensory cortex engages cyclooxygenase 2, which catalyzes the synthesis of the powerful vasodilator, prostaglandin E2 (PGE2), which acts through EP2 and EP4 receptors on VSMCs [27,28]. Intriguingly, NO plays a permissive role in PGE2-dependent vasodilation by maintaining the hemodynamic response during sustained NA [25,38]. Long-lasting synaptic activity could lead to an increase in $[Ca^{2+}]_i$ within perisynaptic astrocytic processes (Figure 2), which lags behind the onset of CBF, but is able to activate phospholipase A2 (PLA2) and cleave AA from the plasma membrane. AA, in turn, diffuses to adjacent VSMCs and is converted into the vasoconstricting messenger, 20-hydroxyeicosatetraenoic (20-HETE), by cytochrome P450 4A (CYP4A) [33,38]. However, neuronal-derived NO inhibits CYP4A, thereby preventing 20-HETE formation and maintaining PGE2-dependent vasodilation [9,33,38]. The mechanism(s) whereby NA induces astrocytic Ca^{2+} signals is still unclear. Glutamate has been predicted to increase astrocyte $[Ca^{2+}]_i$ by binding to metabotropic glutamate receptors (mGluRs) 1 and 5 (mGluR1 and mGluR5, respectively), which stimulate phospholipase Cβ (PLCβ) through Gqα monomer and induce inositol-1,4,5-trisphosphate ($InsP_3$)-dependent Ca^{2+} release from the endoplasmic reticulum (ER) [9,39]. However, mGluR1 and mGluR5 are lacking in adult astrocytes, and the genetic deletion of type 2 $InsP_3$ receptor ($InsP_3R2$), which represents the primary $InsP_3R$ isoform in glial cells, does not prevent NVC [2,30,39]. Nevertheless, there is indisputable evidence that astrocytes require mGluRs to drive the hemodynamic response to sensory stimulation also in adult mice [24,40,41]. Furthermore, alternative mechanisms may drive astrocyte Ca^{2+} signaling, including NMDARs, purinergic receptors and multiple transient receptor potential (TRP) channels [14,39]. We refer the reader to a number of exhaustive and recent reviews about the controversial role of astrocyte $[Ca^{2+}]_i$ in NVC [14,30,39,42]. Intriguingly, astrocytes are able to sense transmural pressure across the vascular wall through vanilloid TRP 4 (TRPV4) channels [43,44], which are located on their perivascular end feet. It has, therefore, been proposed that the initial hemodynamic response to NA activates these mechanosensitive Ca^{2+}-permeable channels, thereby causing an increase in astrocyte $[Ca^{2+}]_i$ and recruiting the Ca^{2+}-dependent PLA2 [14]. In addition, synaptically-released ATP could mobilize ER Ca^{2+} by activating P2Y2 and P2Y4 receptors, which are located on astrocytic end feet wrapped around cerebral vessels [45].

Although it has long been thought that CBF regulation occurs at the arteriole level [6], recent studies have convincingly shown that most of the hydraulic resistance that must be decreased in order to increase cortical perfusion is located in the capillary bed, which are wrapped by contractile pericytes [30,46,47]. This model makes physiological sense as, on average, firing neurons are remarkably closer to capillaries than to arterioles (8–23 μm away versus 70–160 μm) [47]. Therefore, capillaries are located in a much more suitable position to rapidly detect NA and initiate the hemodynamic response that ultimately generates BOLD signals. In addition, the regulation of CBF at the capillary level could selectively increase CBF only in active areas, thereby finely matching the local tissue O_2 supply to local cerebral demand [30,46,47]. The cerebellar cortex is composed of three layers: molecular layer, Purkinje cells and granular layer, respectively, from outermost to innermost [48]. Recent studies demonstrated that, at the molecular layer, synaptic activity caused an increase in astrocytic end feet Ca^{2+} concentration by inducing Ca^{2+} entry through P2X1 channels. This spatially-restricted Ca^{2+} signal, in turn, recruited phospholipase D2 (PLD2) and diacylglycerol lipase to synthesize AA, which was then metabolized by cyclooxygenase 1 into PGE2. Finally, PGE2 evoked capillary dilation by binding to EP4 receptors, which were presumably located on pericytes [13]. Conversely, in the granular layer, synaptic activity induced robust NO release by promoting NMDARs-mediated Ca^{2+} entry in granule cells without astrocyte involvement [11].

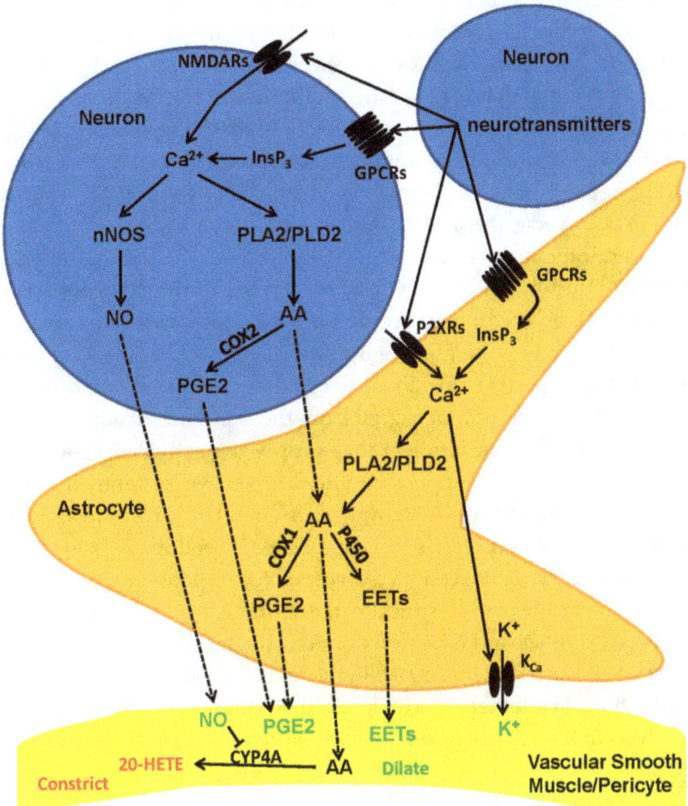

Figure 2. The mechanisms by which neurons and astrocytes stimulate arteriole and capillary dilation in response to synaptic activity. Synaptic activity increases intracellular Ca^{2+} levels within the postsynaptic neuron by stimulating metabotropic G-protein-coupled receptors (GPCRs), ionotropic receptors (e.g., NMDARs) or L-type voltage-operated Ca^{2+} channels (VOCs). This increase in $[Ca^{2+}]_i$ leads to the synthesis of NO and PGE2, which may relax both smooth muscle cells (arterioles) and pericytes (capillaries). Synaptically-released neurotransmitters may also increase $[Ca^{2+}]_i$ in perisynaptic astrocytes, thereby triggering NO release and PGE2/EET production. AA, which may be synthesized by PLA2 in both neurons and astrocytes, may be converted in the vasoconstricting factor, 20-HETE, in perivascular cells. Abbreviations: 20-HETE: 20-hydroxyeicosatetraenoic acid; AA: arachidonic acid; COX1: cyclooxygenase 1; COX2: cyclooxygenase 2; EETs: epoxyeicosatrienoic acids; GPCRs; G-protein-coupled receptors; $InsP_3$: inositol-1,4,5-trisphosphate; K_{Ca}: Ca^{2+}-activated intermediate and small conductance K^+ channels; NO: nitric oxide; nNOS: neuronal NO synthase; P450: cytochrome P450; PGE2: prostaglandin E2; PLA2: phospholipase A2; PLD2: phospholipase D2; VOCs: L-type voltage-operated Ca^{2+} channels.

2.3. *Is There a Role for Endothelial Ca^{2+} Signaling in Neurovascular Coupling?*

Vascular endothelial cells control vascular tone by releasing a myriad of vasoactive mediators in response to an increase in $[Ca^{2+}]_i$ evoked by either chemical or mechanical inputs [16,18–20,49]. For instance, endothelial Ca^{2+} signals trigger the selective increase in blood flow to skeletal, respiratory and cardiac muscles induced by physical exercise/training [15]. Curiously, it is still unclear whether and how endothelial Ca^{2+} signals play any role in translating NA into vasoactive signals in the brain [7,24]. Luminal perfusion of neurotransmitters and neuromodulators, such as acetylcholine, ATP, ADP and bradykinin, results in the dilation of cerebral arterioles by inducing the activation of specific Gq-protein-coupled receptors (GPCRs), which are located on the endothelial membrane [29,50–53]. GPCRs stimulate PLCβ to synthesize the ER Ca^{2+}-releasing messenger, $InsP_3$, thereby triggering the Ca^{2+}-mediated signaling cascade that leads to the synthesis and release of most endothelial-derived vasoactive messengers [16,18–20,49,54]. Nevertheless, only scarce information is available about the role of endothelial Ca^{2+} signaling in the hemodynamic response to NA. As described elsewhere [50], synaptically-activated neurons and/or astrocytes could directly stimulate brain microvessels by releasing mediators that traverse the VSMC or pericyte layers and bind to receptors located on the abluminal side of endothelium. Interestingly, brain microvascular endothelial cells may also express NMDARs [55,56] and mGluR1 [57–60], although their potential contribution to NVC has been barely appreciated [4]. Accordingly, recent studies demonstrated that endothelial NMDARs participate in glutamate-induced vasodilation of intraparenchymal arterioles by mediating Ca^{2+} entry and subsequent eNOS activation [55,56]. Finally, brain endothelium could actively mediate the retrograde propagation of the vasodilation signal from the capillaries feeding the sites of NA to the upstream pial vessels (arteries and arterioles) that supply the activated area [6,7,61]. Retrograde vasodilation into the proximal arterial supply is required to achieve an optimally-localized increase in blood flow to active neurons and to avoid a "flow steal" from interconnected vascular networks [4,62]. Retrograde propagation in peripheral vessels is accomplished by endothelial Ca^{2+} signals that impinge on two distinct components to drive the conducted vasomotor response: (1) the Ca^{2+}-dependent rapid activation of EDH, which is restricted to stimulated endothelial cells, but is rapidly transmitted to more remote sites along the vessel and (2) a slower intercellular Ca^{2+} wave that spreads vasodilation via endothelial release of NO and PGE2 [7,63,64]. In the following section, we will discuss the evidence supporting the contribution of endothelial Ca^{2+} signals in NVC and the possibility that interendothelial Ca^{2+} waves are involved in the control of cerebrovessel diameters and microhemodynamics during intense synaptic activity.

3. The Role of Endothelial Ca^{2+} Signaling in Neurovascular Coupling

3.1. *Endothelial Ca^{2+} Signaling in Brief*

It has long been known that an increase in endothelial $[Ca^{2+}]_i$ delivers the crucial signal to induce the synthesis of multiple vasoactive mediators [16–20]. For instance, intracellular Ca^{2+} signals recruit the Ca^{2+}/CaM-dependent endothelial NOS (eNOS) and the Ca^{2+}-dependent phospholipase A2 (PLA_2) to generate NO and prostaglandin I2 (PGI_2 or prostacyclin), respectively [16,19]. The Ca^{2+} response to extracellular autacoids typically consists of an initial Ca^{2+} peak, which is due to $InsP_3$-dependent ER Ca^{2+} release, followed by sustained Ca^{2+} entry through store-operated Ca^{2+} channels (Figure 3) [65–67]. Store-operated Ca^{2+} entry (SOCE) is a major Ca^{2+} entry pathway in endothelial cells, being activated by any stimulus leading to the depletion of the ER Ca^{2+} pool [68–70]. The dynamic interplay between $InsP_3$-dependent Ca^{2+} release, which could be amplified by adjoining ryanodine receptors (RyRs) through the process of Ca^{2+}-induced Ca^{2+} release (CICR) (Figure 3), and SOCE results in biphasic Ca^{2+} signals [71,72] or repetitive $[Ca^{2+}]_i$ oscillations in brain vascular endothelium [73–75].

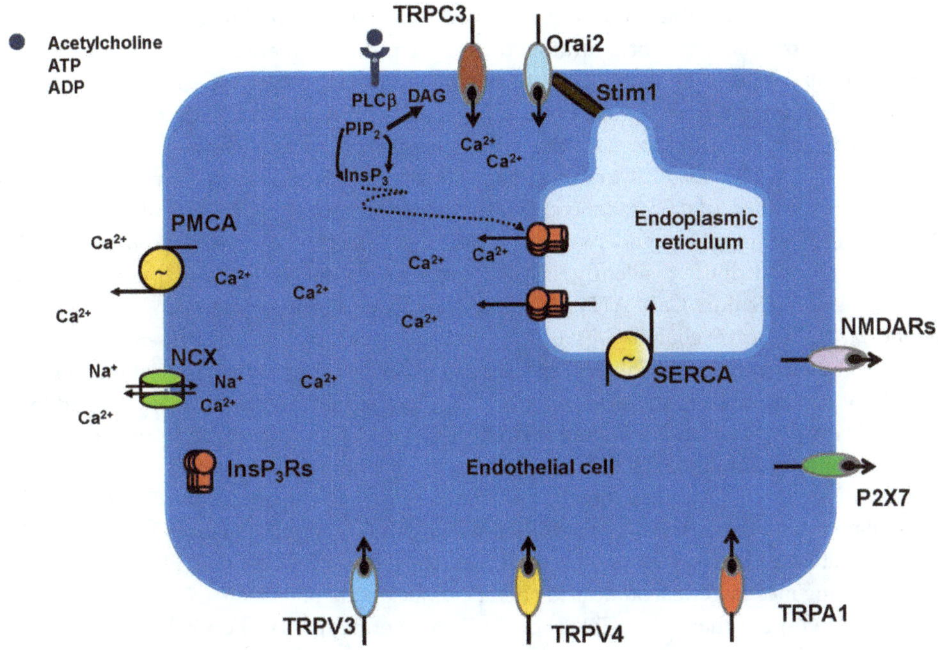

Figure 3. The Ca^{2+} signaling toolkit in brain microvascular endothelial cells. There is scarce information available regarding the molecular components of the Ca^{2+} signaling toolkit in brain microvascular endothelial cells. A recent investigation, however, provided a thorough characterization of the Ca^{2+} machinery in bEND5 cells [76], which represent an established mouse brain microvascular endothelial cell line. Further information was obtained by the analysis of endothelial Ca^{2+} signals in rodent parenchymal arterioles and in the human hCMEC/D3 cell line. Extracellular autacoids bind to specific G-protein-coupled receptors, such as M-AchRs and P2Y1 receptors, thereby activating PLCβ, which in turn cleaves PIP_2 into $InsP_3$ and DAG. $InsP_3$ triggers ER-dependent Ca^{2+} releasing by gating $InsP_3Rs$, while DAG could activate TRPC3. The $InsP_3$-dependent drop in ER Ca^{2+} levels induces SOCE, which is mediated by the interaction between Stim1 and Orai2 in bEND5 cells. Moreover, extracellular Ca^{2+} entry may occur through TRPV3, TRPV4 and TRPA1, which are coupled to either eNOS or EDH [77,78]. Finally, brain microvascular endothelial cells may express Ca^{2+}-permeable ionotropic receptors, such as NMDARs and P2X7 receptors. The elevation in $[Ca^{2+}]_i$ decays to the baseline via the concerted interaction between SERCA and PMCA pumps, as well as through NCX [72,79,80]. Abbreviations: $InsP_3$, inositol-1,4,5-trisphosphate; DAG, diacylglycerol; $InsP_3Rs$, $InsP_3$ receptors; NCX, Na^+–Ca^{2+} exchanger; PMCA, plasma membrane Ca^{2+} ATPase; PIP_2, phosphatidylinositol-4,5-bisphosphate; PLCβ, phospholipase Cβ; SERCA, sarco-endoplasmic reticulum Ca^{2+}-ATPase. The thicker line connecting PIP_2 to $InsP_3$ indicates a high amount of second messenger produced upon PLCβ activation.

3.1.1. The Endothelial Ca^{2+} Toolkit in Brain Microvascular Endothelial Cells: Endogenous Ca^{2+} Release

Scarce information is available regarding the composition of the Ca^{2+} toolkit in brain endothelial cells [74]. For instance, human brain microvascular endothelial cells expressed $InsP_3R1$ and displayed ER-dependent Ca^{2+} release in response to an increase in $InsP_3$ levels [81]. Conversely, only $InsP_3R2$ was expressed in capillary endothelium of rat hippocampus, while RyRs could not be detected in the same study [82]. Nevertheless, RyRs sustained $InsP_3$-dependent Ca^{2+} release in rat brain microvascular endothelial cells in vitro [71], which suggests that the brain endothelial Ca^{2+} toolkit could undergo substantial remodeling in cell cultures [19]. A recent investigation carried out a thorough investigation of the Ca^{2+} toolkit in bEND5 cells [76], a widely-employed mouse brain microvascular endothelial cell line [83]. This analysis revealed that bEND5 cells expressed both

$InsP_3R1$ and $InsP_3R2$, while they lacked $InsP_3R2$ and RyRs [76]. ER-mobilized Ca^{2+} may lead to an increase in mitochondrial Ca^{2+} concentration due to the physical interaction between $InsP_3Rs$ and the Ca^{2+}-permeable voltage-dependent anion channels (VDACs), which are embedded in the outer mitochondrial membrane. $InsP_3Rs$-released Ca^{2+} is transferred by VDAC1 into the intermembrane space, from which it is routed towards the mitochondrial matrix though the mitochondrial Ca^{2+} uniporter (MCU) [84,85]. This ER-to-mitochondria Ca^{2+} transfer boosts cellular bioenergetics by stimulating intramitochondrial Ca^{2+}-dependent dehydrogenases, such as oxoglutarate dehydrogenase, NAD-isocitrate dehydrogenase and pyruvate dehydrogenase [86–88]. Subsequently, endothelial mitochondria may contribute to silently (i.e., without a global increase in $[Ca^{2+}]_i$) refill the ER in a sarco-endoplasmic reticulum Ca^{2+}-ATPase (SERCA)-mediated fashion by releasing Ca^{2+} though the mitochondrial Na^+/Ca^{2+} exchanger [89,90]. Ca^{2+} entry through the MCU is driven by the negative (i.e., -180 mV) membrane potential ($\Delta\Psi$) that exists across the inner mitochondrial membrane [86]. Mitochondrial content in cerebrovascular endothelium is significantly higher as compared to other vascular districts [91], and $InsP_3Rs$-mediated mitochondrial Ca^{2+} signals arise in rat [92,93] and human [94] brain capillary endothelial cells. Moreover, mitochondrial depolarization hampers the ER-to-mitochondrial Ca^{2+} shuttle, thereby causing a remarkable increase in $[Ca^{2+}]_i$ in rat brain microvascular endothelial cells [95]. An additional mode of Ca^{2+}-mediated cross-talk in endothelial cells may be established between ER and the acidic vesicles of the endolysosomal (EL) Ca^{2+} store [96,97]. The Ca^{2+}-releasing messenger, nicotinic acid adenine dinucleotide phosphate (NAADP), mobilizes EL Ca^{2+} by gating two-pore channels 1 (TPC1) or TPC2 in several types of endothelial cells [54,98–100]. Lysosomal Ca^2 release may be amplified by adjacent ER-embedded $InsP_3Rs$ through CICR [97]. The role of lysosomal Ca^{2+} signaling in brain microvascular endothelial cells remains, however, elusive.

3.1.2. The Endothelial Ca^{2+} Toolkit in Brain Microvascular Endothelial Cells: Endogenous Ca^{2+} Release

Likewise, the molecular structure of SOCE in brain vascular endothelium remains to be fully elucidated. Typically, endothelial SOCE is comprised of Stromal interaction molecule 1 and 2 (Stim1 and Stim2, respectively), which sense the drop in ER Ca^{2+} concentration ($[Ca^{2+}]_{ER}$) and Orai1-2 channels, which provide the Ca^{2+}-permeable channel-forming subunits on the plasma membrane [65, 76,101–104]. More specifically, Stim2 is activated by small fluctuations in $[Ca^{2+}]_{ER}$, drives resting Ca^{2+} entry and maintains basal Ca^{2+} levels in endothelial cells [103], while Stim1 is engaged by massive ER Ca^{2+} depletion and sustains agonists-induced extracellular Ca^{2+} entry [65,76,101,102, 104]. However, exceptions to this widespread model may exist. For instance, Stim2 expression is rather modest, and Orai2 represents the only Orai isoform endowed to mouse brain microvascular cells [76]. Of note, Orai2 constitutes the prominent pore-forming subunit of store-operated channels also in mouse neurons [105], which is consistent with the notion that endothelial cells are sensitive to both environmental cues and epigenetic modifications [106]. As widely illustrated in [106,107], the endothelial phenotype is unique in its plasticity and depends on both site-specific signal inputs delivered by the surrounding milieu (which may be diluted in cell culture) and by site-specific epigenetic modifications (i.e., DNA methylation, histone methylation and histone acetylation), which persist under in vitro culture conditions. Similar to other endothelial cells types [108,109], SOCE is constitutively activated to refill the $InsP_3$-dependent ER Ca^{2+} pool and maintains basal Ca^{2+} levels in mouse brain microvascular endothelial cells [76]. AA may induce extracellular Ca^{2+} entry in vascular endothelial cells by prompting Orai1 to interact with Orai3 independently on Stim1 [110]. Nevertheless, Orai3 is not expressed in bEND5 cells, and therefore, this mechanism is unlikely to work in NVC [76]. The endothelial SOCE machinery could involve additional components, such as members of the canonical TRP (TRPC) sub-family of non-selective cation channels, of which seven isoforms exist (TRPC1-7) [68,111]. For instance, Stim1 could also recruit TRPC1 and TRPC4 to assemble into a ternary complex [67,112], whose Na^+/Ca^{2+} permeability is determined by Orai1 [113,114]. TRPC1 was expressed in bEND5 cells, while TRPC4 was absent [76]. However, TRPC1 requires Orai1 to be recruited into the SOCE complex upon ER Ca^{2+} depletion [112,115]. It is, therefore, unlikely that it contributes to SOCE in mouse brain microvascular endothelial cells, which lack Orai1 [76], while it may assemble with polycystic TRP2 (TRPP2) to form a stretch-sensitive Ca^{2+}-permeable channel [116]. In addition to SOCE, PLCβ-dependent signaling may lead to the activation of diacylglycerol (DAG)-sensitive Ca^{2+} channels, such as TRPC3 [117,118] and TRPC6 [119,120]. The TRP superfamily of non-selective cation channels consists of 28 members that are classified into six sub-families based on their amino acid sequence homology and structural homology [121,122]. These subfamilies are designated as canonical (TRPC1-7), vanilloid (TRPV1-6), melastatin (TRPM1-8), ankyrin (TRPA1), mucolipin (TRPML1-3) and polycystin (TRPP; TRPP2, TRPP3 and TRPP5) [78,121]. Endothelial cells from different vascular beds may dispose of distinct complements of TRP channels to respond to a myriad of different chemical, mechanical and thermal stimuli [19,78,123,124]. Vascular tone in the brain is specifically regulated by a restricted number of TRP channels, including TRPC3, TRPV3, TRPV4 and TRPA1 (Figure 3), which may stimulate NO release and/or activate IK_{Ca} and SK_{Ca} channels to trigger endothelial-dependent hyperpolarization (EDH) [19,77,78,124]. Endothelial TRPV4 has a potentially relevant role in NVC as it can be activated by AA and its cytochrome P450 epoxygenases-metabolites, i.e., epoxyeicosatrienoic acids (EETs), which evoke vasodilation in intraparenchymal arterioles [9,125]. Finally, human [126], mouse [55,56,127] and rat [128] brain endothelial cells express functional NMDARs (Figure 3) [55], which may mediate glutamate-induced extracellular Ca^{2+} entry. Likewise, P2X7 receptors were recently found in both hCMEC/D3 cells (Figure 3) [129], an immortalized human brain endothelial cell line, and in rat brain endothelial cells in situ [130]. Therefore, the multifaceted endothelial Ca^{2+} toolkit, being located at the interface between neuronal projections and flowing blood, is in an ideal position to trigger and/or modulate NVC by sensing synaptically-released neurotransmitters and blood-borne autacoids.

3.2. Endothelial NMDA Receptors Trigger Glutamate-Induced Nitric Oxide-Mediated Vasodilation

NO represents the major mediator whereby endothelial cells control the vascular tone in large conduit arteries (up to 100%), while its contribution to agonists and/or flow-induced vasodilation decreases (up to 20–50%) as the vascular tree branches into the network of arterioles and capillaries that locally control blood flow [131,132]. NO release is mainly sustained by SOCE rather than by ER-dependent Ca^{2+} release (Figure 4) [133–135], and an oscillatory increase in $[Ca^{2+}]_i$ is the typical waveform that leads to extracellular autacoids-induced NO liberation from vascular endothelial cells [136,137]. In addition to SOCE, TRPC3 [138] and TRPV4 [139] may also evoke extracellular Ca^{2+} entry-mediated eNOS activation, NO release and endothelium-dependent vasodilation. Once released, NO diffuses to adjoining VSMCs to stimulate soluble guanylate cyclase and induce cyclic guanosine-3′,5′-monophosphate (cGMP) production (Figure 4). Then, cGMP activates protein kinase (PKG), which phosphorylates multiple targets to prevent the Ca^{2+}-dependent recruitment of myosin light chain kinase and induce VSMC relaxation. For instance, PKG-dependent phosphorylation inhibits the increase in VSMC $[Ca^{2+}]_i$ induced by L-type voltage-operated Ca^{2+} channels and $InsP_3Rs$; in addition, PKG phosphorylates SERCA, thereby boosting cytosolic Ca^{2+} sequestration into the ER lumen [16,140]. In addition, PKG-dependent phosphorylation stimulates large-conductance Ca^{2+}-activated K^+ channels (BK_{Ca}), thereby inducing VSMC hyperpolarization and vessel dilation [16,140]. Finally, NO accelerates SERCA-mediated Ca^{2+} reuptake through S-nitrosylation of its cysteine thiols [140]. It is generally assumed that neuronal-derived NO induces vasodilation in the hippocampus and cerebellum [10,25,26,28,33,35–37] and plays a permissive role in PGE2-induced vasorelaxation [9,33,38]. However, alternative sources, e.g., eNOS, have been implicated in glutamate-induced NVC [141,142]. Moreover, glutamate evoked endothelium-dependent vasodilation in the presence of D-serine, a NMDAR co-agonist, in isolated middle cerebral arteries and in brain slice parenchymal arterioles [55,143,144]. NMDARs are heterotetramers comprising seven distinct subunits (GluN1, GluN2A–D and GluN3A and B); GluN1 is strictly required for the assembly of a functional channel and for correct trafficking of the other subunits [144–146].

NMDAR activation requires binding of glutamate to GluN1 and of a co-agonist, D-serine or glycine, to GluN2 [145]. Two GluN1 subunits associate with GluN2A and GluN2B to form neuronal NMDARs, which are therefore sensitive to extracellular Mg^{2+} block [145]. As a consequence, NMDAR activation requires simultaneous binding of synaptically-released glutamate and release of Mg^{2+} inhibition by AMPARs-dependent membrane depolarization [145]. However, in non-neuronal cells, GluN1 subunits assemble with GluN2C and GluN2D, which confer a lower sensitivity to Mg^{2+}, while incorporation of GluN3 further decreases the inhibitory effect of extracellular Mg^{2+} and limits Ca^{2+} permeability [144,145]. NMDAR subunits (i.e., GluN1 and GluN2A-D) have been detected in brain endothelial cells in vitro [126–128,143,147,148] and in cerebral cortex in situ [127] (Figure 3). Intriguingly, NMDARs are more abundant on the basolateral endothelial membrane, which place them in the most suitable position to mediate direct neuronal-to-vascular communication [127]. Recently, Anderson's group demonstrated that glutamate and NMDA induced vasodilation in isolated middle cerebral arteries and in brain slices penetrating arterioles only in the presence of the NMDAR co-agonist, D-serine [55]. This feature could explain why previous investigations, which omitted D-serine or glycine from the bathing solution, failed to observe NMDA-induced dilation of cerebral vessels [149]. Subsequently, the same group showed that astrocytic Ca^{2+} signals induced D-serine release, which was in turn able to activate endothelial NMDARs, thereby activating eNOS in a Ca^{2+}-dependent manner [56,127]. Interestingly, NO evoked dilation of cortical penetrating arterioles by suppressing 20-hydroxyeicosatetraenoic acid (20-HETE) synthesis and boosting PGE2-induced vasorelaxation [56,127]. Future work will have to assess whether eNOS is also activated in response to somatosensory stimulation in vivo and to elucidate the physiological transmitter that induces the Ca^{2+} response in astrocytes. However, these findings clearly show that endothelial NMDARs may control CBF by engaging eNOS at the arteriole level.

3.3. Intracellular Ca^{2+} Signals Drive Acetylcholine-Induced Nitric Oxide Release from Brain Microvascular Endothelial Cells

The cortex receives a widespread acetylcholine innervation mainly arising from the basal forebrain nucleus [28,150]. Basal forebrain acetylcholine neurons broadly projects on intraparenchymal arterioles and capillaries and are, therefore, optimally suited to directly control CBF [28,29,150]. Accordingly, electrical stimulation of basal forebrain neurons results in dilation of intracortical arterioles and increases local CBF [53,151,152]. Acetylcholine-induced cerebral vasodilation is mediated by Gq-coupled muscarinic M5 receptors (M5-AchRs) [53,153], which activate eNOS and induce NO release from cerebrovascular endothelium, in mouse and pig [53,150]. Acetylcholine evokes NO release by triggering repetitive $[Ca^{2+}]_i$ oscillations in several types of endothelial cells [137,154,155]. A recent study focused on bEND5 cells to unravel how acetylcholine induces Ca^{2+}-dependent NO release from cerebrovascular endothelium [76,156,157]. Acetylcholine triggered intracellular oscillations in $[Ca^{2+}]_i$, which were driven by rhythmical $InsP_3$-dependent ER Ca^{2+} release and maintained by SOCE (Figure 3) [76]. The Ca^{2+} response to acetylcholine was mediated by M3-AchRs, which represent another PLCβ-coupled M-AchR isoform. Conversely, nicotine did not cause any detectable increase in $[Ca^{2+}]_i$ [76], as also observed in other endothelial cell types [158]. Acetylcholine-induced Ca^{2+} oscillations led to robust NO release, with eNOS requiring both intracellular Ca^{2+} release and extracellular Ca^{2+} entry to be fully activated [76]. These data were partially confirmed in hCMEC/D3 cells, which displayed a biphasic increase in $[Ca^{2+}]_i$ in response to acetylcholine [76]; of note, human brain microvascular endothelial cells express M5-AchRs [52]. Our preliminary data suggest that the distinct waveforms of acetylcholine-induced Ca^{2+} signals in human vs. mouse brain endothelial cells reflect crucial differences in their Ca^{2+} toolkit. Accordingly, hCMEC/D3 only express $InsP_3R3$, while they lack $InsP_3R1$ and $InsP_3R2$ [159]). $InsP_3R2$, which shows the sharpest dependence on ambient Ca^{2+} and is the most sensitive $InsP_3R$ isoform to $InsP_3$, has long been known as the main oscillatory Ca^{2+} unit [160,161]. Conversely, $InsP_3R3$, which is not inhibited by surrounding Ca^{2+}, tends to suppress intracellular Ca^{2+} oscillations [160,161]. Therefore, endothelial Ca^{2+} signaling can be truly regarded as a crucial determinant for acetylcholine-induced NVC.

3.4. *Endothelial Ca^{2+} Signals Could Mediate ATP-Induced Vasodilation*

Following synaptic activity, neurons and astrocytes release ATP, which serves as a modulator of cellular excitability, synaptic strength and plasticity [162]. ATP is rapidly (200 ms) hydrolyzed by extracellular ectonucleotidases into ADP and adenosine [162], all these mediators being able to physiologically increase CBF [52,163]. Luminal application of ATP and ADP has long been known to induce endothelium-dependent vasodilation in isolated cerebral vessels. For instance, ATP and ADP promoted vasodilation in rat middle cerebral arteries by stimulating NO release, although ATP could also act through cytochrome P450-metabolites [51]. Of note, extraluminal administration of ATP evoked a biphasic vasomotor response in rat penetrating arterioles, consisting of an initial transient vasoconstriction followed by local vasodilation, which was subsequently propagated to upstream locations ($\approx$500 μm) along the vascular wall [164,165]. ATP-induced local vasoconstriction was mediated by ionotropic P2X receptors on VSMCs, while endothelial P2Y1 receptors triggered local vasodilation [164]. ATP induced vasodilation by stimulating NO release and stimulating EET production to activate EDH (see below); EETs were also responsible for the upstream propagation of the vasomotor response [164]. These findings were supported by a recent study, showing that the hemodynamic response to whisker stimulation in the mouse somatosensory cortex required P2Y1 receptor-dependent eNOS activation. Previous administration of fluoroacetate, a glial-specific metabolic toxin, prevented eNOS-dependent functional hyperemia, thereby suggesting that ATP was mainly released by perivascular astrocytes [166]. P2Y1 receptors are GPCRs, which bind to both ATP and ADP and control the vascular tone by inducing intracellular Ca^{2+} signals in endothelial cells through distinct signaling pathways [167–169]. For instance, P2Y1 receptors induced $InsP_3$-dependent ER Ca^{2+} release in rat cardiac microvascular endothelial cells [167], whereas they stimulated cyclic nucleotide-gated channels in the H5V endothelial cell line, in primary cultures of bovine aortic endothelial cells and in mouse aorta endothelial cells in situ [169]. Moreover, ATP induced local and conducted vasodilation in hamster cheek pouch arterioles by triggering an increase in endothelial $[Ca^{2+}]_i$ both locally and $\approx$1200 μm upstream along the same vessel (see also below) [170]. Therefore, it is conceivable that endothelial Ca^{2+} signals mediate P2Y1 receptor-dependent NO release and functional hyperemia in the somatosensory cortex.

3.5. *TRP Channels Trigger Endothelial-Dependent Hyperpolarization in Cerebrovascular Endothelial Cells*

EDH provides the largest contribution to endothelial vasorelaxing mechanisms in resistance-sized arteries and arterioles, as shown in coronary, renal and mesenteric circulation [22,23,131,171]. EDH is initiated by an increase in endothelial $[Ca^{2+}]_i$, which stimulates IK_{Ca} ($KCa_{3.1}$) and SK_{Ca} ($KCa_{2.3}$) channels to hyperpolarize the endothelial cell membrane (Figure 4). Hyperpolarizing current spreads from vascular endothelium to overlying smooth muscle cells to trigger VSMC relaxation and vessel dilation by inhibiting voltage-dependent Ca^{2+} entry (Figure 4) [22,23,77]. This vasorelaxing mechanism requires a tightly-regulated disposition of the Ca^{2+} sources and the Ca^{2+}-sensitive decoders, i.e., SK_{Ca} and IK_{Ca} channels, which effect membrane hyperpolarization. To achieve such a precise spatial arrangement, endothelial cells extend cellular protrusions through the internal elastic lamina, which establish a heterocellular coupling with adjacent smooth muscle cells through connexin-based myo-endothelial gap-junctions (MEGJs) [22,23,77]. The endothelial ER also protrudes into these myo-endothelial microdomain sites and forms spatially-discrete $InsP_3$-sensitive Ca^{2+} pools, which are juxtaposed with IK_{Ca} channels, which are located on the endothelial protrusions traversing the holes in the vascular wall [23,172]. Conversely, SK_{Ca} channels are distributed throughout the endothelial cell membrane, although they are enriched at MEGJs and may, therefore, sense ER-released Ca^{2+} [173]. $InsP_3Rs$ are constitutively activated to produce repetitive, spatially-restricted Ca^{2+} release events, termed Ca^{2+} pulsars, whose frequency can be increased by extracellular vasoactive agonists, such as acetylcholine. Spontaneous $InsP_3$-driven Ca^{2+} pulsars are selectively coupled to IK_{Ca} and SK_{Ca} channels, thereby hyperpolarizing the endothelial membrane potential at myo-endothelial projections and inducing dilation in adjoining VSMCs [172]. K^+ signaling in the myo-endothelial space could then be boosted by the stimulation of endothelial inward rectifying K^+ (K_{ir}) channels or of Na^+/K^+ ATPase in VSMCs. In addition to $InsP_3$-induced ER-dependent Ca^{2+} release, IK_{Ca} and SK_{Ca} channels can be activated by extracellular Ca^{2+} entry through TRPV4 channels, which is also largely expressed at MEGJs [174]. EDH does not only evoke local vasodilation at the site of endothelial stimulation; the hyperpolarizing current spreads along vascular endothelium to upstream arteries (up to $\approx$2 mm) and drives the retrograde vasodilation, which is ultimately responsible for the drop in vascular resistance that increases blood supply to active regions [175–177]. Preliminary data revealed that the same clustered architecture of TRPV4 and IK_{Ca} channels is maintained at MEGJs of rodent cerebral arteries [24,78,178–180], which suggests that discrete $InsP_3$Rs-mediated Ca^{2+} pulsars arise also in parenchymal vessels [24]. Moreover, cerebral MEGJs are enriched with TRPA1 channels, which may also support EDH in cortical circulation [178,181]. EDH mediates ATP-, UTP- and SLIGR (a selective agonist of protease-activated receptor 2)-induced vasodilation in rat middle cerebral arteries [164,180,182–185] and acetylcholine-induced vasodilation in mouse posterior cerebral arteries [186].

3.5.1. TRPV4

TRPV4 is emerging as a major regulator of CBF [24,39,163] and represents one of the most important Ca^{2+}-entry pathways in vascular endothelial cells [111,187–191]. TRPV4 channels are polymodal non-selective cation channels that mediate Ca^{2+} entry in response to distinct chemical, thermal and mechanical stimuli [54,125,187,191–194], thereby integrating the diverse surrounding cues acting on the vascular wall. For instance, TRPV4 may be activated by EETs, which are produced via AA epoxygenation by cytochrome P450 epoxygenase enzymes. Briefly, an increase in endothelial $[Ca^{2+}]_i$ induced by either mechanical or chemical stimuli may stimulate the Ca^{2+}-dependent phospholipase A2 (PLA2), which cleaves AA from membrane phospholipids [77,125]. AA is, turn, is converted into EETs, such as 5,6-EET, 8,9-EET, 11,12-EET and 14,15-EET, by cytochrome P450 2C9 or cytochrome P450 2J2 [125]. In particular, 5,6-EET, 8,9-EET and 11,12-EET were shown to stimulate TRPV4-mediated non-selective cation channels and intracellular Ca^{2+} signals in endothelial cells from several vascular beds [191,195–198]. Luminal application of UTP, which is a selective agonist of P2Y2 receptors, in rat middle cerebral arteries induced TRPV4-mediated Ca^{2+} entry across the luminal and abluminal face of the endothelial monolayer. TRPV4-mediated Ca^{2+} entry, in turn, activated PLA2, whose activation was polarized to the abluminal side, to activate EDH and induce vasodilation [199,200]. Moreover, TRPV4-mediated Ca^{2+} entry was necessary to activate IK_{Ca} and SK_{Ca} channels and induce the vasodilatory response to acetylcholine in mouse posterior cerebral arteries [186]. TRPV4 is, therefore, the most likely candidate to trigger ATP-induced retrograde vasodilation in rat parenchymal arterioles, which is initiated by EETs and involves IK_{Ca}, but not SK_{Ca}, channels [164]. Intriguingly, EETs mediate vasodilation in rodent cortical arterioles [27,42,201,202]. It has been proposed that NA-induced elevation in astrocyte $[Ca^{2+}]_i$ engages PLA2 to liberate AA, thereby resulting in EET synthesis, either within astrocytes or adjacent smooth muscle cells [33,42]. Future work is necessary to assess whether: (1) astrocyte-derived EETs also activate endothelial TRPV4 within the NVU; and/or (2) neural activity directly stimulates brain endothelial cells to produce EET synthesis and activate TRPV4. A recent investigation suggested that EDH was responsible for propagating the hemodynamic response to somatosensory stimulation from the capillary bed to upstream penetrating arterioles and pial arteries [7,203]. However, IK_{Ca} and IK_{Ca} currents could not be recorded in mouse brain capillary endothelial cells [204]. These findings strongly suggest that: (1) EDH is restricted to pial arteries and arterioles, but it cannot be activated in the capillary bed; and (2) an alternative mechanism spreads the vasomotor signal from brain capillaries to upstream vessels (see below) [204].

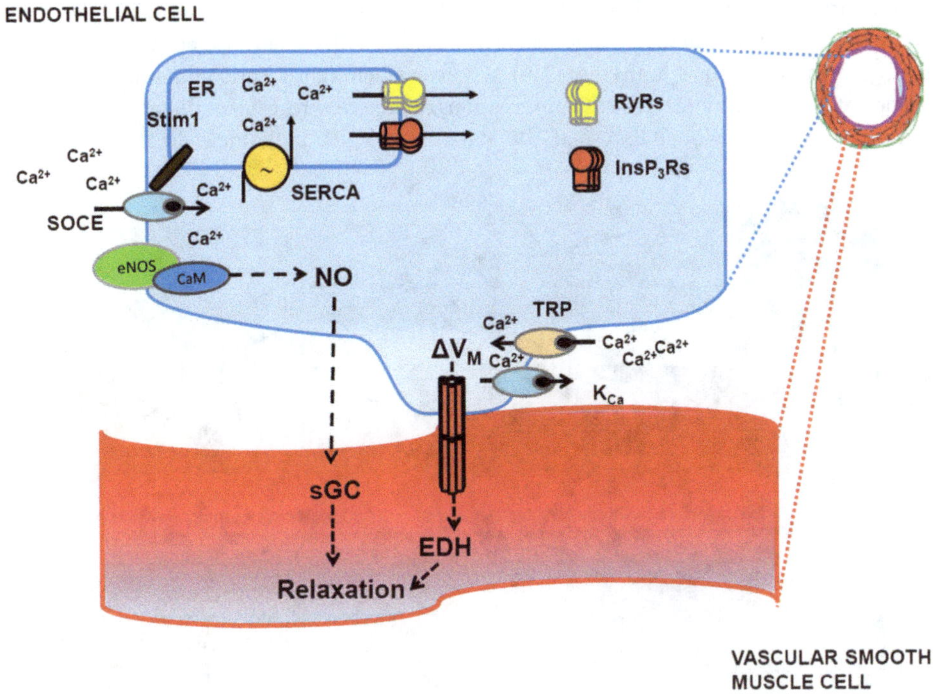

Figure 4. Ca^{2+}-regulated endothelium-dependent vasodilation. Agonists-induced $InsP_3$-dependent ER Ca^{2+} depletion in vascular endothelial cells leads to store-operated Ca^{2+} entry (SOCE). SOCE is tightly coupled to eNOS, thereby triggering robust NO release. NO, in turn, diffuses towards adjacent vascular smooth muscle cells (VSMCs) at myo-endothelial projections and activates soluble guanylyl cyclase (sGC) to induce vasorelaxation. Extracellular agonists may also activate TRP channels (e.g., TRPC3, TRPV3, TRPV4 and TRPA1), which is preferentially coupled to Ca^{2+}-activated K^+ channels (K_{Ca}), such as IK_{Ca} and SK_{Ca}. Endothelial hyperpolarization spreads through myo-endothelial gap junctions to adjoining VSMCs to induce vasorelaxation according to a mechanism known endothelial-dependent hyperpolarization (EDH).

3.5.2. TRPA1, TRPC3 and TRPV3

In addition to TRPV4, TRPA1 channels are also enriched at MEGJs in rat cerebral arteries and may induce EDH in isolated vessels [78]. Accordingly, allyl isothiocyanate (AITC) (mustard oil), a selective TRPA1 agonist, induced local Ca^{2+} entry events, IK_{Ca} channel activation and vasodilation in rat cerebral arteries [205,206]. TRPA1 was physiologically activated by superoxide anions generated by the reactive oxygen species-generating enzyme NADPH oxidase isoform 2 (NOX2), which colocalizes with TRPA1 in cerebral endothelium, but not in other vascular districts [178]. Furthermore, EDH in cerebral arteries may be elicited by Ca^{2+} influx through TRPC3 and TRPV3 channels [78]. TRPC3 is gated by DAG to conduct extracellular Ca^{2+} into vascular endothelial cells upon PLC activation [117,118,138,207]. A recent investigation revealed that TRPC3 contributed to ATP-induced vasodilation in mouse middle cerebral arteries and posterior cerebral arteries by delivering the Ca^{2+} necessary for IK_{Ca} and SK_{Ca} channel activation. IK_{Ca} channels mediated the initial phase of ATP-induced hyperpolarization, whereas SK_{Ca} channels sustained the delayed phase of endothelial hyperpolarization [208]. IK_{Ca} and SK_{Ca} channels in rat brain vessels may also be activated by TRPV3. Accordingly, carvacrol, a monoterpenoid phenol compound highly concentrated in the essential oil of oregano, caused vasodilation by selectively activating TRPV3 in rat isolated posterior cerebral and superior cerebellar arteries. TRPV3-mediated extracellular Ca^{2+} entry, in turn, engaged IK_{Ca} and SK_{Ca} channels to trigger EDH. Unlike TRPV4 and TRPA1, TRPV3 was not specifically localized to MEGJs, but was distributed throughout the endothelial membrane [209]. In addition to carvacrol, TRPV3, as well as TRPA1 are sensitive to several dietary agonists. For instance, TRPV3 may also be activated by eugenol (clove oil) and thymol (found in thyme) [210], whereas allicin (garlic) and cinnamaldehyde induce TRPA1-mediated Ca^{2+} entry [211]. These observations, therefore, suggest that dietary manipulation of TRP channels could be exploited to rescue CBF in cerebrovascular pathologies [6,212].

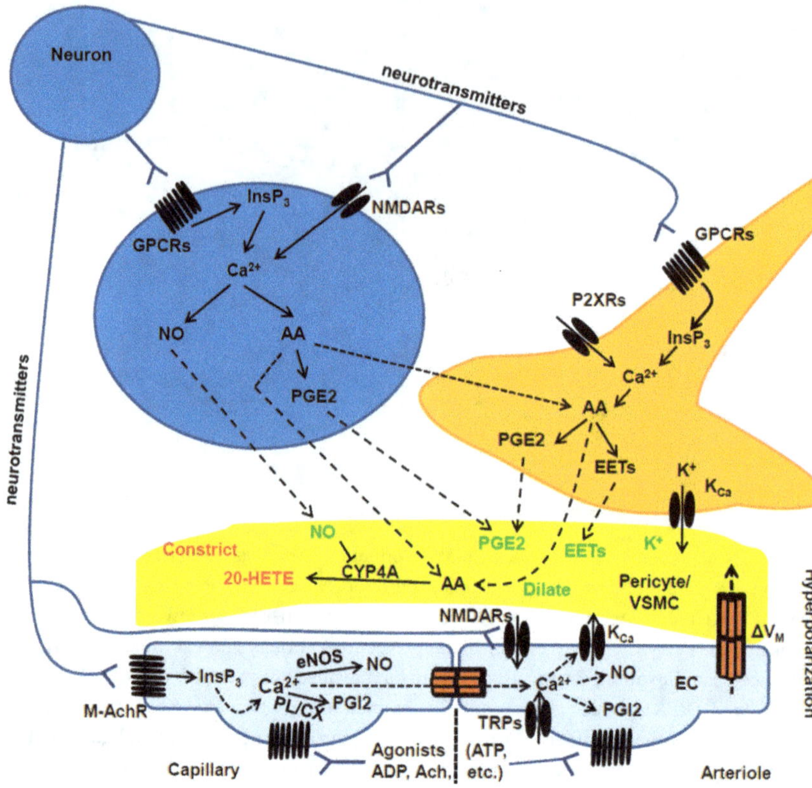

Figure 5. The putative role of endothelial Ca^{2+} signaling in neurovascular coupling. Synaptic activity leads to arteriole and capillary vasodilation by inducing an increase in $[Ca^{2+}]_i$ in postsynaptic neurons and perisynaptic astrocytes, as shown in Figure 2. Recent evidence indicated that synaptically-released glutamate may activate endothelial NMDARs, thereby eliciting the Ca^{2+}-dependent activation of eNOS, in intraparenchymal arterioles. Moreover, acetylcholine may induce Ca^{2+}-dependent NO release from brain endothelial cells by initiating the concerted interplay between $InsP_3Rs$ and SOCE [76]. This local vasodilation may be spread to more remote sites ($\approx$ 500 μm) through the initiation of an interendothelial Ca^{2+} wave, which ignites NO release and PGI2 production as long as it travels along the endothelial monolayer. Moreover, this propagating Ca^{2+} sweep could induce vasodilation by also stimulating K_{Ca} channels and evoking EDH in arterioles. Finally, blood-borne autacoids and dietary agonists could induce vasodilation by, respectively, binding to their specific GPCRs and stimulating multiple TRP channels (TRPC3, TRPV3, TRPV4, TRPA1) to initiate EDH. Abbreviations: 20-HETE: 20-hydroxyeicosatetraenoic acid; AA: arachidonic acid; CX: cyclooxygenases 1 and 2; EETs: epoxyeicosatrienoic acids; GPCRs; G-protein-coupled receptors; $InsP_3$: inositol-1,4,5-trisphosphate; K_{Ca}: Ca^{2+}-activated intermediate and small conductance K^+ channels; NO: nitric oxide; nNOS: neuronal NO synthase; P450: cytochrome P450; PGE2: prostaglandin E2; PL: phospholipase A2; PLD2: phospholipase D2; TRPs: TRP channels; VSMC: vascular smooth muscle cell.

4. How to Integrate Endothelial Ca^{2+} Signals in the Current Models of Neurovascular Coupling and Propagated Vasodilation

The role of endothelial Ca^{2+} signaling in NVC has largely been neglected as most authors focused on other cellular components of the NVU, such as neurons, interneurons, astrocytes, smooth muscle cells and, more recently, pericytes [2,4,6,9,30]. However, the literature discussed in the previous paragraphs suggests that the endothelial Ca^{2+} toolkit could play a crucial role in detecting and translating NA into a local vasoactive signal that is then propagated to upstream vessels [7,24]. In order to interpret the precise function of the ion channel network that controls CBF, it is mandatory to understand when and where the hemodynamic response starts. As mentioned earlier, synaptic activity-evoked NVC could be initiated at the capillary level and conducted upstream [12,47,203,213–217]. This hypothesis makes physiological sense as capillaries are closer to active neurons compared to arterioles and could represent the earliest vascular component to detect NA. In addition, endothelial signaling underpins long-range and almost unattenuated propagation (up to 2 mm) of vasomotor responses along peripheral vessels [22,62,63,176,177,218]. Endothelial engagement during NVC could, therefore, fulfill the same function observed in peripheral circulation, i.e., initiating (or contributing to initiate) vasodilation in proximity of the most metabolically active areas and conducting the vasomotor signal to upstream feeding arteries and arterioles to ensure an adequate increase in local blood supply (Figure 5) [62,177].

We believe that there is wide evidence to conclude that acetylcholine, which is liberated by cholinergic afferents emanated from the basal forebrain neurons during alert wakefulness, arousal, learning and attentional effort [28,150], increases CBF by inducing an increase in endothelial $[Ca^{2+}]_I$, which results in robust NO release [53,76,150,151,153,157]. This model does not rule out the possibility that the sub-population of acetylcholine and NO-synthesizing basal forebrain neurons that send projections onto cortical microvessels could directly control CBF though NO liberation [150,219]; in addition, acetylcholine and NO-synthesizing fibers may also contact NO cortical interneurons, which could contribute to NO-dependent vasodilation [150,220]. Similar to acetylcholine, glutamate could induce NMDARs-mediated NO release from brain endothelium [55,56,127]. Endothelial-derived NO could represent the alternative, i.e., non-neuronal, source of NO supporting vasodilation in cerebellar [141] and cortical [142] parenchymal arterioles. Future work will have to assess whether capillary brain endothelial cells express functional NMDARs and generate NO in response to glutamate stimulation. Intriguingly, the role of glial Ca^{2+} signals [13], which lead to the indispensable release of D-serine in arterioles, and of NO in glutamate-induced capillary dilation have been recently reported in cerebellum [11]. Alternatively, glutamate-induced endothelial-dependent vasodilation could contribute to pre-dilate upstream pial arteries and parenchymal arterioles prior to the retrograde propagation of the vasomotor signal from the capillary bed. Retrograde propagation of the initial vasodilation to upstream cortical vessels still represents a matter of hot debate [4,7,221]. Hillman's group demonstrated that light-dye disruption of endothelial lining dampens propagation of stimulus (electrical hind paw stimulation)-induced vasodilation in pial arteries in vivo, thereby blunting the increase in CBF [203]. In Section 2.3, we anticipated that endothelial Ca^{2+} signals mediate retrograde propagation of the vasomotor response by engaging two distinct mechanisms: (1) IK_{Ca} and SK_{Ca} channels, which effect the fast component of conducted vasodilation to upstream feeding vessels (i.e., pial arteries and arterioles); and (2) an intercellular Ca^{2+} wave that mediates the slower component of conducted vasodilation by inducing the release of NO and PGI2 from vascular endothelial cells [7,63,64]. EDH cannot, however, be initiated by brain capillary endothelial cells, which lack functional IK_{Ca} and SK_{Ca} channels [204]. Nelson's group recently revealed that a modest increase in the extracellular K^+ concentration ($\approx$ 10 mM), which is likely to truly reflect NA, activated brain capillary cell endothelial inward rectifier K^+ ($K_{IR}2.1$) channels to generate a hyperpolarizing signal that rapidly propagated to upstream arterioles (monitored up to $\approx$500 μm) to cause vasodilation and increase CBF in vivo [204]. Therefore, it appears that $K_{IR}2.1$, rather than IK_{Ca} and SK_{Ca}, channels mediate the fast component of conducted vasodilation in cerebral circulation. Of note, the hyperpolarization conduction velocity was in the same range, ≈2 mm/s, as the propagation speed of retrograde pial artery dilation [204,213]. The secondary slow component of conducted vasodilation is sustained by interendothelial Ca^{2+} waves, as discussed elsewhere [4,7,62]. Earlier work carried out by monitoring endothelial Ca^{2+} with a Ca^{2+}-sensitive dye, i.e., Fura-2, showed that local delivery of acetylcholine triggered a Ca^{2+} wave travelling bidirectionally along the endothelium of hamster feed arteries for no less than 1 mm [222]. This intercellular Ca^{2+} wave, in turn, sustained the slow component of conducted vasodilation by promoting NO and PGI2 release [223]. The use of a genetic Ca^{2+} indicator, GCaMP2, confirmed these data in cremaster muscle arterioles in vivo. Accordingly, acetylcholine induced a local increase in endothelial $[Ca^{2+}]_i$ that activated K_{Ca} channels to induce rapidly-conducting vasodilation at distances >1 mm and travelled along the endothelium as an intercellular Ca^{2+} wave for 300–400 μm. As shown in previous studies [222,223], the intercellular Ca^{2+} wave preceded a secondary vasodilation that was mediated by NO and PGE2 [63]. These findings led to the suggestion that the initial hyperpolarization conducts vasodilation far away (> several millimeters) from the local site of stimulation (including daughter and parent branches), whereas the slower Ca^{2+} wave encompasses only the vessel segments closer to the active region, thereby finely tuning the magnitude and duration of the vasodilatory response [62,63]. A recent study showed that removal of extracellular Ca^{2+} may induce intercellular Ca^{2+} waves in immortalized (RBE4) and primary (from bovine origin) brain microvascular endothelial cells [224]. Although this finding is not sufficient to confirm that interendothelial Ca^{2+} waves may propagate NA-induced local vasodilation to upstream arteries and arterioles, it confirms that brain endothelial cells are able to generate this type of propagating Ca^{2+} signal, which is likely to impinge on $InsP_3Rs$ [225–227]. Future work will have to assess whether: (1) an intercellular Ca^{2+} wave contributes to propagating the hemodynamic response from the capillary bed to feeding vessels; and (2) if so, whether EDH is stimulated by the incoming Ca^{2+} wave in parenchymal arterioles to sustain the vasomotor response.

5. Conclusions

Neurovascular coupling is the crucial process to adjust local CBF to the metabolic requirements of active neurons and maintain brain function. Although functional magnetic resonance imaging is routinely employed to monitor the changes in neuronal spiking activity, BOLD signals actually reflect the increases in CBF induced by NA. Therefore, understanding the cellular and molecular bases of NVC is mandatory to interpret the complex relationship between neuronal firing, metabolism and blood flow in both physiological and pathological conditions [6–8]. Intracellular Ca^{2+} signaling has long been known to drive NVC by coupling synaptic activity with the production of vasoactive messengers. Emerging evidence, however, suggests that the endothelial Ca^{2+} toolkit could also be recruited by neurotransmitters (i.e., glutamate and acetylcholine) and neuromodulators (e.g., ATP) to induce NO and PGE2 release or activate EDH. Future work will benefit from the availability of transgenic mice selectively expressing a genetic Ca^{2+} indicator, e.g., GCaMP2, in vascular endothelial cells [63,174]. The combination of endothelial GCaMP2-expressing mice with the novel advances in imaging techniques (i.e., two- or three-photon microscopy) will permit assessing whether synaptic activity increases endothelial $[Ca^{2+}]_i$ in vivo. It will also be important to decipher the role played by IK_{Ca} and SK_{Ca} channels in the hemodynamic response to NA. Earlier work showed that EDH is involved in local and retrograde vasodilation in parenchymal arterioles and middle cerebral arteries, but is absent in capillaries, where functional hyperemia is likely to initiate in response to sensory stimulation. The local vasodilation induced by NA increases the strength of laminar shear stress acting on vascular endothelium, a mechanism that leads to TRPV4-dependent NO release and EDH in peripheral circulation [191,228]. There is, however, conflicting evidence regarding flow-induced vasodilation in pial arteries and parenchymal arteries in the brain [229–231]. Nevertheless, EDH could be exploited by dietary manipulation to treat the vascular dysfunctions associated with aging and neurodegenerative disorders, which cause cognitive impairment by halting CBF [2,3,6,212]. For instance, NVC is severely impaired in Alzheimer's disease due to a defect in NO release [217,232,233]. A recent study revealed that the intracellular Ca^{2+} toolkit is severely compromised in rat brain microvascular endothelial cells exposed to amyloid-beta (Aβ) peptide [93], whose accumulation in brain parenchyma and in the cerebrovasculature represents a major pathogenic factor in AD [4]. Future work will have to assess whether dietary and/or pharmacological manipulation is able to rescue or halt functional hyperemia in these subjects by activating the TRP channels (i.e., TRPC3, TRPV3, TRPV4 and TRPA1), which are coupled to EDH in the brain. An emerging area of research for NVC is represented by the vasodilating role of mitochondrial Ca^{2+} in brain microvascular endothelial cells. A recent series of studies revealed that mitochondrial depolarization, induced by BMS-291095 (BMS), an opener of mitochondrial ATP-dependent K^+ (K_{ATP}) channels, prevented Ca^{2+} accumulation within the mitochondrial matrix, thereby resulting in a large increase in $[Ca^{2+}]_i$ in rat cerebral arteries endothelial cells. This mitochondrial-derived Ca^{2+} signal, in turn, induced eNOS activation, NO release and endothelium-dependent vasodilation [95]. Intriguingly, subsequent work demonstrated that mitochondrial-dependent NO release and vasodilation were impaired in cerebral arteries of obese Zucker rats [234]. The physiological stimulus responsible for mitochondrial-induced NO release in cerebral endothelium is yet to be elucidated, but it could be implicated in many other cerebrovascular disorders. Finally, astrocyte-released AA is the precursor of multiple vasoactive messengers, such as PGE2, EETs and 20-HETE, which are all involved in NVC, although their contributions depend on the vascular segment and/or the brain area. However, AA was shown to induce Ca^{2+}-dependent NO release by directly activating TRPV4 in several types of endothelial cells [54,235]. Some evidence suggested that AA was able to induce Ca^{2+} signals in human brain microvascular endothelial cells [94]. Given the key role of this lipid mediator within the NVU, we predict that future work will unveil that AA-dependent endothelial Ca^{2+} signals may also contribute to NVC.

Abbreviations

20-HETE	20-hydroxyeicosatetraenoic acid
AA	arachidonic acid
AMPA	a-amino-3-hydroxy-5-methyl-4-isoxazol epropionic acid
BK_{Ca}	large-conductance Ca^{2+}-activated K^+ channels
BOLD	blood oxygenation level dependent
CaM	Calmodulin
CBF	cerebral blood flow
cGMP	cyclic guanosine-3′,5′-monophosphate
CICR	Ca^{2+}-induced Ca^{2+} release
CYP4A	cytochrome P450 4A
DAG	Diacylglycerol
EDH	endothelial-dependent hyperpolarization
EETs	epoxyeicosatrienoic acids
EL	Endolysosomal
eNOS	neuronal NO synthase
ER	endoplasmic reticulum
GPCRs	G-protein-coupled receptors
IK_{Ca}	intermediate-conductance Ca^{2+}-activated K^+ channels
$InsP_3$	inositol-1,4,5-trisphosphate
$InsP_3Rs$	$InsP_3$ receptors
K_{Ca}	Ca^{2+}-activated K^+ channels
K_{ir}	inward rectifying K^+ channels
M-AchRs	muscarinic acetylcholine receptors
MCU	mitochondrial Ca^{2+} uniporter

MEGJs	myo-endothelial gap-junctions
mGluRs	metabotropic glutamate receptors
NA	neuronal activity
NAADP	nicotinic acid adenine dinucleotide phosphate
NMDA	*N*-methyl-D-aspartate
NMDARs	NMDA receptors
NO	nitric oxide
nNOS	neuronal NO synthase
NOX2	NADPH oxidase isoform 2
NVC	neurovascular coupling
NVU	neurovascular unit
O_2	Oxygen
P450	cytochrome P450
PGE2	prostaglandin E2
PIP_2	phosphatidylinositol-4,5-bisphosphate
PLA2	phospholipase A2
PLD2	phospholipase D2
RyRs	ryanodine receptors
SERCA	sarco-endoplasmic reticulum Ca^{2+}-ATPase
SK_{Ca}	small-conductance Ca^{2+}-activated K^+ channels
SOCE	store-operated Ca^{2+} entry
Stim	stromal interaction molecule
TPCs	two-pore channels
TRPs	TRP channels
TRPA1	ankyrin TRP 1
TRPC	canonical TRP
TRPV4	vanilloid TRP 4
VDACs	voltage-dependent anion channels
VSMC	vascular smooth muscle cells

References

1. Attwell, D.; Laughlin, S.B. An energy budget for signaling in the grey matter of the brain. *J. Cereb. Blood Flow Metab.* **2001**, *21*, 1133–1145. [CrossRef] [PubMed]
2. Kisler, K.; Nelson, A.R.; Montagne, A.; Zlokovic, B.V. Cerebral blood flow regulation and neurovascular dysfunction in Alzheimer disease. *Nat. Rev. Neurosci.* **2017**, *18*, 419–434. [CrossRef] [PubMed]
3. Zlokovic, B.V. Neurovascular pathways to neurodegeneration in Alzheimer's disease and other disorders. *Nat. Rev. Neurosci.* **2011**, *12*, 723–738. [CrossRef] [PubMed]
4. Iadecola, C. The Neurovascular Unit Coming of Age: A Journey through Neurovascular Coupling in Health and Disease. *Neuron* **2017**, *96*, 17–42. [CrossRef] [PubMed]
5. Koller, A.; Toth, P. Contribution of flow-dependent vasomotor mechanisms to the autoregulation of cerebral blood flow. *J. Vasc. Res.* **2012**, *49*, 375–389. [CrossRef] [PubMed]
6. Iadecola, C. Neurovascular regulation in the normal brain and in Alzheimer's disease. *Nat. Rev. Neurosci.* **2004**, *5*, 347–360. [CrossRef] [PubMed]
7. Hillman, E.M. Coupling mechanism and significance of the BOLD signal: A status report. *Annu. Rev. Neurosci.* **2014**, *37*, 161–181. [CrossRef] [PubMed]
8. Attwell, D.; Iadecola, C. The neural basis of functional brain imaging signals. *Trends Neurosci.* **2002**, *25*, 621–625. [CrossRef]
9. Attwell, D.; Buchan, A.M.; Charpak, S.; Lauritzen, M.; Macvicar, B.A.; Newman, E.A. Glial and neuronal control of brain blood flow. *Nature* **2010**, *468*, 232–243. [CrossRef] [PubMed]
10. Rancillac, A.; Rossier, J.; Guille, M.; Tong, X.K.; Geoffroy, H.; Amatore, C.; Arbault, S.; Hamel, E.; Cauli, B. Glutamatergic Control of Microvascular Tone by Distinct GABA Neurons in the Cerebellum. *J. Neurosci.* **2006**, *26*, 6997–7006. [CrossRef] [PubMed]

11. Mapelli, L.; Gagliano, G.; Soda, T.; Laforenza, U.; Moccia, F.; D'Angelo, E.U. Granular Layer Neurons Control Cerebellar Neurovascular Coupling Through an NMDA Receptor/NO-Dependent System. *J. Neurosci.* **2017**, *37*, 1340–1351. [CrossRef] [PubMed]
12. Hall, C.N.; Reynell, C.; Gesslein, B.; Hamilton, N.B.; Mishra, A.; Sutherland, B.A.; O'Farrell, F.M.; Buchan, A.M.; Lauritzen, M.; Attwell, D. Capillary pericytes regulate cerebral blood flow in health and disease. *Nature* **2014**, *508*, 55–60. [CrossRef] [PubMed]
13. Mishra, A.; Reynolds, J.P.; Chen, Y.; Gourine, A.V.; Rusakov, D.A.; Attwell, D. Astrocytes mediate neurovascular signaling to capillary pericytes but not to arterioles. *Nat. Neurosci.* **2016**, *19*, 1619–1627. [CrossRef] [PubMed]
14. Rosenegger, D.G.; Gordon, G.R. A slow or modulatory role of astrocytes in neurovascular coupling. *Microcirculation* **2015**, *22*, 197–203. [CrossRef] [PubMed]
15. Laughlin, M.H.; Davis, M.J.; Secher, N.H.; van Lieshout, J.J.; Arce-Esquivel, A.A.; Simmons, G.H.; Bender, S.B.; Padilla, J.; Bache, R.J.; Merkus, D.; et al. Peripheral circulation. *Compr. Physiol.* **2012**, *2*, 321–447. [PubMed]
16. Khaddaj Mallat, R.; Mathew John, C.; Kendrick, D.J.; Braun, A.P. The vascular endothelium: A regulator of arterial tone and interface for the immune system. *Crit. Rev. Clin. Lab. Sci.* **2017**, *54*, 458–470. [CrossRef] [PubMed]
17. Mancardi, D.; Pla, A.F.; Moccia, F.; Tanzi, F.; Munaron, L. Old and new gasotransmitters in the cardiovascular system: Focus on the role of nitric oxide and hydrogen sulfide in endothelial cells and cardiomyocytes. *Curr. Pharm. Biotechnol.* **2011**, *12*, 1406–1415. [CrossRef] [PubMed]
18. Vanhoutte, P.M.; Tang, E.H. Endothelium-dependent contractions: When a good guy turns bad! *J. Physiol.* **2008**, *586*, 5295–5304. [CrossRef] [PubMed]
19. Moccia, F.; Berra-Romani, R.; Tanzi, F. Update on vascular endothelial Ca^{2+} signalling: A tale of ion channels, pumps and transporters. *World J. Biol. Chem.* **2012**, *3*, 127–158. [CrossRef] [PubMed]
20. Moccia, F.; Tanzi, F.; Munaron, L. Endothelial remodelling and intracellular calcium machinery. *Curr. Mol. Med.* **2014**, *14*, 457–480. [CrossRef] [PubMed]
21. Altaany, Z.; Moccia, F.; Munaron, L.; Mancardi, D.; Wang, R. Hydrogen sulfide and endothelial dysfunction: Relationship with nitric oxide. *Curr. Med. Chem.* **2014**, *21*, 3646–3661. [CrossRef] [PubMed]
22. Behringer, E.J. Calcium and electrical signaling in arterial endothelial tubes: New insights into cellular physiology and cardiovascular function. *Microcirculation* **2017**, *24*. [CrossRef] [PubMed]
23. Garland, C.J.; Dora, K.A. EDH: Endothelium-dependent hyperpolarization and microvascular signalling. *Acta Physiologica* **2017**, *219*, 152–161. [CrossRef] [PubMed]
24. Longden, T.A.; Hill-Eubanks, D.C.; Nelson, M.T. Ion channel networks in the control of cerebral blood flow. *J. Cereb. Blood Flow Metab.* **2016**, *36*, 492–512. [CrossRef] [PubMed]
25. Cauli, B.; Hamel, E. Revisiting the role of neurons in neurovascular coupling. *Front. Neuroenerg.* **2010**, *2*, 9. [CrossRef] [PubMed]
26. Cauli, B.; Tong, X.K.; Rancillac, A.; Serluca, N.; Lambolez, B.; Rossier, J.; Hamel, E. Cortical GABA interneurons in neurovascular coupling: Relays for subcortical vasoactive pathways. *J. Neurosci.* **2004**, *24*, 8940–8949. [CrossRef] [PubMed]
27. Lecrux, C.; Toussay, X.; Kocharyan, A.; Fernandes, P.; Neupane, S.; Levesque, M.; Plaisier, F.; Shmuel, A.; Cauli, B.; Hamel, E. Pyramidal neurons are "neurogenic hubs" in the neurovascular coupling response to whisker stimulation. *J. Neurosci.* **2011**, *31*, 9836–9847. [CrossRef] [PubMed]
28. Lecrux, C.; Hamel, E. Neuronal networks and mediators of cortical neurovascular coupling responses in normal and altered brain states. *Philos. Trans. R. Soc. Lond. B Biol. Sci.* **2016**, *371*. pii: 20150350. [CrossRef] [PubMed]
29. Hamel, E. Perivascular nerves and the regulation of cerebrovascular tone. *J. Appl. Physiol. (1985)* **2006**, *100*, 1059–1064. [CrossRef] [PubMed]
30. Nortley, R.; Attwell, D. Control of brain energy supply by astrocytes. *Curr. Opin. Neurobiol.* **2017**, *47*, 80–85. [CrossRef] [PubMed]
31. Mathiisen, T.M.; Lehre, K.P.; Danbolt, N.C.; Ottersen, O.P. The perivascular astroglial sheath provides a complete covering of the brain microvessels: An electron microscopic 3D reconstruction. *Glia* **2010**, *58*, 1094–1103. [CrossRef] [PubMed]
32. Zlokovic, B.V. Neurovascular mechanisms of Alzheimer's neurodegeneration. *Trends Neurosci.* **2005**, *28*, 202–208. [CrossRef] [PubMed]

33. Dalkara, T.; Alarcon-Martinez, L. Cerebral microvascular pericytes and neurogliovascular signaling in health and disease. *Brain Res.* **2015**, *1623*, 3–17. [CrossRef] [PubMed]
34. Lauritzen, M. Reading vascular changes in brain imaging: Is dendritic calcium the key? *Nat. Rev. Neurosci.* **2005**, *6*, 77–85. [CrossRef] [PubMed]
35. Lourenco, C.F.; Ledo, A.; Barbosa, R.M.; Laranjinha, J. Neurovascular-neuroenergetic coupling axis in the brain: Master regulation by nitric oxide and consequences in aging and neurodegeneration. *Free Radic. Biol. Med.* **2017**, *108*, 668–682. [CrossRef] [PubMed]
36. Yang, G.; Chen, G.; Ebner, T.J.; Iadecola, C. Nitric oxide is the predominant mediator of cerebellar hyperemia during somatosensory activation in rats. *Am. J. Physiol.* **1999**, *277 (Pt 2)*, R1760–R1770. [CrossRef] [PubMed]
37. Lovick, T.A.; Brown, L.A.; Key, B.J. Neurovascular relationships in hippocampal slices: Physiological and anatomical studies of mechanisms underlying flow-metabolism coupling in intraparenchymal microvessels. *Neuroscience* **1999**, *92*, 47–60. [CrossRef]
38. Duchemin, S.; Boily, M.; Sadekova, N.; Girouard, H. The complex contribution of NOS interneurons in the physiology of cerebrovascular regulation. *Front. Neural Circ.* **2012**, *6*, 51. [CrossRef] [PubMed]
39. Filosa, J.A.; Morrison, H.W.; Iddings, J.A.; Du, W.; Kim, K.J. Beyond neurovascular coupling, role of astrocytes in the regulation of vascular tone. *Neuroscience* **2016**, *323*, 96–109. [CrossRef] [PubMed]
40. Wang, X.; Lou, N.; Xu, Q.; Tian, G.F.; Peng, W.G.; Han, X.; Kang, J.; Takano, T.; Nedergaard, M. Astrocytic Ca^{2+} signaling evoked by sensory stimulation in vivo. *Nat. Neurosci.* **2006**, *9*, 816–823. [CrossRef] [PubMed]
41. Takano, T.; Tian, G.F.; Peng, W.; Lou, N.; Libionka, W.; Han, X.; Nedergaard, M. Astrocyte-mediated control of cerebral blood flow. *Nat. Neurosci.* **2006**, *9*, 260–267. [CrossRef] [PubMed]
42. Mishra, A. Binaural blood flow control by astrocytes: Listening to synapses and the vasculature. *J. Physiol.* **2017**, *595*, 1885–1902. [CrossRef] [PubMed]
43. Kim, K.J.; Iddings, J.A.; Stern, J.E.; Blanco, V.M.; Croom, D.; Kirov, S.A.; Filosa, J.A. Astrocyte contributions to flow/pressure-evoked parenchymal arteriole vasoconstriction. *J. Neurosci.* **2015**, *35*, 8245–8257. [CrossRef] [PubMed]
44. Kim, K.J.; Ramiro Diaz, J.; Iddings, J.A.; Filosa, J.A. Vasculo-Neuronal Coupling: Retrograde Vascular Communication to Brain Neurons. *J. Neurosci.* **2016**, *36*, 12624–12639. [CrossRef] [PubMed]
45. Simard, M.; Arcuino, G.; Takano, T.; Liu, Q.S.; Nedergaard, M. Signaling at the gliovascular interface. *J. Neurosci.* **2003**, *23*, 9254–9262. [PubMed]
46. Attwell, D.; Mishra, A.; Hall, C.N.; O'Farrell, F.M.; Dalkara, T. What is a pericyte? *J. Cereb. Blood Flow Metab.* **2016**, *36*, 451–455. [CrossRef] [PubMed]
47. Hamilton, N.B.; Attwell, D.; Hall, C.N. Pericyte-mediated regulation of capillary diameter: A component of neurovascular coupling in health and disease. *Front. Neuroenerg.* **2010**, *2*. pii: 5. [CrossRef] [PubMed]
48. D'Angelo, E. The organization of plasticity in the cerebellar cortex: From synapses to control. *Prog. Brain Res.* **2014**, *210*, 31–58. [PubMed]
49. Berra-Romani, R.; Avelino-Cruz, J.E.; Raqeeb, A.; Della Corte, A.; Cinelli, M.; Montagnani, S.; Guerra, G.; Moccia, F.; Tanzi, F. Ca(2)(+)-dependent nitric oxide release in the injured endothelium of excised rat aorta: A promising mechanism applying in vascular prosthetic devices in aging patients. *BMC Surg.* **2013**, *13* (Suppl. 2), S40. [CrossRef] [PubMed]
50. Andresen, J.; Shafi, N.I.; Bryan, R.M., Jr. Endothelial influences on cerebrovascular tone. *J. Appl. Physiol. (1985)* **2006**, *100*, 318–327. [CrossRef] [PubMed]
51. You, J.; Johnson, T.D.; Childres, W.F.; Bryan, R.M., Jr. Endothelial-mediated dilations of rat middle cerebral arteries by ATP and ADP. *Am. J. Physiol.* **1997**, *273 (Pt 2)*, H1472–H1477. [CrossRef] [PubMed]
52. Elhusseiny, A.; Cohen, Z.; Olivier, A.; Stanimirovic, D.B.; Hamel, E. Functional acetylcholine muscarinic receptor subtypes in human brain microcirculation: Identification and cellular localization. *J. Cereb. Blood Flow Metab.* **1999**, *19*, 794–802. [CrossRef] [PubMed]
53. Elhusseiny, A.; Hamel, E. Muscarinic–but not nicotinic–acetylcholine receptors mediate a nitric oxide-dependent dilation in brain cortical arterioles: A possible role for the M5 receptor subtype. *J. Cereb. Blood Flow Metab.* **2000**, *20*, 298–305. [CrossRef] [PubMed]
54. Zuccolo, E.; Dragoni, S.; Poletto, V.; Catarsi, P.; Guido, D.; Rappa, A.; Reforgiato, M.; Lodola, F.; Lim, D.; Rosti, V.; et al. Arachidonic acid-evoked Ca^{2+} signals promote nitric oxide release and proliferation in human endothelial colony forming cells. *Vascul. Pharmacol.* **2016**, *87*, 159–171. [CrossRef] [PubMed]

55. LeMaistre, J.L.; Sanders, S.A.; Stobart, M.J.; Lu, L.; Knox, J.D.; Anderson, H.D.; Anderson, C.M. Coactivation of NMDA receptors by glutamate and D-serine induces dilation of isolated middle cerebral arteries. *J. Cereb. Blood Flow Metab.* **2012**, *32*, 537–547. [CrossRef] [PubMed]
56. Stobart, J.L.; Lu, L.; Anderson, H.D.; Mori, H.; Anderson, C.M. Astrocyte-induced cortical vasodilation is mediated by D-serine and endothelial nitric oxide synthase. *Proc. Natl. Acad. Sci. USA* **2013**, *110*, 3149–3154. [CrossRef] [PubMed]
57. Beard, R.S., Jr.; Reynolds, J.J.; Bearden, S.E. Metabotropic glutamate receptor 5 mediates phosphorylation of vascular endothelial cadherin and nuclear localization of beta-catenin in response to homocysteine. *Vascul. Pharmacol.* **2012**, *56*, 159–167. [CrossRef] [PubMed]
58. Lv, J.M.; Guo, X.M.; Chen, B.; Lei, Q.; Pan, Y.J.; Yang, Q. The Noncompetitive AMPAR Antagonist Perampanel Abrogates Brain Endothelial Cell Permeability in Response to Ischemia: Involvement of Claudin-5. *Cell. Mol. Neurobiol.* **2016**, *36*, 745–753. [CrossRef] [PubMed]
59. Collard, C.D.; Park, K.A.; Montalto, M.C.; Alapati, S.; Buras, J.A.; Stahl, G.L.; Colgan, S.P. Neutrophil-derived glutamate regulates vascular endothelial barrier function. *J. Biol. Chem.* **2002**, *277*, 14801–14811. [CrossRef] [PubMed]
60. Gillard, S.E.; Tzaferis, J.; Tsui, H.C.; Kingston, A.E. Expression of metabotropic glutamate receptors in rat meningeal and brain microvasculature and choroid plexus. *J. Comp. Neurol.* **2003**, *461*, 317–332. [CrossRef] [PubMed]
61. Itoh, Y.; Suzuki, N. Control of brain capillary blood flow. *J. Cereb. Blood Flow Metab.* **2012**, *32*, 1167–1176. [CrossRef] [PubMed]
62. Segal, S.S. Integration and Modulation of Intercellular Signaling Underlying Blood Flow Control. *J. Vasc. Res.* **2015**, *52*, 136–157. [CrossRef] [PubMed]
63. Tallini, Y.N.; Brekke, J.F.; Shui, B.; Doran, R.; Hwang, S.M.; Nakai, J.; Salama, G.; Segal, S.S.; Kotlikoff, M.I. Propagated endothelial Ca^{2+} waves and arteriolar dilation in vivo: Measurements in Cx40BAC GCaMP2 transgenic mice. *Circ. Res.* **2007**, *101*, 1300–1309. [CrossRef] [PubMed]
64. Wolfle, S.E.; Chaston, D.J.; Goto, K.; Sandow, S.L.; Edwards, F.R.; Hill, C.E. Non-linear relationship between hyperpolarisation and relaxation enables long distance propagation of vasodilatation. *J. Physiol.* **2011**, *589 (Pt 10)*, 2607–2623. [CrossRef] [PubMed]
65. Abdullaev, I.F.; Bisaillon, J.M.; Potier, M.; Gonzalez, J.C.; Motiani, R.K.; Trebak, M. STIM1 and ORAI1 mediate CRAC currents and store-operated calcium entry important for endothelial cell proliferation. *Circ. Res.* **2008**, *103*, 1289–1299. [CrossRef] [PubMed]
66. Zuccolo, E.; Di Buduo, C.; Lodola, F.; Orecchioni, S.; Scarpellino, G.; Kheder, D.A.; Poletto, V.; Guerra, G.; Bertolini, F.; Balduini, A.; et al. Stromal Cell-Derived Factor-1alpha Promotes Endothelial Colony-Forming Cell Migration Through the Ca(2+)-Dependent Activation of the Extracellular Signal-Regulated Kinase 1/2 and Phosphoinositide 3-Kinase/AKT Pathways. *Stem Cells Dev.* **2018**, *27*, 23–34. [CrossRef] [PubMed]
67. Freichel, M.; Suh, S.H.; Pfeifer, A.; Schweig, U.; Trost, C.; Weissgerber, P.; Biel, M.; Philipp, S.; Freise, D.; Droogmans, G.; et al. Lack of an endothelial store-operated Ca^{2+} current impairs agonist-dependent vasorelaxation in TRP4-/- mice. *Nat. Cell Biol.* **2001**, *3*, 121–127. [CrossRef] [PubMed]
68. Blatter, L.A. Tissue Specificity: SOCE: Implications for Ca^{2+} Handling in Endothelial Cells. *Adv. Exp. Med. Biol.* **2017**, *993*, 343–361. [PubMed]
69. Moccia, F.; Lodola, F.; Dragoni, S.; Bonetti, E.; Bottino, C.; Guerra, G.; Laforenza, U.; Rosti, V.; Tanzi, F. Ca^{2+} signalling in endothelial progenitor cells: A novel means to improve cell-based therapy and impair tumour vascularisation. *Curr. Vasc. Pharmacol.* **2014**, *12*, 87–105. [CrossRef] [PubMed]
70. Moccia, F.; Dragoni, S.; Lodola, F.; Bonetti, E.; Bottino, C.; Guerra, G.; Laforenza, U.; Rosti, V.; Tanzi, F. Store-dependent Ca(2+) entry in endothelial progenitor cells as a perspective tool to enhance cell-based therapy and adverse tumour vascularization. *Curr. Med. Chem.* **2012**, *19*, 5802–5018. [CrossRef] [PubMed]
71. Brailoiu, E.; Shipsky, M.M.; Yan, G.; Abood, M.E.; Brailoiu, G.C. Mechanisms of modulation of brain microvascular endothelial cells function by thrombin. *Brain Res.* **2017**, *1657*, 167–175. [CrossRef] [PubMed]
72. Domotor, E.; Abbott, N.J.; Adam-Vizi, V. Na^{+}-Ca^{2+} exchange and its implications for calcium homeostasis in primary cultured rat brain microvascular endothelial cells. *J. Physiol.* **1999**, *515 (Pt 1)*, 147–155. [CrossRef] [PubMed]

73. De Bock, M.; Culot, M.; Wang, N.; Bol, M.; Decrock, E.; De Vuyst, E.; da Costa, A.; Dauwe, I.; Vinken, M.; Simon, A.M.; et al. Connexin channels provide a target to manipulate brain endothelial calcium dynamics and blood-brain barrier permeability. *J. Cereb. Blood Flow Metab.* **2011**, *31*, 1942–1957. [CrossRef] [PubMed]
74. De Bock, M.; Wang, N.; Decrock, E.; Bol, M.; Gadicherla, A.K.; Culot, M.; Cecchelli, R.; Bultynck, G.; Leybaert, L. Endothelial calcium dynamics, connexin channels and blood-brain barrier function. *Prog. Neurobiol.* **2013**, *108*, 1–20. [CrossRef] [PubMed]
75. Scharbrodt, W.; Abdallah, Y.; Kasseckert, S.A.; Gligorievski, D.; Piper, H.M.; Boker, D.K.; Deinsberger, W.; Oertel, M.F. Cytosolic Ca^{2+} oscillations in human cerebrovascular endothelial cells after subarachnoid hemorrhage. *J. Cereb. Blood Flow Metab.* **2009**, *29*, 57–65. [CrossRef] [PubMed]
76. Zuccolo, E.; Lim, D.; Kheder, D.A.; Perna, A.; Catarsi, P.; Botta, L.; Rosti, V.; Riboni, L.; Sancini, G.; Tanzi, F.; et al. Acetylcholine induces intracellular Ca_{2+} oscillations and nitric oxide release in mouse brain endothelial cells. *Cell Calcium* **2017**, *66*, 33–47. [CrossRef] [PubMed]
77. Earley, S. Endothelium-dependent cerebral artery dilation mediated by transient receptor potential and Ca^{2+}-activated K^+ channels. *J. Cardiovasc. Pharmacol.* **2011**, *57*, 148–153. [CrossRef] [PubMed]
78. Earley, S.; Brayden, J.E. Transient receptor potential channels in the vasculature. *Physiol. Rev.* **2015**, *95*, 645–690. [CrossRef] [PubMed]
79. Manoonkitiwongsa, P.S.; Whitter, E.F.; Wareesangtip, W.; McMillan, P.J.; Nava, P.B.; Schultz, R.L. Calcium-dependent ATPase unlike ecto-ATPase is located primarily on the luminal surface of brain endothelial cells. *Histochem. J.* **2000**, *32*, 313–324. [CrossRef] [PubMed]
80. Moccia, F.; Berra-Romani, R.; Baruffi, S.; Spaggiari, S.; Signorelli, S.; Castelli, L.; Magistretti, J.; Taglietti, V.; Tanzi, F. Ca^{2+} uptake by the endoplasmic reticulum Ca^{2+}-ATPase in rat microvascular endothelial cells. *Biochem. J.* **2002**, *364 (Pt 1)*, 235–244. [CrossRef] [PubMed]
81. Haorah, J.; Knipe, B.; Gorantla, S.; Zheng, J.; Persidsky, Y. Alcohol-induced blood-brain barrier dysfunction is mediated via inositol 1,4,5-triphosphate receptor (IP3R)-gated intracellular calcium release. *J. Neurochem.* **2007**, *100*, 324–336. [CrossRef] [PubMed]
82. Hertle, D.N.; Yeckel, M.F. Distribution of inositol-1,4,5-trisphosphate receptor isotypes and ryanodine receptor isotypes during maturation of the rat hippocampus. *Neuroscience* **2007**, *150*, 625–638. [CrossRef] [PubMed]
83. Czupalla, C.J.; Liebner, S.; Devraj, K. In vitro models of the blood-brain barrier. *Methods Mol. Biol.* **2014**, *1135*, 415–437. [PubMed]
84. Pedriali, G.; Rimessi, A.; Sbano, L.; Giorgi, C.; Wieckowski, M.R.; Previati, M.; Pinton, P. Regulation of Endoplasmic Reticulum-Mitochondria Ca^{2+} Transfer and Its Importance for Anti-Cancer Therapies. *Front. Oncol.* **2017**, *7*, 180. [CrossRef] [PubMed]
85. Dong, Z.; Shanmughapriya, S.; Tomar, D.; Siddiqui, N.; Lynch, S.; Nemani, N.; Breves, S.L.; Zhang, X.; Tripathi, A.; Palaniappan, P.; et al. Mitochondrial Ca^{2+} Uniporter Is a Mitochondrial Luminal Redox Sensor that Augments MCU Channel Activity. *Mol. Cell* **2017**, *65*, 1014–1028.e7. [CrossRef] [PubMed]
86. De Stefani, D.; Rizzuto, R.; Pozzan, T. Enjoy the Trip: Calcium in Mitochondria Back and Forth. *Annu. Rev. Biochem.* **2016**, *85*, 161–192. [CrossRef] [PubMed]
87. Moccia, F. Remodelling of the Ca^{2+} Toolkit in Tumor Endothelium as a Crucial Responsible for the Resistance to Anticancer Therapies. *Curr. Signal Transduct. Ther.* **2017**, *12*. [CrossRef]
88. Kluge, M.A.; Fetterman, J.L.; Vita, J.A. Mitochondria and endothelial function. *Circ. Res.* **2013**, *112*, 1171–1188. [CrossRef] [PubMed]
89. Malli, R.; Frieden, M.; Hunkova, M.; Trenker, M.; Graier, W.F. Ca^{2+} refilling of the endoplasmic reticulum is largely preserved albeit reduced Ca^{2+} entry in endothelial cells. *Cell Calcium* **2007**, *41*, 63–76. [CrossRef] [PubMed]
90. Malli, R.; Frieden, M.; Osibow, K.; Zoratti, C.; Mayer, M.; Demaurex, N.; Graier, W.F. Sustained Ca^{2+} transfer across mitochondria is Essential for mitochondrial Ca^{2+} buffering, sore-operated Ca^{2+} entry, and Ca^{2+} store refilling. *J. Biol. Chem.* **2003**, *278*, 44769–44779. [CrossRef] [PubMed]
91. Busija, D.W.; Rutkai, I.; Dutta, S.; Katakam, P.V. Role of Mitochondria in Cerebral Vascular Function: Energy Production, Cellular Protection, and Regulation of Vascular Tone. *Compr. Physiol.* **2016**, *6*, 1529–1548. [PubMed]
92. Gerencser, A.A.; Adam-Vizi, V. Selective, high-resolution fluorescence imaging of mitochondrial Ca^{2+} concentration. *Cell Calcium* **2001**, *30*, 311–321. [CrossRef] [PubMed]

93. Fonseca, A.C.; Moreira, P.I.; Oliveira, C.R.; Cardoso, S.M.; Pinton, P.; Pereira, C.F. Amyloid-beta disrupts calcium and redox homeostasis in brain endothelial cells. *Mol. Neurobiol.* **2015**, *51*, 610–622. [CrossRef] [PubMed]
94. Evans, J.; Ko, Y.; Mata, W.; Saquib, M.; Eldridge, J.; Cohen-Gadol, A.; Leaver, H.A.; Wang, S.; Rizzo, M.T. Arachidonic acid induces brain endothelial cell apoptosis via p38-MAPK and intracellular calcium signaling. *Microvasc. Res.* **2015**, *98*, 145–158. [CrossRef] [PubMed]
95. Katakam, P.V.; Wappler, E.A.; Katz, P.S.; Rutkai, I.; Institoris, A.; Domoki, F.; Gaspar, T.; Grovenburg, S.M.; Snipes, J.A.; Busija, D.W. Depolarization of mitochondria in endothelial cells promotes cerebral artery vasodilation by activation of nitric oxide synthase. *Arterioscler. Thromb. Vasc. Biol.* **2013**, *33*, 752–759. [CrossRef] [PubMed]
96. Ronco, V.; Potenza, D.M.; Denti, F.; Vullo, S.; Gagliano, G.; Tognolina, M.; Guerra, G.; Pinton, P.; Genazzani, A.A.; Mapelli, L.; et al. A novel Ca(2)(+)-mediated cross-talk between endoplasmic reticulum and acidic organelles: Implications for NAADP-dependent Ca(2)(+) signalling. *Cell Calcium* **2015**, *57*, 89–100. [CrossRef] [PubMed]
97. Zuccolo, E.; Lim, D.; Poletto, V.; Guerra, G.; Tanzi, F.; Rosti, V.; Moccia, F. Acidic Ca_{2+} stores interact with the endoplasmic reticulum to shape intracellular Ca^{2+} signals in human endothelial progenitor cells. *Vascul. Pharmacol.* **2015**, *75*, 70–71. [CrossRef]
98. Di Nezza, F.; Zuccolo, E.; Poletto, V.; Rosti, V.; De Luca, A.; Moccia, F.; Guerra, G.; Ambrosone, L. Liposomes as a Putative Tool to Investigate NAADP Signaling in Vasculogenesis. *J. Cell. Biochem.* **2017**, *118*, 3722–3729. [CrossRef] [PubMed]
99. Favia, A.; Desideri, M.; Gambara, G.; D'Alessio, A.; Ruas, M.; Esposito, B.; Del Bufalo, D.; Parrington, J.; Ziparo, E.; Palombi, F.; et al. VEGF-induced neoangiogenesis is mediated by NAADP and two-pore channel-2-dependent Ca^{2+} signaling. *Proc. Natl. Acad. Sci. USA* **2014**, *111*, E4706–E4715. [CrossRef] [PubMed]
100. Favia, A.; Pafumi, I.; Desideri, M.; Padula, F.; Montesano, C.; Passeri, D.; Nicoletti, C.; Orlandi, A.; Del Bufalo, D.; Sergi, M.; et al. NAADP-Dependent Ca(2+) Signaling Controls Melanoma Progression, Metastatic Dissemination and Neoangiogenesis. *Sci. Rep.* **2016**, *6*, 18925. [CrossRef] [PubMed]
101. Lodola, F.; Laforenza, U.; Bonetti, E.; Lim, D.; Dragoni, S.; Bottino, C.; Ong, H.L.; Guerra, G.; Ganini, C.; Massa, M.; et al. Store-operated Ca^{2+} entry is remodelled and controls in vitro angiogenesis in endothelial progenitor cells isolated from tumoral patients. *PLoS ONE* **2012**, *7*, e42541. [CrossRef] [PubMed]
102. Li, J.; Cubbon, R.M.; Wilson, L.A.; Amer, M.S.; McKeown, L.; Hou, B.; Majeed, Y.; Tumova, S.; Seymour, V.A.L.; Taylor, H.; et al. ORAI1 and CRAC channel dependence of VEGF-activated Ca^{2+} entry and endothelial tube formation. *Circ. Res.* **2011**, *108*, 1190–1198. [CrossRef] [PubMed]
103. Brandman, O.; Liou, J.; Park, W.S.; Meyer, T. STIM2 is a feedback regulator that stabilizes basal cytosolic and endoplasmic reticulum Ca^{2+} levels. *Cell* **2007**, *131*, 1327–1339. [CrossRef] [PubMed]
104. Antigny, F.; Jousset, H.; Konig, S.; Frieden, M. Thapsigargin activates Ca(2)+ entry both by store-dependent, STIM1/ORAI1-mediated, and store-independent, TRPC3/PLC/PKC-mediated pathways in human endothelial cells. *Cell Calcium* **2011**, *49*, 115–127. [CrossRef] [PubMed]
105. Moccia, F.; Zuccolo, E.; Soda, T.; Tanzi, F.; Guerra, G.; Mapelli, L.; Lodola, F.; D'Angelo, E. Stim and Orai proteins in neuronal Ca(2+) signaling and excitability. *Front. Cell. Neurosci.* **2015**, *9*, 153. [CrossRef] [PubMed]
106. Aird, W.C. Endothelial cell heterogeneity. *Cold Spring Harb. Perspect. Med.* **2012**, *2*, a006429. [CrossRef] [PubMed]
107. Regan, E.R.; Aird, W.C. Dynamical systems approach to endothelial heterogeneity. *Circ. Res.* **2012**, *111*, 110–130. [CrossRef] [PubMed]
108. Moccia, F.; Berra-Romani, R.; Baruffi, S.; Spaggiari, S.; Adams, D.J.; Taglietti, V.; Tanzi, F. Basal nonselective cation permeability in rat cardiac microvascular endothelial cells. *Microvasc. Res.* **2002**, *64*, 187–197. [CrossRef] [PubMed]
109. Zuccolo, E.; Bottino, C.; Diofano, F.; Poletto, V.; Codazzi, A.C.; Mannarino, S.; Campanelli, R.; Fois, G.; Marseglia, G.L.; Guerra, G.; et al. Constitutive Store-Operated Ca(2+) Entry Leads to Enhanced Nitric Oxide Production and Proliferation in Infantile Hemangioma-Derived Endothelial Colony-Forming Cells. *Stem Cells Dev.* **2016**, *25*, 301–319. [CrossRef] [PubMed]

110. Li, J.; Bruns, A.F.; Hou, B.; Rode, B.; Webster, P.J.; Bailey, M.A.; Appleby, H.L.; Moss, N.K.; Ritchie, J.E.; Yuldasheva, N.Y.; et al. ORAI3 Surface Accumulation and Calcium Entry Evoked by Vascular Endothelial Growth Factor. *Arterioscler. Thromb. Vasc. Biol.* **2015**, *35*, 1987–1994. [CrossRef] [PubMed]
111. Moccia, F. Endothelial Ca(2+) Signaling and the Resistance to Anticancer Treatments: Partners in Crime. *Int. J. Mol. Sci.* **2018**, *19*. pii: E217. [CrossRef] [PubMed]
112. Sundivakkam, P.C.; Freichel, M.; Singh, V.; Yuan, J.P.; Vogel, S.M.; Flockerzi, V.; Malik, A.B.; Tiruppathi, C. The Ca(2+) sensor stromal interaction molecule 1 (STIM1) is necessary and sufficient for the store-operated Ca(2+) entry function of transient receptor potential canonical (TRPC) 1 and 4 channels in endothelial cells. *Mol. Pharmacol.* **2012**, *81*, 510–526. [CrossRef] [PubMed]
113. Cioffi, D.L.; Wu, S.; Chen, H.; Alexeyev, M.; St Croix, C.M.; Pitt, B.R.; Uhlig, S.; Stevens, T. ORAI1 determines calcium selectivity of an endogenous TRPC heterotetramer channel. *Circ. Res.* **2012**, *110*, 1435–1444. [CrossRef] [PubMed]
114. Xu, N.; Cioffi, D.L.; Alexeyev, M.; Rich, T.C.; Stevens, T. Sodium entry through endothelial store-operated calcium entry channels: Regulation by ORAI1. *Am. J. Physiol. Cell Physiol.* **2015**, *308*, C277–C288. [CrossRef] [PubMed]
115. Cioffi, D.L.; Barry, C.; Stevens, T. Store-operated calcium entry channels in pulmonary endothelium: The emerging story of TRPCS and Orai1. *Adv. Exp. Med. Biol.* **2010**, *661*, 137–154. [PubMed]
116. Berrout, J.; Jin, M.; O'Neil, R.G. Critical role of TRPP2 and TRPC1 channels in stretch-induced injury of blood-brain barrier endothelial cells. *Brain Res.* **2012**, *1436*, 1–12. [CrossRef] [PubMed]
117. Dragoni, S.; Laforenza, U.; Bonetti, E.; Lodola, F.; Bottino, C.; Guerra, G.; Borghesi, A.; Stronati, M.; Rosti, V.; Tanzi, F.; et al. Canonical transient receptor potential 3 channel triggers vascular endothelial growth factor-induced intracellular Ca^{2+} oscillations in endothelial progenitor cells isolated from umbilical cord blood. *Stem Cells Dev.* **2013**, *22*, 2561–2580. [CrossRef] [PubMed]
118. Moccia, F.; Lucariello, A.; Guerra, G. TRPC3-mediated Ca(2+) signals as a promising strategy to boost therapeutic angiogenesis in failing hearts: The role of autologous endothelial colony forming cells. *J. Cell. Physiol.* **2017**. [CrossRef]
119. Hamdollah Zadeh, M.A.; Glass, C.A.; Magnussen, A.; Hancox, J.C.; Bates, D.O. VEGF-mediated elevated intracellular calcium and angiogenesis in human microvascular endothelial cells in vitro are inhibited by dominant negative TRPC6. *Microcirculation* **2008**, *15*, 605–614. [CrossRef] [PubMed]
120. Chaudhuri, P.; Rosenbaum, M.A.; Birnbaumer, L.; Graham, L.M. Integration of TRPC6 and NADPH oxidase activation in lysophosphatidylcholine-induced TRPC5 externalization. *Am. J. Physiol. Cell Physiol.* **2017**, *313*, C541–C555. [CrossRef] [PubMed]
121. Gees, M.; Colsoul, B.; Nilius, B. The role of transient receptor potential cation channels in Ca^{2+} signaling. *Cold Spring Harb. Perspect. Biol.* **2010**, 2, a003962. [CrossRef] [PubMed]
122. Parenti, A.; De Logu, F.; Geppetti, P.; Benemei, S. What is the evidence for the role of TRP channels in inflammatory and immune cells? *Br. J. Pharmacol.* **2016**, *173*, 953–969. [CrossRef] [PubMed]
123. Di, A.; Malik, A.B. TRP channels and the control of vascular function. *Curr. Opin. Pharmacol.* **2010**, *10*, 127–132. [CrossRef] [PubMed]
124. Wong, C.O.; Yao, X. TRP channels in vascular endothelial cells. *Adv. Exp. Med. Biol.* **2011**, *704*, 759–780. [PubMed]
125. Ellinsworth, D.C.; Earley, S.; Murphy, T.V.; Sandow, S.L. Endothelial control of vasodilation: Integration of myoendothelial microdomain signalling and modulation by epoxyeicosatrienoic acids. *Pflugers Arch.* **2014**, *466*, 389–405. [CrossRef] [PubMed]
126. Sharp, C.D.; Hines, I.; Houghton, J.; Warren, A.; Jackson, T.H. t.; Jawahar, A.; Nanda, A.; Elrod, J.W.; Long, A.; Chi, A.; et al. Glutamate causes a loss in human cerebral endothelial barrier integrity through activation of NMDA receptor. *Am. J. Physiol. Heart Circ. Physiol.* **2003**, *285*, H2592–H2598. [CrossRef] [PubMed]
127. Lu, L.; Hogan-Cann, A.D.; Globa, A.K.; Lu, P.; Nagy, J.I.; Bamji, S.X.; Anderson, C.M. Astrocytes drive cortical vasodilatory signaling by activating endothelial NMDA receptors. *J. Cereb. Blood Flow Metab.* **2017**. [CrossRef] [PubMed]
128. Krizbai, I.A.; Deli, M.A.; Pestenacz, A.; Siklos, L.; Szabo, C.A.; Andras, I.; Joo, F. Expression of glutamate receptors on cultured cerebral endothelial cells. *J. Neurosci. Res.* **1998**, *54*, 814–819. [CrossRef]

129. Yang, F.; Zhao, K.; Zhang, X.; Zhang, J.; Xu, B. ATP Induces Disruption of Tight Junction Proteins via IL-1 Beta-Dependent MMP-9 Activation of Human Blood-Brain Barrier In Vitro. *Neural Plast.* **2016**, *2016*, 8928530. [CrossRef] [PubMed]
130. Zhao, H.; Zhang, X.; Dai, Z.; Feng, Y.; Li, Q.; Zhang, J.H.; Liu, X.; Chen, Y.; Feng, H. P2X7 Receptor Suppression Preserves Blood-Brain Barrier through Inhibiting RhoA Activation after Experimental Intracerebral Hemorrhage in Rats. *Sci. Rep.* **2016**, *6*, 23286. [CrossRef] [PubMed]
131. De Wit, C.; Wolfle, S.E. EDHF and gap junctions: Important regulators of vascular tone within the microcirculation. *Curr. Pharm. Biotechnol.* **2007**, *8*, 11–25. [CrossRef] [PubMed]
132. Shu, X.; Keller, T.C. t.; Begandt, D.; Butcher, J.T.; Biwer, L.; Keller, A.S.; Columbus, L.; Isakson, B.E. Endothelial nitric oxide synthase in the microcirculation. *Cell. Mol. Life Sci.* **2015**, *72*, 4561–4575. [CrossRef] [PubMed]
133. Dedkova, E.N.; Blatter, L.A. Nitric oxide inhibits capacitative Ca_{2+} entry and enhances endoplasmic reticulum Ca^{2+} uptake in bovine vascular endothelial cells. *J. Physiol.* **2002**, *539 (Pt 1)*, 77–91. [CrossRef] [PubMed]
134. Parekh, A.B. Functional consequences of activating store-operated CRAC channels. *Cell Calcium* **2007**, *42*, 111–121. [CrossRef] [PubMed]
135. Lin, S.; Fagan, K.A.; Li, K.X.; Shaul, P.W.; Cooper, D.M.; Rodman, D.M. Sustained endothelial nitric-oxide synthase activation requires capacitative Ca^{2+} entry. *J. Biol. Chem.* **2000**, *275*, 17979–17985. [CrossRef] [PubMed]
136. Kimura, C.; Oike, M.; Ohnaka, K.; Nose, Y.; Ito, Y. Constitutive nitric oxide production in bovine aortic and brain microvascular endothelial cells: A comparative study. *J. Physiol.* **2004**, *554 (Pt 3)*, 721–730. [CrossRef] [PubMed]
137. Boittin, F.X.; Alonso, F.; Le Gal, L.; Allagnat, F.; Beny, J.L.; Haefliger, J.A. Connexins and M3 muscarinic receptors contribute to heterogeneous Ca(2+) signaling in mouse aortic endothelium. *Cell. Physiol. Biochem.* **2013**, *31*, 166–178. [CrossRef] [PubMed]
138. Yeon, S.I.; Kim, J.Y.; Yeon, D.S.; Abramowitz, J.; Birnbaumer, L.; Muallem, S.; Lee, Y.H. Transient receptor potential canonical type 3 channels control the vascular contractility of mouse mesenteric arteries. *PLoS ONE* **2014**, *9*, e110413. [CrossRef] [PubMed]
139. Zhang, D.X.; Mendoza, S.A.; Bubolz, A.H.; Mizuno, A.; Ge, Z.D.; Li, R.; Warltier, D.C.; Suzuki, M.; Gutterman, D.D. Transient receptor potential vanilloid type 4-deficient mice exhibit impaired endothelium-dependent relaxation induced by acetylcholine in vitro and in vivo. *Hypertension* **2009**, *53*, 532–538. [CrossRef] [PubMed]
140. Zhao, Y.; Vanhoutte, P.M.; Leung, S.W. Vascular nitric oxide: Beyond eNOS. *J. Pharmacol. Sci.* **2015**, *129*, 83–94. [CrossRef] [PubMed]
141. Iadecola, C.; Zhang, F.; Xu, X. Role of nitric oxide synthase-containing vascular nerves in cerebrovasodilation elicited from cerebellum. *Am. J. Physiol.* **1993**, *264 (Pt 2)*, R738–R346. [CrossRef] [PubMed]
142. De Labra, C.; Rivadulla, C.; Espinosa, N.; Dasilva, M.; Cao, R.; Cudeiro, J. Different sources of nitric oxide mediate neurovascular coupling in the lateral geniculate nucleus of the cat. *Front. Syst. Neurosci.* **2009**, *3*, 9. [CrossRef] [PubMed]
143. Fiumana, E.; Parfenova, H.; Jaggar, J.H.; Leffler, C.W. Carbon monoxide mediates vasodilator effects of glutamate in isolated pressurized cerebral arterioles of newborn pigs. *Am. J. Physiol. Heart Circ. Physiol.* **2003**, *284*, H1073–H1079. [CrossRef] [PubMed]
144. Hogan-Cann, A.D.; Anderson, C.M. Physiological Roles of Non-Neuronal NMDA Receptors. *Trends Pharmacol. Sci.* **2016**, *37*, 750–767. [CrossRef] [PubMed]
145. Paoletti, P.; Bellone, C.; Zhou, Q. NMDA receptor subunit diversity: Impact on receptor properties, synaptic plasticity and disease. *Nat. Rev. Neurosci.* **2013**, *14*, 383–400. [CrossRef] [PubMed]
146. Fukaya, M.; Kato, A.; Lovett, C.; Tonegawa, S.; Watanabe, M. Retention of NMDA receptor NR2 subunits in the lumen of endoplasmic reticulum in targeted NR1 knockout mice. *Proc. Natl. Acad. Sci. USA* **2003**, *100*, 4855–4860. [CrossRef] [PubMed]
147. Neuhaus, W.; Freidl, M.; Szkokan, P.; Berger, M.; Wirth, M.; Winkler, J.; Gabor, F.; Pifl, C.; Noe, C.R. Effects of NMDA receptor modulators on a blood-brain barrier in vitro model. *Brain Res.* **2011**, *1394*, 49–61. [CrossRef] [PubMed]
148. Kamat, P.K.; Kalani, A.; Tyagi, S.C.; Tyagi, N. Hydrogen Sulfide Epigenetically Attenuates Homocysteine-Induced Mitochondrial Toxicity Mediated Through NMDA Receptor in Mouse Brain Endothelial (bEnd3) Cells. *J. Cell. Physiol.* **2015**, *230*, 378–394. [CrossRef] [PubMed]

149. Busija, D.W.; Bari, F.; Domoki, F.; Louis, T. Mechanisms involved in the cerebrovascular dilator effects of *N*-methyl-D-aspartate in cerebral cortex. *Brain Res. Rev.* **2007**, *56*, 89–100. [CrossRef] [PubMed]
150. Hamel, E. Cholinergic modulation of the cortical microvascular bed. *Prog. Brain Res.* **2004**, *145*, 171–178. [PubMed]
151. Zhang, F.; Xu, S.; Iadecola, C. Role of nitric oxide and acetylcholine in neocortical hyperemia elicited by basal forebrain stimulation: Evidence for an involvement of endothelial nitric oxide. *Neuroscience* **1995**, *69*, 1195–1204. [CrossRef]
152. Iadecola, C.; Zhang, F. Permissive and obligatory roles of NO in cerebrovascular responses to hypercapnia and acetylcholine. *Am. J. Physiol.* **1996**, *271 (Pt 2)*, R990–R1001. [CrossRef] [PubMed]
153. Yamada, M.; Lamping, K.G.; Duttaroy, A.; Zhang, W.; Cui, Y.; Bymaster, F.P.; McKinzie, D.L.; Felder, C.C.; Deng, C.X.; Faraci, F.M.; et al. Cholinergic dilation of cerebral blood vessels is abolished in M(5) muscarinic acetylcholine receptor knockout mice. *Proc. Natl. Acad. Sci. USA* **2001**, *98*, 14096–14101. [CrossRef] [PubMed]
154. Sandow, S.L.; Senadheera, S.; Grayson, T.H.; Welsh, D.G.; Murphy, T.V. Calcium and endothelium-mediated vasodilator signaling. *Adv. Exp. Med. Biol.* **2012**, *740*, 811–831. [PubMed]
155. Kasai, Y.; Yamazawa, T.; Sakurai, T.; Taketani, Y.; Iino, M. Endothelium-dependent frequency modulation of Ca^{2+} signalling in individual vascular smooth muscle cells of the rat. *J. Physiol.* **1997**, *504 (Pt 2)*, 349–357. [CrossRef] [PubMed]
156. Weikert, S.; Freyer, D.; Weih, M.; Isaev, N.; Busch, C.; Schultze, J.; Megow, D.; Dirnagl, U. Rapid Ca^{2+}-dependent NO-production from central nervous system cells in culture measured by NO-nitrite/ozone chemoluminescence. *Brain Res.* **1997**, *748*, 1–11. [CrossRef]
157. Radu, B.M.; Osculati, A.M.M.; Suku, E.; Banciu, A.; Tsenov, G.; Merigo, F.; Di Chio, M.; Banciu, D.D.; Tognoli, C.; Kacer, P.; et al. All muscarinic acetylcholine receptors (M1–M5) are expressed in murine brain microvascular endothelium. *Sci. Rep.* **2017**, *7*, 5083. [CrossRef] [PubMed]
158. Moccia, F.; Frost, C.; Berra-Romani, R.; Tanzi, F.; Adams, D.J. Expression and function of neuronal nicotinic ACh receptors in rat microvascular endothelial cells. *Am. J. Physiol. Heart Circ. Physiol.* **2004**, *286*, H486–H491. [CrossRef] [PubMed]
159. Laforenza, U.; Pellavio, G.; Moccia, F. Muscarinic M5 receptors trigger acetylcholine-induced Ca^{2+} signals and nitric oxide release in human brain microvascular endothelial cells. 2017; manuscript in preparation.
160. Uhlen, P.; Fritz, N. Biochemistry of calcium oscillations. *Biochem. Biophys. Res. Commun.* **2010**, *396*, 28–32. [CrossRef] [PubMed]
161. Mikoshiba, K. IP3 receptor/Ca^{2+} channel: From discovery to new signaling concepts. *J. Neurochem.* **2007**, *102*, 1426–1046. [CrossRef] [PubMed]
162. Guzman, S.J.; Gerevich, Z. P2Y Receptors in Synaptic Transmission and Plasticity: Therapeutic Potential in Cognitive Dysfunction. *Neural Plast.* **2016**, *2016*, 1207393. [CrossRef] [PubMed]
163. Munoz, M.F.; Puebla, M.; Figueroa, X.F. Control of the neurovascular coupling by nitric oxide-dependent regulation of astrocytic Ca(2+) signaling. *Front. Cell. Neurosci.* **2015**, *9*, 59. [CrossRef] [PubMed]
164. Dietrich, H.H.; Horiuchi, T.; Xiang, C.; Hongo, K.; Falck, J.R.; Dacey, R.G., Jr. Mechanism of ATP-induced local and conducted vasomotor responses in isolated rat cerebral penetrating arterioles. *J. Vasc. Res.* **2009**, *46*, 253–264. [CrossRef] [PubMed]
165. Dietrich, H.H.; Kajita, Y.; Dacey, R.G., Jr. Local and conducted vasomotor responses in isolated rat cerebral arterioles. *Am. J. Physiol.* **1996**, *71*, H1109–H1116. [CrossRef] [PubMed]
166. Toth, P.; Tarantini, S.; Davila, A.; Valcarcel-Ares, M.N.; Tucsek, Z.; Varamini, B.; Ballabh, P.; Sonntag, W.E.; Baur, J.A.; Csiszar, A.; et al. Purinergic glio-endothelial coupling during neuronal activity: Role of P2Y1 receptors and eNOS in functional hyperemia in the mouse somatosensory cortex. *Am. J. Physiol. Heart Circ. Physiol.* **2015**, *309*, H1837–H1845. [CrossRef] [PubMed]
167. Moccia, F.; Baruffi, S.; Spaggiari, S.; Coltrini, D.; Berra-Romani, R.; Signorelli, S.; Castelli, L.; Taglietti, V.; Tanzi, F. P2y1 and P2y2 receptor-operated Ca^{2+} signals in primary cultures of cardiac microvascular endothelial cells. *Microvasc. Res.* **2001**, *61*, 240–252. [CrossRef] [PubMed]
168. Marrelli, S.P. Mechanisms of endothelial P2Y(1)- and P2Y(2)-mediated vasodilatation involve differential $[Ca^{2+}]$i responses. *Am. J. Physiol. Heart Circ. Physiol.* **2001**, *281*, H1759–H1766. [CrossRef] [PubMed]
169. Kwan, H.Y.; Cheng, K.T.; Ma, Y.; Huang, Y.; Tang, N.L.; Yu, S.; Yao, X. CNGA2 contributes to ATP-induced noncapacitative Ca^{2+} influx in vascular endothelial cells. *J. Vasc. Res.* **2010**, *47*, 148–156. [CrossRef] [PubMed]

170. Duza, T.; Sarelius, I.H. Conducted dilations initiated by purines in arterioles are endothelium dependent and require endothelial Ca^{2+}. *Am. J. Physiol. Heart Circ. Physiol.* **2003**, *285*, H26–H37. [CrossRef] [PubMed]
171. Kohler, R.; Hoyer, J. The endothelium-derived hyperpolarizing factor: Insights from genetic animal models. *Kidney Int.* **2007**, *72*, 145–150. [CrossRef] [PubMed]
172. Ledoux, J.; Taylor, M.S.; Bonev, A.D.; Hannah, R.M.; Solodushko, V.; Shui, B.; Tallini, Y.; Kotlikoff, M.I.; Nelson, M.T. Functional architecture of inositol 1,4,5-trisphosphate signaling in restricted spaces of myoendothelial projections. *Proc. Natl. Acad. Sci. USA* **2008**, *105*, 9627–9632. [CrossRef] [PubMed]
173. Dora, K.A.; Gallagher, N.T.; McNeish, A.; Garland, C.J. Modulation of endothelial cell KCa3.1 channels during endothelium-derived hyperpolarizing factor signaling in mesenteric resistance arteries. *Circ. Res.* **2008**, *102*, 1247–1255. [CrossRef] [PubMed]
174. Sonkusare, S.K.; Bonev, A.D.; Ledoux, J.; Liedtke, W.; Kotlikoff, M.I.; Heppner, T.J.; Hill-Eubanks, D.C.; Nelson, M.T. Elementary Ca^{2+} signals through endothelial TRPV4 channels regulate vascular function. *Science* **2012**, *336*, 597–601. [CrossRef] [PubMed]
175. Behringer, E.J.; Segal, S.S. Spreading the signal for vasodilatation: Implications for skeletal muscle blood flow control and the effects of ageing. *J. Physiol.* **2012**, *590*, 6277–6284. [CrossRef] [PubMed]
176. Behringer, E.J.; Segal, S.S. Tuning electrical conduction along endothelial tubes of resistance arteries through Ca(2+)-activated K(+) channels. *Circ. Res.* **2012**, *110*, 1311–1321. [CrossRef] [PubMed]
177. Kapela, A.; Behringer, E.J.; Segal, S.S.; Tsoukias, N.M. Biophysical properties of microvascular endothelium: Requirements for initiating and conducting electrical signals. *Microcirculation* **2017**, *25*. [CrossRef] [PubMed]
178. Sullivan, M.N.; Gonzales, A.L.; Pires, P.W.; Bruhl, A.; Leo, M.D.; Li, W.; Oulidi, A.; Boop, F.A.; Feng, Y.; Jaggar, J.H.; et al. Localized TRPA1 channel Ca^{2+} signals stimulated by reactive oxygen species promote cerebral artery dilation. *Science Signal.* **2015**, *8*, ra2. [CrossRef] [PubMed]
179. Hannah, R.M.; Dunn, K.M.; Bonev, A.D.; Nelson, M.T. Endothelial SK(Ca) and IK(Ca) channels regulate brain parenchymal arteriolar diameter and cortical cerebral blood flow. *J. Cereb. Blood Flow Metab.* **2011**, *31*, 1175–1186. [CrossRef] [PubMed]
180. Sokoya, E.M.; Burns, A.R.; Setiawan, C.T.; Coleman, H.A.; Parkington, H.C.; Tare, M. Evidence for the involvement of myoendothelial gap junctions in EDHF-mediated relaxation in the rat middle cerebral artery. *Am. J. Physiol. Heart Circ. Physiol.* **2006**, *291*, H385–H393. [CrossRef] [PubMed]
181. Bagher, P.; Beleznai, T.; Kansui, Y.; Mitchell, R.; Garland, C.J.; Dora, K.A. Low intravascular pressure activates endothelial cell TRPV4 channels, local Ca^{2+} events, and IKCa channels, reducing arteriolar tone. *Proc. Natl. Acad. Sci. USA* **2012**, *109*, 18174–18179. [CrossRef] [PubMed]
182. Golding, E.M.; Marrelli, S.P.; You, J.; Bryan, R.M., Jr. Endothelium-derived hyperpolarizing factor in the brain: A new regulator of cerebral blood flow? *Stroke* **2002**, *33*, 661–663. [PubMed]
183. McNeish, A.J.; Dora, K.A.; Garland, C.J. Possible role for K^+ in endothelium-derived hyperpolarizing factor-linked dilatation in rat middle cerebral artery. *Stroke* **2005**, *36*, 1526–1532. [CrossRef] [PubMed]
184. Marrelli, S.P.; Eckmann, M.S.; Hunte, M.S. Role of endothelial intermediate conductance KCa channels in cerebral EDHF-mediated dilations. *Am. J. Physiol. Heart Circ. Physiol.* **2003**, *285*, H1590–H1599. [CrossRef] [PubMed]
185. McNeish, A.J.; Sandow, S.L.; Neylon, C.B.; Chen, M.X.; Dora, K.A.; Garland, C.J. Evidence for involvement of both IKCa and SKCa channels in hyperpolarizing responses of the rat middle cerebral artery. *Stroke* **2006**, *37*, 1277–1282. [CrossRef] [PubMed]
186. Zhang, L.; Papadopoulos, P.; Hamel, E. Endothelial TRPV4 channels mediate dilation of cerebral arteries: Impairment and recovery in cerebrovascular pathologies related to Alzheimer's disease. *Br. J. Pharmacol.* **2013**, *170*, 661–670. [CrossRef] [PubMed]
187. Dragoni, S.; Guerra, G.; Fiorio Pla, A.; Bertoni, G.; Rappa, A.; Poletto, V.; Bottino, C.; Aronica, A.; Lodola, F.; Cinelli, M.P.; et al. A functional Transient Receptor Potential Vanilloid 4 (TRPV4) channel is expressed in human endothelial progenitor cells. *J. Cell. Physiol.* **2015**, *230*, 95–104. [CrossRef] [PubMed]
188. Thodeti, C.K.; Matthews, B.; Ravi, A.; Mammoto, A.; Ghosh, K.; Bracha, A.L.; Ingber, D.E. TRPV4 channels mediate cyclic strain-induced endothelial cell reorientation through integrin-to-integrin signaling. *Circ. Res.* **2009**, *104*, 1123–1130. [CrossRef] [PubMed]
189. He, D.; Pan, Q.; Chen, Z.; Sun, C.; Zhang, P.; Mao, A.; Zhu, Y.; Li, H.; Lu, C.; Xie, M.; et al. Treatment of hypertension by increasing impaired endothelial TRPV4-KCa2.3 interaction. *EMBO Mol. Med.* **2017**, *9*, 1491–1503. [CrossRef] [PubMed]

190. Ho, W.S.; Zheng, X.; Zhang, D.X. Role of endothelial TRPV4 channels in vascular actions of the endocannabinoid, 2-arachidonoylglycerol. *Br. J. Pharmacol.* **2015**, *172*, 5251–5264. [CrossRef] [PubMed]
191. Mendoza, S.A.; Fang, J.; Gutterman, D.D.; Wilcox, D.A.; Bubolz, A.H.; Li, R.; Suzuki, M.; Zhang, D.X. TRPV4-mediated endothelial Ca^{2+} influx and vasodilation in response to shear stress. *Am. J. Physiol. Heart Circ. Physiol.* **2010**, *298*, H466–H476. [CrossRef] [PubMed]
192. Zheng, X.; Zinkevich, N.S.; Gebremedhin, D.; Gauthier, K.M.; Nishijima, Y.; Fang, J.; Wilcox, D.A.; Campbell, W.B.; Gutterman, D.D.; Zhang, D.X. Arachidonic acid-induced dilation in human coronary arterioles: Convergence of signaling mechanisms on endothelial TRPV4-mediated Ca^{2+} entry. *J. Am. Heart Assoc.* **2013**, *2*, e000080. [CrossRef] [PubMed]
193. Watanabe, H.; Vriens, J.; Suh, S.H.; Benham, C.D.; Droogmans, G.; Nilius, B. Heat-evoked activation of TRPV4 channels in a HEK293 cell expression system and in native mouse aorta endothelial cells. *J. Biol. Chem.* **2002**, *277*, 47044–47051. [CrossRef] [PubMed]
194. White, J.P.; Cibelli, M.; Urban, L.; Nilius, B.; McGeown, J.G.; Nagy, I. TRPV4: Molecular Conductor of a Diverse Orchestra. *Physiol. Rev.* **2016**, *96*, 911–973. [CrossRef] [PubMed]
195. Watanabe, H.; Vriens, J.; Prenen, J.; Droogmans, G.; Voets, T.; Nilius, B. Anandamide and arachidonic acid use epoxyeicosatrienoic acids to activate TRPV4 channels. *Nature* **2003**, *424*, 434–438. [CrossRef] [PubMed]
196. Graier, W.F.; Simecek, S.; Sturek, M. Cytochrome P450 mono-oxygenase-regulated signalling of Ca^{2+} entry in human and bovine endothelial cells. *J. Physiol.* **1995**, *482 (Pt 2)*, 259–274. [CrossRef] [PubMed]
197. Vriens, J.; Owsianik, G.; Fisslthaler, B.; Suzuki, M.; Janssens, A.; Voets, T.; Morisseau, C.; Hammock, B.D.; Fleming, I.; Busse, R.; et al. Modulation of the Ca^2 permeable cation channel TRPV4 by cytochrome P450 epoxygenases in vascular endothelium. *Circ. Res.* **2005**, *97*, 908–915. [CrossRef] [PubMed]
198. Earley, S.; Pauyo, T.; Drapp, R.; Tavares, M.J.; Liedtke, W.; Brayden, J.E. TRPV4-dependent dilation of peripheral resistance arteries influences arterial pressure. *Am. J. Physiol. Heart Circ. Physiol.* **2009**, *297*, H1096–H1102. [CrossRef] [PubMed]
199. Marrelli, S.P.; O'Neil R, G.; Brown, R.C.; Bryan, R.M., Jr. PLA2 and TRPV4 channels regulate endothelial calcium in cerebral arteries. *Am. J. Physiol. Heart Circ. Physiol.* **2007**, *292*, H1390–H1397. [CrossRef] [PubMed]
200. You, J.; Marrelli, S.P.; Bryan, R.M., Jr. Role of cytoplasmic phospholipase A2 in endothelium-derived hyperpolarizing factor dilations of rat middle cerebral arteries. *J. Cereb. Blood Flow Metab.* **2002**, *22*, 1239–1247. [CrossRef] [PubMed]
201. Liu, X.; Li, C.; Gebremedhin, D.; Hwang, S.H.; Hammock, B.D.; Falck, J.R.; Roman, R.J.; Harder, D.R.; Koehler, R.C. Epoxyeicosatrienoic acid-dependent cerebral vasodilation evoked by metabotropic glutamate receptor activation in vivo. *Am. J. Physiol. Heart Circ. Physiol.* **2011**, *301*, H373–H381. [CrossRef] [PubMed]
202. Peng, X.; Zhang, C.; Alkayed, N.J.; Harder, D.R.; Koehler, R.C. Dependency of cortical functional hyperemia to forepaw stimulation on epoxygenase and nitric oxide synthase activities in rats. *J. Cereb. Blood Flow Metab.* **2004**, *24*, 509–517. [CrossRef] [PubMed]
203. Chen, B.R.; Kozberg, M.G.; Bouchard, M.B.; Shaik, M.A.; Hillman, E.M. A critical role for the vascular endothelium in functional neurovascular coupling in the brain. *J. Am. Heart Assoc.* **2014**, *3*, e000787. [CrossRef] [PubMed]
204. Longden, T.A.; Dabertrand, F.; Koide, M.; Gonzales, A.L.; Tykocki, N.R.; Brayden, J.E.; Hill-Eubanks, D.; Nelson, M.T. Capillary K(+)-sensing initiates retrograde hyperpolarization to increase local cerebral blood flow. *Nat. Neurosci.* **2017**, *20*, 717–726. [CrossRef] [PubMed]
205. Earley, S.; Gonzales, A.L.; Crnich, R. Endothelium-dependent cerebral artery dilation mediated by TRPA1 and Ca^{2+}-Activated K^+ channels. *Circ. Res.* **2009**, *104*, 987–994. [CrossRef] [PubMed]
206. Qian, X.; Francis, M.; Solodushko, V.; Earley, S.; Taylor, M.S. Recruitment of dynamic endothelial Ca^{2+} signals by the TRPA1 channel activator AITC in rat cerebral arteries. *Microcirculation* **2013**, *20*, 138–148. [CrossRef] [PubMed]
207. Senadheera, S.; Kim, Y.; Grayson, T.H.; Toemoe, S.; Kochukov, M.Y.; Abramowitz, J.; Housley, G.D.; Bertrand, R.L.; Chadha, P.S.; Bertrand, P.P.; et al. Transient receptor potential canonical type 3 channels facilitate endothelium-derived hyperpolarization-mediated resistance artery vasodilator activity. *Cardiovasc. Res.* **2012**, *95*, 439–447. [CrossRef] [PubMed]
208. Kochukov, M.Y.; Balasubramanian, A.; Abramowitz, J.; Birnbaumer, L.; Marrelli, S.P. Activation of endothelial transient receptor potential C3 channel is required for small conductance calcium-activated potassium

channel activation and sustained endothelial hyperpolarization and vasodilation of cerebral artery. *J. Am. Heart Assoc.* **2014**, *3*, e000913. [CrossRef] [PubMed]

209. Pires, P.W.; Sullivan, M.N.; Pritchard, H.A.; Robinson, J.J.; Earley, S. Unitary TRPV3 channel Ca^{2+} influx events elicit endothelium-dependent dilation of cerebral parenchymal arterioles. *Am. J. Physiol. Heart Circ. Physiol.* **2015**, *309*, H2031–H2041. [CrossRef] [PubMed]
210. Yang, P.; Zhu, M.X. Trpv3. In *Handbook of Experimental Pharmacology*; Springer: Berlin, Germany, 2014; Volume 222, pp. 273–291.
211. Zygmunt, P.M.; Hogestatt, E.D. Trpa1. In *Handbook of Experimental Pharmacology*; Springer: Berlin, Germany, 2014; Volume 222, pp. 583–630.
212. Iadecola, C. The pathobiology of vascular dementia. *Neuron* **2013**, *80*, 844–866. [CrossRef] [PubMed]
213. Chen, B.R.; Bouchard, M.B.; McCaslin, A.F.; Burgess, S.A.; Hillman, E.M. High-speed vascular dynamics of the hemodynamic response. *NeuroImage* **2011**, *54*, 1021–1030. [CrossRef] [PubMed]
214. Hillman, E.M.; Devor, A.; Bouchard, M.B.; Dunn, A.K.; Krauss, G.W.; Skoch, J.; Bacskai, B.J.; Dale, A.M.; Boas, D.A. Depth-resolved optical imaging and microscopy of vascular compartment dynamics during somatosensory stimulation. *NeuroImage* **2007**, *35*, 89–104. [CrossRef] [PubMed]
215. Stefanovic, B.; Hutchinson, E.; Yakovleva, V.; Schram, V.; Russell, J.T.; Belluscio, L.; Koretsky, A.P.; Silva, A.C. Functional reactivity of cerebral capillaries. *J. Cereb. Blood Flow Metab.* **2008**, *28*, 961–972. [CrossRef] [PubMed]
216. O'Herron, P.; Chhatbar, P.Y.; Levy, M.; Shen, Z.; Schramm, A.E.; Lu, Z.; Kara, P. Neural correlates of single-vessel haemodynamic responses in vivo. *Nature* **2016**, *534*, 378–382. [CrossRef] [PubMed]
217. Kisler, K.; Nelson, A.R.; Rege, S.V.; Ramanathan, A.; Wang, Y.; Ahuja, A.; Lazic, D.; Tsai, P.S.; Zhao, Z.; Zhou, Y.; et al. Pericyte degeneration leads to neurovascular uncoupling and limits oxygen supply to brain. *Nat. Neurosci.* **2017**, *20*, 406–416. [CrossRef] [PubMed]
218. Behringer, E.J.; Socha, M.J.; Polo-Parada, L.; Segal, S.S. Electrical conduction along endothelial cell tubes from mouse feed arteries: Confounding actions of glycyrrhetinic acid derivatives. *Br. J. Pharmacol.* **2012**, *166*, 774–787. [CrossRef] [PubMed]
219. Vaucher, E.; Hamel, E. Cholinergic basal forebrain neurons project to cortical microvessels in the rat: Electron microscopic study with anterogradely transported Phaseolus vulgaris leucoagglutinin and choline acetyltransferase immunocytochemistry. *J. Neurosci.* **1995**, *15*, 7427–7441. [PubMed]
220. Vaucher, E.; Linville, D.; Hamel, E. Cholinergic basal forebrain projections to nitric oxide synthase-containing neurons in the rat cerebral cortex. *Neuroscience* **1997**, *79*, 827–836. [CrossRef]
221. Jensen, L.J.; Holstein-Rathlou, N.H. The vascular conducted response in cerebral blood flow regulation. *J. Cereb. Blood Flow Metab.* **2013**, *33*, 649–656. [CrossRef] [PubMed]
222. Uhrenholt, T.R.; Domeier, T.L.; Segal, S.S. Propagation of calcium waves along endothelium of hamster feed arteries. *Am. J. Physiol. Heart Circ. Physiol.* **2007**, *292*, H1634–H1640. [CrossRef] [PubMed]
223. Domeier, T.L.; Segal, S.S. Electromechanical and pharmacomechanical signalling pathways for conducted vasodilatation along endothelium of hamster feed arteries. *J. Physiol.* **2007**, *579*, 175–186. [CrossRef] [PubMed]
224. De Bock, M.; Culot, M.; Wang, N.; da Costa, A.; Decrock, E.; Bol, M.; Bultynck, G.; Cecchelli, R.; Leybaert, L. Low extracellular Ca^{2+} conditions induce an increase in brain endothelial permeability that involves intercellular Ca^{2+} waves. *Brain Res.* **2012**, *1487*, 78–87. [CrossRef] [PubMed]
225. Berra-Romani, R.; Raqeeb, A.; Torres-Jácome, J.; Guzman-Silva, A.; Guerra, G.; Tanzi, F.; Moccia, F. The mechanism of injury-induced intracellular calcium concentration oscillations in the endothelium of excised rat aorta. *J. Vasc. Res.* **2012**, *49*, 65–76. [CrossRef] [PubMed]
226. Francis, M.; Waldrup, J.R.; Qian, X.; Solodushko, V.; Meriwether, J.; Taylor, M.S. Functional Tuning of Intrinsic Endothelial Ca^{2+} Dynamics in Swine Coronary Arteries. *Circ. Res.* **2016**, *118*, 1078–1090. [CrossRef] [PubMed]
227. Wilson, C.; Saunter, C.D.; Girkin, J.M.; McCarron, J.G. Clusters of specialized detector cells provide sensitive and high fidelity receptor signaling in the intact endothelium. *FASEB J.* **2016**, *30*, 2000–2013. [CrossRef] [PubMed]
228. Kohler, R.; Hoyer, J. Role of TRPV4 in the Mechanotransduction of Shear Stress in Endothelial Cells. In *TRP Ion Channel Function in Sensory Transduction and Cellular Signaling Cascades*; Liedtke, W.B., Heller, S., Eds.; CRC Press: Boca Raton, FL, USA, 2007.

229. Bryan, R.M., Jr.; Marrelli, S.P.; Steenberg, M.L.; Schildmeyer, L.A.; Johnson, T.D. Effects of luminal shear stress on cerebral arteries and arterioles. *Am. J. Physiol. Heart Circ. Physiol.* **2001**, *280*, H2011–H2022. [CrossRef] [PubMed]
230. Bryan, R.M., Jr.; Steenberg, M.L.; Marrelli, S.P. Role of endothelium in shear stress-induced constrictions in rat middle cerebral artery. *Stroke* **2001**, *32*, 1394–1400. [CrossRef] [PubMed]
231. Ngai, A.C.; Winn, H.R. Modulation of cerebral arteriolar diameter by intraluminal flow and pressure. *Circ. Res.* **1995**, *77*, 832–840. [CrossRef] [PubMed]
232. Lourenco, C.F.; Ledo, A.; Barbosa, R.M.; Laranjinha, J. Neurovascular uncoupling in the triple transgenic model of Alzheimer's disease: Impaired cerebral blood flow response to neuronal-derived nitric oxide signaling. *Exp. Neurol.* **2017**, *291*, 36–43. [CrossRef] [PubMed]
233. Lourenco, C.F.; Ledo, A.; Dias, C.; Barbosa, R.M.; Laranjinha, J. Neurovascular and neurometabolic derailment in aging and Alzheimer's disease. *Front. Aging Neurosci.* **2015**, *7*, 103. [CrossRef] [PubMed]
234. Katakam, P.V.; Domoki, F.; Snipes, J.A.; Busija, A.R.; Jarajapu, Y.P.; Busija, D.W. Impaired mitochondria-dependent vasodilation in cerebral arteries of Zucker obese rats with insulin resistance. *Am. J. Physiol. Regul. Integr. Comp. Physiol.* **2009**, *296*, R289–R298. [CrossRef] [PubMed]
235. Mottola, A.; Antoniotti, S.; Lovisolo, D.; Munaron, L. Regulation of noncapacitative calcium entry by arachidonic acid and nitric oxide in endothelial cells. *FASEB J.* **2005**, *19*, 2075–2077. [CrossRef] [PubMed]

4

Expression, Subcellular Localization, and Interactions of CPK Family Genes in Maize

Muhammad Hayder Bin Khalid [1,†], Muhammad Ali Raza [2,†], Hao Qiang Yu [1,†], Imran Khan [3], Fu Ai Sun [1], Ling Yang Feng [2], Jing Tao Qu [1], Feng Ling Fu [1,*] and Wan Chen Li [1,*]

1 Maize Research Institute, Sichuan Agricultural University, Chengdu 611130, China; haider2323@gmail.com (M.H.B.K.); yhq1801@sicau.edu.cn (H.Q.Y.); andapanda86@outlook.com (F.A.S.); pandaanda86@outlook.com (J.T.Q.)

2 College of Agronomy, Sichuan Agricultural University, Chengdu 611130, China; razaali0784@yahoo.com (M.A.R.); lilymalik87@hotmail.com (L.Y.F.)

3 Department of Grassland Science, Sichuan Agricultural University, Chengdu 611130, China; imran.62k@gmail.com

* Correspondence: ffl@sicau.edu.cn (F.L.F.); aumdyms@sicau.edu.cn (W.C.L.)

† These authors contributed equally to this work.

Abstract: Calcium-dependent protein kinase (CPKs) is a key player in the calcium signaling pathway to decode calcium signals into various physiological responses. cDNA sequences of 9 *Zm*CPK genes were successfully cloned from all four phylogenetic groups in maize. qRT-PCR analysis showed the expression variation of these selected genes under abscisic acid (ABA) and calcium chloride ($CaCl_2$) treatment. Due to the presence of N-myristoylation/palmitoylation sites, the selected *Zm*CPK members were localized in a plasma membrane. To clarify whether *Zm*CPK, a key player in calcium signaling, interacts with key players of ABA, protein phosphatase 2Cs (PP2Cs) and the SNF1-related protein kinase 2s (SnRK2s) and mitogen-activated protein kinase (MAPK) signaling pathways in maize, we examined the interaction between 9 CPKs, 8 PP2Cs, 5 SnRKs, and 20 members of the MPK family in maize by using yeast two-hybrid assay. Our results showed that three *Zm*CPKs interact with three different members of *Zm*SnRKs while four *Zm*CPK members had a positive interaction with 13 members of *Zm*MPKs in different combinations. These four *Zm*CPK proteins are from three different groups in maize. These findings of physical interactions between *Zm*CPKs, *Zm*SnRKs, and *Zm*MPKs suggested that these signaling pathways do not only have indirect influence but also have direct crosstalk that may involve the defense mechanism in maize. The present study may improve the understanding of signal transduction in plants.

Keywords: calcium dependent protein kinases; PP2C; SnRK; MAPK; ABA; signal transduction; Y2H; BiFC

---

1. Introduction

Plants suffer from various environmental stresses, including drought, high salinity, and extreme temperatures [1–3]. To cope with these stresses, plants have developed sophisticated signal transduction pathways and a series of survival mechanisms. Among those pathways and mechanisms, transitory calcium is an important messenger of signal transduction that responds to hormonal and biotic or abiotic stresses [4–6]. Calcium signals are deciphered by calcium sensors, including calcium dependent protein kinases (CPKs). Genetics and biochemical studies have shown that CPKs are the key players in biological processes and plant signaling pathways, including innate immunity, oxidative burst, abiotic stress responses, and hormone signaling [7–15].

Abscisic acid (ABA) signaling pathway regulate many processes of plant growth and development under abiotic stress conditions, such as dehydration or high salinity [3,16,17]. Initially, it was thought that Ca^{2+}- and ABA-mediating signaling pathways were independent of each other [18,19], or that the latter regulates the former [20], but later it was found that these two pathways are interdependent [21–24]. Phosphorylation of two ABA-responsive transcription factors (ABF1 and ABF4) by AtCPK4, 11, and 32, suggested the role of kinases in the regulation of ABA signaling through these transcription factors under stress conditions [11,25]. *AtCPK3*, 6, 10, 21, and 23 of the *Arabidopsis* CPK family control stomatal closure by two signaling pathways of ABA and Ca^{2+} [26–29]. In maize leaf protoplasts, two CPKs are able to transactivate the expression of an ABA-inducible gene in the absence of the hormone [21]. In seed germination and seedling growth of *Arabidopsis*, ABA sensitivity was enhanced by the overexpression of *ZmCPK4* [30]. The mutants of CPK1 and CPK1a inhibited the activity of protein phosphatases type 2Cs (PP2C), which is capable of blocking the responses to ABA in *Arabidopsis* [21]. All this evidence suggests that CPK plays a critical role at the intersection of ABA and Ca^{2+} signaling pathways [31]. *AtCPK4* and *AtCPK11* catalyze the ABF1 and ABF4 (ABRE2), and meanwhile, ABF1 is also a target of *SnRK2*.2 and *SnRK2*.3, indicating that two kinases can share a single substrate. On the other hand, MAPK signaling cascade is another pathway which proceeds beside calcium signaling and is triggered by the same environmental stimuli. Most of the MAPK family members in *Arabidopsis* are related to the pathogen response, while *At*CPKs were found to play their roles under abiotic stress as well as pathogen response [15,32]. MAPKs and CPKs are thought to be the key components arbitrating immunity in plants and are activated in response to almost the same stimuli [33–36]. Crosstalk between CPK and MAPK signaling pathways have been studied in animals, but have hardly been focused in plant systems [37,38]. Reports have suggested that MAPKs and CPKs act differentially in salt signaling and innate immunity without direct crosstalk [12,39]. In tobacco, it has been suggested that MAPK and CPK signaling mechanisms proceed in parallel, to provide backup and fine-tuning of the overlapping reactions underway [40]. With regard to hindrance in the expression of *ZmMPK5* due to *ZmCPK11* silencing, the role of *ZmCPK11* upstream of *ZmMPK5* has been proposed [41]. Phosphorylation of *OsMPK5* by *OsCPK18* proposed the direct crosstalk between CPK and MAPK signaling pathways [42].

Although CPKs seem to play their role in ABA signaling, it is not clear whether there is a link between the CPK pathway and the core ABA signaling components in ABA signaling. Moreover, in addition to the classical MAPK cascade model, plants may have developed a calcium-mediated pathway for the activation of MAPK. Taking into account the importance of CPKs, we isolated nine CPKs distributed in four groups on the basis of phylogenetic analysis (Figure 1). Expression analysis of the selected CPKs under ABA and $CaCl_2$ treatment explained their sensitivity towards these treatments. Because of the importance of location of protein expression, we visualized subcellular localization of selected members in tobacco leaves. Yeast two-hybrid assay revealed the interaction between *Zm*CPKs, *Zm*SnRKs, and *Zm*MPKs.

It has become clear that signaling pathways have to be envisaged as complex networks rather than as linear or branched signaling pathways.

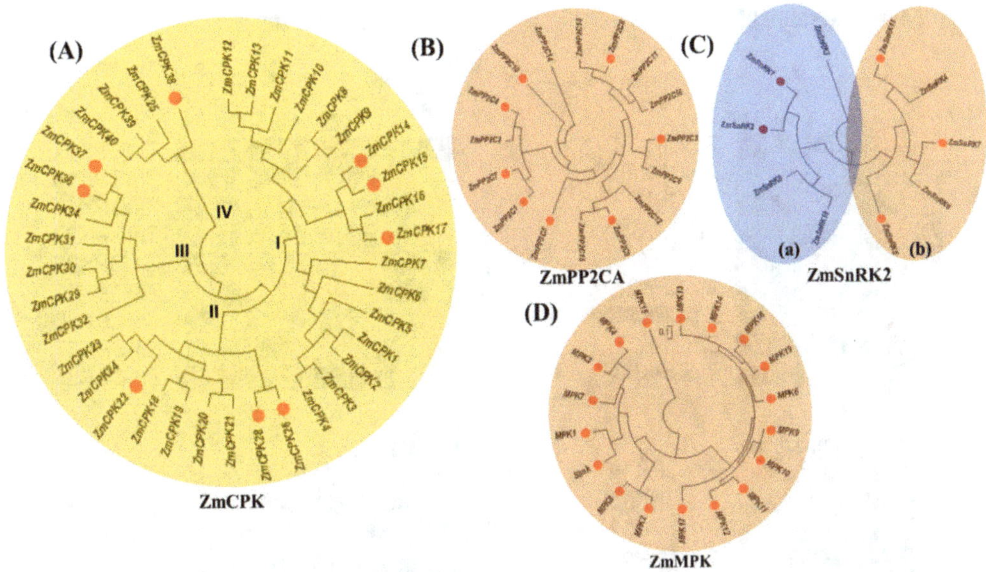

Figure 1. Phylogenetic tree of CPKs, PP2Cs, SnRKs, and MPKs from maize. The phylogenetic trees were constructed after alignment of full-length sequences of (**A**) ZmCPKs, (**B**) *Zm*PP2Cs clade A, (**C**) *Zm*SnRK2s, and (**D**) ZmMAPKs from maize using the neighbor-joining method. MEGA7 was used with 1000 bootstraps. Members marked in red were used in this study.

2. Results

2.1. Expression Analysis of ZmCPK Family Members

Gene expression patterns can provide important information about gene function. In this study, the expression patterns of the nine *Zm*CPKs were detected by qRT-PCR, and the results showed variations under ABA and $CaCl_2$ as calcium treatments. The analysis with qRT–PCR showed that the expressions of these nine *Zm*CPK genes were regulated by these treatments. Under ABA and calcium treatment, we noticed that *Zm*CPK expressed differentially at different time points (Figure 2). *ZmCPK14*, 22, 36, and 37 appeared to be significantly down-regulated at 6 h of post-ABA treatment relative to the control leaf samples. On the other hand, *ZmCPK14*, 15, and 38 seemed to be upregulated at 6 h of post ABA-treated root samples. Besides *ZmCPK14*, 15, and 22, all other genes were highly expressed at 12 h of ABA-treated leaf samples and expression of most of the genes was changed into downregulation at 24 h, i.e., *ZmCPK26*, 28, 36, 37, and 38. Unlike leaf samples, root samples had a downregulation of the *ZmCPK17*, 26, 28, and 36 genes. *ZmCPK37* and *ZmCPK38* genes were expressed differentially at 48 h in leaf and root samples after ABA treatment.

Moreover, the expression of nominated nine CPKs was evaluated both in leaves and roots at different time points under $CaCl_2$ treatment. Compared to control samples of leaves, sudden expression change was observed in all selected genes at 10 min of post-$CaCl_2$ treatment. However, the expressions of *ZmCPK17*, 22, and 36 were identified to be downregulated in $CaCl_2$-treated leaves. Although the expression patterns of these nine genes were also evaluated under root samples. *ZmCPK14*, 17, 26, 36, and 37 had upregulation in the first hour, i.e., at 20, 30, and 60 min, in selected root samples. The different response of the transcripts of these *Zm*CPK genes to ABA and $CaCl_2$ treatments suggested their potential role in resistance against abiotic stress. Given the expression pattern of the genes, all the genes had a heartbeat-like expression pattern at different time points. We can predict per expression levels of these genes that the first hour is most important after abiotic stress treatment, i.e., ABA and $CaCl_2$, and these findings could be useful for researchers for detailed functional analysis.

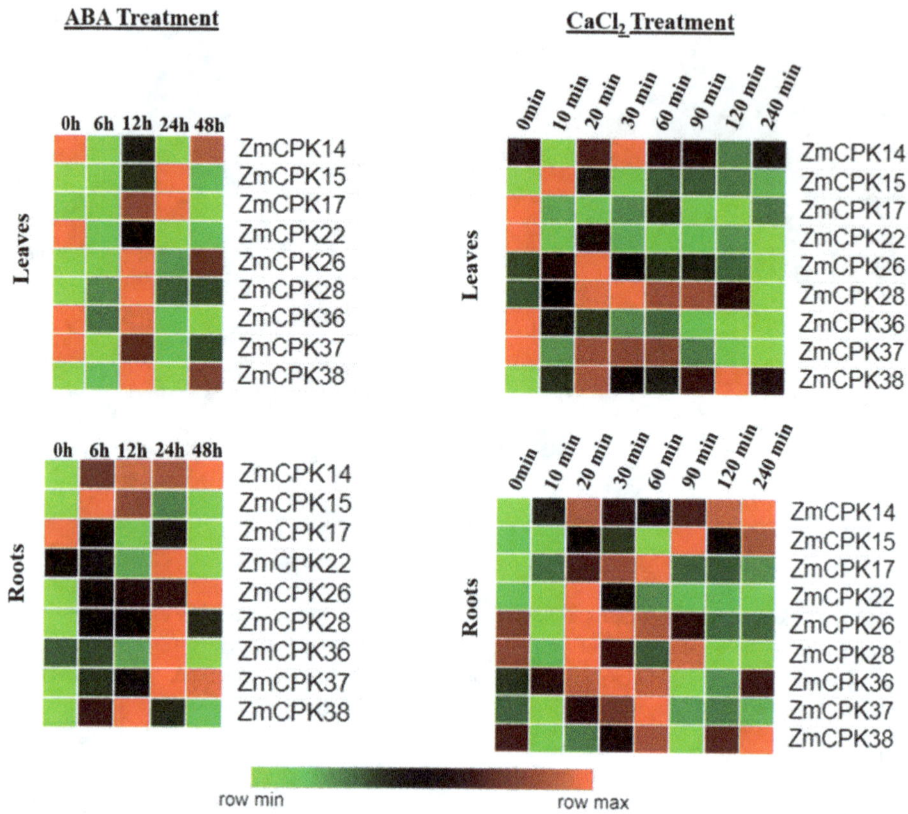

Figure 2. Heat maps illustrating the expression pattern of nine *ZmCPK* genes in maize B73 leaf and root under abscisic acid (ABA) and $CaCl_2$ treatment. The *ZmGAPDH* gene was used as an internal control. Both leaf and root tissues were collected at time points of 0, 6, 12, 24, and 48 h after ABA and at 0, 10, 20, 30, 60, 90, 120, and 240 min after $CaCl_2$ treatment.

2.2. Subcellular Localization of Selected ZmCPK Proteins

The expression site of the protein is closely related to its physiological function. Therefore, nine maize CPKs were fused with green fluorescent proteins regarding their subcellular localization in tobacco leaves. Among the tested *Zm*CPKs, *ZmCPK17*, 22, 28, and 38 were predicted to have a myristoylation motif, while ZmCPK14, 15, 26, 36, and 37 do not have myristoylation motif. In addition to the N-myristoylation motif, the palmitoylation site is also important for the membrane localization of CPK proteins. In this experiment, the non-myristoylated *Zm*CPKs were predicted to have palmitoylation sites. To verify the subcellular localization of these CPKs, eGFP fusion of these proteins was transiently expressed in tobacco leaves. The subcellular localization was analyzed by laser confocal microscopy. As shown in Figure 3, all nine *Zm*CPKs were localized to the cell membrane.

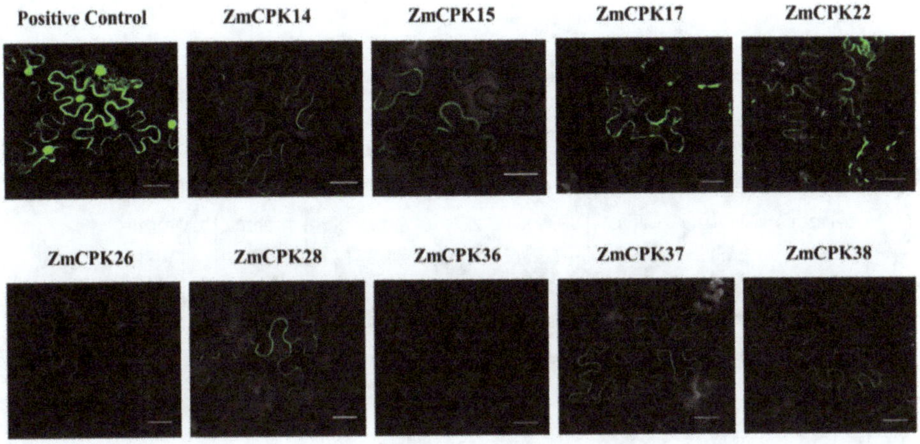

Figure 3. Subcellular localization of selected *Zm*CPK-GFP fusion proteins in *N. benthamiana* leaf epidermal cells. All the members of *Zm*CPKs were localized to the plasma membrane. eGFP was used as a positive control. Bars = 20 μm.

2.3. *Interaction Between ZmCPK Family and A Subclass ZmPP2C Family*

Prior to the yeast two-hybrid assay, all recombinant vectors involved in this experiment were first tested by a self-activation assay, in which the gene fragment was inserted into the vector pGADT7, and the pGBDT7 empty vector was co-transformed into yeast strain Y2HGold, and the gene fragment was inserted. The recombinant vector pGBDT7 was co-transformed with the pGADT7 empty vector into the yeast strain Y2HGold, and the growth target was used to observe whether the inserted target fragment has the function of activating the yeast two-hybrid system alone (Figure 4). All the vectors of the experiment had no self-activation effect. False positive results caused by self-activation are excluded.

There were 72 yeast double-hybrid interactions between nine *Zm*CPK genes and eight members of class A of *Zm*PP2C. *ZmPP2C1/3/4/5/7/8/9* and 10 were tested in these interactions in different ways and combinations (Figure 5). Seven combinations (*Zm*CPK15×*Zm*PP2C7, *Zm*CPK36 × *ZmPP2C3/4/7*and 9 and *ZmCPK38* × *ZmPP2C5* and 9) showed possible interaction by growing on a selective QDO (−Leu/−Trp/−His/−Ade, X-α-Gal) medium, but had no effect of X-α-Gal application, which raised a question on the authenticity and strength of these interactions, which could be due to inaccessibility of prey protein to the UAS region that is related to Gal4 binding site in rare cases. In this case, we needed to confirm the interaction using additional techniques, e.g., coimmunoprecipitation (coIP) or Bimolecular fluorescence complementation (BiFC). To confirm the interaction between the pairs of possibly interacted proteins in yeast, we performed BiFC in tobacco leaves. The results of BiFC confirmed that the interaction showed by pairs in Y2H were false positives. We replicated the BiFC experiment three times to exclude any mistake and found negative results, indicating that there is no interaction between *Zm*CPKs and *Zm*PP2Cs in tobacco leaves.

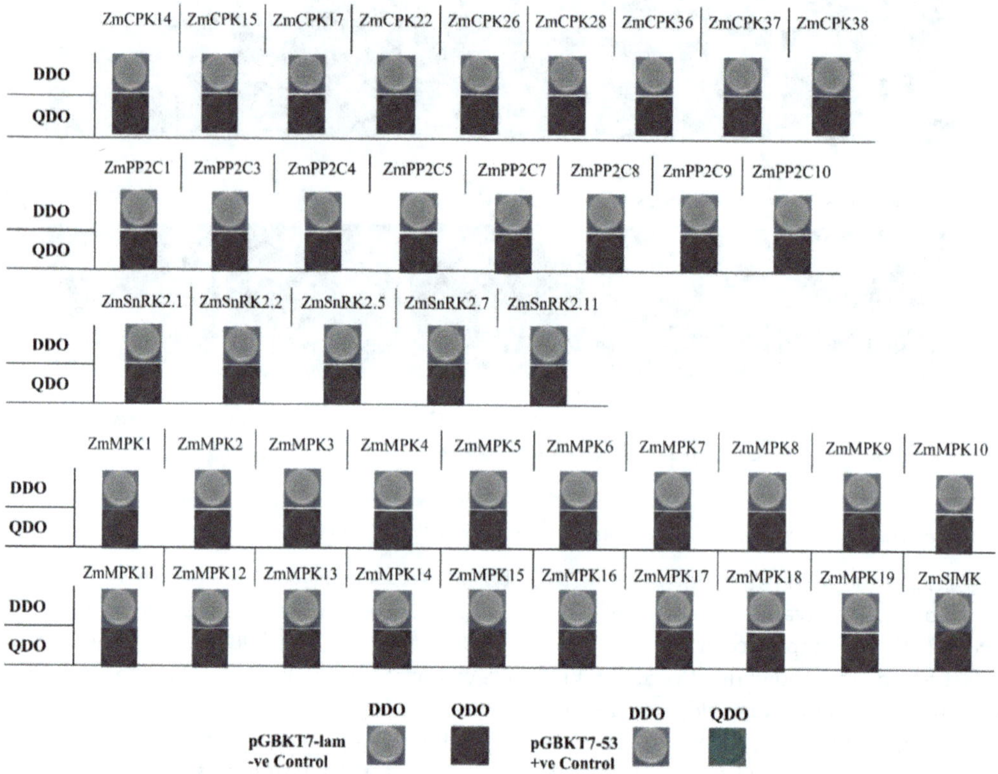

Figure 4. Self-activation detection of *Zm*CPK, *Zm*PP2C, *Zm*SnRK2, and *Zm*MPK family members for yeast two-hybrid assay. pGBKT7-lam and pGBKT7-53 were used as a negative and positive control, respectively. DDO medium = −Leu/−Trp, QDO = −Leu/−Trp/−His/−Ade/ + X-α-Gal.

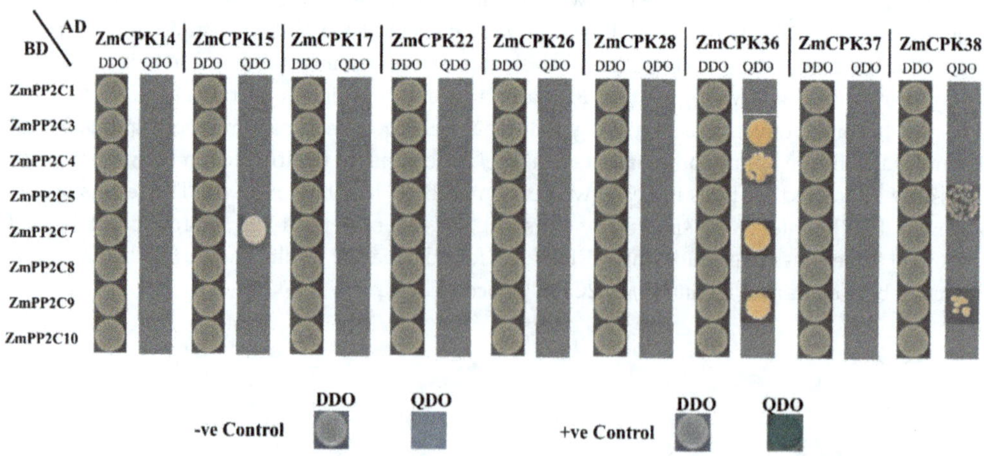

Figure 5. Y2H interaction assay between CPK and clade A *Zm*PP2C members. DDO medium lacking Leu/Trp, QDO medium lacking Leu/Trp/His/Ade with X-α-Gal. pGBKT7-Lam and pGBKT7-53 were used as −ve and +ve control, respectively. Pseudo interacting pairs grown on QDO are in brown color due to adenine deficiency and have no effect of X-α-Gal on them, indicating that these could be false positives.

2.4. Interaction of the ZmCPK Family and the ZmSnRK2 Family

There are 11 members in the *Zm*SnRK2 subclass. In this study, yeast two-hybrid assay was carried out with five *Zm*SnRK2s and nine *Zm*CPKs of maize. Based on the predicted interactions of the BrCPK-BrSnRK superfamily in Chinese cabbage [43], we used different combinations to verify the CPK-SnRK interaction in maize.

Yeast two-hybrid experiments were performed between *Zm*SnRK2 and nine *Zm*CPKs in 42 combinations. Out of all these combinations, only three combinations showed interactions. Those combinations *wereZmCPK38* × *ZmSnRK2*.1, *ZmCPK36* × *ZmSnRK2*.2, and *ZmCPK17* × *ZmSnRK2*.5 (Figure 6). All these combinations showed interaction with each other as they grew in the form of blue dots on QDO medium.

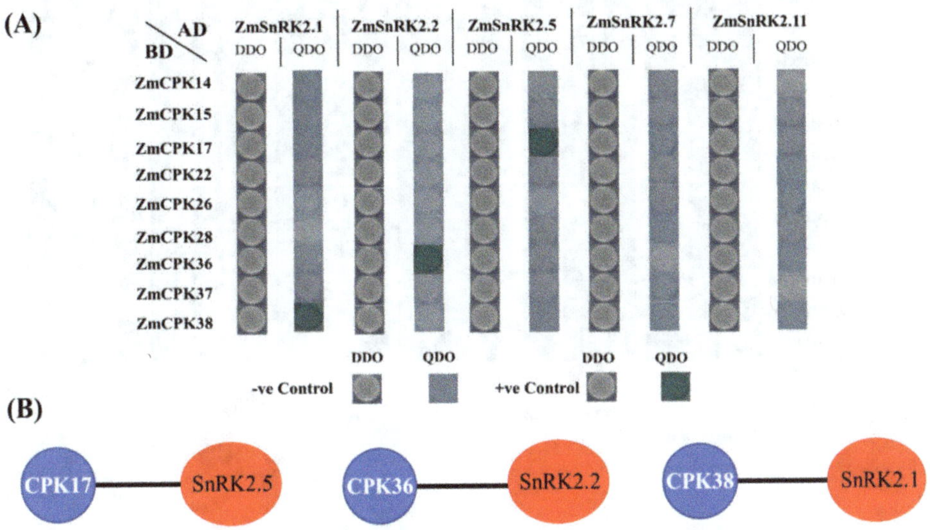

Figure 6. Y2H assay between *Zm*CPK and *Zm*SnRK2s. (**A**) Interaction was determined by the colony growth of the yeast strain Y2HGold co-transformed with these indicated plasmid combinations on DDO medium lacking Leu/Trp, QDO medium lacking–Leu/–Trp/–His/–Ade with X-α-Gal, pGBKT7-Lam and pGBKT7-53 were used as −ve and +ve controls, respectively. (**B**) Interactome of *Zm*CPK and *Zm*SnRK2 family members. The data are compiled from results presented in **A**.

2.5. Physical Interaction Between ZmCPK and ZmMPK Gene Families

In order to test the intersection of the calcium signaling pathway and MAPK signaling pathway, we tested the interaction between family members of *Zm*CPK and *Zm*MPK. Our results showed that there is a strong interaction between these two family members, as they activated in response to the same stimuli. In this study, nine members were cloned and constructed into a pGADT7 vector, and yeast two-hybrid assay was carried out with 20 members of *Zm*MPK of maize. *ZmCPK15*,17,28 and 38 interacted with *Zm*MPK family members in different combinations. Interactional pairs were *ZmCPK15* × *ZmMPK3/4/6/7/8/9/11/17/SIMK*, *ZmCPK17* × *ZmMPK3/6/13/14/19/SIMK*, *ZmCPK28* × *ZmMPK2/4/8*, and *ZmCPK38* × *ZmMPK3/4* (Figure 7).

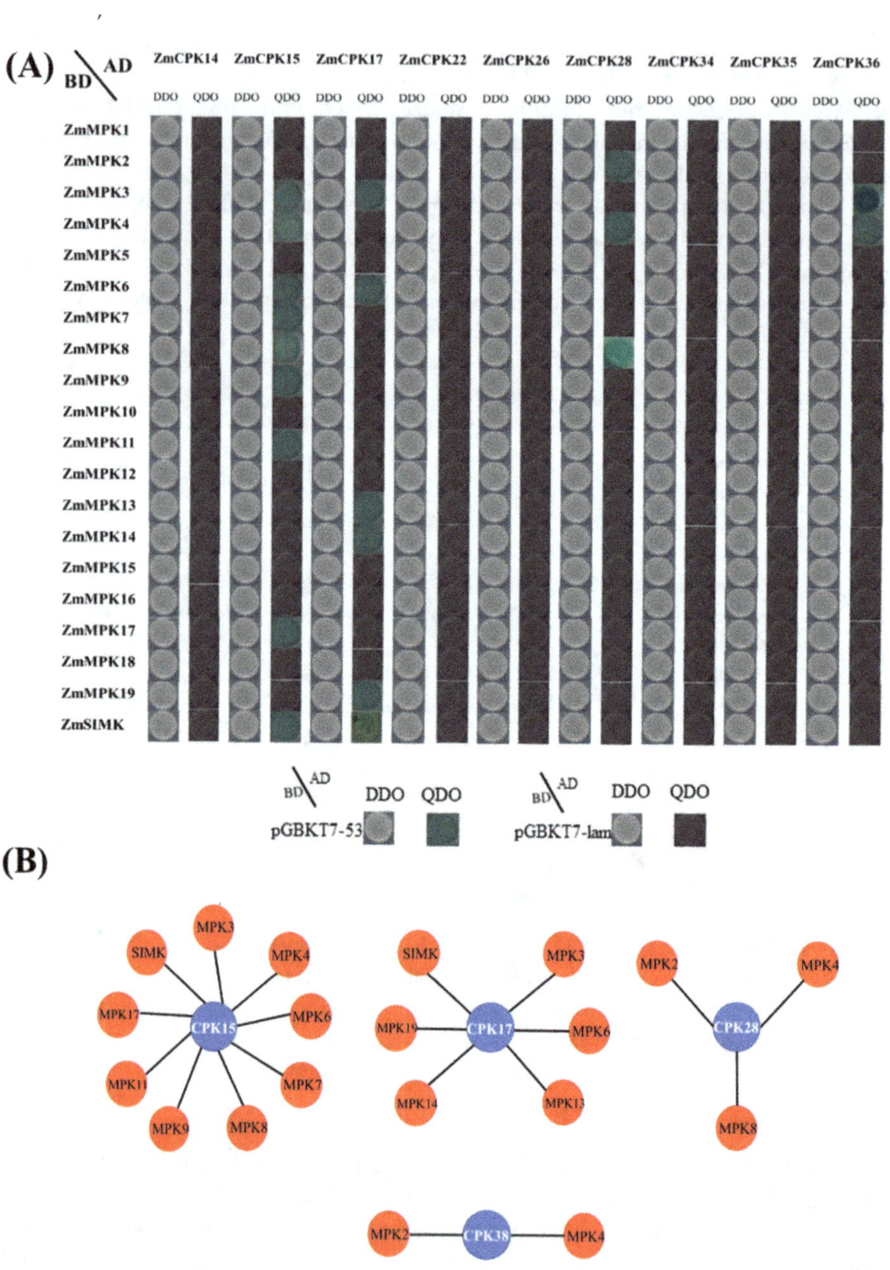

Figure 7. Y2H interaction assay between *Zm*CPK and *Zm*MPK members. (**A**) Interaction was determined by the colony growth of the yeast strain Y2HGold co-transformed with these indicated plasmid combinations on DDO medium lacking Leu/Trp and QDO medium lacking Leu/Trp/His/Ade with X-α-Gal. pGBKT7-Lam and pGBKT7-53 were used as −ve and +ve controls, respectively. (**B**) Interactome of *Zm*CPK and *Zm*MPK family members. The data were compiled from results presented in A.

3. Discussion

3.1. Expression Profiles of ZmCPKs

An increasing body of evidence has shown that CPKs regulate ABA signal transduction in plants [11,26,28]. *ZmCPK17* was expressed at high levels in ABA treatment, but downregulated in calcium treatment, which indicated that these two treatments have antagonistic effects on the expression of *ZmCPK17* in leaves, while in roots, they showed the same trend but were lower in transcription levels, as described in an earlier study [44]. *ZmCPK14*, 22, and 37 showed downregulation in leaves under ABA treatment and upregulation in calcium treatment, depicting the sensitivity of these genes towards calcium and ABA treatments. *ZmCPK15* belongs to group A of the *Zm*CPK family and is closely related to *AtCPK4* and *AtCPK11* [45]. Biphasic fluctuation in the expression of *ZmCPK15* under calcium and ABA treatments, indicating its role in defense mechanism through increasing the production of superoxide dismutase (SOD) and ascorbate peroxidase (APX) in maize protoplasts [41]. *ZmCPK26* and *ZmCPK28* showed slight fluctuation under these treatments, indicating a possible role similar to their ortholog in rice, and suggesting that these two genes are not sensitive to calcium treatment or their affinity towards calcium is not enough for expression alterations as reported in *Arabidopsis* [46,47]. This minute fluctuation in expression time-course analysis showed that calcium treatment and ABA treatment led to a significant increase and decrease of the expressions of these members, indicating their positive and negative roles in the downstream signaling process.

Given the expression pattern of the genes, all the genes have a heartbeat-like expression pattern at different time points. We can predict per expression levels of these genes that the first hour is the most important after abiotic stress treatments, i.e., ABA and $CaCl_2$, and these findings could be useful for researchers for detailed functional analysis.

3.2. *Sub-Cellular Localization of CPK Genes in Maize*

In plants, CPKs are widely distributed, having ubiquitous expression in roots, flowers, leaves, and siliques, etc. Some CPKs are expressed in multiple places, while others are specific with respect to their expression. [48–50] The widespread distribution of calcium-dependent protein kinases (CPKs) in plants indicates that they play significant roles in signaling transduction pathways by activating numerous substrates [51]. To examine the subcellular localizations of *Zm*CPKs, we used translational fusions with (GFP) green fluorescence protein. All the selected *Zm*CPKs were located in a plasma membrane (Figure 3). Most of the members of CPKs have predicted a N-myristoylation motif (asite with a gly residue at position 2), which is involved in membrane targeting. This co-translational acylation is irreversible and involves a second post-translational signal, such as reversible palmitoylation sites (a motif with cys residue at 4/5 position) to retain the membrane association [30,52]. *OsCPK18* showed that the N-terminus myristoylation site is vital for plasma membrane targeting [53]. *NtCPK5* lost its plasma membrane localization after three mutations (Gly2Ala or Cys4Ala or a double mutant of Gly2Ala/Cys4Ala) and showed distribution throughout the cell, suggesting the importance of N-myristoylation and palmitoylation sites for CPKs' plasma membrane localization. [8,54] Supportively, *AtCPK34* and *AtCPK2* were localized on the plasma membrane, but after point mutation at myristoylation/palmitoylation sites, abolished the plasma membrane localization and caused cytoplasmic distribution [55]. In soybeans, *GmCPK3* and GmCPK31 had N-Myristoylation sites and were localized in the plasma membrane of *Arabidopsis* mesophyll protoplasts [56]. Another biochemical study, which was in line with our study, revealed that CPK is highly Ca2+ sensitive and associated with plasma membranes [47]. As calcium-dependent protein kinases take part in various aspects of plant biology, from transcription to ion transport [57], the localization of these *Zm*CPKs indicated their possible role in signal transduction or ion transportation in or out of the cell.

3.3. *Interaction Analysis of ZmCPKs, ZmPP2Cs, and ZmSnRKs*

In plants, protein-protein interactions are influenced by many factors, such as tissue and organ specificity, stress response expression patterns, subcellular localization, protein-to-protein preference, and phosphorylation levels, etc. [58].

*Zm*CPK, *Zm*PP2C, and *Zm*SnRK family members are numerous, their expression patterns and subcellular localizations are different, and interactions show variability in different crops, resulting in the complexity of the upstream calcium and ABA signal network of maize [59]. This diversity and complexity are also a favorable regulation mechanism for maize to cope with the external adverse environment. In ABA signaling, members of the PP2C and SnRK family were found to have interactions in maize [60], and in contrast to our study, members of PP2C and CPK were found to interact in canola [61]. In this study, possible points of interaction between Calcium and ABA signaling were studied in maize. In contrast to the results in *Brassica napus* and *Arabidopsis*, selected candidates of *Zm*CPK and *Zm*PP2C had no interaction in maize, suggesting that they may or may not be involved in other cellular functions [61,62]. The SnRK2 family is the major kinase involved in ABA signaling, and is classified into two subclasses of SnRK2a (*ZmSnRK*2.1/2.2/2.3/2.8/2.10) and SnRK2b (*ZmSnRK*2.4/2.5/2.6/2.7/2.11) according to the degree of enrichment of its C-terminal amino acids [63–65]. Among them, the SnRK2a subclass was confirmed to be mainly involved in the binding to PP2C and the transmission of ABA signals [66–68]. Along with the interaction and phosphorylation of other substrates, two kinases can phosphorylate each other, or they could have a common substrate [67]. Our results suggested that two kinases (CPKs and SnRKs) can interact with each other in a way that may activate downstream signaling. In the present study, both the *Zm*SnRK2a and *Zm*SnRK2b subclasses were found to interact with *Zm*CPK. Subcellular localization of SnRK2s [60] and our results of subcellular localization support these results of Y2H interactions. It implies that ZmSnRK2 members might be the intersectional point of calcium and ABA signal transmission. Further functional dissection of these individual interactions between *Zm*CPKs and *Zm*SnRKs will be needed to completely understand their specific role in the growth and development of maize.

3.4. *Interaction of Calcium and MAPK Signaling Pathways*

Plant protein interactions are controlled by many factors, including stress and tissue or organ specificity. Traditional MAPK cascade recognizes MAPK kinase as the only kinase for signal transmission in yeast, animal, and plant systems. Efforts have been made to study calcium/CPK and MAPK signaling in plants, but the crosstalk between these important kinase signaling pathways have not been studied. Previous studies suggest that these two pathways work parallel to each other without having direct crosstalk in *Arabidopsis* and tobacco [39]. In this study, interactional analysis of CPK and MAPK members in maize indicated that the calcium signaling pathway can directly influence the MAPK signaling mechanism through CPKs. Our results suggested that *Zm*CPK 15, 17, 28 and 38 make direct contributions to the crosstalk between calcium and MAPK signaling mechanisms in rice as reported by [42]. Generally, MAPKs are characterized by the conserved TXY motifs in their T-loop and activated by dual phosphorylation. *OsCPK18* (an ortholog of *ZmCPK38*) interacted with *OsMPK5* (an ortholog of *ZmMPK4*) and was identified as activating OsMPK5 by phosphorylating two threonine residues without affecting the phosphorylation of conserved TXY (TEY & TDY) motifs [42]. This scheme may be followed in maize, but it is difficult to determine the interactional activity from the primary amino acid sequence, because it largely depends upon conformational diversity. Our results implied that the crosstalk between calcium and MAPK signaling is not only carried out indirectly, but also directly, through the physical interaction of CPK and MAPK in maize. Further intensive research on the function of the CPK-MAPK complex will help us to explore the roles of these individual complexes and improve our understanding of crosstalk between these pathways in plant growth and development, as well as responses in biotic and abiotic stresses. These findings suggest that MAPKs are not only phosphorylated by MAP kinase kinase (MKK) but that plants might evolve another pathway to phosphorylate MAPKs via CPKs.

4. Materials and Methods

4.1. *Plant Materials and Growth Conditions*

The seeds of the model maize inbred line B73 were sterilized, germinated in petri dishes, and transplanted into 1/4 concentration Hoagland medium for hydroponic culture, and the nutrient solution was replaced once every 3–4 days. The culture temperature was 18–25 °C, the photoperiod was 12 h (light)/12 h (dark), and the light intensity was 400 μmol/s·m^2.

4.2. ABA/$CaCl_2$ Treatment and RNA Extraction

As described in the literature, 100uM ABA and 10 mM $CaCl_2$ can significantly stimulate and mimic stress responses [41,60,69–71]. At the three-leaf stage, plants were treated with ABA and $CaCl_2$ by soaking roots in nutrient solution with 100 μM (±) ABA ((Sigma-Aldrich, Foster City, CA, USA)) and 10 mM $CaCl_2$. After 0 (Control), 6, 12, 24 and 48 h of ABA treatment and 0 (Control), 10, 20, 30, 60, 90, 120 and 240 min of calcium chloride treatment, roots and leaves were sampled and three biological replicates were set for each set of samples. Roots and leaves were sampled separately and frozen at −80 °C immediately after collection. RNAiso Plus (TaKaRa, Kusatsu, Japan) and the PrimeScript™II1st strand cDNA synthesis kit (Takara, Kusatsu, Japan) were used for total RNA extraction and reverse transcription for cDNA synthesis, respectively.

4.3. Quantitative Real-Time PCR

The Primer-BLAST tool (http://www.ncbi.nlm.nih.gov/tools/primer-blast) was used to design gene-specific primers to amplify a 100–300 bp fragment of each selected member of the *Zm*CPK gene family (Supplementary Table S1). The cDNA samples prepared above were used as templates for SYBR green-based qRT-PCR analysis. Each reaction mixture (10 μL) contained SYBR Premix Ex Taq II (Tli RNaseH Plus) 5 μL, F/R primer 0.5 μL, diluted 1 μL of cDNA, and ddH2O supplemented to 10 μL. The thermal cycling conditions were 95 °C for 30 sec followed by 39 cycles of 95 °C for 5 s, 55–60 °C for 30 s and 72 °C for 15 s. Referring to Lin et al. (2014) [72], the ZmGAPDH gene was used as an internal control. The $2^{-\Delta\Delta CT}$ method of the CFX Manger™ software version 2.0 (Bio-Rad, Philadelphia, PA, USA) was used to normalize the differential gene expression among the multiple internal controls and the target genes [72].

4.4. Subcellular Localization

Gene-specific primers were used to amplify the targeted coding sequences of nine *Zm*CPKs with appropriate restriction sites and without terminal codons from cDNA samples (Supplementary Table S2). Amplified fragments were ligated in-frame to the 5′-terminus with the expression vector pCambia2300-eGFP. Constructed plasmids were infiltrated into abaxial air space of three-week-old *N. benthamiana* leaves with the help of transformed *Agrobacterium* strain GV3101. Infiltrated parts of the leaves were marked and fluorescence was observed under the confocal laser scanning microscope model A1 (Nikon, Tokyo, Japan) after 48 h of infiltration.

4.5. Yeast Two-Hybrid Assay

The complete coding sequences of 9 *Zm*CPK family members from different phylogenetic groups, 8 *Zm*PP2C family, 5 genes of *Zm*SnRK family and 20 genes from the *Zm*MPK family (Figure 1) were amplified by using gene specific primers from the cDNA samples (primer information in Supplementary Table S3). These amplified fragments were inserted into bait (pGBKT7) and pray (pGADT7) vectors of the yeast two-hybrid assay after sequence confirmation by sequencing. Yeast strain Y2HGold was co-transferred in all possible combinations of pGBKT7-*Zm*CPK14, 15, 17, 22, 26, 28, 36, 37, 38 with pGADT7-*ZmSnRK*2.1, 2.2, 2.5, 2.7, 2.11, and pGADT7-*ZmCPK14*, 15, 17, 22, 26, 28, 36, 37, and 38 with pGBKT7-*ZmPP2C1*, 3–5, and 7–10, pGBKT7-*ZmMPK1*-19, and ZmSIMK (primer information in Supplementary Tables S4 and S5, MAPK identifiers in Supplementary Table S8), according to the protocol provided by Yeastmaker™ Yeast Transformation System 2 (Clontech, Japan). pGBKT7-53 and pGBKT7-lam were used as positive and negative controls, respectively. To exclude the possible autoactivation of *Zm*CPK members, a control experiment was carried out by co-transformation of loaded bait and prey plasmids with empty prey and bait plasmids, respectively. After screening on solid DDO (−Leu/−Trp) medium for 3 days at 30 °C, selected monoclones were inoculated in liquid DDO (−Leu/−Trp) medium until OD600 = 0.7−1.0. These cultures were inoculated on QDO (−Leu/−h−His/-Ade/ + X-α-Gal) medium after ten-fold dilution. Results were observed after 4–5 days of incubation at 30 °C.

4.6. Bimolecular Fluorescence Complementation

Coding sequences of *ZmCPK15/36/38* and *ZmPP2C3/4/5/7/9* were amplified by using two sets of primers (Supplementary Tables S6 and S7) and inserted into pSAT6-nEYFP-N1_(E2913) and pSAT6A-cEYPF-N1_(pE3086) yellow fluorescent protein plasmids without stop codon, respectively. These plasmids were transformed into Agrobacterium tumefaciens strain GV3101. As described by Wang et al. 2018 [60], transformed A. tumefaciens strains loaded with expression plasmids pSAT6-CPKs-nEYFP and pSAT6A-*Zm*PP2Cs-cEYPF were co-infiltrated into the abaxial air space of 22 days old *Nicotiana benthamiana* leaves at OD600 = 0.7:0.7. The epidermal cells of the tobacco leaves near the infiltrated sites were observed for fluorescence after 36 h of the co-infiltration under confocal laser scanning microscope model A1 (Nikon, Tokyo, Japan).

5. Conclusions

In this study, 9 members of the *Zm*CPK family were amplified from different groups on the basis of phylogenetic analysis. qRT-PCR analysis showed that most of the *Zm*CPK genes are responsive to both ABA and $CaCl_2$ treatments inferred by heat maps. Furthermore, subcellular localization of the candidate genes provided preliminary information for future investigation of the CPK family in maize. Moreover, three members of the *Zm*CPK family interacted with the members of *Zm*SnRK2 family that are active members in the core ABA signaling cascade. Four *Zm*CPK members were found to interact with thirteen *Zm*MPK members, depicting the complexity of signaling networks. The current investigation lays the groundwork for a functional characterization of the *Zm*CPK gene family to improve our understanding of complex signaling pathways.

Author Contributions: M.H.B.K. and W.C.L. conceived and designed the experiments; M.H.B.K. and H.Q.Y. performed the experiments; M.A.R. and I.K. analyzed the data; F.A.S., J.T.Q. and L.Y.F. contributed reagents/materials/analysis tools; M.H.B.K. wrote the paper; F.L.F. and W.C.L. supplied funds; M.H.B.K., W.C.L. and M.A.R. revised the paper.

Acknowledgments: We thank Dongtao Ren at China Agriculture University for providing pGBKT7-*Zm*MPK plasmids and technical support from the National Key Laboratory of Biology and Genetic Improvement of Maize in the Southwest region.

References

1. Zhu, J.K. Plant salt tolerance. *Trends Plant Sci.* **2001**, *6*, 66–71. [CrossRef]
2. Xiong, L.; Schumaker, K.S.; Zhu, J.K. Cell signaling during cold, drought, and salt stress. *Plant Cell* **2002**, *14*, S165–S183. [CrossRef]
3. Zhu, J.K. Salt and drought stress signal transduction in plants. *Annu. Rev. Plant Biol.* **2002**, *53*, 247–273. [CrossRef] [PubMed]
4. Evans, N.H.; McAinsh, M.R.; Hetherington, A.M. Calcium oscillations in higher plants. *Curr. Opin. Plant Biol.* **2001**, *4*, 415–420. [CrossRef]
5. Tuteja, N.; Mahajan, S. Calcium signaling network in plants: An overview. *Plant Signal. Behav.* **2007**, *2*, 79–85. [CrossRef] [PubMed]
6. Kudla, J.; Batistič, O.; Hashimoto, K. Calcium signals: The lead currency of plant information processing. *Plant Cell* **2010**, *22*, 541–563. [CrossRef] [PubMed]

7. Saijo, Y.; Hata, S.; Kyozuka, J.; Shimamoto, K.; Izui, K. Over-expression of a single Ca2+-dependent protein kinase confers both cold and salt/drought tolerance on rice plants. *Plant J.* **2000**, *23*, 319–327. [CrossRef]
8. Cheng, S.H.; Willmann, M.R.; Chen, H.C.; Sheen, J. Calcium Signaling through Protein Kinases. The Arabidopsis Calcium-Dependent Protein Kinase Gene Family. *Plant Physiol.* **2002**, *129*, 469–485. [CrossRef]
9. Harper, J.F.; Breton, G.; Harmon, A. Decoding Ca2+ signals through plant protein kinases. *Annu. Rev. Plant Biol.* **2004**, *55*, 263–288. [CrossRef]
10. Xiong, T.C.; Bourque, S.; Lecourieux, D.; Amelot, N.; Grat, S.; Brière, C.; Mazars, C.; Pugin, A.; Ranjeva, R. Calcium signaling in plant cell organelles delimited by a double membrane. *Biochim. Biophys. Acta Mol. Cell Res.* **2006**, *1763*, 1209–1215. [CrossRef]
11. Zhu, S.Y.; Yu, X.C.; Wang, X.J.; Rui, Z.; Yan, L.; Fan, R.C.; Yi, S.; Du, S.Y.; Wang, X.F.; Wu, F.Q.; et al. Two Calcium-Dependent Protein Kinases, CPK4 and CPK11, Regulate Abscisic Acid Signal Transduction in Arabidopsis. *Plant Cell* **2007**, *19*, 3019–3036. [CrossRef] [PubMed]
12. Mehlmer, N.; Wurzinger, B.; Stael, S.; Hofmannrodrigues, D.; Csaszar, E.; Pfister, B.; Bayer, R.; Teige, M. The Ca2+-dependent protein kinase CPK3 is required for MAPK-independent salt-stress acclimation in Arabidopsis. *Plant J.* **2010**, *63*, 484–498. [CrossRef] [PubMed]
13. Asano, T.; Hakata, M.; Nakamura, H.; Aoki, N.; Komatsu, S.; Ichikawa, H.; Hirochika, H.; Ohsugi, R. Functional characterisation of OsCPK21, a calcium-dependent protein kinase that confers salt tolerance in rice. *Plant Mol. Biol.* **2011**, *75*, 179–191. [CrossRef] [PubMed]
14. Kobayashi, M.; Yoshioka, M.; Asai, S.; Nomura, H.; Kuchimura, K.; Mori, H.; Doke, N.; Yoshioka, H. StCDPK5 confers resistance to late blight pathogen but increases susceptibility to early blight pathogen in potato via reactive oxygen species burst. *N. Phytol.* **2012**, *196*, 223–237. [CrossRef]
15. Ranty, B.; Aldon, D.; Cotelle, V.; Galaud, J.P.; Thuleau, P.; Mazars, C. Calcium sensors as key hubs in plant responses to biotic and abiotic stresses. *Front. Plant Sci.* **2016**, *7*, 327. [CrossRef]
16. Pennisi, E. *Stressed Out over a Stress Hormone*; American Association for the Advancement of Science: Washington, DC, USA, 2009.
17. Cutler, S.R.; Rodriguez, P.L.; Finkelstein, R.R.; Abrams, S.R. Abscisic acid: Emergence of a core signaling network. *Annu. Rev. Plant Biol.* **2010**, *61*, 651–679. [CrossRef]
18. Allan, A.C.; Fricker, M.D.; Ward, J.L.; Beale, M.H.; Trewavas, A.J. Two transduction pathways mediate rapid effects of abscisic acid in Commelina guard cells. *Plant Cell* **1994**, *6*, 1319–1328. [CrossRef]
19. Levchenko, V.; Konrad, K.R.; Dietrich, P.; Roelfsema, M.R.G.; Hedrich, R. Cytosolic abscisic acid activates guard cell anion channels without preceding Ca2+ signals. *Proc. Natl. Acad. Sci. USA* **2005**, *102*, 4203–4208. [CrossRef]
20. McAinsh, M.R.; Brownlee, C.; Hetherington, A.M. Abscisic acid-induced elevation of guard cell cytosolic Ca2+ precedes stomatal closure. *Nature* **1990**, *343*, 186–188. [CrossRef]
21. Sheen, J. Ca2+-dependent protein kinases and stress signal transduction in plants. *Science* **1996**, *274*, 1900–1902. [CrossRef]
22. MacRobbie, E.A.C. ABA activates multiple Ca2+ fluxes in stomatal guard cells, triggering vacuolar K+ (Rb+) release. *Proc. Natl. Acad. Sci. USA* **2000**, *97*, 12361–12368. [CrossRef] [PubMed]
23. Köhler, B.; Blatt, M.R. Protein phosphorylation activates the guard cell Ca2+ channel and is a prerequisite for gating by abscisic acid. *Plant J.* **2002**, *32*, 185–194. [CrossRef] [PubMed]
24. Siegel, R.S.; Xue, S.; Murata, Y.; Yang, Y.; Nishimura, N.; Wang, A.; Schroeder, J.I. Calcium elevation-dependent and attenuated resting calcium-dependent abscisic acid induction of stomatal closure and abscisic acid-induced enhancement of calcium sensitivities of S-type anion and inward-rectifying K+ channels in Arabidopsis guard cells. *Plant J.* **2009**, *59*, 207–220. [CrossRef] [PubMed]
25. Choi, H.I.; Park, H.J.; Park, J.H.; Kim, S.; Im, M.-Y.; Seo, H.H.; Kim, Y.W.; Hwang, I.; Kim, S.Y. Arabidopsis calcium-dependent protein kinase AtCPK32 interacts with ABF4, a transcriptional regulator of abscisic acid-responsive gene expression, and modulates its activity. *Plant Physiol.* **2005**, *139*, 1750–1761. [CrossRef]
26. Mori, I.C.; Murata, Y.; Yang, Y.; Munemasa, S.; Wang, Y.F.; Andreoli, S.; Tiriac, H.; Alonso, J.M.; Harper, J.F.; Ecker, J.R. CDPKs CPK6 and CPK3 function in ABA regulation of guard cell S-type anion-and Ca2+-permeable channels and stomatal closure. *PLoS Biol.* **2006**, *4*, e327. [CrossRef]
27. Ma, S.Y.; Wu, W.H. AtCPK23 functions in Arabidopsis responses to drought and salt stresses. *Plant Mol. Biol.* **2007**, *65*, 511–518. [CrossRef]

28. Zou, J.J.; Wei, F.J.; Wang, C.; Wu, J.J.; Ratnasekera, D.; Liu, W.X.; Wu, W.H. Arabidopsis calcium-dependent protein kinase AtCPK10 functions in ABA and Ca2+-mediated stomatal regulation in response to drought stress. *Plant Physiol.* **2010**, *154*, 1232–1243. [CrossRef]
29. Franz, S.; Ehlert, B.; Liese, A.; Kurth, J.; Cazalé, A.C.; Romeis, T. Calcium-dependent protein kinase CPK21 functions in abiotic stress response in *Arabidopsis thaliana. Mol. Plant* **2011**, *4*, 83–96. [CrossRef]
30. Jiang, S.; Zhang, D.; Wang, L.; Pan, J.; Liu, Y.; Kong, X.; Zhou, Y.; Li, D. Biochemistry. A maize calcium-dependent protein kinase gene, ZmCPK4, positively regulated abscisic acid signaling and enhanced drought stress tolerance in transgenic Arabidopsis. *Plant Physiol. Biochem.* **2013**, *71*, 112–120. [CrossRef]
31. Sanders, D.; Pelloux, J.; Brownlee, C.; Harper, J.F. Calcium at the crossroads of signaling. *Plant Cell* **2002**, *14*, S401–S417. [CrossRef]
32. Wurzinger, B.; Mair, A.; Pfister, B.; Teige, M. Cross-talk of calcium-dependent protein kinase and MAP kinase signaling. *Plant Signal. Behav.* **2011**, *6*, 8–12. [CrossRef] [PubMed]
33. Petersen, M.; Brodersen, P.; Naested, H.; Andreasson, E.; Lindhart, U.; Bo, J.; Nielsen, H.B.; Lacy, M.; Austin, M.J.; Parker, J.E. Arabidopsis MAP Kinase 4 Negatively Regulates Systemic Acquired Resistance. *Cell* **2000**, *103*, 1111–1120. [CrossRef]
34. Asai, T.; Tena, G.; Plotnikova, J.; Willmann, M.R.; Chiu, W.L.; Gomez-Gomez, L.; Boller, T.; Ausubel, F.M.; Sheen, J. MAP kinase signalling cascade in Arabidopsis innate immunity. *Nature* **2002**, *415*, 977–983. [CrossRef] [PubMed]
35. Mészáros, T.; Helfer, A.; Hatzimasoura, E.; Magyar, Z.; Serazetdinova, L.; Rios, G.; Bardóczy, V.; Teige, M.; Koncz, C.; Peck, S. The Arabidopsis MAP kinase kinase MKK1 participates in defence responses to the bacterial elicitor flagellin. *Plant J.* **2006**, *48*, 485–498. [CrossRef] [PubMed]
36. Dóczi, R.; Brader, G.; Pettkószandtner, A.; Rajh, I.; Djamei, A.; Pitzschke, A.; Teige, M.; Hirt, H. The Arabidopsis Mitogen-Activated Protein Kinase Kinase MKK3 Is Upstream of Group C Mitogen-Activated Protein Kinases and Participates in Pathogen Signaling. *Plant Cell* **2007**, *19*, 3266–3279. [CrossRef] [PubMed]
37. Agell, N.; Bachs, O.; Rocamora, N.; Villalonga, P. Modulation of the Ras/Raf/MEK/ERK pathway by Ca 2+, and Calmodulin. *Cell. Signal.* **2002**, *14*, 649–654. [CrossRef]
38. Rozengurt, E. Mitogenic signaling pathways induced by G protein-coupled receptors. *J. Cell. Physiol.* **2007**, *213*, 589–602. [CrossRef]
39. Boudsocq, M.; Willmann, M.; McCormack, M.; Lee, H.; Shan, L.; He, P.; Bush, J.; Cheng, S.; Sheen, J. Differential innate immune signalling via Ca2+ sensor protein kinases in plants. *Nature* **2010**, *464*, 418–422. [CrossRef]
40. Ludwig, A.A.; Saitoh, H.; Felix, G.; Freymark, G.; Miersch, O.; Wasternack, C.; Boller, T.; Jones, J.D.G.; Romeis, T. Ethylene-mediated cross-talk between calcium-dependent protein kinase and MAPK signaling controls stress responses in plants. *Proc. Natl. Acad. Sci. USA* **2005**, *102*, 10736–10741. [CrossRef]
41. Ding, Y.; Cao, J.; Ni, L.; Zhu, Y.; Zhang, A.; Tan, M.; Jiang, M. ZmCPK11 is involved in abscisic acid-induced antioxidant defence and functions upstream of ZmMPK5 in abscisic acid signalling in maize. *J. Exp. Bot.* **2013**, *64*, 871–884. [CrossRef]
42. Xie, K.; Yang, Y. Direct phosphorylation and activation of a mitogen-activated protein kinase by a calcium-dependent protein kinase in rice. *Plant Cell* **2014**, *26*, 3077–3089. [CrossRef] [PubMed]
43. Wu, P.; Wang, W.; Duan, W.; Li, Y.; Hou, X. Comprehensive Analysis of the CDPK-SnRK Superfamily Genes in Chinese Cabbage and Its Evolutionary Implications in Plants. *Front. Plant Sci.* **2017**, *8*, 162. [CrossRef] [PubMed]
44. Kong, X.; Lv, W.; Jiang, S.; Zhang, D.; Cai, G.; Pan, J.; Li, D. Genome-wide identification and expression analysis of calcium-dependent protein kinase in maize. *BMC Genom.* **2013**, *14*, 433. [CrossRef] [PubMed]
45. Boudsocq, M.; Sheen, J. CDPKs in immune and stress signaling. *Trends Plant Sci.* **2013**, *18*, 30–40. [CrossRef] [PubMed]
46. Morello, L.; Bardini, M.; Cricrì, M.; Sala, F.; Breviario, D. Functional analysis of DNA sequences controlling the expression of the rice OsCDPK2 gene. *Planta* **2006**, *223*, 479–491. [CrossRef] [PubMed]
47. Boudsocq, M.; Droillard, M.J.; Regad, L.; Laurière, C. Characterization of Arabidopsis calcium-dependent protein kinases: Activated or not by calcium? *Biochem. J.* **2012**, *447*, 291–299. [CrossRef]
48. Hong, Y.; Takano, M.; Liu, C.M.; Gasch, A.; Chye, M.L.; Chua, N.H. Expression of three members of the calcium-dependent protein kinase gene family in *Arabidopsis thaliana. Plant Mol. Biol.* **1996**, *30*, 1259–1275. [CrossRef]

49. Ye, S.; Wang, L.; Xie, W.; Wan, B.; Li, X.; Lin, Y. Expression profile of calcium-dependent protein kinase (CDPKs) genes during the whole lifespan and under phytohormone treatment conditions in rice (*Oryza sativa* L. ssp. indica). *Plant Mol. Biol.* **2009**, *70*, 311–325. [CrossRef]
50. Monaghan, J.; Matschi, S.; Shorinola, O.; Rovenich, H.; Matei, A.; Segonzac, C.; Malinovsky, F.G.; Rathjen, J.P.; Maclean, D.; Romeis, T.; et al. The calcium-dependent protein kinase CPK28 buffers plant immunity and regulates BIK1 turnover. *Cell Host Microbe* **2014**, *16*, 605–615. [CrossRef]
51. Rudd, J.J.; Franklin Tong, V.E. Unravelling response-specificity in Ca2+ signalling pathways in plant cells. *N. Phytol.* **2001**, *151*, 7–33. [CrossRef]
52. Campo, S.; Baldrich, P.; Messeguer, J.; Lalanne, E.; Coca, M.; San, S.B. Overexpression of a Calcium-Dependent Protein Kinase Confers Salt and Drought Tolerance in Rice by Preventing Membrane Lipid Peroxidation. *Plant Physiol.* **2014**, *165*, 688–704. [CrossRef] [PubMed]
53. Campos-Soriano, L.; Gómez-Ariza, J.; Bonfante, P.; San Segundo, B. A rice calcium-dependent protein kinase is expressed in cortical root cells during the presymbiotic phase of the arbuscular mycorrhizal symbiosis. *BMC Plant Biol.* **2011**, *11*, 90. [CrossRef] [PubMed]
54. Yun, W.; Zhang, M.; Ke, K.; Lu, Y.T. Cellular localization and biochemical characterization of a novel calcium-dependent protein kinase from tobacco. *Cell Res.* **2005**, *15*, 604–612.
55. Gutermuth, T.; Lassig, R.; Portes, M.T.; Maierhofer, T.; Romeis, T.; Borst, J.W.; Hedrich, R.; Feijó, J.A.; Konrad, K.R. Pollen tube growth regulation by free anions depends on the interaction between the anion channel SLAH3 and calcium-dependent protein kinases CPK2 and CPK20. *Plant Cell* **2013**, *25*, 4525–4543. [CrossRef]
56. Liu, H.; Che, Z.; Zeng, X.; Zhou, X.; Sitoe, H.M.; Wang, H.; Yu, D.J.F.; Genomics, I. Genome-wide analysis of calcium-dependent protein kinases and their expression patterns in response to herbivore and wounding stresses in soybean. *Funct. Integr. Genom.* **2016**, *16*, 481–493. [CrossRef]
57. Curran, A.; Chang, I.F.; Chang, C.L.; Garg, S.; Miguel, R.M.; Barron, Y.D.; Li, Y.; Romanowsky, S.; Cushman, J.C.; Gribskov, M. Calcium-dependent protein kinases from Arabidopsis show substrate specificity differences in an analysis of 103 substrates. *Front. Plant Sci.* **2011**, *2*, 36. [CrossRef]
58. Klimecka, M.; Muszynska, G. Structure and functions of plant calcium-dependent protein kinases. *Acta Biochim. Pol. Engl. Ed.* **2007**, *54*, 219–233. [CrossRef]
59. Soon, F.F.; Ng, L.M.; Zhou, X.E.; West, G.M.; Kovach, A.; Tan, M.E.; Suino-Powell, K.M.; He, Y.; Xu, Y.; Chalmers, M.J.; et al. Molecular mimicry regulates ABA signaling by SnRK2 kinases and PP2C phosphatases. *Science* **2012**, *335*, 85–88. [CrossRef]
60. Wang, Y.G.; Fu, F.L.; Yu, H.Q.; Hu, T.; Zhang, Y.Y.; Tao, Y.; Zhu, J.K.; Zhao, Y.; Li, W.C. Interaction network of core ABA signaling components in maize. *Plant Mol. Biol.* **2018**, *96*, 245–263. [CrossRef]
61. Zhang, H.; Liu, W.Z.; Zhang, Y.; Deng, M.; Niu, F.; Yang, B.; Wang, X.; Wang, B.; Liang, W.; Deyholos, M.K.; et al. Identification, expression and interaction analyses of calcium-dependent protein kinase (CPK) genes in canola (*Brassica napus* L.). *BMC Genom.* **2014**, *15*, 211. [CrossRef]
62. Lynch, T.; Erickson, B.J.; Finkelstein, R.R. Direct interactions of ABA-insensitive (ABI)-clade protein phosphatase (PP) 2Cs with calcium-dependent protein kinases and ABA response element-binding bZIPs may contribute to turning off ABA response. *Plant Mol. Biol.* **2012**, *80*, 647–658. [CrossRef] [PubMed]
63. Belin, C.; De Franco, P.O.; Bourbousse, C.; Chaignepain, S.; Schmitter, J.M.; Vavasseur, A.; Giraudat, J.; Barbier-Brygoo, H.; Thomine, S. Identification of features regulating OST1 kinase activity and OST1 function in guard cells. *Plant Physiol.* **2006**, *141*, 1316–1327. [CrossRef] [PubMed]
64. Yoshida, R.; Umezawa, T.; Mizoguchi, T.; Takahashi, S.; Takahashi, F.; Shinozaki, K. The regulatory domain of SRK2E/OST1/SnRK2. 6 interacts with ABI1 and integrates abscisic acid (ABA) and osmotic stress signals controlling stomatal closure in Arabidopsis. *J. Biol. Chem.* **2006**, *281*, 5310–5318. [CrossRef] [PubMed]
65. Yoshida, T.; Nishimura, N.; Kitahata, N.; Kuromori, T.; Ito, T.; Asami, T.; Shinozaki, K.; Hirayama, T. ABA-hypersensitive germination3 encodes a protein phosphatase 2C (AtPP2CA) that strongly regulates abscisic acid signaling during germination among Arabidopsis protein phosphatase 2Cs. *Plant Physiol.* **2006**, *140*, 115–126. [CrossRef] [PubMed]
66. Mustilli, A.C.; Merlot, S.; Vavasseur, A.; Fenzi, F.; Giraudat, J. Arabidopsis OST1 Protein Kinase Mediates the Regulation of Stomatal Aperture by Abscisic Acid and Acts Upstream of Reactive Oxygen Species Production. *Plant Cell* **2002**, *14*, 3089–3099. [CrossRef]

67. Vlad, F.; Rubio, S.; Rodrigues, A.; Sirichandra, C.; Belin, C.; Robert, N.; Leung, J.; Rodriguez, P.L.; Laurière, C.; Merlot, S. Protein Phosphatases 2C Regulate the Activation of the Snf1-Related Kinase OST1 by Abscisic Acid in Arabidopsis. *Plant Cell* **2009**, *21*, 3170–3184. [CrossRef]
68. Kulik, A.; Wawer, I.; Krzywińska, E.; Bucholc, M.; Dobrowolska, G. SnRK2 Protein Kinases—Key Regulators of Plant Response to Abiotic Stresses. *OMICS J. Integr. Biol.* **2011**, *15*, 859–872. [CrossRef]
69. Xu, C.; Li, X.; Zhang, L. The effect of calcium chloride on growth, photosynthesis, and antioxidant responses of Zoysia japonica under drought conditions. *PLoS ONE* **2013**, *8*, e68214. [CrossRef]
70. Yang, C.; Liu, J.; Dong, X.; Cai, Z.; Tian, W.; Wang, X. Short-term and continuing stresses differentially interplay with multiple hormones to regulate plant survival and growth. *Mol. Plant* **2014**, *7*, 841–855. [CrossRef]
71. Patykowski, J.; Kołodziejek, J.; Wala, M. Biochemical and growth responses of silver maple (*Acer saccharinum* L.) to sodium chloride and calcium chloride. *PeerJ* **2018**, *6*, e5958. [CrossRef]
72. Lin, F.; Jiang, L.; Liu, Y.; Lv, Y.; Dai, H.; Zhao, H. Genome-wide identification of housekeeping genes in maize. *Plant Mol. Biol.* **2014**, *86*, 543–554. [CrossRef] [PubMed]

5

Calcium Signaling in Vertebrate Development and its Role in Disease

Sudip Paudel †, Regan Sindelar † and Margaret Saha *

College of William and Mary, Williamsburg, VA 23187, USA; spaudel@email.wm.edu (S.P.); rssindelar@email.wm.edu (R.S.)

* Correspondence: mssaha@wm.edu

† These authors contributed equally to this work.

Abstract: Accumulating evidence over the past three decades suggests that altered calcium signaling during development may be a major driving force for adult pathophysiological events. Well over a hundred human genes encode proteins that are specifically dedicated to calcium homeostasis and calcium signaling, and the majority of these are expressed during embryonic development. Recent advances in molecular techniques have identified impaired calcium signaling during development due to either mutations or dysregulation of these proteins. This impaired signaling has been implicated in various human diseases ranging from cardiac malformations to epilepsy. Although the molecular basis of these and other diseases have been well studied in adult systems, the potential developmental origins of such diseases are less well characterized. In this review, we will discuss the recent evidence that examines different patterns of calcium activity during early development, as well as potential medical conditions associated with its dysregulation. Studies performed using various model organisms, including zebrafish, *Xenopus*, and mouse, have underscored the critical role of calcium activity in infertility, abortive pregnancy, developmental defects, and a range of diseases which manifest later in life. Understanding the underlying mechanisms by which calcium regulates these diverse developmental processes remains a challenge; however, this knowledge will potentially enable calcium signaling to be used as a therapeutic target in regenerative and personalized medicine.

Keywords: calcium; development; embryo; human disease; animal model; *Xenopus*; zebrafish; mouse

---

1. Introduction

Calcium is an ancient and ubiquitous signaling ion involved in a wide array of physiological processes throughout the entire lifespan of the individual, from fertilization through senescence. Given its central position in virtually every cellular process and tissue type, it is not surprising that its dysregulation leads to a wide spectrum of pathophysiological conditions. These conditions include, but are not limited to, infertility, miscarriage, developmental defects, neuropathic pain, epilepsy and seizures [1].

Under normal conditions, the basal intracellular concentration of calcium ion (Ca^{2+}) is maintained at very low levels compared to the extracellular spaces by a number of proteins, including pumps and transporters that have Ca^{2+} binding ability. Calcium pumps, using ATP, remove Ca^{2+} from the cytosol and help cells maintain roughly hundred to thousand-fold concentration gradients between intracellular and extracellular environments. Ca^{2+}-ATPases are an example of these pumps, which are localized in various cellular organelles; for example, P2A ATPase is localized in the endoplasmic reticulum, P2B ATPase is localized in the plasma membrane, and P2C ATPase is localized in the Golgi apparatus [2–4]. In addition, Ca^{2+} enters the cytosol passively from the extracellular matrix, neighboring cells, and cytoplasmic Ca^{2+} stores; the opening and closing of these channels is driven by both chemical and electrical gradients. Cells create these gradients in order to use calcium as a signaling molecule [3]. During calcium-mediated signaling events, Ca^{2+} permeant channels open transiently allowing spontaneous influx of Ca^{2+} into the cytosol. This influx of Ca^{2+} increases intracellular concentration of Ca^{2+}, and is often referred to generically as a calcium transient [5–7]. The short transient fluctuation in cytoplasmic Ca^{2+} within a given (typically isolated) cell due to these influxes are often defined as calcium spikes [8]. When a calcium spike originates from one or a few cells and then propagates along several other neighboring cells via gap junctions or paracrine signaling, it is called an intercellular wave [5,8–10]. The change in concentration of cytoplasmic Ca^{2+} is sensed by a number of calcium sensing proteins with calcium binding affinities ranging from hundreds to million-fold [11]. These calcium sensing proteins transduce calcium signals, decoding spatial and temporal changes in amplitude, duration and localization of calcium activity [7,11,12].

In turn, these signaling cascades drive a wide array of physiological responses throughout the life of the organism from fertilization onwards. While the role of calcium and its associated pathologies have been intensively studied in the adult organism, less attention has been focused on the role of calcium and its relationship to disease during embryonic development. Given the importance of calcium activity as a regulator of embryonic development from fertilization through organogenesis, dysregulation of these calcium dynamics, due to either mutations or epigenetic developmental dysregulation of some element or component of the calcium signaling process, results in various diseases and disorders (Supplementary Table S1). For example, Timothy syndrome is caused by constitutive opening of voltage gated calcium channel CaV1.2 due to mutations in the *cacna1c* gene. This gene is expressed embryonically and the syndrome is characterized by syndactyly, intellectual disability, congenital heart defects, distinctive facial features and developmental delay. Similarly, mutations in the gene ATP2A1 (sarco(endo)plasmic reticulum calcium-ATPase 1 (SERCA1)) on chromosome 16p11 result in Brody myopathy, which is characterized by a decrease or loss of sarcoplasmic reticulum Ca^{2+}-ATPase activity and problems with muscle contraction [13]. While some of these mutations and dysregulated processes are embryonic lethal, many manifest their effects at birth, and others may not show symptoms until later in life due to their indirect effects within a complex genetic network [14].

Calcium activity during development is varied and complex with embryos exhibiting different patterns of spikes and waves. Animal model studies during early stages of development have provided a broad understanding of human developmental defects and diseases related to the dysregulation of calcium activity. In this review, we will provide an overview of the current state of knowledge regarding the role of calcium activity in embryonic and fetal development and disease. Given the obvious challenges of studying calcium activity in human embryonic development, much of the information we will discuss derives from model systems, particularly frogs, fish, and mice. We will discuss each stage of development from fertilization through organogenesis chronologically. Each section will begin with a brief overview of the key developmental events that occur during that particular stage and then proceed to analyze the role of calcium in those processes, including how dysregulation of calcium dynamics can, and does, lead to disease.

2. Calcium Activity during Development and Its Role in Disease

2.1. *Fertilization and Egg Activation*

Fertilization is the process by which DNA of the sperm and egg unite to give rise to a new diploid organism. Sperm entry then triggers the oocyte to transition into a developing embryo in a process known as egg activation. Egg activation is characterized by the occurrence of a number of sequential events in the oocyte during fertilization: recruitment of maternal mRNA and formation of polysomes, completion of meiosis, modification of the plasma membrane and zona pellucida in order to prevent polyspermy, cortical granule exocytosis, formation of male and female proneuclei, and syngamy, the fusion of two genomes [15,16]. While species differences exist, the process of egg activation is a relatively conserved mechanism that is mediated and coordinated by calcium; failure in any step of this process typically results in infertility.

The importance of calcium activity in the process of fertilization and egg activation cannot be underestimated. Fertilization initiates elevations of intracellular Ca^{2+} concentration in all vertebrate oocytes studied to date [17]. These elevations are initiated from the site of sperm-egg fusion, and are caused by transient influxes of Ca^{2+} from both the extracellular milieu and intracellular calcium stores. The patterns of these influxes do vary somewhat across species. For example, the oocytes from some lower vertebrates such as *Xenopus* and zebrafish achieve this elevation via a single calcium transient, while mammalian oocytes exhibit an initial transient increase within a few minutes of the sperm binding to the egg surface, followed by subsequent oscillations in cytoplasmic Ca^{2+} concentration at 20 to 30 min intervals [17–19]. This calcium activity was visualized for the first time in a mammalian egg by imaging zona-free mouse oocytes using aequorin during in vitro fertilization [20]. Similar calcium behavior was observed in mouse and human oocytes during in vitro fertilization and intracytoplasmic sperm injection using aequorin or various other calcium sensitive dyes [21,22]. Inhibition of this calcium activity results in fertilization failure. For example, when extracellular Ca^{2+} was restricted from entering the oocyte cytoplasm using bivalent cation chelators such as BAPTA or EGTA, calcium insulators such as gadolinium, or a Ca^{2+} free culture medium, *Xenopus* oocytes failed to develop normal cleavage furrows, and mouse oocytes failed to resume meiosis during the egg activation process [18,23,24]. Similarly, wild type sperm was unable to activate zona-free mouse eggs lacking a gene (cacna1h$^{-/-}$) that codes for a T-type voltage gated calcium channel, which mediates extracellular calcium entry [25]. Pharmacological inhibition of this channel, using mibefradil or pimozide, also resulted in a phenotype similar to cacna1h$^{-/-}$ following in vitro fertilization [25]. These observations suggest that impaired entry of calcium from extracellular matrix to the oocyte cytoplasm may result in fertilization failure and lead to infertility.

Sperm specific phospholipase C, such as phospholipase C isozyme zeta, on its own is sufficient to trigger calcium release from intracellular calcium stores via channels, including STIM1, in order to accomplish egg activation in different species [17]. Mouse eggs injected with sperm-supernatants treated with an anti-phospholipase C zeta antibody failed to show Ca^{2+} elevation and subsequent egg activation events. However, when the eggs were subsequently microinjected with only sperm specific phospholipase C zeta, normal elevation of Ca^{2+} was observed and egg activation was rescued [26, 27]. These observations have been further strengthened by recent gene editing technology. For example, intracytoplasmic sperm injection or in vivo fertilization with phospholipase C zeta knockout sperm failed to trigger egg activation in wild type mouse eggs, but subsequent injection of wild type phospholipase C zeta-RNA successfully rescued egg activation and zygotic development [28]. Various measures, including histological analysis, sperm viability, sperm motility and hyperactivity assays, showed that phospholipase C zeta gene knockout using CRISPR/Cas9 results in no defects in spermatogenesis [28]. Similarly, the knock down of STIM1 from pig oocytes using STIM1 specific siRNA resulted in failure to generate Ca^{2+} oscillations, decreased oocyte survival rate, and decreased efficiency of cleavage and blastocyst development [29]. Taken together, these observations suggest that impaired function of sperm-specific phospholipase C zeta also results in impaired egg activation leading to mammalian infertility. Inhibition of these Ca^{2+} elevations results in failure to initiate and complete functional fertilization in humans and other model organisms studied to date, including mice, *Xenopus* and zebrafish [16,30,31].

Recent molecular advances, including genomics, transcriptomics, proteomics, gene editing technology, and imaging technology, are opening new avenues of exploration for the potential therapeutic use of calcium in infertility treatments. For instance, assisted oocyte activation using mechanical, chemical, or electrical artificial triggers that cause calcium influx into the oocyte cytoplasm have been shown to trigger egg activation [30]. Therefore, manipulation of calcium activity could serve as a potential therapy for severe infertility associated with egg activation failure.

2.2. Cleavage and Blastula Stages

Following egg activation, vertebrate embryos undergo a rapid series of cell divisions leading to a relatively unstructured and undifferentiated mass of fluid-filled cells termed the blastula. It is during this stage that the maternal to zygotic transition occurs in which the zygotic genome is activated, an event that happens very early in mammals. It is also during this stage of mammalian development that the first important developmental lineage decisions occur, namely the determination of the inner cell mass that will give rise to the embryo proper and the trophoblast that will eventually give rise to the extraembryonic tissues. The blastula stage occurs during the first few weeks of embryonic development in humans; developmental defects during this stage of development affect germ layer formation and morphogenetic movements and will typically result in a non-viable embryo [32].

Calcium plays a key role in the regulation of the cell divisions during this early stage of development. During the initial cleavage stages, the embryo exhibits slowly moving intracellular waves of calcium. When imaging transgenic zebrafish that express genetically encoded calcium indicators, including cameleon YC2.60 and GCaMP6 driven by promoters of constitutive genes such as hspa8 and βactin, these waves were localized exclusively to the cleavage furrow [19,33]. Immunocytochemistry assays in zebrafish embryos have shown that this furrow-associated calcium activity is mediated by store-operated Ca^{2+} entry channels, including STIM1 and ORAI1 [34,35]. Furrow-associated waves have been shown to remodel microtubules into a furrow microtubule array that is required for cytokinesis in zebrafish. The furrow microtubule array is analogous to the mammalian midbody during the final stage of cytokinesis. The furrow microtubule array remodeling process is mediated by *nebel*, a cytokinesis regulator gene. When *nebel* mutant embryos were imaged using Oregon-Green BAPTA dextran, they not only showed reductions in both amplitude and frequency of slowly moving intracellular waves of calcium, but also failed to undergo cleavage. However, when calcium activity was enhanced by injecting IP3, NAADP, and $CaCl_2$ solution into *nebel* mutant eggs, furrow microtubule array formation was increased and the eggs progressed towards pseudocleavage [36]. These observations underscore the importance of calcium during the first divisions of zygotic development.

As embryos develop, furrow-associated Ca^{2+} activity accompanies successive cleavages until the 64-cell stage and then becomes undetectable, at least in zebrafish [37]. These results were consistent with findings from previous studies in wild type zebrafish [38]. During later blastula stages, namely the 128 cell-stage in zebrafish, both forms of calcium activity (spikes and waves) are observed. During a calcium spike, cytoplasmic calcium activity spreads over a whole cell, whereas in a calcium wave, it propagates across a group of contiguous or nearly contiguous cells in the blastoderm [37,39]. In *Xenopus*, this activity has been shown to be enriched in anterior ectoderm at late blastula stages [40]. Dysregulation of this calcium activity results in misalignment of microtubules, which leads ultimately to gastrulation failure [41]. These stages are very challenging to image in mammalian development, so it remains unknown if similar mechanisms are at play in human embryogenesis.

2.3. Gastrulation

During gastrulation, embryos exhibit characteristic calcium activity patterns which regulate well-coordinated cell movements that result in the formation of the characteristic vertebrate body plan and the induction of the nervous system. The three germ layers began to form during blastula stages, but become well established and exhibit regional differences during gastrulation [42]. The massive degree of cell movement leads to significant tissue rearrangements and inductive interactions that will ultimately give rise to specific organ systems [43,44]. The ectoderm gives rise to the nervous system and epidermis, while the mesoderm predominantly gives rise to the circulatory system, skeletal-muscular system, connective tissue and urinary system. The digestive, respiratory and glandular systems including organs such as the liver and the pancreas are derived from the endoderm. Typically, failure of gastrulation is embryonic lethal; however, some rare birth defects such as caudal dysgenesis and conjoined twinning are considered to be associated with impaired gastrulation [45].

Not surprisingly, calcium activity is essential for virtually every aspect of gastrulation. Because of the size and accessibility of the embryos, the majority of our knowledge of the critical role of calcium derives from studying amphibian and fish embryos. As embryos progress from the mid-blastula transition to early gastrula stage of development, calcium activity drastically decreases in its frequency of occurrence and its amplitude in zebrafish and *Xenopus* [37,39]. This period of decreased activity, in *Xenopus* and zebrafish embryos, is referred to as a "quiet period" [37,40,46,47]. After mid-blastula transition, however, calcium activity is more complex and tissue specific, that is, different tissue domains of the embryo exhibit differential propagating multicellular waves and non-propagating spikes [39]. For example, the dorsal quadrant of embryos, in comparison with ventral and lateral quadrants, exhibits higher calcium activity in terms of both frequency of occurrence and amplitude in *Xenopus* and zebrafish [37,40,48]. Similarly, calcium activity is enriched in the frequency of occurrence and amplitude in Blastula Chordin and Noggin-expressing region which is located in dorsal animal cells and presumptive neural tissue in *Xenopus* and Zebrafish [40,47,49–51]. Additionally, a gradual elevation of Ca^{2+} in the dorsal quadrant is superimposed at early gastrula stages and continues until late neurula in zebrafish and *Xenopus* [40,52]. Immunohistochemical assays have shown that newt embryos express various calcium channels, including L-type voltage-gated calcium channels and ryanodine receptors, throughout the presumptive ectoderm during gastrulation. Inhibition of this calcium activity, using low concentrations of potent inhibitors of the sarco/endoplasmic reticulum Ca^{2+} ATPase pumps, including thapsigargin and cyclopiazonic acid, during late blastula/early gastrula stages of development resulted in severe developmental defects including cyclopia and tail defects in zebrafish [53].

Calcium activity has been shown to regulate the extensive tissue movements that occur during gastrulation; disruption of calcium activity severely impedes cell and tissue movements. During massive tissue rearrangement processes, intercellular calcium waves are generated, which are considered an extension of the spikes and waves observed in the blastula stage [54]. The leading cells of this rearrangement initiate gastrulation by sensing and translating the signal from the ectoderm. Interestingly, these leader cells exhibit remarkably high calcium activity compared to other cells from second and third rows throughout gastrulation in *Xenopus* explants [44,55]. Disruption of calcium activity during gastrulation using calcium chelators such as BAPTA-AM, calcium channel blockers such as R(+)BayK and nicardipine, and morpholino antisense constructs which inhibit translation of calcium channels such as L-type calcium channels and purinergic receptor P2Y11, significantly reduces leader cells' protrusion activity and migration speed, thus impairing gastrulation and convergent extension [43,44]. When calcium activity was increased by ionomycin treatment, higher migratory activity of the leading edge mesoderm was induced during gastrulation in *Xenopus* [44]. In humans, some extremely rare congenital defects such as sirenomelia and caudal dysgenesis are thought to occur due to impaired gastrulation, however, the relationship to calcium activity remains unclear. In general, defects in gastrulation will result in embryonic lethality [45,56].

2.4. Neural Induction

The tissue rearrangements that occur during gastrulation form the basis for neural induction, as the overlying dorsal ectoderm comes into contact with the dorsal mesendoderm. This process commences in the dorsal ectoderm near the onset of gastrulation, from embryonic day 6.0–8.5 in mice, due to a complex interaction between the blastula-chordin and noggin-expressing center and Nieuwkoop center in *Xenopus* [42,57]. This interaction triggers the secretion of neural inducing factors, such as noggin, chordin, follistatin, Xnr3 and Cerberus. These factors inhibit a potent epidermal inducer, bone morphogenetic protein (BMP), and also coordinate regulation of fibroblast growth factor (FGF), Ca^{2+} and Wnt signaling. Therefore, neural induction has been considered the default pathway during development as it occurs in the absence of the suppressive influence of BMP [40,51]. Calcium activity is strongly implicated in the process of neural induction.

Barth and Barth, using amphibian embryos, first introduced the concept that Ca^{2+} treatment can activate neural induction [58]. Stableford (1967) supported this idea by showing an eight-fold increase in Ca^{2+} concentration in the internal fluid of amphibian neurula when compared with the blastula [59]. Recently, further detailed analysis of calcium activity has been achieved with both spatial and temporal resolution during early neural development. These studies have shown that the increase in cytoplasmic Ca^{2+} concentration in the dorsal ectoderm is critical in enabling cross-talk between different signaling pathways, including BMP and FGF/Erk, required for neural induction in *Xenopus* and zebrafish [51,60]. Similar results in mice are consistent with the findings in lower vertebrates. For example, pharmacological inhibition of calcineurin, a calcium-calmodulin-dependent serine/threonine phosphatase, using cyclosporin A and FK506 on day 3 to day 8 of gestation in mice results in reduced expression of neuroectodermal markers *FezF1*, *FezF2*, and *Six3*. This inhibition also increased expression of mesodermal marker Brachyury and BMP responsive genes including BMP4-Smad1/5 target genes, and induced phosphorylation of Smad1/5. However, FGF- or ionophore elevated Ca^{2+} dependent activation of calcineurin rescued these effects [51]. Similarly, when elevation of cytoplasmic Ca^{2+} from neuroectoderm is inhibited using L-type voltage gated calcium channel antagonists, such as R(+)BayK8644, and calcium chelators, such as BAPTA, neuralization of the dorsal ectoderm is prevented. This inhibition blocks the expression of neural markers, including geminin, *Zic3*, and *NCAM* and also results in developmental defects, including head and CNS deformation, absence of eyes, abnormal notochord formation, and spina bifida in *Xenopus* [40]. Similar results have been obtained by inhibiting calcium influx from both the extracellular matrix via other channels including TRP and intracellular stores via mediators such as IP3R and RyR [40,49,51,61]. Taken together, these experiments indicate that calcium signaling is essential for neural induction in a wide range of vertebrate animals, including mammals, and that deregulated calcium activity may result in severe developmental defects.

2.5. Organogenesis

Gastrulation establishes the basic vertebrate body plan with recognizable anterior-posterior, dorsal-ventral, and left-right axes. During organogenesis, the regional differences in the three germ layers that developed during the blastula and gastrula stages give rise to different organ systems. While calcium activity is critical for the development of all organ systems, in this review, we will focus on the importance of calcium activity during nervous system, heart, kidney, muscle and immune system development, as well as diseases associated with abnormal calcium activity.

2.5.1. Nervous System Development

Neural Tube Closure

Following gastrulation and neural induction, the neural tube forms the rudimentary basis of the central nervous system. By the end of gastrulation, undifferentiated neural progenitor cells form the neural plate on the dorsal surface of the embryo as a flat neuro-epithelial sheet. As embryonic development progresses, this plate lengthens along the anterior-posterior axis in a process called rostro-caudal extension, the lateral borders narrow via a process called mediolateral convergence, elevate, and bend to form ridged/grooved neural folds. The edges of the neural fold meet and fuse to form the hollow neuroepithelial tube. This process is characterized by apical constriction, cell shape change, and morphogenetic movements. Apical constriction is believed to drive convergent extension, movement, and transformation of two-dimensional planar neural tissue to a three-dimensional structure with dorsoventral, mediolateral and rostrocaudal axes in the neural tube between weeks three and four of human gestation [62,63]. This seemingly simple process of neural tube formation is orchestrated by a complex gene expression network, which is tightly regulated by spatiotemporal interactions between different tissue types as well as a number of molecular processes involving ~300 genes, epigenetic modifications, and multiple signaling pathways [63,64].

Calcium activity has been shown to regulate biomechanical processes and epithelial re-modeling during neural tube closure. Smedley and Stanisstreet (1986) cultured rat embryos at 10.4 days in papaverine and D-600, which are pharmacological inhibitors of calcium channels, and found that this inhibition results in severe neural tube defects [65]. This experiment provided the first evidence that calcium is necessary for neural tube closure. Recent studies using vertebrate model organisms including zebrafish, *Xenopus*, and mice have supported the idea that calcium is a major regulator of neural tube closure [4,66–68]. Visualization of calcium influxes during neural tube closure using genetically encoded calcium indicator showed distinct spikes and waves. Spikes were short, only lasting less than 40 s, and limited to single cells, while waves were long-lasting transients that originate from a single cell or a group of a few cells and propagate across several to hundreds of cells in *Xenopus* [4,66]. As neural tube closure progresses, intracellular calcium activity increases in terms of both incidence and frequency of occurrence, becoming extremely frequent during late stages of closure. Interestingly, the frequency of calcium activity correlates with a number of biomechanical processes that are necessary for neural tube closure: apical contraction, cell polarization and intercalation, morphogenetic movement, and cell shape change [4,66,68]. The apical regions of neuroepithelial cells are highly enriched with motor proteins that mediate contractions of actin filaments in the apical regions of these cells. When this actin activity is imaged using F-actin marker Lifeact-EGFP at the same time as calcium activity, F-actin mesh-like structures develop in the center of the cells following the calcium transients during neural tube closure in *Xenopus* [4]. In addition, calcium transients trigger non-muscle myosin II activation within a minute of their occurrence in *Xenopus* and *Drosophila* [4,64,67].

Pharmacologically disrupting calcium transients by blocking IP3R and other membrane-localized Ca^{2+} channels, using 2ABP and nifedipine, abolished both apical constriction and movements of the neural plate towards closing. Also, this disruption reduced the width of the neural gene *Sox2* expression zone and ultimately resulted in neural tube closure defects; these defects included impaired forebrain, hindbrain and midbrain in *Xenopus* and *Ciona* [4,66,68–71]. These results suggest that dysregulation of calcium during convergent extension and neural tube closure results in severe central nervous system defects. Central nervous system defects, including neural tube closure defects, accounted for about 8% of all birth defects in the United States from 2004–2006; examples include anencephaly, spina bifida without anencephaly, encephalocele, and anophthalmia/microphthalmia [72]. The causes of these defects are not completely understood; however, impaired calcium activity is likely to be a potential factor.

Neurotransmitter Phenotype Specification

A functional brain is composed of heterogeneous populations of cells that express different neurotransmitters and neurotransmitter receptors. The process of neural differentiation and acquisition of functional identity, also known as neurotransmitter phenotype specification, is achieved by complex interactions between transcription factors, morphogenetic proteins such as BMP and Shh, and other signaling molecules [73]. Common examples of neurotransmitter phenotypes include cholinergic and GABAergic. Interestingly, the neurotransmitter phenotype for a single neuron is plastic and can change depending upon its extrinsic and intrinsic signaling environment [74].

Calcium signaling has been established as a key regulator of neural differentiation and phenotype specification. In 1993, Spitzer and colleagues first introduced the concept of calcium activity-mediated neurotransmitter specification. When presumptive spinal cord neurons of *Xenopus* were cultured in calcium free medium, the total number of GABA immunoreactive GABAergic neurons decreased by 30% [75]. Similarly, when calcium activity of primary neural cells isolated from 3 different stages of development, namely neural plate, neural fold and neural tube, in *Xenopus* was enhanced via exposure to (−)BayK 8644, a pharmacological activator of voltage gated calcium channels, the total number of excitatory neurons was found to be decreased [76]. Furthermore, these results were strengthened by evidence that enhancing calcium spike frequency, either by overexpression of the voltage-gated rat brain sodium channels $rNa_v2a\alpha$ and $rNav_2a\beta$ or application of the sodium channel antagonist veratridine, resulted in increased GABA and Glycine immunoreactive inhibitory neurons in *Xenopus* embryonic spinal cord in vitro as well as in vivo. Conversely, chronic induction of hyperpolarization, either by overexpression of mRNA encoding the human inward rectifier potassium channel hKir2.1 or pharmacological blockers, including EGTA or a mixture of GVIA w-conotoxin, calcicludine, flunarizine and tetrodotoxin, has been shown to suppress calcium activity. This suppression of calcium activity increased the number of glutamatergic and cholinergic immunoreactive excitatory neurons in *Xenopus* spinal cord, in vivo as well as in tissue culture [77,78]. These changes in the ratio between excitatory neurons and inhibitory neurons lead to neurodevelopmental diseases, including autism spectrum disorder, Rett syndrome, and fragile X syndrome [79,80]. For example, reduction in the number of inhibitory synapses and over-activation of class I metabotropic glutamate receptor in the cerebral cortex and hippocampus due to lack of a modulator of voltage gated calcium channel Cav2.2 called Fragile X Mental Retardation 1(*fmr1*) gene results in fragile X syndrome in mice. Strikingly, when these mice were treated with the GABA agonist gaboxadol, the phenotype was rescued [80].

Furthermore, different cell types have been shown to exhibit their signature frequencies and incidence patterns of calcium activity during a 10-h period of *Xenopus* spinal cord development after neural tube closure. Dorsal sensory Rohon–Beard neurons exhibit spontaneous calcium activity characterized by a low and constant frequency, dorsolateral interneurons exhibit monotonically increasing frequency, ventral motoneurons exhibit low to high stepping frequency, and ventral interneurons maintain a high frequency throughout the larval period [77]. Differential proportions of these different cell types have been implicated in developmental diseases: for example, decreased GABAergic, glutamate decarboxylase 65 KDa isoform immunoreactive, neurons in the cerebellar dentate nuclei and increased GABAergic interneurons in the hippocampus have been implicated in idiopathic autism. Similarly, in vivo studies have associated reduced $GABA_A$ receptor density in the fronto-temporal cortex in Rett Syndrome [81]. Involvement of calcium channels in a wide array of disorders associated with impaired neurotransmitter phenotype specification suggests that dysregulation of calcium homeostasis might play a role in the etiology of various pathophysiological conditions, including chronic pain, cerebellar ataxia, autism schizophrenia and migraine [82]. Therefore, dysregulation of calcium signaling could be a potential cause of neurodevelopmental diseases.

2.5.2. Muscle Development

Muscle tissue, both skeletal and cardiac, is the derivative of mesoderm, the middle of the three germ layers. Paraxial mesoderm gives rise to subcutaneous tissue and to somitomeres, which develop into head mesenchyme, skeletal muscles and the axial skeleton, intermediate mesoderm gives rise to urogenital structures, and lateral plate mesoderm gives rise to components of limbs and the circulatory system, including the heart. It is apparent that derivation of these structures from a primitive germ layer is the product of a complex spatio-temporally regulated interaction network which is affected by multiple factors, calcium being one of them. Disruption of calcium activity during muscle development results in severe developmental defects, including but not limited to reduced and amorphous skeletal muscle [83,84]. Differentiating muscle cells, both in culture and in situ, exhibit spontaneous calcium transients as a first step of muscle development (myogenesis) in vertebrates. Chernoff and Hilfer for the first time cultured chick embryo trunk in media supplemented with and without calcium, in addition to various other calcium transport agonists and antagonists, including caffeine, ionophore A23187, papaverine, and verapamil, and determined that calcium activity was required for somitogenesis [85]. In addition, somite maturation and calcium activity have been shown to correlate in *Xenopus*. When somites of early tailbud stage *Xenopus* embryos, namely anterior somites, maturing somites, segmenting somites and unsegmented paraxial mesoderm, were exposed and imaged using a calcium sensitive dye (fluo-3) along the AP-axis, calcium activity, measured by incidence, frequency and duration of transients, was inversely correlated with the maturity of somites. For instance, anterior somites were more mature than unsegmented paraxial mesoderm as reflected in the expression of sarcomeric myosin, and anterior somites exhibited no calcium activity while unsegmented paraxial mesoderm exhibited the highest activity [86]. Interestingly, two distinct calcium active periods with a frequency ranging from 0.02 Hz to 0.12 Hz and a quiet period lasting about 3.5 h at the 28-somite stage have been shown to occur during slow muscle cell development after the 17-somite stage in transgenic zebrafish. This transgenic zebrafish constitutively expressed apoaequorin under an actin promoter only in developing muscle cells [87–89]. All of these observations underscore the importance of calcium during early muscle development.

Calcium activity observed during myogenesis, as opposed to neurogenesis, mainly relies upon store-operated calcium entry, which is mediated by RyR, TPC2, IP3R, ORAI/STIM, VGCC, and TRPC, among others [90–92]. Dysregulation of cytoplasmic Ca^{2+} homeostasis due to mutations or pharmacological manipulation of these mediators during myogenesis results in severe fetal neuromuscular disorders [82,93,94]. For instance, mice lacking STIM1 die perinatally [95]. Loss of function mutations in *STIM1* and *Orai* have also been identified in human patients suffering from hypotonia [96,97]. Additionally, blocking the the RyR channel using ryanodine resulted in inhibited differentiation of fetal myoblasts in E9.5 mice [98], hydrocephaly and scoliokyphosis in *Xenopus* [86]; and disruption of the vertical myoseptae and smooth muscle cell spanning in zebrafish [87,99]. Also, a homozygous Ile4898 to I4898T mutation in RYR1 in E9.5 mice causes severe asphyxia characterized by reduced and amorphous skeletal muscle, disorganized myofibrils, delayed cardiovascular development and perinatal death [92]. Similarly, when TPC2 expression was blocked using a TPC2 specific morpholino or a CRISPR/Cas system, decreased myotome width, U-shaped somites in the anterior trunk instead of the usual chevron-shaped, and smooth muscle cell myofibrils that were not aligned into bundles were observed in zebrafish [88,89]. These phenotypic effects were rescued when additional calcium release from intracellular stores was triggered using caffeine, IP3 or expression of mRNA construct of TPC2 [88,89]. These observations demonstrate the crucial role of calcium signaling in diseases of the muscle.

2.5.3. Heart Development

The heart and other components of the circulatory system derive from lateral plate mesoderm during the third week of pregnancy in humans. A plate of promyocardial cells intermixed with endothelial strands on either side of the neural fold represents the developing heart at the neural fold stage of development. As embryos progress, two massive growth events, a massive growth of the anterior portion of the neural tube and an endodermal invagination to form the foregut, drive the folding of the promyocardial plate to form endocardial tubes in either side of the neural tube. At this stage, the heart is bilaterally symmetrical, with an inverted Y shape. Although establishment of the pulmonary system happens later in development, the endothelial strand ensures the presence of the circulatory system. These two tubes elongate and converge towards each other to form the primitive heart tube, which quickly forms different cardiac structures including the truncus arteriosus and sinus venosus [100–102]. It is indisputable that a genetic blueprint is required to regulate all the developmental processes and necessary cell differentiations in the developing heart, including formation of cardiomyocytes and valvular interstitial cells. For example, NKX2.5 is expressed in early heart progenitor cells, and is considered a master controller of cardiac development. In addition, various members of the TGF-β superfamily including nodal and activin, BMPs, and other signaling molecules including calcium have been shown to play an indispensable role during cardiogenesis [100].

Dysregulation of calcium activity results in failure in every step of cardiac development, including differentiation of cardiac progenitor cells and cardiac tube formation. Two decades ago, using aequorin, Creton et al. found a 10-fold increase in cytoplasmic Ca^{2+} in the cardiac region during cardiogenesis at the segmentation period of zebrafish development. This increase was achieved by a large number of calcium spikes that occurred every 10–20 min for several hours [103]. When this calcium activity was blocked by injecting low concentrations of BAPTA buffer into the zygote (higher concentrations stalled fertilized embryos in the one cell stage), the heart defects were apparent.

The heart was smaller and stretched longitudinally, and it was not able to pump the blood [103]. Recent findings using a diverse group of organisms, both in vivo and in vitro, have strengthened our understanding of calcium's role in cardiac development. For example, elevated levels of calcium were observed while imaging the heart region in both transgenic zebrafish and mice at comparable stages of development. These transgenic animals expressed genetically encoded calcium markers GCaMP6 and GCaMP2 respectively, which express under the promoter of ubiquitously expressing genes (Tg[βactin2:GCaMP6s]stl351, Tg[ubi:GCaMP6s]stl352 in zebrafish and Tet-Off αMHC-CaMP2 in mice) in order to visualize cytosolic Ca^{2+} ions [37,104]. When these calcium spikes during cardiogenesis are blocked by culturing mouse embryos at E7.5 to E8.5 in media supplemented with the L-type voltage gated calcium channel blockers, nifedipine and verapamil, various heart defects, including lack of a right ventricle and a large left ventricle, were apparent [105]. In addition, these blockers reduced DNA synthesis, (assayed using tritiated [3H]-thymidine incorporation), cell division (via reduction of the mRNA expression of a positive cell cycle regulator), cyclin B1 (detected using qPCR), and differentiation of cardiac progenitors (detected using immunolabeling of differentiation marker sarcomeric myosin), in cardiac cells from E11.5 mice in vitro [106]. Similarly, mouse embryos treated with bivalent cation chelators such as BAPTA or EGTA, or embryos that lack calreticulin, which is a Ca^{2+}-binding chaperone of the endoplasmic reticulum, die in utero due to defective heart development. However, the calcium ionophore ionomycin restored myofibrillogenesis in cardiomyocytes [107,108]. These observations underscore the importance of calcium activity during cardiogenesis and associated developmental heart defects.

2.5.4. Kidney Development

Disruptions to calcium signaling pathways have also been implicated in a number of developmental dysfunctions and diseases in the kidney. Calcium activity mediates the initiation and formation of the kidney field; the kidney field is defined as the mesodermal territory where expression of *pax8* and *lhx1* overlap during organogenesis. Overexpression of *pax8* and/or *lhx1* leads to the formation of enlarged and ectopic pronephroi [52,109]. In addition, calcium regulates the transcription of genes that are involved in the formation of renal structures, including *pax8*, *lhx1*, *osr1*, and *osr2*. Loss of function of any of these genes leads to impaired pronephros development in *Drosophila* and *Xenopus* [109–111]. Using aequorin imaging, it was shown that calcium transients occur in intact *Xenopus* embryos in the lateral mesoderm during kidney tubule formation. When these transients were inhibited with a calcium chelator during the late gastrula or mid neurula stage, kidney tubule development was disrupted. Interestingly, incubating *Xenopus* ectoderm with activin A, which is a IP3 modulator/mesodermal inducer, followed by retinoic acid, which is a voltage-gated calcium channels modulator, rescued pronephric tubule formation [112]. Note that retinoic acid stimulates calcium transients, while activin A alone does not. However, treating *Xenopus* ectoderm with activin A followed by caffeine or ionomycin not only increased cytoplasmic Ca^{2+}, but also triggered normal tubule differentiation [112]. This evidence suggests that calcium signaling is necessary and sufficient to regulate the tubule differentiation process.

Additionally, several calcium-dependent proteins have been linked to pronephros development and function. For example, the proteins polycystin-1 and polycystin-2, encoded by the genes PKD1 and PKD2 respectively, are expressed in renal tissue of both humans and mice during renal development and have been shown to function together to create a calcium-permeable channel [107]. Mutations in PKD1 and PKD2 result in autosomal dominant polycystic kidney disease in humans [113]. The protein products of these mutant forms of PKD1 and PKD2 do not form their normal calcium permeable channel during embryogenesis, which results in the development of large fluid-filled cysts within the kidney tubules and collecting ducts. Collectively these observations suggest that calcium activity is instrumental to multiple aspects of kidney development and associated diseases.

2.5.5. Immune System

Although the immune system does not appear until later stages of development, for example until 12 weeks of development in humans, immune cell differentiation and migration, which are mediated by calcium signaling, are hallmarks of development. Boucek and Snyderman first demonstrated the requirement for calcium activity during neutrophil functioning using lanthanum chloride, which is a calcium influx inhibitor [114]. More recent studies have strengthened the idea that a transient increase in cytoplasmic Ca^{2+} is required for immune cell differentiation and migration. *Stim1* and *Stim2* double-knockout mice at 5–6 weeks showed reduced regulatory T cells in the thymus, spleen, and lymph nodes, and these T-cells showed no Ca^{2+} influx when imaged using calcium indicator Fura-2-AM even after thapsigargin treatment [115]. These knockout mice exhibited a decrease in regulatory T-cells in all immune organs, leading to autoimmune disease symptoms [115,116]. Similarly, imaging transgenic zebrafish embryos that expressed GCaMP3 specifically in neutrophils under the promoter *LysC* showed elevated cytoplasmic Ca^{2+} as well as enhanced calcium spiking behavior during migration and phagocytosis at wound sites in vivo. Inhibition of this calcium activity using the calcium channel antagonist SKF 96365 resulted in impaired recruitment of these neutrophils due to their undirected movement [117]. Likewise, when fura-2 loaded human neutrophils were cultured in media with and without calcium supplement, cells cultured in calcium free medium were unable to migrate on poly-D-lysine-coated glass [118].

In addition, cytoplasmic Ca^{2+} elevation is also facilitated by store-operated calcium entry through calcium release activated channels in immune cells, including lymphocytes and mast cells [115,119]. Dysregulation of calcium release activated channels results in a severely compromised immune system. For example, mice lacking STIM1 or with a gene-trap insertion in the *Orai1* gene show defective mast cell degranulation and hereditary immunodeficiencies [120,121]. Patients with these mutations displayed symptoms such as severe T cell immunodeficiency, impaired T cell activation and proliferation leading to recurrent viral, bacterial, and fungal infections, and muscular hypotonia [119]. Similarly, alterations in *STIM1* expression affected the sensitivity of immunoglobin E-mediated immediate-phase anaphylactic responses in vivo in mice [120]. These observations underscore the importance of calcium homeostasis and signaling behaviors in immune system development and function.

3. Wound Healing and Regeneration

3.1. Wound Healing before Formation of Immune System

In order to reach maturity, embryos must be able to respond to a wide variety of perturbations that are likely to occur during development. Embryos in the early stages of development exhibit inflammatory-response-free and scar-free wound closure with a mechanism that closely parallels neural tube closure during morphogenesis [122]. This healing is often called regenerative healing, and does not occur in most adult organisms [123]. The mechanisms of these abilities rely mostly on calcium signaling. Imaging the optic tectum of *Xenopus* larva using Oregon Green BAPTA-1 showed that fine micropipette mediated mechanical insults and targeted induction of cell death by a high-voltage electrical stimulus triggered a rapid influx of Ca^{2+} that generated further spontaneous Ca^{2+} influxes, which propagated through multiple rows of cells surrounding the injury site within seconds. These rapid calcium waves were accompanied by contractions of the neuroepithelium and expulsion of potentially damaged cells [124]. When this calcium activity was blocked and/or inhibited using BAPTA, thapsigargin, 2-aminoethoxydiphenyl borate, which is an IP3R blocker, and an array of purinergic receptor blockers, tissue contractions and potentially damaged cell expulsions were hindered. This blockage also negatively impacted wound healing [124]. In addition, damaged cells also release ATP into the extracellular milieu, which is recognized by the metabotropic purinergic receptors P2Ys on adjacent undamaged cells [125]. Activation of P2Y receptors further increases the cytoplasmic Ca^{2+} and activated metalloproteinase. This leads to heparin-binding EGF-like growth factor activation and subsequent cell proliferation, as well as formation of actomyosin cables and actin-rich protrusions, which guide tissue contraction, cell elongation, and migration in order to close the wound [125]. Therefore, calcium activity plays an indispensable role in embryonic scar-free regenerative wound healing.

3.2. Wound Healing Following the Formation of Immune System

At later stages of development, following the formation of functional immune cells, wound healing differs mechanistically from early embryonic wound healing, however, calcium activity continues to play a critical role. During this later stage of development, immune cells are also involved in wound healing and healing is no longer inflammation-response-free. Interestingly, this mechanism of healing is also regulated by calcium as a first responder. In this mechanism, calcium regulates both modulation of actomyosin and recruitment of immune cells. In *Drosophila* embryos, elevated calcium induced by laser wounding was shown to activate hydrogen peroxide synthase, resulting in a rapid accumulation of H_2O_2 around wound sites as indicated by the fluorigenic reporter Amplex Ultrared. A reduction in H_2O_2 signals at the wound site was observed in TRPM2 channel and innexin 2 mutant embryos, in which calcium elevation did not occur, indicating that calcium was necessary to this inflammatory response [126]. Likewise, using a genetically encoded H_2O_2 ratiometric sensor HyPer, Niethammer et al. for the first time showed that H_2O_2 is required for rapid recruitment of leukocytes to the wound in zebrafish larva [127]. Inhibition of this calcium activity using thapsigargin or EGTA not only inhibits H_2O_2 release healing efficiency, but also reduces the average hemocyte response during laser-induced epithelial wound healing in *Drosophila* embryos [126]. Even after the immune system develops and matures, calcium continues to play a critical role in embryonic wound healing.

3.3. Regeneration

Equally importantly, embryos also show calcium-mediated tissue regenerative capabilities during the early stages of development. Spontaneous calcium activity has been observed in the regenerating tail in *Xenopus* during the first hours of recovery. Inhibiting this calcium activity using ryanodine reduced the number of activated muscle progenitor cells and skeletal muscle stem cells, also known as muscle satellite cells, in the regenerating tissue. This, in turn, inhibited the regeneration process [128]. During regeneration, calcium activity increases secretion and mobilization of growth factors including insulin-like growth factor, interleukins, including interleukin-1, interleukin-6 and interleukin-8, parathyroid hormone, transforming growth factor-β and platelet-derived growth factor. Platelets are widely recognized as inducers of cell proliferation, and are also involved in stem cell differentiation, cell migration, and revascularization of damaged tissue via increased release of growth factors during tissue regeneration. Therefore, platelet-rich plasma derivatives have been used in regenerative medicine [129]. This increase in growth factor release has been proven- to be mediated by calcium activity. This calcium activity has been shown to be critical for platelet functioning using double knockouts of various calcium channels, including TRPM7, TRPC6, Orai1 and SERCA, and pharmacological agonists and antagonists in mice [130–132]. These factors facilitate changes in the expression of specific genes that are responsible for proper cell proliferation and migration during regeneration [122,133]. Using microarrays, Patterson et al. reported 624 upregulated and 826 downregulated genes associated with epidermal regeneration in *Drosophila* embryos, including genes related to cellular component organization and stress response [134]. Similarly, using an RNA-Seq approach, Tsujioka et al. reported 25 candidate genes involved in tail regeneration in *Xenopus*, including genes related to cell proliferation, for example, CDK1 and cyclin B2 [135]. These results indicate that calcium activates expression of a number of gene cascades that are responsible for tissue regeneration. Understanding the underlying mechanisms of embryonic regeneration could lead to new developments in the fields of regenerative clinical medicine and tissue engineering.

4. Conclusions and Future Directions

Over the last three decades, studies have shown that dysregulation of calcium signaling results in severe medical conditions, including neural tube and other developmental defects. Given the versatility of calcium as a signaling molecule and its near ubiquitous involvement with virtually every aspect of embryonic development (Figure 1), this is scarcely surprising. Yet despite its importance in development and the genesis of human disease, there is much that we do not understand about calcium and many avenues for future research. For example, currently, a comprehensive analysis of the regulation of spatiotemporal expression patterns of calcium-associated genes and their interactions is lacking. Such an analysis would suggest specific tissues and developmental timepoints where calcium-regulated processes can go awry. New gene editing and tissue engineering techniques as well as novel next generation sequencing approaches may aid in understanding the etiology and mechanisms of medical conditions associated with dysregulation of calcium, as well as potential therapeutic solutions for these conditions. In addition, determination of the embryonic expression patterns, both at the mRNA and protein level, of genes known to be involved in calcium activity and screens for novel genes implicated in calcium activity are warranted.

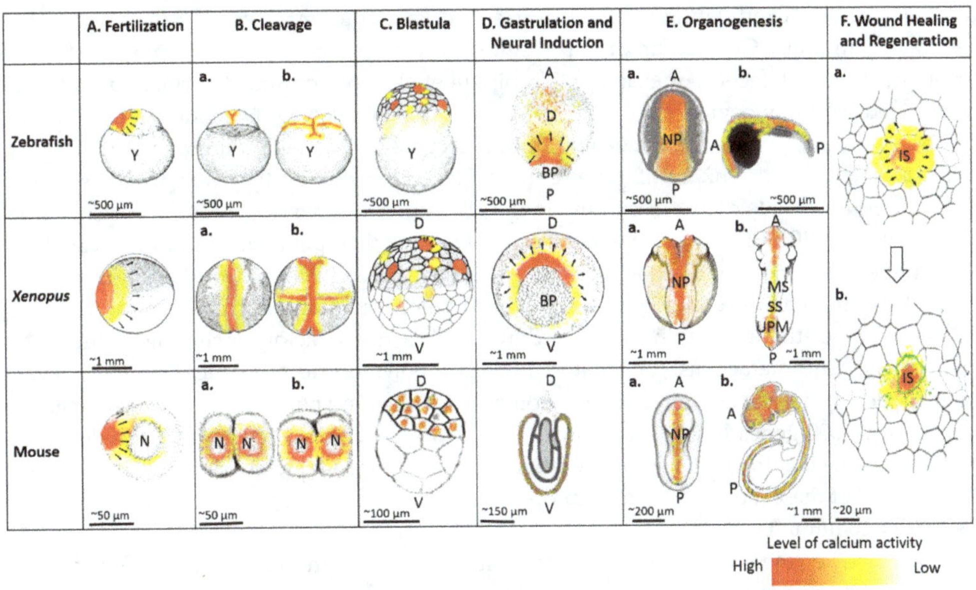

Figure 1. Comparative calcium activity among zebrafish, frogs, and mice during egg activation and fertilization (**A**), first cleavage (**Ba**), second cleavage (**Bb**), blastula (**C**), gastrulation and neural induction (**D**) and organogenesis (**E**), including neural tube closure (**Ea**) and muscle development (**Eb**) in zebrafish, *Xenopus*, and mouse; also, depiction of calcium dynamics immediately after tissue damage (**Fa**) and calcium mediated actomyosin filament (green lining) during wound healing and regeneration (**Fb**). Black arrows show direction of propagation of calcium waves. Y, yolk; A, anterior; P, posterior; D, dorsal; V, ventral; BP, blastopore; NP, neural plate; UPM, unsegmented paraxial mesoderm; MS, matured somites; SS, segmenting somites; IS, injury site; N, nucleus. Images were adapted and re-drawn from [136–138].

In addition to many avenues for further research, there is also a huge potential to utilize this versatile signaling molecule therapeutically and in regenerative medicine. It is well known that developing embryos up to the blastocyst stage of development and embryonic stem cells exhibit the ability to either maintain stem cell identity or differentiate into the lineages of all three germ layers. This opens new avenues for the exploration of these cells' potential uses in regenerative medicine [139–141]. The ultimate goal of regenerative medicine is to restore or substitute damaged tissue so the tissue can maintain its function(s); this may require boosting natural defense systems, which could harness and accelerate innate healing processes in order to cure previously unmanageable medical problems [142]. Towards this end, Jergova et al. engineered neural progenitor cells to generate transgenic GABAergic cells that release an analgesic peptide serine-histogranin, which antagonizes NMDA receptors. These cells were then transplanted onto an injury site in order to assess their effect on peripheral neuropathic pain after 1 week of induced injury in rat spinal cord. This transplant resulted in improved pain sensitivity based upon a pain-related behavioral assessment with four pain parameters: mechanical, heat, cold, and tactile [143]. Similarly, Mery et al. engineered pacemaker cells from embryonic stem cells that induce InsP3 mediated oscillatory calcium release from the endoplasmic reticulum; these cells had a potential to serve as pacemaker cells prior to the development of pacemaker ionic channels in early embryos. This oscillatory calcium release was required for pacemaker activity. When InsP3 receptor antisense cDNA was expressed constitutively in these cells, expression of type I InsP3 receptors and spontaneous calcium activity, both frequency of occurrence and amplitude, were reduced, and beating activity was also impaired [144]. These observations underscore the potential use of calcium signaling toolkits in medical cell and tissue engineering.

Current approaches used to understand calcium-related defects include mapping and cloning causative gene(s), recapitulating clinical features, and using pharmacological reagents in order to ameliorate symptoms. Gene editing is also beginning to be explored as a way to understand and possibly treat these defects. For example, Limpitikul et al. reprogrammed dermal fibroblasts into iPSC-derived cardiomyocytes, then they used these cardiomyocytes to model calm2-associated Long-QT syndrome, which is a congenital birth defect associated with a mutation in the calcium binding protein calmodulin2, *calm2* gene. Using CRISPRi technology, they were able to selectively correct the mutated allele and rescue the long QT syndrome phenotype [145]. Molecular gene editing tools like CRISPR have tremendous capabilities for therapeutic use, and these capabilities will continue to expand as molecular technology improves even further. However, off-target effects of these tools and complex etiologies of some birth defects continue to create critical therapeutic challenges. These etiologies can be influenced by a wide array of factors, including multiple genes, impaired gene-gene interactions, combinatorial effects of polymorphisms, changes in protein levels in time and space, epigenetic modifications, environmental effects, and mutations in epigenetic regulators. For example, ~300 genes have been shown to be critically required for neural tube closure [63], and RNAseq data identified at least 30 causative genes for congenital fetal akinesia deformation [83]. Addressing all of these factors during the process of therapeutic development remains a challenge in the scientific community.

Taken together, these results indicate that calcium is involved in a diverse array of physiological processes, and dysregulation of calcium activity results in wide array of developmental defects. There is a tremendous demand for scar-free healing after major surgical procedures, an ability the embryo possesses following unique patterns of calcium activity. The capacity of the embryo to either maintain stem cell identity or differentiate into specific lineages such as neural cells could also be used to treat neurodegenerative diseases. While great progress has been made in recent decades, translation of these fetal abilities into adult systems still remains a challenge. Understanding how an ion as simple as Ca^{2+} plays diverse physiological roles may help solve these medical problems. Some aspects of calcium signaling are understood in a variety of processes already, and recent advances in genomics, transcriptomics, proteomics, and gene editing technology may help expand our knowledge about the underlying mechanisms of calcium function and enable its potential therapeutic use.

Author Contributions: All authors read the primary literature and participated in writing and reviewing the manuscript.

Acknowledgments: This work was funded by the National Institutes of Health (1R15HD077624-01) and the National Science Foundation (IOS-1257895) to MSS.

References

1. Database GeneCards. GeneCards—Human Genes | Gene Database | Gene Search. Available online: https://www.genecards.org/ (accessed on 23 September 2018).
2. Lew, V.L.; Tsien, R.Y.; Miner, C.; Bookchin, R.M. Physiological [Ca^{2+}]Ilevel and Pump-Leak Turnover in Intact Red Cells Measured Using an Incorporated Ca Chelator. *Nature* **1982**, *298*, 478–481. [CrossRef] [PubMed]
3. Kramer, I. Intracellular Calcium. In *Signal Transduction*; Elsevier: Amsterdam, The Netherlands, 2016; pp. 381–439. [CrossRef]
4. Suzuki, M.; Sato, M.; Koyama, H.; Hara, Y.; Hayashi, K.; Yasue, N.; Imamura, H.; Fujimori, T.; Nagai, T.; Campbell, R.E.; et al. Distinct Intracellular Ca^{2+} Dynamics Regulate Apical Constriction and Differentially Contribute to Neural Tube Closure. *Development* **2017**, *144*, 1307–1316. [CrossRef] [PubMed]
5. Balaji, R.; Bielmeier, C.; Harz, H.; Bates, J.; Stadler, C.; Hildebrand, A.; Classen, A.K. Calcium Spikes, Waves and Oscillations in a Large, Patterned Epithelial Tissue. *Sci. Rep.* **2017**, *7*, 42786. [CrossRef] [PubMed]
6. Suzuki, J.; Kanemaru, K.; Ishii, K.; Ohkura, M.; Okubo, Y.; Iino, M. Imaging Intraorganellar Ca^{2+} at Subcellular Resolution Using CEPIA. *Nat. Commun.* **2014**, *5*, 4153. [CrossRef] [PubMed]
7. Berridge, M.J.; Lipp, P.; Bootman, M.D. The Versatility and Universality of Calcium Signalling. *Nat. Rev. Mol. Cell Biol.* **2000**, *1*, 11–21. [CrossRef] [PubMed]
8. Dupont, G.; Combettes, L.; Leybaert, L. Calcium Dynamics: Spatio-Temporal Organization from the Subcellular to the Organ Level. *Int. Rev. Cytol.* **2007**, *261*, 193–245. [PubMed]
9. Markova, O.; Sénatore, S.; Chardès, C.; Lenne, P.F. Calcium Spikes in Epithelium: Study on Drosophila Early Embryos. *Sci. Rep.* **2015**, *5*, 11379. [CrossRef] [PubMed]
10. Gu, X.; Olson, E.C.; Spitzer, N.C. Spontaneous Neuronal Calcium Spikes and Waves during Early Differentiation. *J. Neurosci. Off. J. Soc. Neurosci.* **1994**, *14*, 6325–6335. [CrossRef]
11. Clapham, D.E. Calcium Signaling. *Cell* **2007**, *131*, 1047–1058. [CrossRef] [PubMed]
12. Berridge, M.J.; Bootman, M.D.; Roderick, H.L. Calcium Signalling: Dynamics, Homeostasis and Remodelling. *Nat. Rev. Mol. Cell Biol.* **2003**, *4*, 517–529. [CrossRef] [PubMed]
13. National Institutes of Health. National Center for Advancing Translational Sciences (NCATS). Available online: https://ncats.nih.gov/ (accessed on 23 September 2018).
14. Zhaurova, K. Genetic Causes of Adult-Onset Disorders. *Nat. Educ.* **2008**, *1*, 49.
15. Robertson, S.; Lin, R. Oocyte-to-Zygote Transition. *Semin. Cell Dev. Biol.* **2018**. [CrossRef] [PubMed]
16. Kuroda, K.; Brosens, J.J.; Quenby, S.; Takeda, S. (Eds.) *Treatment Strategy for Unexplained Infertility and Recurrent Miscarriage*; Springer: Midtown Manhattan, NY, USA, 2018.
17. Kashir, J.; Deguchi, R.; Jones, C.; Coward, K.; Stricker, S.A. Comparative Biology of Sperm Factors and Fertilization-Induced Calcium Signals across the Animal Kingdom. *Mol. Reprod. Dev.* **2013**. [CrossRef] [PubMed]
18. Wozniak, K.L.; Mayfield, B.L.; Duray, A.M.; Tembo, M.; Beleny, D.O.; Napolitano, M.A.; Sauer, M.L.; Wisner, B.W.; Carlson, A.E. Extracellular Ca^{2+} Is Required for Fertilization in the African Clawed Frog, Xenopus Laevis. *PLoS ONE* **2017**, *12*, e0170405. [CrossRef] [PubMed]
19. Mizuno, H.; Sassa, T.; Higashijima, S.; Okamoto, H.; Miyawaki, A. Transgenic Zebrafish for Ratiometric Imaging of Cytosolic and Mitochondrial Ca^{2+} response in Teleost Embryo. *Cell Calcium* **2013**, *54*, 236–245. [CrossRef] [PubMed]
20. Cuthbertson, K.S.; Whittingham, D.G.; Cobbold, P.H. Free Ca^{2+} Increases in Exponential Phases during Mouse Oocyte Activation. *Nature* **1981**, *294*, 754–757. [CrossRef] [PubMed]
21. Deguchi, R.; Shirakawa, H.; Oda, S.; Mohri, T.; Miyazaki, S. Spatiotemporal Analysis of Ca(2+) Waves in Relation to the Sperm Entry Site and Animal-Vegetal Axis during Ca(2+) Oscillations in Fertilized Mouse Eggs. *Dev. Biol.* **2000**, *218*, 299–313. [CrossRef] [PubMed]
22. Ferrer-Buitrago, M.; Bonte, D.; De Sutter, P.; Leybaert, L.; Heindryckx, B. Single Ca^{2+} Transients vs. Oscillatory Ca^{2+} Signaling for Assisted Oocyte Activation: Limitations and Benefits. *Reproduction* **2018**, *155*, R105–R119. [CrossRef] [PubMed]

23. Miao, Y.-L.; Stein, P.; Jefferson, W.N.; Padilla-Banks, E.; Williams, C.J. Calcium Influx-Mediated Signaling Is Required for Complete Mouse Egg Activation. *Proc. Natl. Acad. Sci. USA* **2012**, *109*, 4169–4174. [CrossRef] [PubMed]
24. Miao, Y.L.; Williams, C.J. Calcium Signaling in Mammalian Egg Activation and Embryo Development: The Influence of Subcellular Localization. *Mol. Reprod. Dev.* **2012**, *79*, 742–756. [CrossRef] [PubMed]
25. Bernhardt, M.L.; Zhang, Y.; Erxleben, C.F.; Padilla-Banks, E.; McDonough, C.E.; Miao, Y.-L.; Armstrong, D.L.; Williams, C.J. CaV3.2 T-Type Channels Mediate Ca^{2+} Entry during Oocyte Maturation and Following Fertilization. *J. Cell Sci.* **2015**, *128*, 4442–4452. [CrossRef] [PubMed]
26. Tang, T.S.; Dong, J.B.; Huang, X.Y.; Sun, F.Z.; Royse, J.; Blayney, L.M.; Swann, K.; Lai, F.A. Ca(2+) Oscillations Induced by a Cytosolic Sperm Protein Factor Are Mediated by a Maternal Machinery That Functions only Once in Mammalian Eggs. *Development* **2000**, *127*, 1141–1150. [PubMed]
27. Saunders, C.M. Ca^{2+} Oscillations Triggered by Sperm PLCz. 2002.
28. Hachem, A.; Godwin, J.; Ruas, M.; Lee, H.C.; Ferrer Buitrago, M.; Ardestani, G.; Bassett, A.; Fox, S.; Navarrete, F.; de Sutter, P.; et al. PLCζ Is the Physiological Trigger of the Ca^{2+} Oscillations That Induce Embryogenesis in Mammals but Conception Can Occur in Its Absence. *Development* **2017**, *144*, 2914–2924. [CrossRef] [PubMed]
29. Lee, K.; Wang, C.; Machaty, Z. STIM1 Is Required for Ca^{2+} Signaling during Mammalian Fertilization. *Dev. Biol.* **2012**, *367*, 154–162. [CrossRef] [PubMed]
30. Yamaguchi, T.; Kuroda, K.; Tanaka, A.; Watanabe, S. Fertilization Failure. In *Treatment Strategy for Unexplained Infertility and Recurrent Miscarriage*; Springer: Singapore, 2018; pp. 7–17.
31. Ferrer-Buitrago, M.; Dhaenens, L.; Lu, Y.; Bonte, D.; Vanden Meerschaut, F.; De Sutter, P.; Leybaert, L.; Heindryckx, B. Human Oocyte Calcium Analysis Predicts the Response to Assisted Oocyte Activation in Patients Experiencing Fertilization Failure after ICSI. *Hum. Reprod.* **2018**, *33*, 416–425. [CrossRef] [PubMed]
32. Halliday, J.L.; Ukoumunne, O.C.; Baker, H.W.G.; Breheny, S.; Jaques, A.M.; Garrett, C.; Healy, D.; Amor, D. Increased Risk of Blastogenesis Birth Defects, Arising in the First 4 Weeks of Pregnancy, after Assisted Reproductive Technologies. *Hum. Reprod.* **2010**, *25*, 59–65. [CrossRef] [PubMed]
33. Chen, C.; Jin, J.; Lee, G.A.; Silva, E.; Donoghue, M. Cross-Species Functional Analyses Reveal Shared and Separate Roles for Sox11 in Frog Primary Neurogenesis and Mouse Cortical Neuronal Differentiation. *Biol. Open* **2016**, *5*, 409–417. [CrossRef] [PubMed]
34. Chan, C.M.; Chen, Y.; Hung, T.S.; Miller, A.L.; Shipley, A.M.; Webb, S.E. Inhibition of SOCE Disrupts Cytokinesis in Zebrafish Embryos via Inhibition of Cleavage Furrow Deepening. *Int. J. Dev. Biol.* **2015**, *59*, 289–301. [CrossRef] [PubMed]
35. Chan, C.M.; Aw, J.T.M.; Webb, S.E.; Miller, A.L. SOCE Proteins, STIM1 and Orai1, Are Localized to the Cleavage Furrow during Cytokinesis of the First and Second Cell Division Cycles in Zebrafish Embryos. *Zygote* **2016**, *24*, 880–889. [CrossRef] [PubMed]
36. Eno, C.; Gomez, T.; Slusarski, D.C.; Pelegri, F. Slow Calcium Waves Mediate Furrow Microtubule Reorganization and Germ Plasm Compaction in the Early Zebrafish Embryo. *Development* **2018**, *145*, dev156604. [CrossRef] [PubMed]
37. Chen, J.; Xia, L.; Bruchas, M.R.; Solnica-Krezel, L. Imaging Early Embryonic Calcium Activity with GCaMP6s Transgenic Zebrafish. *Dev. Biol.* **2017**, *430*, 385–396. [CrossRef] [PubMed]
38. Webb, S.E.; Miller, A.L. Ca^{2+} Signaling during Vertebrate Somitogenesis. *Acta Pharmacol. Sin.* **2006**, *27*, 781–790. [CrossRef] [PubMed]
39. Ma, L.H.; Webb, S.E.; Chan, C.M.; Zhang, J.; Miller, A.L. Establishment of a Transitory Dorsal-Biased Window of Localized Ca^{2+} Signaling in the Superficial Epithelium Following the Mid-Blastula Transition in Zebrafish Embryos. *Dev. Biol.* **2009**, *327*, 143–157. [CrossRef] [PubMed]
40. Leclerc, C.; Webb, S.E.; Daguzan, C.; Moreau, M.; Miller, A.L. Imaging Patterns of Calcium Transients during Neural Induction in Xenopus Laevis Embryos. *J. Cell Sci.* **2000**, *113*, 3519–3529. [PubMed]
41. Hara, Y.; Nagayama, K.; Yamamoto, T.S.; Matsumoto, T.; Suzuki, M.; Ueno, N. Directional Migration of Leading-Edge Mesoderm Generates Physical Forces: Implication in Xenopus Notochord Formation during Gastrulation. *Dev. Biol.* **2013**, *382*, 482–495. [CrossRef] [PubMed]
42. Levine, A.J.; Brivanlou, A.H. Proposal of a Model of Mammalian Neural Induction. *Dev. Biol.* **2007**, *308*, 247–256. [CrossRef] [PubMed]

43. Shindo, A.; Hara, Y.; Yamamoto, T.S.; Ohkura, M.; Nakai, J.; Ueno, N. Tissue-Tissue Interaction-Triggered Calcium Elevation Is Required for Cell Polarization during Xenopus Gastrulation. *PLoS ONE* **2010**, *5*, e8897. [CrossRef] [PubMed]
44. Hayashi, K.; Yamamoto, T.S.; Ueno, N. Intracellular Calcium Signal at the Leading Edge Regulates Mesodermal Sheet Migration during Xenopus Gastrulation. *Sci. Rep.* **2018**, *8*, 2433. [CrossRef] [PubMed]
45. Ferrer-Vaquer, A.; Hadjantonakis, A.K. Birth Defects Associated with Perturbations in Preimplantation, Gastrulation, and Axis Extension: From Conjoined Twinning to Caudal Dysgenesis. *Wiley Interdiscip. Rev. Dev. Biol.* **2013**, *2*, 427–442. [CrossRef] [PubMed]
46. Gilland, E.; Miller, A.L.; Karplus, E.; Baker, R.; Webb, S.E. Imaging of Multicellular Large-Scale Rhythmic Calcium Waves during Zebrafish Gastrulation. *Proc. Natl. Acad. Sci. USA* **1999**, *96*, 157–161. [CrossRef] [PubMed]
47. Webb, S.E.; Moreau, M.; Leclerc, C.; Miller, A.L. Calcium Transients and Neural Induction in Vertebrates. *Cell Calcium* **2005**, *37*, 375–385. [CrossRef] [PubMed]
48. Yuen, M.Y.F.; Webb, S.E.; Chan, C.M.; Thisse, B.; Thisse, C.; Miller, A.L. Characterization of Ca^{2+} signaling in the External Yolk Syncytial Layer during the Late Blastula and Early Gastrula Periods of Zebrafish Development. *Biochim. Biophys. Acta* **2013**, *1833*, 1641–1656. [CrossRef] [PubMed]
49. Moreau, M.; Neant, I.; Webb, S.E.; Miller, A.L.; Leclerc, C. Calcium Signalling during Neural Induction in Xenopus Laevis Embryos. *Philos. Trans. R. Soc. B Biol. Sci.* **2008**, *363*, 1371–1375. [CrossRef] [PubMed]
50. Leclerc, C.; Néant, I.; Moreau, M. Early Neural Development in Vertebrates Is also a Matter of Calcium. *Biochimie* **2011**, *93*, 2102–2111. [CrossRef] [PubMed]
51. Cho, A.; Tang, Y.; Davila, J.; Deng, S.; Chen, L.; Miller, E.; Wernig, M.; Graef, I.A. Calcineurin Signaling Regulates Neural Induction through Antagonizing the BMP Pathway. *Neuron* **2014**, *82*, 109–124. [CrossRef] [PubMed]
52. Moreau, M.; Néant, I.; Webb, S.E.; Miller, A.L.; Riou, J.F.; Leclerc, C. Ca^{2+} coding and Decoding Strategies for the Specification of Neural and Renal Precursor Cells during Development. *Cell Calcium* **2016**, *59*, 75–83. [CrossRef] [PubMed]
53. Creton, R. The Calcium Pump of the Endoplasmic Reticulum Plays a Role in Midline Signaling during Early Zebrafish Development. *Dev. Brain Res.* **2004**, *151*, 33–41. [CrossRef] [PubMed]
54. Webb, S.E.; Miller, A.L. Ca^{2+} signaling and Early Embryonic Patterning during the Blastula and Gastrula Periods of Zebrafish and Xenopus Development. *Biochim. Biophys. Acta* **2006**, *1763*, 1192–1208. [CrossRef] [PubMed]
55. Wallingford, J.B.; Ewald, A.J.; Harland, R.M.; Fraser, S.E. Calcium Signaling during Convergent Extension in Xenopus. *Curr. Biol.* **2001**, *11*, 652–661. [CrossRef]
56. Lange, L.; Marks, M.; Liu, J.; Wittler, L.; Bauer, H.; Piehl, S.; Bläß, G.; Timmermann, B.; Herrmann, B.G. Patterning and Gastrulation Defects Caused by the Tw18 Lethal Are Due to Loss of Ppp2r1a. *Biol. Open* **2017**, *6*, 752–764. [CrossRef] [PubMed]
57. Yasuoka, Y.; Taira, M. The Molecular Basis of the Gastrula Organizer in Amphibians and Cnidarians. In *Reproductive and Developmental Strategies. Diversity and Commonality in Animals*; Springer: Tokyo, Japan, 2018; pp. 667–708.
58. Barth, L.G.; Barth, L.J. Sequential Induction of the Presumptive Epidermis of the Rana Pipiens Gastrula. *Biol. Bull.* **1964**, *127*, 413–427. [CrossRef]
59. Stableford, L.T. A Study of Calcium in the Early Development of the Amphibian Embryo. *Dev. Biol.* **1967**, *16*, 303–314. [CrossRef]
60. Leclerc, C.; Néant, I.; Moreau, M. The Calcium: An Early Signal That Initiates the Formation of the Nervous System during Embryogenesis. *Front. Mol. Neurosci.* **2012**, *5*, 1–12. [CrossRef] [PubMed]
61. Lee, H.-K.; Lee, H.-S.; Moody, S.A. Neural Transcription Factors: From Embryos to Neural Stem Cells. *Mol. Cells* **2014**, *37*, 705–712. [CrossRef] [PubMed]
62. Flanagan, M.; Sonnen, J.A.; Keene, C.D.; Hevner, R.F.; Montine, T.J. *Molecular Basis of Diseases of the Nervous System*, 2nd ed.; Elsevier Inc.: Amsterdam, The Netherlands, 2018.
63. Wilde, J.J.; Petersen, J.R.; Niswander, L. Genetic, Epigenetic, and Environmental Contributions to Neural Tube Closure. *Annu. Rev. Genet.* **2014**, *48*, 583–611. [CrossRef] [PubMed]
64. Nikolopoulou, E.; Galea, G.L.; Rolo, A.; Greene, N.D.E.; Copp, A.J. Neural Tube Closure: Cellular, Molecular and Biomechanical Mechanisms. *Development* **2017**, *144*, 552–566. [CrossRef] [PubMed]

65. Smedley, M.J.; Stanisstreet, M. Calcium and Neurulation in Mammalian Embryos II. Effects of Cytoskeletal Inhibitors and Calcium Antagonists on the Neural Folds of Rat Embryos. *Development* **1986**, *93*, 167–178.
66. Christodoulou, N.; Skourides, P.A.A. Cell-Autonomous Ca^{2+} Flashes Elicit Pulsed Contractions of an Apical Actin Network to Drive Apical Constriction during Neural Tube Closure. *Cell Rep.* **2015**, *13*, 2189–2202. [CrossRef] [PubMed]
67. Kong, D.; Wolf, F.; Großhans, J. In Vivo Optochemical Control of Cell Contractility at Single Cell Resolution by Ca^{2+} Induced Myosin Activation. *bioRxiv* **2018**. [CrossRef]
68. Sahu, S.U.; Visetsouk, M.R.; Garde, R.J.; Hennes, L.; Kwas, C.; Gutzman, J.H. Calcium Signals Drive Cell Shape Changes during Zebrafish Midbrain–hindbrain Boundary Formation. *Mol. Biol. Cell* **2017**, *28*, 875–882. [CrossRef] [PubMed]
69. Abdul-Wajid, S.; Morales-Diaz, H.; Khairallah, S.M.; Smith, W.C. T-Type Calcium Channel Regulation of Neural Tube Closure and EphrinA/EPHA Expression. *Cell Rep.* **2015**, *13*, 829–839. [CrossRef] [PubMed]
70. Sequerra, E.B.; Goyal, R.; Castro, P.A.; Levin, J.B.; Borodinsky, L.N. NMDA Receptor Signaling Is Important for Neural Tube Formation and for Preventing Antiepileptic Drug-Induced Neural Tube Defects NMDA Receptor Signaling Is Important for Neural Tube Formation and for Preventing Antiepileptic Drug-Induced Neural Tube Defec. *J. Neurosci.* **2018**. [CrossRef] [PubMed]
71. Borodinsky, L.N. Xenopus Laevis as a Model Organism for the Study of Spinal Cord Formation, Development, Function and Regeneration. *Front. Neural Circuits* **2017**, *11*, 90. [CrossRef] [PubMed]
72. Parker, S.E.; Mai, C.T.; Canfield, M.A.; Rickard, R.; Wang, Y.; Meyer, R.E.; Anderson, P.; Mason, C.A.; Collins, J.S.; Kirby, R.S.; et al. Updated National Birth Prevalence Estimates for Selected Birth Defects in the United States, 2004–2006. *Birth Defects Res. Part A—Clin. Mol. Teratol.* **2010**, *88*, 1008–1016. [CrossRef] [PubMed]
73. Borodinsky, L.N.; Belgacem, Y.H. Crosstalk among Electrical Activity, Trophic Factors and Morphogenetic Proteins in the Regulation of Neurotransmitter Phenotype Specification. *J. Chem. Neuroanat.* **2016**, *73*, 3–8. [CrossRef] [PubMed]
74. Spitzer, N.C. Neurotransmitter Switching? No Surprise. *Neuron* **2015**, *86*, 1131–1144. [CrossRef] [PubMed]
75. Spitzer, N.C.; Debaca, R.C.; Allen, K.A.; Holliday, J. Calcium Dependence of Differentiation of GABA Immunoreactivity in Spinal Neurons. *J. Comp. Neurol.* **1993**, *337*, 168–175. [CrossRef] [PubMed]
76. Lewis, B.B.; Miller, L.E.; Herbst, W.A.; Saha, M.S. The Role of Voltage-Gated Calcium Channels in Neurotransmitter Phenotype Specification: Coexpression and Functional Analysis in Xenopus Laevis. *J. Comp. Neurol.* **2014**, *522*, 2518–2531. [CrossRef] [PubMed]
77. Borodinsky, L.N.; Root, C.M.; Cronin, J.A.; Sann, S.B.; Gu, X.; Spitzer, N.C. Activity-Dependent Homeostatic Specification of Transmitter Expression in Embryonic Neurons. *Nature* **2004**, *429*, 523–530. [CrossRef] [PubMed]
78. Marek, K.W.; Kurtz, L.M.; Spitzer, N.C. CJun Integrates Calcium Activity and Tlx3 Expression to Regulate Neurotransmitter Specification. *Nat. Neurosci.* **2010**, *13*, 944–950. [CrossRef] [PubMed]
79. Nelson, S.B.; Valakh, V. Excitatory/Inhibitory Balance and Circuit Homeostasis in Autism Spectrum Disorders. *Neuron* **2015**, *87*, 684–698. [CrossRef] [PubMed]
80. Olmos-Serrano, J.L.; Paluszkiewicz, S.M.; Martin, B.S.; Kaufmann, W.E.; Corbin, J.G.; Huntsman, M.M. Defective GABAergic Neurotransmission and Pharmacological Rescue of Neuronal Hyperexcitability in the Amygdala in a Mouse Model of Fragile X. Syndrome. *J. Neurosci.* **2010**, *30*, 9929–9938. [CrossRef] [PubMed]
81. Cellot, G.; Cherubini, E. GABAergic Signaling as Therapeutic Target for Autism Spectrum Disorders. *Front. Pediatr.* **2014**, *2*. [CrossRef] [PubMed]
82. Nanou, E.; Catterall, W.A. Calcium Channels, Synaptic Plasticity, and Neuropsychiatric Disease. *Neuron* **2018**. [CrossRef] [PubMed]
83. Jungbluth, H.; Treves, S.; Zorzato, F.; Sarkozy, A.; Ochala, J.; Sewry, C.; Phadke, R.; Gautel, M.; Muntoni, F. Congenital Myopathies: Disorders of Excitation-Contraction Coupling and Muscle Contraction. *Nat. Rev. Neurol.* **2018**, *14*, 151–167. [CrossRef] [PubMed]
84. Agrawal, A.; Suryakumar, G.; Rathor, R. Role of Defective Ca^{2+} signaling in Skeletal Muscle Weakness: Pharmacological Implications. *J. Cell Commun. Signal.* **2018**, 1–15. [CrossRef] [PubMed]
85. Chernoff, E.A.G.; Hilfer, S.R. Calcium Dependence and Contraction in Somite Formation. *Tissue Cell* **1982**, *14*, 435–449. [CrossRef]

86. Ferrari, M.B.; Spitzer, N.C. Calcium Signaling in the Developing *Xenopus* Myotome. *Dev. Biol.* **1999**, *213*, 269–282. [CrossRef] [PubMed]
87. Cheung, C.Y.; Webb, S.E.; Love, D.R.; Miller, A.L. Visualization, Characterization and Modulation of Calcium Signaling during the Development of Slow Muscle Cells in Intact Zebrafish Embryos. *Int. J. Dev. Biol.* **2011**, *55*, 153–174. [CrossRef] [PubMed]
88. Kelu, J.J.; Webb, S.E.; Parrington, J.; Galione, A.; Miller, A.L. Ca^{2+} release via Two-Pore Channel Type 2 (TPC2) Is Required for Slow Muscle Cell Myofibrillogenesis and Myotomal Patterning in Intact Zebrafish Embryos. *Dev. Biol.* **2017**, *425*, 109–129. [CrossRef] [PubMed]
89. Kelu, J.J.; Chan, H.L.H.; Webb, S.E.; Cheng, A.H.H.; Ruas, M.; Parrington, J.; Galione, A.; Miller, A.L. Two-Pore Channel 2 Activity Is Required for Slow Muscle Cell-Generated Ca^{2+} Signaling during Myogenesis in Intact Zebrafish. *Int. J. Dev. Biol.* **2015**. [CrossRef] [PubMed]
90. Ferrari, M.B.; Rohrbough, J.; Spitzer, N.C. Spontaneous Calcium Transients Regulate Myofibrillogenesis in Embryonic Xenopus Myocytes. *Dev. Biol.* **1996**, *178*, 484–497. [CrossRef] [PubMed]
91. Campbell, N.R.; Podugu, S.P.; Ferrari, M.B. Spatiotemporal Characterization of Short versus Long Duration Calcium Transients in Embryonic Muscle and Their Role in Myofibrillogenesis. *Dev. Biol.* **2006**, *292*, 253–264. [CrossRef] [PubMed]
92. Zvaritch, E.; Depreux, F.; Kraeva, N.; Loy, R.E.; Goonasekera, S.A.; Boncompagni, S.; Kraev, A.; Gramolini, A.O.; Dirksen, R.T.; Franzini-Armstrong, C.; et al. An Ryr1I4895T Mutation Abolishes Ca^{2+} Release Channel Function and Delays Development in Homozygous Offspring of a Mutant Mouse Line. *Proc. Natl. Acad. Sci. USA* **2007**, *104*, 18537–18542. [CrossRef] [PubMed]
93. Farini, A.; Sitzia, C.; Cassinelli, L.; Colleoni, F.; Parolini, D.; Giovanella, U.; Maciotta, S.; Colombo, A.; Meregalli, M.; Torrente, Y. Inositol 1,4,5-Trisphosphate (IP3)-Dependent Ca^{2+} Signaling Mediates Delayed Myogenesis in Duchenne Muscular Dystrophy Fetal Muscle. *Development* **2016**, *143*, 658–669. [CrossRef] [PubMed]
94. Allard, B. From Excitation to Intracellular Ca^{2+} movements in Skeletal Muscle: Basic Aspects and Related Clinical Disorders. *Neuromuscul. Disord.* **2018**. [CrossRef] [PubMed]
95. Stiber, J.A.; Rosenberg, P.B. The Role of Store-Operated Calcium Influx in Skeletal Muscle Signaling. *Cell Calcium* **2011**, *49*, 341–349. [CrossRef] [PubMed]
96. McCarl, C.A.; Picard, C.; Khalil, S.; Kawasaki, T.; Röther, J.; Papolos, A.; Kutok, J.; Hivroz, C.; LeDeist, F.; Plogmann, K.; et al. ORAI1 Deficiency and Lack of Store-Operated Ca^{2+} entry Cause Immunodeficiency, Myopathy, and Ectodermal Dysplasia. *J. Allergy Clin. Immunol.* **2009**, *124*, 1311–1318. [CrossRef] [PubMed]
97. Picard, C.; McCarl, C.-A.; Papolos, A.; Khalil, S.; Lüthy, K.; Hivroz, C.; LeDeist, F.; Rieux-Laucat, F.; Rechavi, G.; Rao, A.; et al. STIM1 Mutation Associated with a Syndrome of Immunodeficiency and Autoimmunity. *N. Engl. J. Med.* **2009**, *360*, 1971–1980. [CrossRef] [PubMed]
98. Pisaniello, A. The Block of Ryanodine Receptors Selectively Inhibits Fetal Myoblast Differentiation. *J. Cell Sci.* **2003**. [CrossRef]
99. Webb, S.E.; Cheung, C.C.Y.; Chan, C.M.; Love, D.R.; Miller, A.L. Application of Complementary Luminescent and Fluorescent Imaging Techniques to Visualize Nuclear and Cytoplasmic Ca^{2+} Signalling during the In Vivo Differentiation of Slow Muscle Cells in Zebrafish Embryos under Normal and Dystrophic Conditions. *Clin. Exp. Pharmacol. Physiol.* **2012**, *39*, 78–86. [CrossRef] [PubMed]
100. Kloesel, B.; Dinardo, J.A.; Body, S.C. Cardiac Embryology and Molecular Mechanisms of Congenital Heart Disease: A Primer for Anesthesiologists. *Anesthesia Analgesia* **2016**. [CrossRef] [PubMed]
101. Schleich, J.M.; Abdulla, T.; Summers, R.; Houyel, L. An Overview of Cardiac Morphogenesis. *Arch. Cardiovasc. Dis.* **2013**. [CrossRef] [PubMed]
102. Moorman, A.; Webb, S.; Brown, N.; Lamers, W.; Anderson, R. Development of the Heart: Formation of the Cardiac Chambers and Arterial Trunks. *Heart* **2003**, *89*, 806–814. [CrossRef] [PubMed]
103. Créton, R.; Speksnijder, J.E.; Jaffe, L.F. Patterns of Free Calcium in Zebrafish Embryos. *J. Cell Sci.* **1998**, *111*, 1613–1622. [CrossRef] [PubMed]
104. Tallini, Y.N.; Ohkura, M.; Choi, B.-R.; Ji, G.; Imoto, K.; Doran, R.; Lee, J.; Plan, P.; Wilson, J.; Xin, H.-B.; et al. Imaging Cellular Signals in the Heart in Vivo: Cardiac Expression of the High-Signal Ca^{2+} Indicator GCaMP2. *Proc. Natl. Acad. Sci. USA* **2006**, *103*, 4753–4758. [CrossRef] [PubMed]
105. Porter, G.A.; Makuck, R.F.; Rivkees, S.A. Intracellular Calcium Plays an Essential Role in Cardiac Development. *Dev. Dyn.* **2003**, *227*, 280–290. [CrossRef] [PubMed]

106. Hotchkiss, A.; Feridooni, T.; Zhang, F.; Pasumarthi, K.B.S. The Effects of Calcium Channel Blockade on Proliferation and Differentiation of Cardiac Progenitor Cells. *Cell Calcium* **2014**, *55*, 238–251. [CrossRef] [PubMed]
107. Webb, S.E.; Miller, A.L. Calcium Signalling during Embryonic Development. *Nat. Rev. Mol. Cell Biol.* **2003**, *4*, 539–551. [CrossRef] [PubMed]
108. Harvey, R.P. Patterning the Vertebrate Heart. *Nat. Rev. Genet.* **2002**, *3*, 544–556. [CrossRef] [PubMed]
109. Carroll, T.J.; Vize, P.D. Synergism between Pax-8 and Lim-1 in Embryonic Kidney Development. *Dev. Biol.* **1999**, *214*, 46–59. [CrossRef] [PubMed]
110. Buisson, I.; Le Bouffant, R.; Futel, M.; Riou, J.F.; Umbhauer, M. Pax8 and Pax2 Are Specifically Required at Different Steps of Xenopus Pronephros Development. *Dev. Biol.* **2015**, *397*, 175–190. [CrossRef] [PubMed]
111. Tena, J.J.; Neto, A.; de la Calle-Mustienes, E.; Bras-Pereira, C.; Casares, F.; Gómez-Skarmeta, J.L. Odd-Skipped Genes Encode Repressors That Control Kidney Development. *Dev. Biol.* **2007**, *301*, 518–531. [CrossRef] [PubMed]
112. Leclerc, C.; Webb, S.E.; Miller, A.L.; Moreau, M. An Increase in Intracellular Ca^{2+} is Involved in Pronephric Tubule Differentiation in the Amphibian Xenopus Laevis. *Dev. Biol.* **2008**, *321*, 357–367. [CrossRef] [PubMed]
113. Gallagher, A.R.; Hidaka, S.; Gretz, N.; Witzgall, R. Molecular Basis of Autosomal-Dominant Polycystic Kidney Disease. *Cell. Mol. Life Sci.* **2002**, *59*, 682–693. [CrossRef] [PubMed]
114. Boucek, M.M.; Snyderman, R. Calcium Influx Requirement for Human Neutrophil Chemotaxis: Inhibition by Lanthanum Chloride. *Science* **1976**, *193*, 905–907. [CrossRef] [PubMed]
115. Oh-hora, M.; Rao, A. Calcium Signaling in Lymphocytes. *Curr. Opin. Immunol.* **2008**, *20*, 250–258. [CrossRef] [PubMed]
116. Oh-Hora, M. Calcium Signaling in the Development and Function of T-Lineage Cells. *Immunol. Rev.* **2009**. [CrossRef] [PubMed]
117. Beerman, R.W.W.; Matty, M.A.A.; Au, G.G.G.; Looger, L.L.L.; Choudhury, K.R.R.; Keller, P.J.J.; Tobin, D.M.M. Direct In Vivo Manipulation and Imaging of Calcium Transients in Neutrophils Identify a Critical Role for Leading-Edge Calcium Flux. *Cell Rep.* **2015**, *13*, 2107–2117. [CrossRef] [PubMed]
118. Marks, P.W.; Maxfield, F.R. Transient Increases in Cytosolic Free Calcium Appear to Be Required for the Migration of Adherent Human Neutrophils. *J. Cell Biol.* **1990**, *110*, 43–52. [CrossRef] [PubMed]
119. Hogan, P.G.; Lewis, R.S.; Rao, A. Molecular Basis of Calcium Signaling in Lymphocytes: STIM and ORAI. *Annu. Rev. Immunol.* **2010**, *28*, 491–533. [CrossRef] [PubMed]
120. Baba, Y.; Nishida, K.; Fujii, Y.; Hirano, T.; Hikida, M.; Kurosaki, T. Essential Function for the Calcium Sensor STIM1 in Mast Cell Activation and Anaphylactic Responses. *Nat. Immunol.* **2008**, *9*, 81–88. [CrossRef] [PubMed]
121. Vig, M.; DeHaven, W.I.; Bird, G.S.; Billingsley, J.M.; Wang, H.; Rao, P.E.; Hutchings, A.B.; Jouvin, M.-H.; Putney, J.W.; Kinet, J.-P. Defective Mast Cell Effector Functions in Mice Lacking the CRACM1 Pore Subunit of Store-Operated Calcium Release–activated Calcium Channels. *Nat. Immunol.* **2008**, *9*, 89–96. [CrossRef] [PubMed]
122. Degen, K.E.; Gourdie, R.G. Embryonic Wound Healing: A Primer for Engineering Novel Therapies for Tissue Repair. *Birth Defects Res. Part C—Embryo Today Rev.* **2012**, *96*, 258–270. [CrossRef] [PubMed]
123. Ud-Din, S.; Volk, S.W.; Bayat, A. Regenerative Healing, Scar-Free Healing and Scar Formation across the Species: Current Concepts and Future Perspectives. *Exp. Dermatol.* **2014**, *23*, 615–619. [CrossRef] [PubMed]
124. Herrgen, L.; Voss, O.P.; Akerman, C.J. Calcium-Dependent Neuroepithelial Contractions Expel Damaged Cells from the Developing Brain. *Dev. Cell* **2014**, *31*, 599–613. [CrossRef] [PubMed]
125. Cordeiro, J.V.; Jacinto, A. The Role of Transcription-Independent Damage Signals in the Initiation of Epithelial Wound Healing. *Nat. Rev. Mol. Cell Biol.* **2013**, *14*, 249–262. [CrossRef] [PubMed]
126. Razzell, W.; Evans, I.R.; Martin, P.; Wood, W. Calcium Flashes Orchestrate the Wound Inflammatory Response through Duox Activation and Hydrogen Peroxide Release. *Curr. Biol.* **2013**, *23*, 424–429. [CrossRef] [PubMed]
127. Niethammer, P.; Grabher, C.; Look, A.T.; Mitchison, T.J. A Tissue-Scale Gradient of Hydrogen Peroxide Mediates Rapid Wound Detection in Zebrafish. *Nature* **2009**, *459*, 996–999. [CrossRef] [PubMed]
128. Tu, M.K.; Borodinsky, L.N. Spontaneous Calcium Transients Manifest in the Regenerating Muscle and Are Necessary for Skeletal Muscle Replenishment. *Cell Calcium* **2014**, *56*, 34–41. [CrossRef] [PubMed]
129. Etulain, J. Platelets in Wound Healing and Regenerative Medicine. *Platelets* **2018**, *15*, 1–3. [CrossRef] [PubMed]

130. Wolf, K.; Braun, A.; Haining, E.J.; Tseng, Y.-L.; Kraft, P.; Schuhmann, M.K.; Gotru, S.K.; Chen, W.; Hermanns, H.M.; Stoll, G.; et al. Partially Defective Store Operated Calcium Entry and Hem(ITAM) Signaling in Platelets of Serotonin Transporter Deficient Mice. *PLoS ONE* **2016**, *11*, e0147664. [CrossRef] [PubMed]
131. Chen, W.; Thielmann, I.; Gupta, S.; Subramanian, H.; Stegner, D.; van Kruchten, R.; Dietrich, A.; Gambaryan, S.; Heemskerk, J.W.M.; Hermanns, H.M.; et al. Orai1-Induced Store-Operated Ca^{2+} Entry Enhances Phospholipase Activity and Modulates Canonical Transient Receptor Potential Channel 6 Function in Murine Platelets. *J. Thromb. Haemost.* **2014**, *12*, 528–539. [CrossRef] [PubMed]
132. Gotru, S.K.; Chen, W.; Kraft, P.; Becker, I.C.; Wolf, K.; Stritt, S.; Zierler, S.; Hermanns, H.M.; Rao, D.; Perraud, A.L.; et al. TRPM7 Kinase Controls Calcium Responses in Arterial Thrombosis and Stroke in Mice. *Arterioscler. Thromb. Vasc. Biol.* **2018**, *38*, 344–352. [CrossRef] [PubMed]
133. Lysaght, M.J.; Jaklenec, A.; Deweerd, E. Great Expectations: Private Sector Activity in Tissue Engineering, Regenerative Medicine, and Stem Cell Therapeutics. *Tissue Eng. Part A* **2008**, *14*, 305–315. [CrossRef] [PubMed]
134. Patterson, R.A.; Juarez, M.T.; Hermann, A.; Sasik, R.; Hardiman, G.; McGinnis, W. Serine Proteolytic Pathway Activation Reveals an Expanded Ensemble of Wound Response Genes in Drosophila. *PLoS ONE* **2013**, *8*. [CrossRef] [PubMed]
135. Tsujioka, H.; Kunieda, T.; Katou, Y.; Shirahige, K.; Kubo, T. Unique Gene Expression Profile of the Proliferating Xenopus Tadpole Tail Blastema Cells Deciphered by RNA-Sequencing Analysis. *PLoS ONE* **2015**. [CrossRef] [PubMed]
136. Mouse Gene Expression Data Search. Available online: http://www.informatics.jax.org/gxd (accessed on 25 September 2018).
137. ZFIN Expression Search. Available online: https://zfin.org/action/expression/search (accessed on 25 September 2018).
138. Gene Expression Search. Available online: http://www.xenbase.org/geneExpression/geneExpressionSearch.do?method=display (accessed on 25 September 2018).
139. Kolios, G.; Moodley, Y. Introduction to Stem Cells and Regenerative Medicine. *Respiration* **2013**, *85*, 3–10. [CrossRef] [PubMed]
140. Ermakov, A.; Daks, A.; Fedorova, O.; Shuvalov, O.; Barlev, N.A. Ca^{2+}-Depended Signaling Pathways Regulate Self-Renewal and Pluripotency of Stem Cells. *Cell Biol. Int.* **2018**, *42*, 1086–1096. [CrossRef] [PubMed]
141. Rahaman, M.N.; Mao, J.J. Stem Cell-Based Composite Tissue Constructs for Regenerative Medicine. *Biotechnol. Bioeng.* **2005**, *91*, 261–284. [CrossRef] [PubMed]
142. Hasan, A. *Tissue Engineering for Artificial Organs: Regenerative Medicine, Smart Diagnostics and Personalized Medicine*; John Wiley & Sons: Hoboken, NJ, USA, 2016; Volume 1–2.
143. Jergova, S.; Gajavelli, S.; Varghese, M.S.; Shekane, P.; Sagen, J. Analgesic Effect of Recombinant GABAergic Cells in a Model of Peripheral Neuropathic Pain. *Cell Transplant.* **2016**, *25*, 629–643. [CrossRef] [PubMed]
144. Méry, A.; Aimond, F.; Ménard, C.; Mikoshiba, K.; Michalak, M.; Pucéat, M. Initiation of Embryonic Cardiac Pacemaker Activity by Inositol 1,4,5-Trisphosphate-Dependent Calcium Signaling. *Mol. Biol. Cell* **2005**, *16*, 2414–2423. [CrossRef] [PubMed]
145. Limpitikul, W.B.; Dick, I.E.; Tester, D.J.; Boczek, N.J.; Limphong, P.; Yang, W.; Choi, M.H.; Babich, J.; Disilvestre, D.; Kanter, R.J.; et al. A Precision Medicine Approach to the Rescue of Function on Malignant Calmodulinopathic Long-QT Syndrome. *Circ. Res.* **2017**, *120*, 39–48. [CrossRef] [PubMed]

6

Role of KCa3.1 Channels in Modulating Ca^{2+} Oscillations during Glioblastoma Cell Migration and Invasion

Luigi Catacuzzeno * and Fabio Franciolini *

Department of Chemistry, Biology and Biotechnology, University of Perugia, 06134 Perugia, Italy
* Correspondence: luigi.catacuzzeno@unipg.it (L.C.); fabio.franciolini@unipg.it (F.F)

Abstract: Cell migration and invasion in glioblastoma (GBM), the most lethal form of primary brain tumors, are critically dependent on Ca^{2+} signaling. Increases of $[Ca^{2+}]_i$ in GBM cells often result from Ca^{2+} release from the endoplasmic reticulum (ER), promoted by a variety of agents present in the tumor microenvironment and able to activate the phospholipase C/inositol 1,4,5-trisphosphate PLC/IP_3 pathway. The Ca^{2+} signaling is further strengthened by the Ca^{2+} influx from the extracellular space through Ca^{2+} release-activated Ca^{2+} (CRAC) currents sustained by Orai/STIM channels, meant to replenish the partially depleted ER. Notably, the elevated cytosolic $[Ca^{2+}]_i$ activates the intermediate conductance Ca^{2+}-activated K (KCa3.1) channels highly expressed in the plasma membrane of GBM cells, and the resulting K^+ efflux hyperpolarizes the cell membrane. This translates to an enhancement of Ca^{2+} entry through Orai/STIM channels as a result of the increased electromotive (driving) force on Ca^{2+} influx, ending with the establishment of a recurrent cycle reinforcing the Ca^{2+} signal. Ca^{2+} signaling in migrating GBM cells often emerges in the form of intracellular Ca^{2+} oscillations, instrumental to promote key processes in the migratory cycle. This has suggested that KCa3.1 channels may promote GBM cell migration by inducing or modulating the shape of Ca^{2+} oscillations. In accordance, we recently built a theoretical model of Ca^{2+} oscillations incorporating the KCa3.1 channel-dependent dynamics of the membrane potential, and found that the KCa3.1 channel activity could significantly affect the IP_3 driven Ca^{2+} oscillations. Here we review our new theoretical model of Ca^{2+} oscillations in GBM, upgraded in the light of better knowledge of the KCa3.1 channel kinetics and Ca^{2+} sensitivity, the dynamics of the Orai/STIM channel modulation, the migration and invasion mechanisms of GBM cells, and their regulation by Ca^{2+} signals.

Keywords: KCa3.1 channels; glioblastoma; cell migration; calcium oscillations; mathematical model

1. The Glioblastoma

The large majority (more than 90%) of cancer deaths are due not to the primary tumor per se, but to relapses arising from new foci established in distant organs via metastasis [1]. Glioblastoma (GBM), the most common and aggressive form of primary brain tumors, is no exception, though it does not metastasize in the classical way (that is, by colonizing other tissues via the bloodstream), but invades brain parenchyma by detaching from the original tumor mass and infiltrating into the healthy tissue by degrading the extracellular matrix or squeezing through the brain interstitial spaces. The urgency of tackling the migration and invasion issues of GBM tumors is clear.

The 2016 World Health Organization classification of the various types of brain tumors regards the presence of isocitrate dehydrogenase gene (*IDH1/2*) mutations as one of the most critical biomarkers. Accordingly, *IDH1/2* wildtype gliomas are categorized as glioblastoma (formerly primary glioblastoma) and *IDH1/2*-mutated gliomas (including formerly classified secondary glioblastoma) as astrocytic glioma and oligodendroglioma. GBMs are further subdivided into four groups (Proneural,

Neural, Classical, and Mesenchymal), mainly based on the abnormally high levels of mutated genes (i.e., *EGFR* is highly upregulated in >98% of Classical GBM, whereas *TP53* (p53), which is most frequently mutated in Proneural GBM (50–60% of patients) is rarely mutated in Classical GBM). In spite of the intensive basic and clinical studies carried out over the past decades, and modern diagnostics and treatments, the average life expectancy for GBM patients is still only around 15 months. The major obstacle with GBM remains its high migratory and invasive potential into healthy brain parenchyma, which prevents complete surgical removal of tumor cells. Even with full clinical treatment (temozolomide-based chemotherapy and radiation therapy), tumors normally recur at some distance from the site of resection, establishing new tumor lesions that are by far the primary cause of mortality in GBM patients. Arguably, at the time of surgery, large numbers of cells have already detached from the original tumor mass and invaded normal brain tissue. Although GBM cell migration and invasion have been deeply investigated, many aspects of these processes are still poorly understood.

GBM cell migration is a highly regulated multistep process that initiates with GBM cells losing adhesion with surrounding elements, avoiding the cell death often associated with extracellular matrix (ECM) disconnection, and acquiring a highly migratory phenotype, which is a critical feature of the invasive process. The basic mechanisms underlying migration of GBM cells are common to most types of migratory cells. Migration is a property of many non-tumor cells, although it is often restricted to specific developmental stages or environmental conditions; the migration of tumor cells could be viewed as the result of mutation-induced dysregulation of specific biochemical pathways that in healthy tissue keep cell migration dormant.

2. Glioblastoma Cell Migration and Ca^{2+} Signaling

2.1. Cell Migration

The basic mechanisms of cell locomotion are now fairly well established. Locomotion can be described as the cyclical repeating of two main processes: (i) protrusion of the cell front due to local gain of cell volume mostly generated by active $Na^+/K^+/2Cl^-$ cotransport accompanied by isoosmotically obliged water, and actin polymerization, with formation of pseudopods; (ii) retraction of the rear cell body in the direction of motion, due to forces produced by actomyosin contraction, accompanied by loss of cell volume generated by passive ion (mainly K^+ and Cl^-) fluxes and osmotic water [2,3]. These two processes involve the coordinated and localized formation of integrin-dependent cell adhesions at the leading edge, and their disassembly at the cell rear [4,5].

Protrusion of the cell front is sustained by localized polymerization of submembrane actin-based cytoskeleton that generates the pushing force and forms flat lamellipodia or needle-like filipodia. A large variety of signaling molecules have been shown to play a leading role in these processes, including the Rho GTPases family (that act as molecular switches to control downstream transduction pathways), and their effector proteins CDC42, RAC1, and RhoA. PI3 kinases have also been deeply implicated in controlling actin polymerization and lamellipodium extension. Activation of PI3 kinase by the pro-invasive signal molecules present in the tumor microenvironment functions as the trigger of the process, in that its activation initiates actin polymerization and generates membrane protrusion [5].

The retraction of the cell rear depends on the contractile forces generated by the activation of the myosin motors along crosslinked actin filaments. The myosin molecule is formed by two heavy chains that make up the two heads and the coiled-coil tail, and four—two essential and two regulatory—myosin light chains (MLCs). The myosin motor is primarily activated by a Ca^{2+}-dependent cascade whereby a cytosolic Ca^{2+} increase activates a MLC kinase (MLCK) that leads to the regulatory MLC phosphorylation, which allows the myosin motor to crosslink with actin and produce tension to pull the rear end of the cell (however, some studies claim that the motor activity of myosin is not required for cell migration [6,7], playing instead a role in the establishment of cell polarity and in the coordination between different cell domains [7,8]). Clearly, to make the rear cell effectively move, the focal adhesions that anchor the actin cytoskeleton to the extracellular matrix (ECM) must be disassembled, normally through a proteolytic process that involves calpain, but also actin microtubules and focal adhesion kinase (FAK)-recruited dynamin that internalizes the integrins. Increasing actomyosin tension disrupts the possible residual resistances of focal adhesions at the cell rear [2]. Myosin can also be modulated by other pathways, including a Rho-dependent cascade (Rho → ROCK → MLC phosphorylation, or MLC phosphatase activation) that does not depend on cytosolic Ca^{2+}.

This locomotion cycle can be initiated by a variety of external stimuli or directional cues which are sensed and decoded by specialized receptors on the plasma membrane and transferred to the cell interior via distinct signaling pathways to actuate the mechanical response. Most common migration stimuli rely on cells sensing (by specific G protein-coupled receptors) local gradients in the concentration of chemical factors (chemoattractants) [9] present in the tumor microenvironment (for instance chemokines IL-8 or CXCL12), or certain growth factors, including the platelet-derived growth factor (PDGF) that plays an important role in tumor metastasis [10]. Cancer cells can also be guided by gradients of bound ligands. Generally attached to the ECM, these ligands are recognized by specific integrins that form oriented focal adhesion to direct the formation of invadopodia, thus the direction of movement. Common is also the observation of cancer cells being guided by mechanical stimuli (mechanotaxis), namely the stiffness of the extracellular matrix they try to penetrate [11]. It is important to note that in vivo cells can be simultaneously subjected to multiple types of cues that need be evaluated as a whole in order to provide appropriate responses.

2.2. *Cell Volume Changes Associated with Migration*

It is important to recognize that these two steps—cell front protrusion and cell rear retraction—normally occur in succession, implying that cells are subjected to major changes in cell volume, especially the cell rear retraction. This requires compensatory ion fluxes (and osmotic water flow) across the plasma membrane in order to maintain proper cytosol osmolarity. Along the line proposed by [12] with regard to cell migration in general, during the protrusion of the cell front the maintenance of a stable osmolarity is sustained by cotransporters and ion exchangers, for instance $Na^+/K^+/2Cl^-$ cotransport, followed by osmotically driven water. By contrast, the control of osmolarity during cell rear retraction critically depends on the activation of K channels and ensuing efflux of K^+ ions (accompanied by Cl^- fluxes, to which the membrane has become highly permeant, and osmotic water). Following the opening of K channels the membrane potential hyperpolarizes to a value between the equilibrium potentials of Cl^- and K^+ (E_{Cl} and E_K), a condition that allows a significant efflux of KCl with consequent loss of water and cell volume decrease that facilitates cell body retraction. This view can explain the slowing of cell locomotion following the inhibition of K and Cl channels [13]. A number of papers indicate that the channel types primarily involved in osmolarity control are the intermediate conductance Ca^{2+}-activated K channel (KCa3.1) and the Cl channel ClC3 [13,14].

2.3. *Ca^{2+} Oscillations*

Much work shows that cell migration is strictly regulated by Ca^{2+}, which is no surprise given that many proteins involved in migration such as myosin, myosin light chain kinase (MLCK), Ca^{2+}/calmodulin-dependent protein kinase II are Ca^{2+} sensitive [15]. In the resting cell Ca^{2+} concentration is kept very low (<100 nM) to prevent activation of Ca^{2+}-sensitive proteins that act as Ca^{2+} sensor and initiation of unwanted biological processes. Ca^{2+} concentration can easily increase more than ten-fold (well over 1 μM) following several types of cell stimulation, and this would bring many of the major cytoplasmic Ca^{2+}-sensitive proteins (calmodulin, PKA, Ca^{2+}-activated channels, etc.) to activate. This opens the question of how Ca^{2+} signaling, under these conditions, can selectively activate a specific Ca^{2+} sensor protein and the associated biochemical cascade. Several strategies have evolved in this respect, but specificity is mostly attained by confining the signal at sub-cellular level, or by regulating kinetics and magnitude of the Ca^{2+} signal. Since in the cytoplasm Ca^{2+} diffusion is extremely low and buffering high, the opening of Ca^{2+} permeant channels results in a very localized

Ca^{2+} increase (the Ca^{2+} microdomain), attaining spatial discrimination of the target proteins and linked downstream pathways. Another strategy relies on the temporal features of Ca^{2+} signals (rising rate, duration, repetition of spikes, etc.) that biological systems are capable to interpret.

In non-excitable cells, including GBM, typical Ca^{2+} signals induced experimentally by robust chemical (hormone) stimulation are slow, large and generally sustained. Hormones binding to their specific G-protein-coupled receptors (GPCRs) often results in the PLC-dependent synthesis of inositol 1,4,5-trisphosphate (IP_3), and consequent Ca^{2+} release from the endoplasmic reticulum (ER) via IP_3 receptors. Eventually, the sustained Ca^{2+} increase subsides as result of the activation of Ca^{2+} pumps placed on the sarco/endoplasmic reticulum (SERCA), or the plasma membrane (PMCA) that transfer Ca^{2+} ions from the cytosol to intracellular stores or the extracellular space.

A physiologically more meaningful type of Ca^{2+} signal, especially in relation to cell migration, where critical Ca^{2+}-dependent processes are cyclical, comes in the form of Ca^{2+} oscillations. It has recently become clear that this type of oscillatory Ca^{2+} increases is what normally occurs when using physiological agonist stimulations. The mechanisms underlying Ca^{2+} oscillations are not fully understood, and often depend on the cell model and agonist used. In some systems Ca^{2+} oscillations are secondary to oscillations of the cytoplasmic IP_3 level, which may be due to various types of feedback control that for brief intervals uncouple the GTP-coupled receptor from PLC. For instance, the reported negative feedback of the PLC products, IP_3 and diacylglycerol (DAG), on PLC itself, or upstream on the GPCR, would result in IP_3 oscillations [16], which would in turn generate cyclical release of Ca^{2+} from intracellular stores, and the Ca^{2+} oscillations [17].

More commonly, however, Ca^{2+} oscillations can be observed in the presence of a constant level of IP_3. The prevailing view is that, under these conditions, the repetitive Ca^{2+} oscillations secondary to moderate (physiological) hormone or agonist activation of membrane PLC and production of a constant amount of IP_3 arises from the biphasic effects of cytosolic Ca^{2+} on the IP_3 receptor gating, with the first phase being embodied by the establishment of the Ca^{2+}-induced Ca^{2+} release (CICR) mechanism, whereby a moderate increase of $[Ca^{2+}]$ in the cytosol via IP_3 receptor causes a positive feedback activation of the IP_3 receptor that results in a higher Ca^{2+} released from internal stores ([18]; Figure 1A).

The second phase occurs when the cytosolic $[Ca^{2+}]$ reaches significantly higher levels, shifting the previous positive feedback on IP_3 receptors into a delayed negative feedback that results in their closing and $[Ca^{2+}]_i$ returning to resting level by the action of both SERCA and PMCA Ca^{2+} pumps (Figure 1A). This view finds support from the observation that at constant levels of IP_3, the activity of the IP_3 receptor and the cytosolic Ca^{2+} concentration display a bell-shaped function, whereby low Ca^{2+} concentrations activate the receptor, whereas high Ca^{2+} concentrations inhibit it [19–22]. Ca^{2+} oscillations generated according to this scheme (i.e., Figure 1A) would however show a rapid decrease of spike amplitude with time, because part of the Ca^{2+} released from the ER during each spike will be pumped out of the cell by PMCA, and be no longer available for refilling the ER and contributing to the next Ca^{2+} spike. This Ca^{2+} oscillations time course is observed experimentally upon removal of external Ca^{2+}, or blockade of Ca^{2+} influx from the extracellular space. To have sustained (or slow decaying) Ca^{2+} oscillations, external Ca^{2+} pumped out of the cell by PMCAs must be allowed to re-enter the cell. This is most commonly accomplished through the mechanism of store-operated Ca^{2+} entry (SOCE).

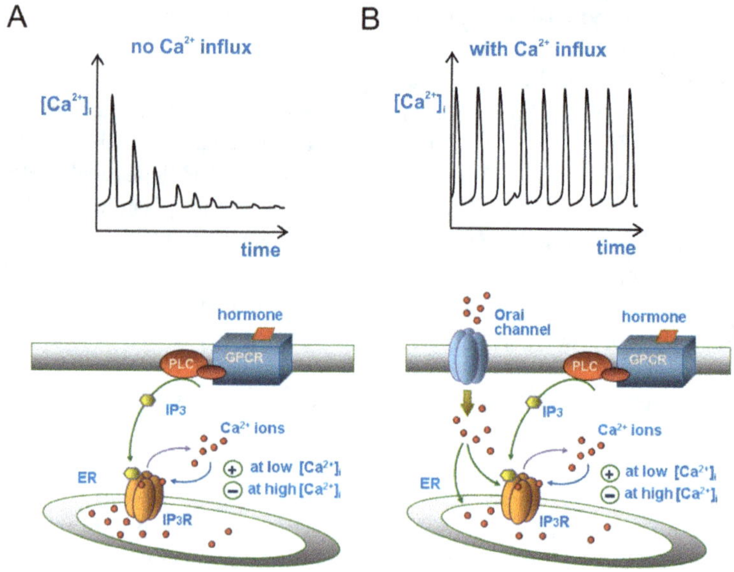

Figure 1. Ca^{2+} oscillations in response to inositol thriphosphate (IP_3) increase, with and without Ca^{2+} influx from extracellular space. (**A**) Bottom, drawing illustrating the hormone-based production of IP_3 that activates the IP_3 receptor to release Ca^{2+} from endoplasmic reticulum (ER). The biphasic effects of cytosolic Ca^{2+} on IP_3 receptor gating (the basic mechanism for Ca^{2+} oscillations), whereby Ca^{2+} modulates positively the receptor at low $[Ca^{2+}]$ but negatively at high $[Ca^{2+}]$, is also illustrated. Top, Ca^{2+} oscillations as produced from the schematics below. Note the decaying trend of Ca^{2+} spikes due to the absence of Ca^{2+} influx from extracellular space; (**B**) Here the drawing has been enriched with a Ca^{2+} influx apparatus from extracellular space through ER-depletion activated Orai channels on the plasma membrane (bottom), which generates sustained Ca^{2+} oscillations (top). For clarity, SERCA and PMCA Ca^{2+} pumps have not been sketched in the drawing, although their activity has always been taken into account. For the same reason, we omitted to draw STIM protein of the ER.

2.4. Store-Operated Ca^{2+} Entry (SOCE)

When Ca^{2+} is released from the ER, i.e., following hormone stimulation, the Ca^{2+} concentration within the organelle decreases. It is this decrease to trigger the process (known as SOCE) whereby Ca^{2+} channels in the plasma membrane are activated, allowing entry of Ca^{2+} into the cell. Although the process of SOCE was described some 30 years ago, its mechanisms and molecular counterparts—the STIM and Orai proteins—have been identified much more recently [23–25]. The stromal interaction molecules (STIMs) are a family of one-passage integral proteins of the ER that act as sensors of the Ca^{2+} concentration within the ER. Orai proteins form instead a group of highly selective Ca^{2+} channels placed in the plasma membrane. Orai channels are neither activated by voltage, nor by agonists (in the classical sense), and are normally silent when the cell is at rest and the ER filled with Ca^{2+} ions. Orai Ca^{2+} channels are opened by the interaction with activated STIM proteins, which occurs following depletion of Ca^{2+} in the ER, and their opening results in Ca^{2+} ions flowing inside the cell (the mechanism has been recently reviewed in depth on neurons by [26]). This Ca^{2+} entry into the cytoplasm enables the refilling of Ca^{2+} depleted ER through the SERCA pumps, and make the Ca^{2+} oscillations amplitude to be kept stable with time (Figure 1B). In the context of our review topic, this feedback mechanism of store-operated Ca^{2+} entry gains further relevance for two more reasons: it contributes to further modulating the IP_3 receptor (this modulation is also dependent on the cell type—for instance, in U87-MG human GBM cells Ca^{2+} oscillations are affected only marginally by removal of Ca^{2+} influx [27], whereas in rat C6 GBM cells this maneuver fully suppresses Ca^{2+}

oscillations [28]), and it represents the target on which the KCa3.1 channel exerts its modulation of Ca^{2+} oscillations, as we will see later.

It needs to be recalled that the store-operated Ca^{2+} entry is not the only suggested mechanism—although it is by far the most widely accepted—that links hormone stimulation to increased plasma membrane Ca^{2+} entry. For instance, heteromers of Orai1 and Orai3 can be activated by arachidonic acid [26], and the Ca^{2+}-permeant TRPC6 by receptor-induced PLC activation via direct action of DAG. Notably, neither pathway involves IP_3 receptor or Ca^{2+} depletion from the ER.

2.5. *Ca^{2+} Oscillations and Cell Migration*

Numerous studies have shown that migrating GBM cells display evident Ca^{2+} oscillations having a period of several tens of seconds, whose presence appears to be essential to the process of migration. For example, in U87-MG cells Ca^{2+} oscillations may be induced by the pro-migratory fetal calf serum (FCS), and trigger focal adhesion disassembly during cell rear retraction [29,30]. Similarly, in D54-MG cells prolonged exposure to bradykinin induces Ca^{2+} oscillations, which significantly enhance cell motility [31]. Mechanistically, the Ca^{2+} signal may translate into actomyosin contraction and rear cell retraction by first binding to calmodulin (CaM). The activated complex, Ca^{2+}/CaM, then activates MLCK, the kinase that phosphorylates the regulatory myosin light chains (MLC), thereby triggering cross-bridge movements and actomyosin contraction. It has been shown that MLCK can be modulated by several other protein kinases including PKA, PKC, and RhoK, indicating that this critical enzyme for cell migration is under the influence and control of many extracellular signals and cytoplasmic pathways. Although the consensus on this mechanism and the role of MLCK is widespread, a little caution is needed as most studies used a pharmacological approach, but really specific MLCK inhibitors were not available [32]. A recent report showed, for instance, that knockout cells for MLCK maintained unaltered (or increased) their migratory ability [33]. This shows that if the critical role of Ca^{2+} in cell motility is undisputed, uncertainties may remain as to how Ca^{2+} is linked to the cell migratory machinery. The role of the Ca^{2+}-activated K channels in cell volume changes and the dynamics of Ca^{2+} signals during migration may be a valid alternative.

3. The Intermediate Conductance Ca^{2+}-Activated K Channel

The intermediate conductance Ca^{2+}-activated K channel, KCa3.1, belongs to the Ca^{2+}-activated K channel (KCa) family, which comprises the large- (KCa1.1), intermediate (KCa3.1), and small conductance (KCa2.1-3) K channels, as originally classified according to their single-channel conductance. Genetic relationship and Ca^{2+} activation mechanisms have later shown that these channels form two well-defined and distantly related groups. One, including only the large-conductance KCa1.1 channel, is gated by the cooperative action of membrane depolarization and $[Ca^{2+}]_i$, while the other group includes both the intermediate conductance KCa3.1 and small-conductance KCa2.1-3 channels, gated solely by cytosolic Ca^{2+} increase.

3.1. *Biophysics, Pharmacology, and Gating of the KCa3.1 Channel*

The biophysical properties of the KCa3.1 channel have been investigated using various experimental approaches and cell models. Work on cultured human glioblastoma cells shows high K^+ selectivity, moderate inward rectification, and single-channel conductance of 20–60 pS in symmetrical 150 mM K^+. The KCa3.1 channel is voltage-insensitive, but highly sensitive to $[Ca^{2+}]_i$ showing an EC_{50} of <200 nM. Ca^{2+} ions activate the KCa3.1 channel via the Ca^{2+}-binding protein CaM, which is constitutively bound to the membrane-proximal region of the intracellular C terminus of the channel. Binding of Ca^{2+} to CaM results in conformational changes of the channel, and its opening. In addition to serving as a Ca^{2+} sensor for channel opening, CaM also regulates the assembly and trafficking of the channel protein to the plasma membrane [34,35].

The KCa3.1 channel can be blocked by peptidic toxins isolated from various scorpions, such as charybdotoxin (ChTx) and maurotoxin (MTx), which display high affinity (IC_{50} in the nM range).

Small synthetic molecules have also been developed, many derived from the clotrimazole template, the classical KCa3.1 channel blocker. The most widely used TRAM-34, developed by Wulff's group, inhibits KCa3.1 channels with an $IC_{50} < 20$ nM and displays high selectivity over the other KCa channels. As for KCa3.1 channel activators, we recall the benzimidazolone 1-EBIO that activates the channels with an $EC_{50} < 30$ μM and DC-EBIO that exhibits 10-fold higher potency. Two structurally similar and still more potent molecules are the oxime NS309 and the benzothiazole SKA-31, both showing an EC_{50} for KCa3.1 channels < 20 nM. We finally mention riluzole, another potent activator of KCa3.1 channels, which is the only FDA-approved drug for treatment of amyotrophic lateral sclerosis (ALS).

3.2. KCa3.1 Channel Expression and Impact on GBM Migration and Invasion

Initial biochemical and electrophysiological studies showed that KCa3.1 channels were diffusely expressed in virtually all cell types investigated. Notably, the KCa3.1 channel was not found in the brain (later the KCa3.1 channel was described in microglia [7], several types of brain neurons [8,36], and the nodes of Ranvier of cerebellar Purkinje neurons [37]), although it was highly expressed in established cell lines from brain tumors, as well as in brain tumors in situ. Moreover, KCa3.1 channel expression was reported in many other tumor types, and implicated in malignancy (increased cell growth, migration, invasion, apoptosis evasion). The KCa3.1 mRNA and protein expression were also found to be significantly enhanced in cancer stem cells derived from both the established cell line U87-MG and primary cell line FCN9 [38]. The REMBRANDT database shows that the gene KCNN4, which encodes the KCa3.1 channel, is overexpressed in more than 30% of gliomas, and its expression is associated with a poor prognosis.

To test the role of the KCa3.1 channel in GBM cells' infiltration into brain parenchyma, human GL-15 GBM cells were xenografted into the brains of SCID mice and later treated with the specific KCa3.1 blocker TRAM-34. Immunofluorescence analyses of cerebral slices after a five-week treatment revealed a significant reduction of tumor infiltration, compared with TRAM-34 untreated mice [39]. Reduction of tumor infiltration was also observed in the brain of mice transplanted with KCa3.1-silenced GL-15 cells, indicating a direct role of KCa3.1 channels in GBM tumor infiltration [40]. Similarly, KCa3.1 channel block with TRAM-34 or silencing by short hairpin RNA (shRNA) completely abolished CXCL12-induced GL-15 cell migration [40]. A strong correlation of KCa3.1 channel expression with migration was reported by Calogero's group in GBM-derived cancer stem cells (CSC) [38]. Blockage of the KCa3.1 channel with TRAM-34 was found to have a much greater impact on the motility of CSCs (75% reduction) that express a high level of KCa3.1 channel than on the FCN9 parental population (32% reduction), where the KCa3.1 channel is expressed at much lower levels. Similar results were also observed with the CSCs derived from U87-MG [38] and with mesenchymal glioblastoma stem cells [41]. On the same line stand the results later obtained by Sontheimer's group showing that pharmacological inhibition of the KCa3.1 channel (with TRAM-34) in U251 glioma cells, or silencing it with inducible siRNA, resulted in a significant reduction of tumor cells' migration in vitro and invasion into surrounding brain parenchyma of SCID mice [42]. On a retrospective study of a patient genomic database, they further showed that KCa3.1 channel expression was inversely correlated with patient survival.

3.3. Basic Functions of KCa3.1 Channel: Regulation of Cell Ca^{2+} Signaling

KCa3.1 channels control a number of basic cellular processes involved in the modulation of several higher-order biological functions critical to brain tumors' malignancy, including migration and invasion. The most relevant and common basic cellular processes controlled by KCa3.1 channels are the modulation of Ca^{2+} signaling and the control of cell volume. Here we will concentrate on the KCa3.1 channel modulation of Ca^{2+} signaling, the focus of this review.

In virtually all cells a robust stimulation of PLC-coupled membrane receptors triggers an initial IP_3-mediated release of Ca^{2+} from intracellular stores, followed by Ca^{2+} influx through store-operated Orai Ca^{2+} channels, which are activated in response to Ca^{2+} depletion of the ER.

One consequence of this Ca^{2+} influx, besides the activation of Ca^{2+}-dependent target proteins (ion channels included), is membrane depolarization, which, if left unchecked, would more and more strongly inhibit further Ca^{2+} influx due to the reduced electrochemical driving force for Ca^{2+} ions. It has been shown that the efflux of K^+ ions following KCa3.1 channels' activation by incoming Ca^{2+} hyperpolarizes the membrane towards E_K, increasing the electromotive force on incoming Ca^{2+} ions [43,44]. This represents an indirect modulatory mechanism of Ca^{2+} entry and, as a result, of cell Ca^{2+} signaling. The Ca^{2+}-dependent and voltage-independent gating of KCa3.1 channels appears to be particularly well suited for this function, since the coupling of Ca^{2+} influx and activation of K^+ channels will create a positive feedback loop whereby more Ca^{2+} influx will increase the K^+ efflux, hyperpolarize the plasma membrane, and further stimulate Ca^{2+} entry, thus amplifying the signal transduction.

This paradigm was first demonstrated in human macrophages, where KCa3.1 channels have been shown to hyperpolarize the membrane and increase the driving force for Ca^{2+} ions during store-operated Ca^{2+} entry [43], and in T cells, where KCa3.1 channels are rapidly upregulated following cell activation, and supposedly used to maximize Ca^{2+} influx during the reactivation of memory T cells [45,46]. Activated T cells isolated from KCa3.1$^{-/-}$ mice show a defective Ca^{2+} response to T cell receptor activation [47]. Also, mast cells appear to use KCa3.1 channels to hyperpolarize the membrane and increase the Ca^{2+} influx following antigen-mediated stimulation. In this case KCa3.1 channels have been found to physically interact with the Orai1 subunit, suggesting that they may be activated by the Ca^{2+} microdomain that forms close to the store-operated Ca^{2+} channel [28,48]. A similar conclusion has been reached for rat microglial cells, where the Orai1-KCa3.1 functional coupling could be interrupted by BAPTA but not by EGTA [49]. Since the two Ca^{2+} chelators have similar Ca^{2+} affinity but different Ca^{2+} binding rates (10 times higher for BAPTA), only a physical coupling between Orai1 and KCa3.1, with a separating distance of few nanometers, would explain these results.

Evidence of cross-talk between the KCa3.1 channel and store depletion-activated Ca^{2+} influx has also been reported for GBM cells. An increase of $[Ca^{2+}]_i$, consisting of a fast peak due to IP_3-dependent Ca^{2+} release from intracellular stores followed by a sustained phase of Ca^{2+} influx across the plasma membrane, supposedly activated by intracellular stores' depletion, was found upon prolonged application of histamine on GL-15 GBM cells [44]. The enhancing role of the KCa3.1 channel in sustaining the protracted influx of external Ca^{2+} was shown by the marked reduction of the sustained histamine-induced $[Ca^{2+}]_i$ increase following application of TRAM-34. This observation could be significant with regard to the KCa3.1 channels' contribution to GBM cell migration exerted through modulation of Ca^{2+} signals.

In another GBM cell line—U87-MG—we also observed that the promigratory FCS is able to promote Ca^{2+} oscillations, and these oscillations cyclically activate the KCa3.1 channels during cell migration [13]. Using a modeling approach, we also found that a channel activity with the properties of KCa3.1 channels could modulate these Ca^{2+} oscillations (it increased the amplitude, duration, and frequency of each Ca^{2+} spike [50]). The results we obtained, which are illustrated in the following section, were unexpectedly interesting, and also able to explain old experiments showing that the KCa3.1 channel inhibition by ChTx abolishes the bradykinin-induced Ca^{2+} oscillations in C6 glioma cells [51].

3.4. Modulation of Ca^{2+} Oscillations by KCa3.1 Channel Activity

We saw earlier that the mechanism underlying Ca^{2+} oscillations is essentially based on the biphasic effects of Ca^{2+} on IP_3 receptor gating—activatory at low and inhibitory at high concentrations. We also discussed the modulation that the KCa3.1 channel could exert on Ca^{2+} oscillations. This modulation relies on the high Ca^{2+} sensitivity and voltage independence of the KCa3.1 channel, and on the output of its activity, that is, the hyperpolarization of the cell membrane potential (V_m) as result of the K^+ efflux. As KCa3.1 channels are activated by Ca^{2+} concentrations well within the range observed during hormone-induced Ca^{2+} oscillations (150–600 nM; cf. above), they are expected to cyclically open (during the Ca^{2+} spikes) and cause cyclic membrane hyperpolarizations in phase with the Ca^{2+} oscillations.

We have observed experimentally such KCa3.1 channel oscillatory activity and associated membrane potential oscillations in U87-MG cells in response to FCS [13], an agent known to induce Ca^{2+} oscillations in these cells [29,30]. Notably, because of their voltage dependence and much lower Ca^{2+} sensitivity, the KCa1.1 channels, also highly expressed in GBM cells, could not be activated under these conditions, even at the peak Ca^{2+} concentration of the oscillations ([13], but see [52] for a case in which the KCa1.1 channel activity may control Ca^{2+} influx). In conclusion, since V_m controls the driving force for external Ca^{2+} entry, KCa3.1-dependent V_m oscillations are expected to cause oscillations in the amplitude of Ca^{2+} influx through the hormone-activated plasma membrane Ca^{2+} channels, which will in turn feedback onto, and modulate the Ca^{2+} oscillations themselves.

We recently implemented basic theoretical models of Ca^{2+} oscillations with the oscillating membrane V_m, as produced by the cyclic activation of KCa3.1 channels. The resulting model was used to predict how the hormone-induced Ca^{2+} oscillations would be influenced by the KCa3.1 channels-induced V_m oscillations in phase with Ca^{2+} oscillations. We found that the cyclic activation of KCa3.1 channels by Ca^{2+} oscillations induces V_m fluctuations that in turn determine oscillations in the Ca^{2+} influx from the extracellular medium, as a result of the oscillating V_m-dependent changes in the driving force for Ca^{2+} ions influx. Since Ca^{2+} influx peaks are in phase with the Ca^{2+} oscillations, the Ca^{2+} spikes will also be increased by the in-phase Ca^{2+} influx. Our model calculations in fact show that KCa3.1-induced V_m oscillations strengthen the hormone-induced $[Ca^{2+}]_i$ signals by increasing the amplitude, duration, and oscillatory frequency of Ca^{2+} spikes (Figure 2; see also [50]).

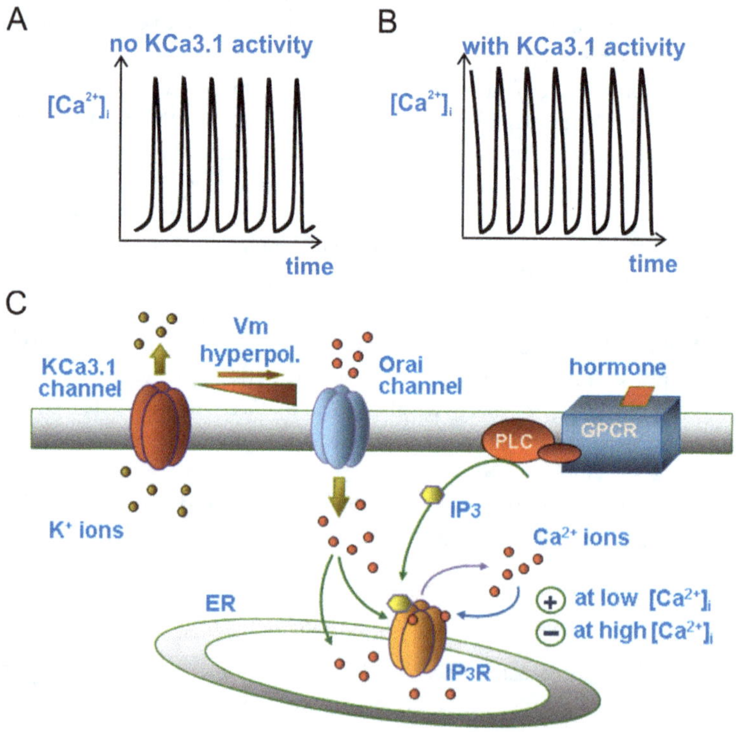

Figure 2. Modulation of Ca^{2+} oscillations by the KCa3.1 channel. Drawing (**C**) illustrating the main ion fluxes generating or modulating the Ca^{2+} oscillations in our model when the KCa3.1 channel is cut off (**A**) or introduced into the system (**B**). Please notice the much longer duration (width) and slightly higher amplitude of the Ca^{2+} oscillations in the presence of KCa3.1 channels.

3.5. KCa3.1 Channels Switch Ca^{2+} Oscillations On and Off

Arguably, the KCa3.1 channels can do more than just modulate the amplitude and frequency of Ca^{2+} oscillations: they can trigger or suppress them. This has been an unexpected prediction of our model that we later found to have been experimentally observed. As experimentally found, our model predicts that Ca^{2+} oscillations can be generated only within a specific range of IP_3 concentrations, while a stable, non-oscillating Ca^{2+} level is present for both lower and higher IP_3 concentrations (Figure 3A). Surprisingly, it was found that when the KCa3.1 conductance is changed in the system, the resulting alteration of the KCa3.1 channels-induced V_m oscillations significantly change the IP_3 concentrations needed to produce Ca^{2+} oscillations. In particular, increasing the activity of KCa3.1 channels shifts leftward the IP_3 range within which Ca^{2+} oscillations are produced. This implies that there are ranges of IP_3 levels where the presence or absence of Ca^{2+} oscillations are only dictated by the activity (conductance) of KCa3.1 channels (Figure 3B bottom, shaded areas).

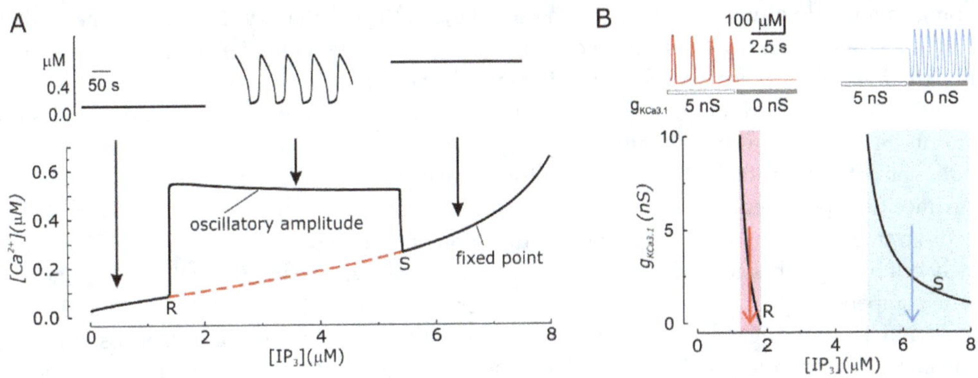

Figure 3. Ca^{2+} oscillation development depends on the IP_3 concentration and the activity of KCa3.1 channels. (**A**) Bottom. Bifurcation diagram showing that IP_3 concentration determines the establishment of the Ca^{2+} oscillations. The red dashed line represents the unstable equilibrium $[Ca^{2+}]_i$. The computation was performed with $g_{KCa3.1}$ = 5 nS. R and S indicate the Hopf bifurcations where Ca^{2+} oscillations appear and disappear, respectively, upon increasing the IP_3 levels. Top. The traces show the temporal changes of the $[Ca^{2+}]_i$ for three different concentrations of IP_3 (indicated by arrows in the bottom part). (**B**) Bottom. Plot of the two Hopf bifurcations (R and S) as a function of the IP_3 concentration and $g_{KCa3.1}$. Three different ranges of IP_3 concentrations, where the KCa3.1-induced V_m oscillations appear to have different effects in the modulation of Ca^{2+} oscillations are evidenced (shaded areas; see text). More specifically in the pink region the presence of KCa3.1 channels is necessary for the existence of Ca^{2+} oscillations, in the light blue region KCa3.1 channels prevents Ca^{2+} oscillations, and finally in the white region in between KCa3.1 channels modulate the shape and duration of oscillation. Top. Simulated Ca^{2+} oscillations obtained with two IP_3 concentrations (1.5 left and 6.2 μM right, as indicated by the red and blue arrows below) within the regions where removing KCa3.1 channels make Ca^{2+} oscillations disappear or appear. (Modified from [48].)

These observations could explain a number of experimental results where KCa3.1 channel modulation was shown to have an effect on Ca^{2+} oscillations. In both activated T lymphocytes and C6 glioma cells, inhibition of V_m oscillations by blocking KCa3.1 channels is able to suppress Ca^{2+} oscillations [51,53]. Moreover *ras*-induced transformation of fibroblasts, which has been reported to upregulate KCa3.1 channels, has also been shown to trigger Ca^{2+} oscillations [54,55]. Finally, KCa3.1 inhibition leads to induction of Ca^{2+} oscillations in about half of the tested glioblastoma cells [52].

3.6. Glioblastoma KCa3.1 Channels Are Activated by Serum-Induced Ca^{2+} Oscillations and Participate to Cell Migration

Glioblastoma cells in vivo are exposed to a variety of tumor microenvironment components and soluble factors that can markedly affect their migratory ability. Among them there are unknown serum components that infiltrate into the tumor area of high-grade gliomas as result of the blood brain barrier breakdown [6,56]. As already shown, FCS enhances migration of U87-MG glioblastoma cells, and does so by inducing Ca^{2+} oscillations that are critically involved in the detachment of focal adhesions (by activating the focal adhesion kinase), and subsequent retraction of the cell rear [29,30]. On the same cell model we further reported that FCS, besides Ca^{2+} oscillations, induces an oscillatory activity of KCa3.1 channels, and this KCa3.1 channel activity is a necessary step to promote U87-MG cell migration by FCS [13]. In the same study, we also found a stable activation of Cl^- currents upon FCS stimulation. Altogether these observations sketch a coherent picture of the cell migratory process, in that, the retraction of the cell rear after the detachment of focal adhesions involves a major cell shape rearrangement and volume reduction. In these instances, the (oscillatory) KCa3.1 channel activity and the consequent cyclical K^+ efflux, in combination with Cl^- efflux (and essential osmotic water), form part of the whole machinery for achieving the cyclical volume reduction needed.

Cell locomotion is thought to be promoted by the cycling replication of four main steps: (i) protrusion of the cell front, associated to asymmetric polymerization of the actin-based cytoskeleton; (ii) adhesion of the protruded cell front to the substratum, mediated by the binding of focal adhesions to the extracellular polymers in contact with the cell; (iii) unbinding of the focal adhesions at the cell rear; (iv) retraction of the rear cell body in the direction of motion (cf. Figure 4). Two of these steps, namely cell front protrusion and cell rear retraction, involve major changes in cell volume and thus require compensatory ion fluxes (and osmotic water flow) across the membrane. Along the lines proposed by [12], during the protrusion of the cell front the maintenance of a stable osmolarity is sustained by cotransporters and ion exchangers—for instance, $Na^+/K^+/2Cl^-$ cotransport—followed by osmotically driven water. During this phase $[Ca^{2+}]_i$ is relatively low (red segment in the Ca^{2+} oscillation, Figure 4a), a condition in which KCa3.1 channels are closed and the resting membrane potential is near E_{Cl}. This condition prevents or strongly limits the loss of KCl via K and Cl channels that would nullify the osmotic work of the $Na^+/K^+/2Cl^-$ cotransport.

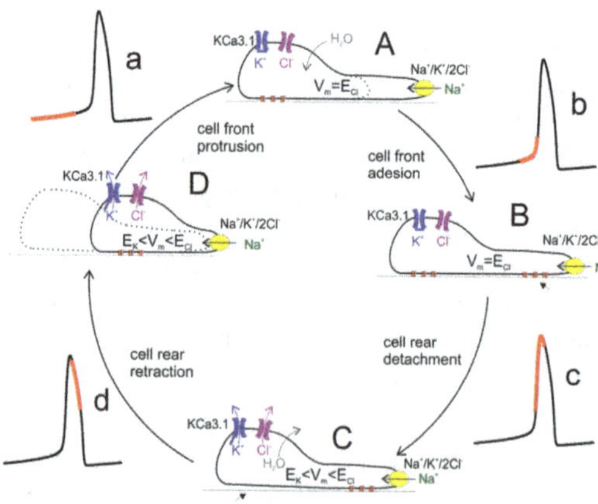

Figure 4. Schematic diagram of cell migratory cycle. The classic view of the cell migration process can be split down into four main cyclical steps. The cycle begins with the cell front protrusion, due to the activity of $Na^+/K^+/2Cl^-$ cotransport (yellow) (**A**) and establishment of adhesion structures (**B**). The elongated cell then removes/weakens the rear adhesions (**C**) so that ensuing contraction can pull the rear cell portion forward (**D**). The concomitant values of $[Ca^{2+}]_i$ is indicated by the red portion on the associated Ca^{2+} oscillation. V_m, ion and water fluxes are also illustrated (see text for details).

By contrast, the control of osmolarity during cell rear retraction critically depends on the activation of KCa3.1 channels and ensuing efflux of K^+ ions (accompanied by Cl^- fluxes, to which membrane has become highly permeant, and osmotic water). Following the increase in $[Ca^{2+}]_i$ towards the peak of the Ca^{2+} oscillation (Figure 4c,d), KCa3.1 channels open and the membrane potential hyperpolarizes to a value between E_{Cl} and E_K, a condition that would allow a significant efflux of KCl with consequent loss of water and cell volume decrease (Figure 4C,D). This would facilitate the process of cell body retraction. This view can explain the slowing of cell locomotion following the inhibition of K and Cl channels [13].

4. Conclusions

Substantial evidence suggests that the oscillatory activity of the KCa3.1 channel during Ca^{2+} oscillations may have both direct and indirect modulatory effects on GBM cell migration. As has long been recognized, the K^+ efflux through KCa3.1 channels directly participates in the cell volume changes during cell rear retraction. The same oscillatory K^+ efflux, however, will also determine the membrane potential oscillations that alter the driving force for Ca^{2+} influx and indirectly modulate the properties and even the presence of Ca^{2+} oscillations, and thus the timing of the migratory machinery. Both modulatory roles make KCa3.1 channels pivotal in GBM cell migration, and thus potential pharmacological targets for this deadly tumor. Although this review concentrates on the role of KCa3.1 and STIM/Orai channels in modulating Ca^{2+} oscillations and cell migration, it needs to be said that other channels including TRPM8 and KCa1.1, also abundantly expressed in glioblastoma cells, may also be important in these processes. Accordingly, TRPM8 [41] and KCa1.1 channel signaling [57] have been demonstrated to be required for basal and radiation-induced migration, respectively, and certainly contribute to Ca^{2+} oscillations.

Author Contributions: Both L.C. and F.F. performed literature search and wrote the paper.

Funding: This research received no external funding.

References

1. Hanahan, D.; Weinberg, R.A. Hallmarks of cancer: The next generation. *Cell* **2011**, *144*, 646–674. [CrossRef] [PubMed]
2. Ridley, A.J.; Schwartz, M.A.; Burridge, K.; Firtel, R.A.; Ginsberg, M.H.; Borisy, G.; Parsons, J.T.; Horwitz, A.R. Cell migration: Integrating signals from front to back. *Science* **2003**, *302*, 1704–1709. [CrossRef] [PubMed]
3. Ridley, A.J. Life at the leading edge. *Cell* **2011**, *145*, 1012–1022. [CrossRef] [PubMed]
4. Carragher, N.O.; Frame, M.C. Focal adhesion and actin dynamics: A place where kinases and proteases meet to promote invasion. *Trends Cell Biol.* **2004**, *14*, 241–249. [CrossRef] [PubMed]
5. Parsons, J.T.; Horwitz, A.R.; Schwartz, M.A. Cell adhesion: Integrating cytoskeletal dynamics and cellular tension. *Nat. Rev. Mol. Cell Biol.* **2010**, *11*, 633–643. [CrossRef] [PubMed]
6. Lund, C.V.; Nguyen, M.T.; Owens, G.C.; Pakchoian, A.J.; Shaterian, A.; Kruse, C.A.; Eliceiri, B.P. Reduced glioma infiltration in Src-deficient mice. *J. Neurooncol.* **2006**, *78*, 19–29. [CrossRef] [PubMed]
7. Wessels, D.; Soll, D.R.; Knecht, D.; Loomis, W.F.; De Lozanne, A.; Spudich, J. Cell motility and chemotaxis in Dictyostelium amebae lacking myosin heavy chain. *Dev. Biol.* **1988**, *128*, 164–177. [CrossRef]
8. Lombardi, M.L.; Knecht, D.A.; Dembo, M.; Lee, J. Traction force microscopy in Dictyostelium reveals distinct roles for myosin II motor and actin-crosslinking activity in polarized cell movement. *J. Cell Sci.* **2007**, *120 Pt 9*, 1624–1634. [CrossRef]
9. Roca-Cusachs, P.; Sunyer, R.; Trepat, X. Mechanical guidance of cell migration: Lessons from chemotaxis. *Curr. Opin. Cell Biol.* **2013**, *25*, 543–549. [CrossRef] [PubMed]
10. Ostman, A.; Heldin, C.H. PDGF receptors as targets in tumor treatment. *Adv. Cancer Res.* **2007**, *97*, 247–274. [PubMed]

11. Lo, C.M.; Wang, H.B.; Dembo, M.; Wang, Y.L. Cell movement is guided by the rigidity of the substrate. *Biophys. J.* **2000**, *79*, 144–152. [CrossRef]
12. Schwab, A.; Fabian, A.; Hanley, P.J.; Stock, C. Role of ion channels and transporters in cell migration. *Physiol. Rev.* **2012**, *92*, 1865–1913. [CrossRef] [PubMed]
13. Catacuzzeno, L.; Aiello, F.; Fioretti, B.; Sforna, L.; Castigli, E.; Ruggieri, P.; Tata, A.M.; Calogero, A.; Franciolini, F. Serum-activated K and Cl currents underlay U87-MG glioblastoma cell migration. *J. Cell Physiol.* **2011**, *226*, 1926–1933. [CrossRef] [PubMed]
14. Cuddapah, V.A.; Habela, C.W.; Watkins, S.; Moore, L.S.; Barclay, T.T.; Sontheimer, H. Kinase activation of ClC-3 accelerates cytoplasmic condensation during mitotic cell rounding. *Am. J. Physiol. Cell Physiol.* **2012**, *302*, C527–C538. [CrossRef] [PubMed]
15. Wei, C.; Wang, X.; Zheng, M.; Cheng, H. Calcium gradients underlying cell migration. *Curr. Opin. Cell Biol.* **2012**, *24*, 254–261. [CrossRef] [PubMed]
16. Cuthbertson, K.S.; Chay, T.R. Modelling receptor-controlled intracellular calcium oscillators. *Cell Calcium* **1991**, *12*, 97–109. [CrossRef]
17. Gaspers, L.D.; Bartlett, P.J.; Politi, A.; Burnett, P.; Metzger, W.; Johnston, J.; Joseph, S.K.; Höfer, T.; Thomas, A.P. Hormone-induced calcium oscillations depend on cross-coupling with inositol 1,4,5-trisphosphate oscillations. *Cell Rep.* **2014**, *9*, 1209–1218. [CrossRef] [PubMed]
18. Dupont, G.; Combettes, L.; Bird, G.S.; Putney, J.W. Calcium oscillations. *Cold Spring Harb. Perspect. Biol.* **2011**, *3*, a004226. [CrossRef] [PubMed]
19. Iino, M. Biphasic Ca^{2+} dependence of inositol 1,4,5-trisphosphate-induced Ca release in smooth muscle cells of the guinea pig taenia caeci. *J. Gen. Physiol.* **1990**, *95*, 1103–1122. [CrossRef] [PubMed]
20. Bezprozvanny, I.; Watras, J.; Ehrlich, B.E. Bell-shaped calcium-response curves of Ins(1,4,5)P3- and calcium-gated channels from endoplasmic reticulum of cerebellum. *Nature* **1991**, *351*, 751–754. [CrossRef] [PubMed]
21. Finch, E.A.; Turner, T.J.; Goldin, S.M. Calcium as a coagonist of inositol 1,4,5-trisphosphate-induced calcium release. *Science* **1991**, *252*, 443–446. [CrossRef] [PubMed]
22. Kaznacheyeva, E.; Lupu, V.D.; Bezprozvanny, I. Single-channel properties of inositol (1,4,5)-trisphosphate receptor heterologously expressed in HEK-293 cells. *J. Gen. Physiol.* **1998**, *111*, 847–856. [CrossRef] [PubMed]
23. Liou, J.; Kim, M.L.; Heo, W.D.; Jones, J.T.; Myers, J.W.; Ferrell, J.E., Jr.; Meyer, T. STIM is a Ca^{2+} sensor essential for Ca^{2+}-store-depletion-triggered Ca^{2+} influx. *Curr. Biol.* **2005**, *15*, 1235–1241. [CrossRef] [PubMed]
24. Roos, J.; DiGregorio, P.J.; Yeromin, A.V.; Ohlsen, K.; Lioudyno, M.; Zhang, S.; Safrina, O.; Kozak, J.A.; Wagner, S.L.; Cahalan, M.D.; et al. STIM1, an essential and conserved component of store-operated Ca^{2+} channel function. *J. Cell Biol.* **2005**, *169*, 435–445. [CrossRef] [PubMed]
25. Peinelt, C.; Vig, M.; Koomoa, D.L.; Beck, A.; Nadler, M.J.; Koblan-Huberson, M.; Lis, A.; Fleig, A.; Penner, R.; Kinet, J.P. Amplification of CRAC current by STIM1 and CRACM1 (Orai1). *Nat. Cell Biol.* **2006**, *8*, 771–773. [CrossRef] [PubMed]
26. Moccia, F.; Ruffinatti, F.A.; Zuccolo, E. Intracellular Ca^{2+} Signals to Reconstruct a Broken Heart: Still a Theoretical Approach? *Curr. Drug Targets* **2015**, *6*, 793–815. [CrossRef]
27. Dubois, C.; Vanden Abeele, F.; Lehen'kyi, V.; Gkika, D.; Guarmit, B.; Lepage, G.; Slomianny, C.; Borowiec, A.S.; Bidaux, G.; Benahmed, M.; et al. Remodeling of channel-forming ORAI proteins determines an oncogenic switch in prostate cancer. *Cancer Cell* **2014**, *26*, 19–32. [CrossRef] [PubMed]
28. Duffy, S.M.; Ashmole, I.; Smallwood, D.T.; Leyland, M.L.; Bradding, P. Orai/CRACM1 and KCa3.1 ion channels interact in the human lung mast cell plasma membrane. *Cell Commun. Signal.* **2015**, *13*, 32. [CrossRef] [PubMed]
29. Rondé, P.; Giannone, G.; Gerasymova, I.; Stoeckel, H.; Takeda, K.; Haiech, J. Mechanism of calcium oscillations in migrating human astrocytoma cells. *Biochim. Biophys. Acta* **2000**, *1498*, 273–280. [CrossRef]
30. Giannone, G.; Rondé, P.; Gaire, M.; Haiech, J.; Takeda, K. Calcium oscillations trigger focal adhesion disassembly in human U87 astrocytoma cells. *J. Biol. Chem.* **2002**, *277*, 26364–26371. [CrossRef] [PubMed]
31. Montana, V.; Sontheimer, H. Bradykinin promotes the chemotactic invasion of primary brain tumors. *J. Neurosci.* **2011**, *31*, 4858–4867. [CrossRef] [PubMed]
32. Totsukawa, G.; Wu, Y.; Sasaki, Y.; Hartshorne, D.J.; Yamakita, Y.; Yamashiro, S.; Matsumura, F. Distinct roles of MLCK and ROCK in the regulation of membrane protrusions and focal adhesion dynamics during cell migration of fibroblasts. *J. Cell Biol.* **2004**, *164*, 427–439. [CrossRef] [PubMed]

33. Chen, C.; Tao, T.; Wen, C.; He, W.Q.; Qiao, Y.N.; Gao, Y.Q.; Chen, X.; Wang, P.; Chen, C.P.; Zhao, W.; et al. Myosin light chain kinase (MLCK) regulates cell migration in a myosin regulatory light chain phosphorylation-independent mechanism. *J. Biol. Chem.* **2014**, *289*, 28478–28488. [CrossRef] [PubMed]
34. Sforna, L.; Megaro, A.; Pessia, M.; Franciolini, F.; Catacuzzeno, L. Structure, Gating and Basic Functions of the Ca^{2+}-activated K Channel of Intermediate Conductance. *Curr. Neuropharmacol.* **2018**, *16*, 608–617. [CrossRef] [PubMed]
35. Catacuzzeno, L.; Franciolini, F. Editorial: The Role of Ca^{2+}-activated K^+ Channels of Intermediate Conductance in Glioblastoma Malignancy. *Curr. Neuropharmacol.* **2018**, *16*, 607. [CrossRef] [PubMed]
36. Vicente-Manzanares, M.; Koach, M.A.; Whitmore, L.; Lamers, M.L.; Horwitz, A.F. Segregation and activation of myosin IIB creates a rear in migrating cells. *J. Cell Biol.* **2008**, *183*, 543–554. [CrossRef] [PubMed]
37. Kaushal, V.; Koeberle, P.D.; Wang, Y.; Schlichter, L.C. The Ca^{2+}-activated K^+ channel KCNN4/KCa3.1 contributes to microglia activation and nitric oxide-dependent neurodegeneration. *J. Neurosci.* **2007**, *27*, 234–244. [CrossRef] [PubMed]
38. Ruggieri, P.; Mangino, G.; Fioretti, B.; Catacuzzeno, L.; Puca, R.; Ponti, D.; Miscusi, M.; Franciolini, F.; Ragona, G.; Calogero, A. The inhibition of KCa3.1 channels activity reduces cell motility in glioblastoma derived cancer stem cells. *PLoS ONE* **2012**, *7*, e47825. [CrossRef] [PubMed]
39. D'Alessandro, G.; Catalano, M.; Sciaccaluga, M.; Chece, G.; Cipriani, R.; Rosito, M.; Grimaldi, A.; Lauro, C.; Cantore, G.; Santoro, A.; et al. KCa3.1 channels are involved in the infiltrative behavior of glioblastoma in vivo. *Cell Death Dis.* **2013**, *4*, e773. [CrossRef] [PubMed]
40. Sciaccaluga, M.; Fioretti, B.; Catacuzzeno, L.; Pagani, F.; Bertollini, C.; Rosito, M.; Catalano, M.; D'Alessandro, G.; Santoro, A.; Cantore, G.; et al. CXCL12-induced glioblastoma cell migration requires intermediate conductance Ca^{2+}-activated K^+ channel activity. *Am. J. Physiol. Cell Physiol.* **2010**, *299*, C175–C184. [CrossRef] [PubMed]
41. Klumpp, L.; Sezgin, E.C.; Skardelly, M.; Eckert, F.; Huber, S.M. KCa3.1 Channels and Glioblastoma: In Vitro Studies. *Curr. Neuropharmacol.* **2018**, *16*, 627–635. [CrossRef] [PubMed]
42. Turner, K.L.; Honasoge, A.; Robert, S.M.; McFerrin, M.M.; Sontheimer, H. A proinvasive role for the Ca(2+)-activated K(+) channel KCa3.1 in malignant glioma. *Glia* **2014**, *62*, 971–981. [CrossRef] [PubMed]
43. Gao, Y.D.; Hanley, P.J.; Rinné, S.; Zuzarte, M.; Daut, J. Calcium-activated K^+ channel ($K_{Ca}3.1$) activity during Ca^{2+} store depletion and store-operated Ca^{2+} entry in human macrophages. *Cell Calcium* **2010**, *48*, 19–27. [CrossRef] [PubMed]
44. Fioretti, B.; Catacuzzeno, L.; Sforna, L.; Aiello, F.; Pagani, F.; Ragozzino, D.; Castigli, E.; Franciolini, F. Histamine hyperpolarizes human glioblastoma cells by activating the intermediate-conductance Ca^{2+}-activated K^+ channel. *Am. J. Physiol. Cell Physiol.* **2009**, *297*, C102–C110. [CrossRef] [PubMed]
45. Ghanshani, S.; Wulff, H.; Miller, M.J.; Rohm, H.; Neben, A.; Gutman, G.A.; Cahalan, M.D.; Chandy, K.G. Up-regulation of the IKCa1 potassium channel during T-cell activation. Molecular mechanism and functional consequences. *J. Biol. Chem.* **2000**, *275*, 37137–37149. [CrossRef] [PubMed]
46. Wulff, H.; Beeton, C.; Chandy, K.G. Potassium channels as therapeutic targets for autoimmune disorders. *Curr. Opin. Drug Discov. Dev.* **2003**, *6*, 640–647.
47. Di, L.; Srivastava, S.; Zhdanova, O.; Ding, Y.; Li, Z.; Wulff, H.; Lafaille, M.; Skolnik, E.Y. Inhibition of the K^+ channel KCa3.1 ameliorates T cell-mediated colitis. *Proc. Natl. Acad. Sci. USA* **2010**, *107*, 1541–1546. [CrossRef] [PubMed]
48. Duffy, M.S.; Berger, P.; Cruse, G.; Yang, W.; Bolton, S.J.; Bradding, P. The K^+ channel iKCA1 potentiates Ca^{2+} influx and degranulation in human lung mast cells. *J. Allergy Clin. Immunol.* **2004**, *114*, 66–72. [CrossRef] [PubMed]
49. Ferreira, R.; Schlichter, L.C. Selective activation of KCa3.1 and CRAC channels by P2Y2 receptors promotes Ca(2+) signaling, store refilling and migration of rat microglial cells. *PLoS ONE* **2013**, *8*, e62345. [CrossRef] [PubMed]
50. Catacuzzeno, L.; Fioretti, B.; Franciolini, F. A theoretical study on the role of Ca^{2+}-activated K^+ channels in the regulation of hormone-induced Ca^{2+} oscillations and their synchronization in adjacent cells. *J. Theor. Biol.* **2012**, *309*, 103–112. [CrossRef] [PubMed]

51. Reetz, G.; Reiser, G. [Ca2+]i oscillations induced by bradykinin in rat glioma cells associated with Ca^{2+} store-dependent Ca^{2+} influx are controlled by cell volume and by membrane potential. *Cell Calcium* **1996**, *19*, 143–156. [CrossRef]
52. Stegen, B.; Klumpp, L.; Misovic, M.; Edalat, L.; Eckert, M.; Klumpp, D.; Ruth, P.; Huber, S.M. K^+ channel signaling in irradiated tumor cells. *Eur. Biophys. J.* **2016**, *45*, 585–598. [CrossRef] [PubMed]
53. Verheugen, J.A.; Vijverberg, H.P. Intracellular Ca^{2+} oscillations and membrane potential fluctuations in intact human T lymphocytes: Role of K^+ channels in Ca^{2+} signaling. *Cell Calcium* **1995**, *17*, 287–300. [CrossRef]
54. Hashii, M.; Nozawa, Y.; Higashida, H. Bradykinin-induced cytosolic Ca^{2+} oscillations and inositol tetrakisphosphate-induced Ca^{2+} influx in voltage-clamped ras-transformed NIH/3T3 fibroblasts. *J. Biol. Chem.* **1993**, *268*, 19403–19410. [PubMed]
55. Rane, S.G. A Ca2(+)-activated K^+ current in ras-transformed fibroblasts is absent from nontransformed cells. *Am. J. Physiol.* **1991**, *260*, C104–C112. [CrossRef] [PubMed]
56. Seitz, R.J.; Wechsler, W. Immunohistochemical demonstration of serum proteins in human cerebral gliomas. *Acta Neuropathol.* **1987**, *73*, 145–152. [CrossRef] [PubMed]
57. Steinle, M.; Palme, D.; Misovic, M.; Rudner, J.; Dittmann, K.; Lukowski, R.; Ruth, P.; Huber, S.M. Ionizing radiation induces migration of glioblastoma cells by activating BK K^+ channels. *Radiother. Oncol.* **2011**, *101*, 122–126. [CrossRef] [PubMed]

7

The Different Facets of Extracellular Calcium Sensors: Old and New Concepts in Calcium-Sensing Receptor Signalling and Pharmacology

Andrea Gerbino * and Matilde Colella *

Department of Biosciences, Biotechnology and Biopharmaceutics, University of Bari, 70121 Bari, Italy; andrea.gerbino@uniba.it (A.G.); matilde.colella@uniba.it (M.C.);

Abstract: The current interest of the scientific community for research in the field of calcium sensing in general and on the calcium-sensing Receptor (CaR) in particular is demonstrated by the still increasing number of papers published on this topic. The extracellular calcium-sensing receptor is the best-known G-protein-coupled receptor (GPCR) able to sense external Ca^{2+} changes. Widely recognized as a fundamental player in systemic Ca^{2+} homeostasis, the CaR is ubiquitously expressed in the human body where it activates multiple signalling pathways. In this review, old and new notions regarding the mechanisms by which extracellular Ca^{2+} microdomains are created and the tools available to measure them are analyzed. After a survey of the main signalling pathways triggered by the CaR, a special attention is reserved for the emerging concepts regarding CaR function in the heart, CaR trafficking and pharmacology. Finally, an overview on other Ca^{2+} sensors is provided.

Keywords: Ca^{2+}-signalling; microelectrode; fluorophore; exocytosis; cAMP signalling; parathyroid extracellular-Ca^{2+}-sensing receptor (CaR); cancer; apoptosis; ischemia/reperfusion; hypertrophy; heart; cardiomyocytes

---

1. Introduction

Over the years, the intracellular Ca^{2+} signal has emerged as the most intensely studied second messenger pathway [1]. Ca^{2+} is the most versatile intracellular messenger, not only in the cytoplasm but also in diverse subcellular compartments, such as mitochondria [2], nuclei [3], lysosomes [4], and the endoplasmic reticulum [5]. The cloning of the extracellular Calcium-sensing receptor (CaR) 25 years ago by Brown et al. [6] paved the way for a heightened understanding of the mechanisms used by Ca^{2+} to signal in the microenvironments outside the cell. Thus, Ca^{2+} acts as a multipurpose messenger, exerting effects at the cellular, subcellular, and extracellular levels. A very interesting and comprehensive list of reviews on the CaR has been recently published by leading experts in the field of extracellular Calcium-sensing and signalling [7]. Other key topics, such as the structure-function relationships [8] and role of the CaR in cancer [9], have also been very recently covered elsewhere and will be named here in the corresponding paragraphs. The readers are thus referred to these fine reviews for a complete understanding of the CaR's function in cell physiology and pathology [10].

The present review will examine the potential conditions under which extracellular Ca^{2+} concentration ($[Ca^{2+}]$) may change in the local microenvironment outside the cell, with a particular emphasis on the experimental approaches utilized to measure ex vivo and in vitro such Ca^{2+} fluctuations. These extracellular Ca^{2+} changes are sensed by the CaR (and other Ca^{2+} sensors) and transduced by means of different signalling pathways that modulate organ-specific physiological processes. Over the last few years, new concepts have emerged regarding the CaR's role in cancer and cardiac physiology and pathology. Also, exciting news have come into view in the fields of CaR signalling, trafficking, and pharmacology. In addition, it is now clear that a number of different Ca^{2+} sensors share important tasks with the CaR. Therefore, the aim of this review is to summarize for the reader of *International Journal of Molecular Science* these new and exciting notions.

2. Extracellular Ca^{2+} Microdomains

Serum-ionized Ca^{2+} concentration (~1.4 mM) is usually precisely regulated by a number of cooperative organs, including parathyroid glands, bones, kidney, and intestine [11]. However, this tight modulation does not prevent local changes in extracellular Ca^{2+}, as those have been identified in the microscopic volume of interstitial fluid surrounding cells of many tissues [12]. This scenario has become increasingly important because the amplitude and the shape of these extracellular Ca^{2+} fluctuations represent an autocrine/paracrine cell-to-cell communication form based on the activity of extracellular Ca^{2+} sensors.

Many factors have been proposed in the onset of such significant extracellular Ca^{2+} changes. At first, we will evaluate how the activity of the cells, in terms of intracellular Ca^{2+} signalling events, can be rapidly translated in asymmetrical Ca^{2+} changes. Then, we will analyze how the exocytosis of Ca^{2+}-loaded vesicles and the synchronous opening of voltage-gated Ca^{2+} conductances perturb Ca^{2+} levels in the close proximity of a cell.

When closely packed to form a functional tissue, cells delimit small diffusion spaces. Within this context, cells themselves represent major sources and/or sinks for Ca^{2+} during intracellular Ca^{2+} signalling events as a result of the activation of Ca^{2+} efflux (e.g., by the plasma membrane Ca^{2+}-ATPase and/or Na+/Ca^{2+} exchanger) and Ca^{2+} influx (e.g., by store-operated channels, SOCs) pathways at the plasma membrane, respectively. Differential dynamics or a polarized distribution of these Ca^{2+} influx/efflux mechanisms [13–16] are required for the onset of significant extracellular Ca^{2+} microdomains. Our group showed that stimulation with carbachol, a Ca^{2+}-mediated agonist, resulted in asymmetrical fluctuations in local extracellular Ca^{2+} levels in a polarized epithelium.

Using Ca^{2+}-selective microelectrodes, we measured a substantial plasma membrane Ca^{2+}-ATPase (PMCA)-dependent local increase in the extracellular Ca^{2+} concentration at the apical face and a similar store-operated channel (SOC)-dependent depletion at the basolateral side of amphibian acid-secreting cells (Figure 1). Thus, in intact epithelia, asymmetrical changes in extracellular Ca^{2+} reflect the polarized distribution of Ca^{2+}-handling mechanisms in the plasma membrane [17]. Of note, PMCA is highly expressed at the apical membrane of these cells where it shares a strategic localization with the extracellular CaR [17]. In fact, these carbachol-induced changes in extracellular Ca^{2+} act as paracrine/autocrine messages sufficient and essential to strictly regulate tissue-specific functions (e.g., alkaline secretion, pepsinogen secretion, and water flux) independently of intracellular $[Ca^{2+}]$ changes [18,19].

Secretory granules are characterized by very high intraluminal Ca^{2+} concentrations [20–22]. Among others, insulin granules of rat insulinoma contain ionized Ca^{2+} in a range between 60 and 120 mM [23]. Thus, it does not come as a surprise that exocytotic events can significantly change extracellular Ca^{2+} in the microenvironment immediately surrounding the cells. Recently, we showed using Ca^{2+}-microelectrodes that stimulation of insulin secretion by high glucose and other secretagogues (glibenclamide, forskolin/IBMX, and ATP) induced a delayed elevation of extracellular Ca^{2+} within rat insulinoma (INS-1E) β-cell pseudoislets [24]. The exocytotic extrusion of Ca^{2+} through Ca^{2+}-loaded vesicles released in the extracellular environment has also been demonstrated in different cell types, such as salivary gland cells [25], bovine adrenal medullary cells [26], neurohypophyseal nerve endings [27], and sea urchin eggs [28].

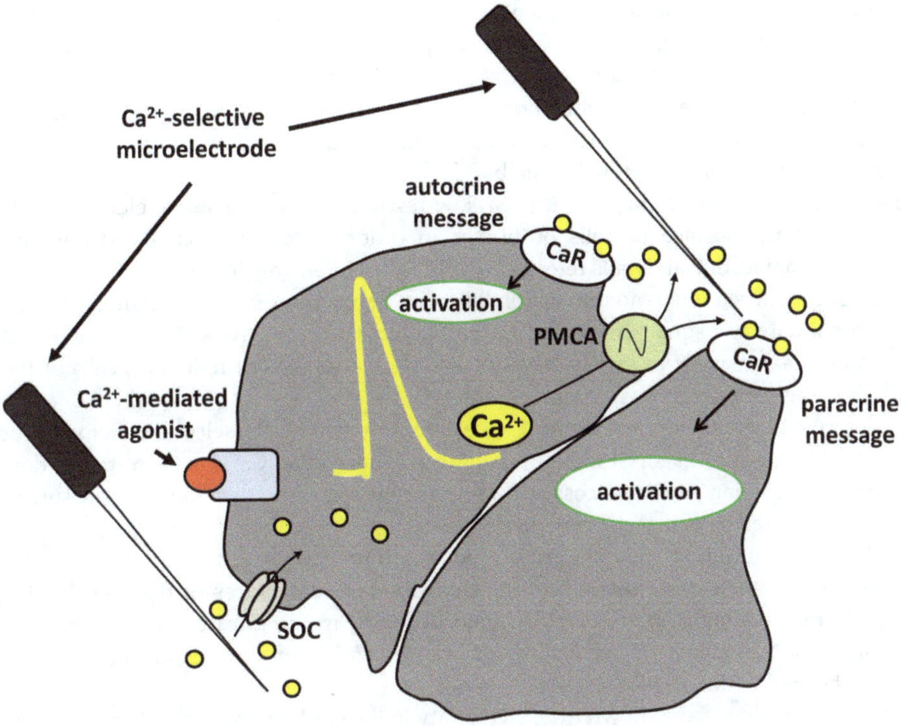

Figure 1. Schematic representation of intercellular communication mediated by the calcium-sensing Receptor (CaR) in polarized epithelial cells. When a Ca^{2+}-mediated agonist (red circle) binds a specific receptor, Ca^{2+} concentration (yellow line) increases within the cytosol. Extrusion of Ca^{2+} via the plasma membrane Ca^{2+}-ATPase (PMCA), localized at the apical membrane, might activate the CaR expressed on cells in close proximity (paracrine message) or on the same cell (autocrine message). The asymmetrical changes in extracellular Ca^{2+} are the result of the polarized localization of PMCA and store-operated channels (SOCs) at the apical and basolateral membranes, respectively. These fluctuations can be recorded with Ca^{2+}-sensitive microelectrodes. Modified from [29].

In addition to the above-mentioned mechanisms, excitable cells express a different kind of voltage-gated Ca^{2+} conductances that, during their physiological activity, might influence extracellular Ca^{2+} levels [30–32]. It has been demonstrated that, in the central nervous system, the synchronized opening of voltage-gated Ca^{2+} channels (VGCCs) elicited significant reductions in extracellular Ca^{2+} levels [33,34]. We and others have shown that in mouse islets of Langerhans and INS-1E pseudoislets, glucose stimulation depleted the concentration of Ca^{2+} in the extracellular milieu (of about 500 μM) as a consequence of Ca^{2+} influx through a VGCC at the plasma membrane [24,35,36]. Transient reductions were also recorded in rat neocortex [33] and in the cardiac muscle [37].

3. Measuring Fluctuations in Extracellular Ca^{2+} Levels

Real-time measurement of extracellular Ca^{2+} changes in the proximity of the plasma membrane (e.g., intercellular spaces) is a technically challenging experimental maneuver. In fact, researchers must deal on the one side with the lack of proper tools to directly access these restricted compartments, and on the other side with the quantification of small Ca^{2+} changes (few hundreds micromolar) over a high extracellular Ca^{2+} background (physiologically around 1.1–1.4 mM). Over the years, many different experimental approaches to measure extracellular Ca^{2+} fluctuations have been proposed in a number of different tissue preparations, although each one of them has its own drawbacks in terms of either sensitivity or spatial resolution.

Microelectrode-based electrophysiological approaches have been applied to diverse cell and tissue types to monitor extracellular Ca^{2+} changes. Early studies directly measured changes in extracellular Ca^{2+} concentration in the rat cerebellum using double-barreled Ca^{2+}-selective micropipettes. During repetitive stimulation of the central nervous system, extracellular Ca^{2+} was significantly reduced, indicating that Ca^{2+} modulation in brain microenvironments may be a significant parameter in the behavior of neuronal ensembles [38].

The vibrating probe technique uses a low-resistance Ca^{2+}-sensitive electrode vibrating in proximity of the surface of cells or tissues in order to measure small extracellular Ca^{2+} gradients (for more technical details refer to [39,40]). Self-referencing, low-resistance vibrating Ca^{2+} microelectrodes were shown to non-invasively quantify, in real time, local extracellular ion gradients with high sensitivity and square micron spatial resolution. Ca^{2+}-selective self-referencing electrodes have been used extensively to show Ca^{2+} flux at different regions of hair cells, cells of the zona pellucida, and the central nervous system [41–44].

As described in the previous section, we extensively used Ca^{2+}-selective microelectrodes to directly record the profile of agonist-induced changes in extracellular $[Ca^{2+}]$ in the restricted domain of both the intact amphibian gastric mucosa [17] and in the microenvironment immediately surrounding INS-1E cells organized as pancreatic pseudoislets [24].

Finally, Mupanomunda and colleagues used in situ microdialysis techniques in anesthetized Wistar rats to monitor extracellular $[Ca^{2+}]$ in the sub-epithelial spaces of the intestine [45] and kidney [46]. In these studies, an increased load of $[Ca^{2+}]$ in the gut lumen or perturbations in whole-animal Ca^{2+} homeostasis were shown to cause significant changes in the external $[Ca^{2+}]$ underlying the epithelium of these tissues.

Ca^{2+}-sensitive dyes have shown extensive utility in the study of the spatio-temporal dynamics of Ca^{2+} fluctuations. These tools, compared with electrophysiological techniques, are more sensitive and provide for a better time resolution and more reliable access to limited spaces. However, these approaches are performed in free extracellular Ca^{2+} (thus, non-physiological conditions) in order to avoid saturation of the fluorophores.

Ca^{2+}-sensitive metallochromic indicators (antipyrylazo III and tetramethylmurexide) were used to estimate extracellular Ca^{2+} dynamics in intact beating hearts. These indicators revealed significant depletion of external Ca^{2+} following the synchronous opening of a Ca^{2+} influx pathway during contraction [37,47,48]. Later on, Tepikin and Petersen introduced the droplet technique [49–51], an elegant approach to measure extracellular Ca^{2+} changes elicited by the activity of the PMCA. This method employs Fluo-3 under Ca^{2+}-free media conditions to quantify extracellular Ca^{2+} changes in small clusters of exocrine gland cells contained in a small droplet of solution (20–50 times the volume of the cells). Extracellular Ca^{2+} was also measured in cell clusters digested from exocrine glands (pancreas, submandibular gland) and placed in a small experimental chamber containing Ca^{2+} green-1 bound to high molecular weight dextran beads in Ca^{2+}-free solution [25,52].

Ca^{2+}-sensitive probes conjugated to a hydrophobic tail have been created to cross the external leaflet of the cell membrane and directly measure extracellular Ca^{2+} levels. A classic example is Ca^{2+}-Green-C-18, the fluorophore conjugated with a lipid derivative that has been successfully used to record Ca^{2+} extrusion and fluctuations $^{2+}$inside the T-tubules [53]. In smooth muscle cells, the Ca^{2+}-sensitive indicator Fura-C18 was used to measure extracellular Ca^{2+} changes near the plasma membrane. The same dye was used to compare the extracellular Ca^{2+} changes recorded with signals measured from the bulk cytosol [54,55]. In addition, Fura-C18 has been used to directly measure near-membrane extracellular $[Ca^{2+}]$ changes in HEK-293 cells, showing that Ca^{2+} extruded to the extracellular space by PMCA, as a consequence of intracellular Ca^{2+} events, activates a positive feedback through the CaR [56].

4. The Extracellular Calcium-Sensing Receptor (CaR)

4.1. Discovery and Cloning of the Extracellular Calcium-Sensing Receptor

The first hints about the existence of an extracellular Ca^{2+} sensor come from a very early study (1966) correlating serum Ca^{2+} concentration and parathyroid hormone (PTH) secretion in whole animals [57,58]. About 10 years later, further evidences corroborated such a correlation in isolated glands and dispersed parathyroid cells [11,59–64].

The direct link between extracellular [Ca^{2+}] changes, PTH secretion, and intracellular Ca^{2+} signals was finally demonstrated by Nemeth and Scarpa [65], while the actual molecular identification of the Ca^{2+} sensor had to wait until 1993, when Brown and colleagues cloned a peculiar G protein coupled receptor (GPCR) from the bovine parathyroid glands [6] and named it "extracellular Calcium-sensing receptor". This receptor, here named CaR, is also referred to as CaSR, after its gene name, or more recently CaS, as recommended by NC-IUPHAR [66].

4.2. Structural Features of the CaR

The extracellular calcium-sensing receptor belongs to class C (or 3) of the GPCR superfamily, which consists of eight metabotropic glutamate receptors (mGlu1-8), two type B metabotropic-aminobutyric acid receptors ($GABA_{B1}$ and $GABA_{B2}$), three taste receptors (T1R1-3), a promiscuous L-α-amino acid (GPRC6A) receptor [67], and seven orphan receptors [68]. The primary structure of human CaR consists of four main parts: a large N-terminal extracellular domain (ECD) (612 amino acids) [69], a cysteine-rich domain, a seven-transmembrane domain (TMD, 250 amino acids), and a 216-residue carboxyterminal (C)-tail [70].

The ECD contains a large bilobed, nutrient-binding Venus Flytrap (VFT) domain related to bacterial periplasmic binding proteins which has been crystallized in several members of GPCR class C (mGlu1 [71], mGlu3 and mGlu7 [72], and the $GABA_{B2}$ subunit [73]). It seems that upon Ca^{2+} (or other agonists) binding, the open cleft of the VFT closes inducing, in turn, conformational changes in both the TMD and the intracellular domains that are believed to initiate signal transduction [74,75].

The functional, highly glycosilated CaR proteins reside on the cell membrane as homodimers [76,77] or hetero-dimers with other class C receptors (mGluR [78] or $GABA_B$ [79]). While non-covalent interactions across the dimer interface [80] are essential for dimer formation, the intermolecular disulfide between two conserved cysteines of the VFTs (C129 and C131 [81–83]) seems not to influence the formation of dimer but to stabilize it [84]. Critical for CaR expression and activity seem to be other three intra-molecular disulfides formed by cysteines in the ECD of the CaR, including C60–C101, C358–C395, and C437–C449 [85]. More recently, Brauner-Osborn showed that one allosteric site per CaR dimer is sufficient for achieving the modulatory effect of positive allosteric modulators, while prevention of activation in both 7TM domains is needed in order to obtain complete inhibition by negative allosteric modulators [86].

The nine-cysteine structure domain (found in all members of the GPCR family C but the GABA B receptor) serves as a linker between the ECD and TMD and is essential for the CaR signalling capacity as demonstrated by the consequences of its removal [87]. The intracellular domain of the CaR shows a low degree of conservation among species [88] except for residues located at the beginning of the C-tail (863–925) that are crucial for surface expression [89] and those (960–984) involved in the interaction with downstream proteins [8,90,91]. As for other GPCRs, CaR activity can be regulated via phosphorylation by different protein kinases, such as PKC [76], which plays a significant role in the negative regulation of the CaR's activity and the generation of CaR-induced intracellular Ca^{2+} oscillations. On the other hand, PKA [92] is involved in the intracellular retention of the receptor [93]. For detailed reviews on the structure of the CaR and new insights on ligand binding pockets, see [8,70,94]. The structural requirements for processes such as bias and allosteric modulation of the CaR have been analyzed by recent works [95]. In addition, a number of papers have highlighted the importance of the potential Ca^{2+} binding sites and their relevance for associated diseases [8,96–102].

Recently, the first high-resolution crystal structure of the ECD of the human CaR bound with Mg^{2+} has been proposed [102]. In this work, the authors found a high-affinity tryptophan derivative in the crystal structure of the CaR, which seems to play a role in potentiating the function of the receptor. The proposed CaR co-activation operated by extracellular Ca^{2+} and amino acids might have a wide impact for the regulation of the receptor in districts of our body that are constantly exposed to both CaR agonists, such as the intestine. An exhaustive review about the recent insights regarding the structure and functional cooperativity of the CaR (and other members of cGPCRs) in health and disease has been published by the group of Jenny Yang [103].

4.3. CaR Promiscuity: Many Orthosteric Agonists and Allosteric Modulators Concur in Modulating CaR Function at Multiple Sites

Although, as the name states, the main ligands of the CaR are extracellular Ca^{2+} ions, which bind to the receptor with a high positive cooperativity, the CaR is also able to respond to other extracellular stimuli (see Table 1), such as inorganic di- and tri-valent cations, including Sr^{2+}, Ba^{2+}, Gd^{3+}, Al^{3+} [11,104], various organic polycations, such as the polyamines spermine and spermidine [105], and basic polypeptides, including polylysine, polyarginine [11,106], and amyloid β-peptides [107]. The CaR has also been shown to be activated by some aminoglycoside antibiotics, such as neomycin and gentamycin [108,109]. These ligands, referred to as type I calcimimetics or orthosteric agonists, can stimulate the receptor also in the absence of Ca^{2+} or any other ligands.

Table 1. Orthosteric agonists and allosteric modulators of the calcium-sensing receptor (modified from [110]).

Orthosteric Agonists (Type I Calcimimetics)	**Class Members**
A. Cations	High potency: Gd^{3+}; Eu^{3+}; Tb^{3+} Intermediate potency: Zn^{2+}; Ni^{2+}; Cd^{2+}; Pb^{2+}; Co^{2+}; Fe^{2+} Low potency: Ca^{2+}; Mg^{2+}; Ba^{2+}; Sr^{2+}; Mn^{2+}
B. Polyamines	spermine, spermidine, putrescine
C. Aminoglycoside antibiotics	neomycin, gentamycin, tobramycin, poromomycin, kanamycin, ribostamycin
D. Basic polypeptides	Poly-L-arginine, poly-L-lysine, protamine, amyloid β-peptides
Allosteric Modulators (Type II Calcimimetics)	
A. L-amino acids	phenylalanine, tryptophan, tyrosine, histidine
B. Glutathione analogs	γ-glutamyl-tripeptides: glutathione, *S*-methylglutathione, *S*-propylglutathione, γ-glutamyl-tripeptides (γ-Glu-Ala, γ-Glu-Cys)
C. Small molecule calcimimetics	The first generation: NPS R-568, NPS R-467, AMG 073, AMG 416 The second generation: cinacalcet The third generation: dibenzylamine calcimimetics, *R,R*-calcimimetic B, AC-265347
D. Small molecule calcilytics	NPS 2143, Calhex 231, ATF396, AXT914, ronacaleret, NPSP795, SB-423557, SB-423562

The other class of CaR agonists is represented by type II calcimimetics (or allosteric modulators), composed by ligands which bind to a site different from that of the orthosteric agonists, thus altering the receptor conformation and, as a consequence, the affinity and/or the signaling capacity of the orthosteric agonist. This action can be exerted in a positive (calcimimetics) or a negative direction (calcilytics).

Belonging to this class of CaR agonists are natural L-amino acids, especially aromatic [111–113] and glutathione analogs [114,115], which synergize with extracellular Ca^{2+}. The receptor's activity is also negatively modulated by low extracellular pH [116] and high ionic strength [117].

A detailed review on the binding sites of orthosteric and allosteric agonists of the CaR has been recently published by Zhang and colleagues [8].

Among the allosteric modulators of the CaR, there are drugs with a high therapeutical potential: first, second, and third generation calcimimetics [118–121].

After the first generation of calcimimetics, which is represented by phenylalkylamine derivatives, such as NPS R-568 and NPS R-467 [122], the second generation reached the clinic: cinacalcet is the only CaR modulator currently used in clinical practice to reduce parathyroid glands activity in patients under dialysis for end-stage renal disease [123–125]. The third generation calcimimetics, which comprise the dibenzylamine calcimimetic, *R,R*-calcimimetic B, and the structurally distinct calcimimetic AC-265347, appear to have enhanced tissue-selective effects on PTH and calcitonin release in rats [126,127] most probably via biased allosteric modulation of the CaR's activity [128] (see below).

Allosteric antagonists or calcilitics, such as NPS-2143 and Calhex 231, antagonize the stimulatory effects of extracellular Ca^{2+} and are currently in clinical trials for treating osteoporosis [129].

In addition to affecting the signaling capacity of the CaR, allosteric CaR modulators have been shown more recently to act as pharmacochaperones that rescue cell surface expression of mutant CaRs that are retained intracellularly (see the paragraphs below).

4.4. Trimeric G Proteins and Signalling Pathways Activated by the CaR

Besides its ability to sense changes in multiple physiological parameters via interaction with different ligands, the CaR has also been shown to activate a wide array of different intracellular pathways (Figure 2) (for comprehensive reviews see [130,131]).

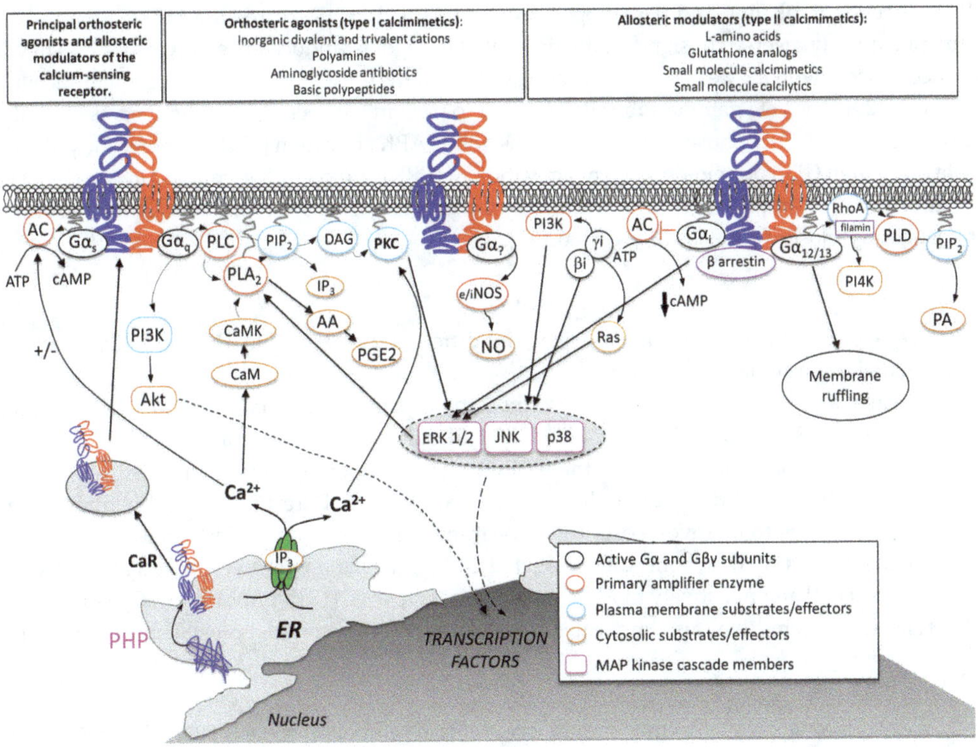

Figure 2. Schematic illustration of intracellular transduction cascades activated by orthosteric agonists or allosteric modulators on the CaR. Modified from [110].

4.4.1. Gq/G11

Most of the studies available in the literature on CaR physiopathological functions in diverse tissues of the body have demonstrated that this receptor, as shown in parathyroids [61,132], mainly interacts with the Gq/11 heterotrimeric G protein, inducing in turn the activation of phospholipase C (PLC)/inositol-1,4,5-trisphosphate (InsP3)/Ca^{2+} mobilization and activation of

PKC and mitogen-activated protein kinase (MAPK) pathways [133]. These intracellular events cause a decrease in parathyroid hormone (PTH) secretion and reduction in renal tubular Ca^{2+} reabsorption [133].

PKC has been described to phosphorylate specific CaR residues and attenuate signalling and/or activate the mitogen-activated protein kinase (MAPK) cascade (ERK1/2, p38 MAPK, JNK), ultimately causing phosphorylation and activation of ERK1/2 [134]. CaR-induced and Gq-mediated activation of phospholipase A2 and phospholipase D have also been reported [135]. Also, activation of the Akt pathway by Gq (via phosphatidylinositol 3-kinase, PI3K) has been described as a determinant in CaR-mediated cell migration [136] and metalloproteases production [137]. The activation of the intracellular Ca^{2+} signalling pathway is also responsible for a Gi/o-independent inhibition of cyclic AMP synthesis mediated by Ca^{2+}-induced inhibition of adenylyl cyclase isoforms 5, 6, and/or 9 [138–140]. For a recent review see also [141].

4.4.2. Gi/o

Early studies have led to the suggestion that the CaR can interact with the pertussis-toxin-sensitive inhibitory G protein, Gαi/o, causing inhibitory control of PTH secretion [142,143] although the matter is still debated [144]. More recent work has clearly demonstrated CaR-mediated inhibition of adenylyl cyclase (AC) by a pertussis toxin (PTX)-sensitive mechanism. In gastric mucosa, CaR activation led to PTX-dependent alkaline and pepsinogen secretion [18]. In CaR-expressing HEK-293 cells, CaR stimulation quickly reversed or significantly prevented the agonist-induced rise in cytosolic cAMP via a dual mechanism involving pertussis-toxin-sensitive Gi and the CaR-stimulated increase in intracellular Ca^{2+} level [139]. The beta gamma subunits of the Gi/o protein have also been described as mediators of CaR -induced ERK1/2 phosphorylation via Ras and MAPK activation [134]. More recently, the role of beta-arrestin in Gi/o protein-dependent activation of (ERK) and c-Jun N-terminal kinases (JNK) has been directly demonstrated [145]. Gi activation by the CaR has also been reported to induce activation of phosphatidylinositol 4-kinase (PI4K) [146,147].

4.4.3. G12/13

G12/13 pathways have been described in different cell types to differentially modulate CaR-mediated intracellular Ca^{2+} kinetics.

Both early and recent studies have described the capacity of aromatic amino acids to induce intracellular Ca^{2+} oscillations via a G12/13 /Rho/ filamin-A pathway, which in turn induce Ca^{2+} influx from the extracellular fluid via the TRPC1 channel opening [148–150]. On the contrary, in pre-osteoblasts, a G12-dependent dephosphorylation of CaR residue T888 by phosphatase PP2A [151,152] seems to be involved in the appearance of the sustained Ca^{2+} increases necessary to activate chemotaxis [153]. In bone, CaR-mediated G12/13 activation positively modulates osteoblast differentiation [154] and negatively modulates osteoclastogenesis [155]. Importantly, CaR-dependent G12/13 signalling seems to be also implicated in the spreading of breast and prostate cancer cells [156].

4.4.4. Gs

In some cell contexts, such as human breast cancer [157] and pituitary cells [158], the CaR has been shown to modulate hormone secretion by activating Gs. Interestingly, a CaR-induced Gs-independent mechanism for the increase of cyclic AMP synthesis has been recently identified in human fetal lung [159].

Finally, a role for nitric oxide (NO) in mediating CaR action on ion channels has been recently demonstrated in the vasculature [160–164].

4.5. Physiopathological Significance of CaR: Old and New Aspects

As the name suggests, the main physiological function of the CaR is to sense subtle changes in serum Ca^{2+} and maintain systemic Ca^{2+} homeostasis within a narrow range via the oppositely directed changes in secretion of PTH, which in turn modulate the equilibrium between Ca^{2+} uptake and conservation via a direct action on kidney, bone, and thyroid C cells [165] and indirect actions on intestine. Also, the CaR controls calcitonin secretion from thyroidal C-cells upon detection of variations in $[Ca^{2+}]$ext (for a review, see [11]).

When the extracellular Ca^{2+} concentration is reduced, CaR-induced parathyroid hormone (PTH) secretion is elicited by the parathyroid glands. Consequences of PTH release are: (a) reduced excretion of Ca^{2+} by the kidney; (b) increased intestinal absorption of Ca^{2+}, which is mediated by PTH-induced synthesis of the active vitamin D metabolite 1,25-dihydroxyvitamin D; and (c) net release of skeletal Ca^{2+} under sustained PTH secretion. All these tasks cooperate to increase serum Ca^{2+} back to physiological levels. On the other hand, when serum Ca^{2+} levels are increased over the physiological range, the receptor is activated inhibiting PTH synthesis and secretion [166].

The physiological significance of the CaR in tissues involved in systemic Ca^{2+} homeostasis is well-highlighted by the identification of mutations of the CaR gene that cause several inherited disorders of extracellular calcium sensing (for a review, see [167]).

Familial hypocalciuric hypercalcemia (FHH) and neonatal severe hyperparathyroidism (NSHPT) [168] are caused by loss-of-function mutations of the CaR gene, whereas autosomal dominant hypocalcemia (ADH) and Bartter Syndrome type V [169] occur from gain-of-function mutations of the CaR [170–172].

Besides its expression in tissues involved in Ca^{2+} homeostasis, such as thyroid glands, kidneys, and bone, the CaR has been identified in an astonishing number of diverse cell types (reviewed in [11] and [173]), i.e., brain [174], pancreas [175], stomach [89], liver [176], and heart [177,178]. In these tissues, the CaR has been shown to regulate a variety of cellular processes, such as secretion, differentiation, and gene expression. For recent reviews on this topic see also [74,179].

4.5.1. CaR Involvement in Cancer

The massive impact of the CaR in such a wide array of cellular processes, including proliferation, differentiation, migration, adhesion, and apoptosis foresees a role for this receptor in many aspects (incidence, progression, recurrence, and lethality) of various cancers (e.g., parathyroid, pancreatic, prostate, breast, colorectal, ovarian, gastric, skin, and neuroblastoma) [9,180]. However, mutations of the CaR are not early events in tumorigenesis. Still, an increasing amount of literature is concordant on CaR participation in various aspects of the neoplastic progression. In this line, a number of papers evaluating the role of CaR polymorphisms in the risk, incidence, recurrence, or lethality of different kinds of cancer has been published [9].

The CaR is associated with both anti- and pro-proliferative signals. Thus, it can either prevent or promote tumorigenesis depending on the cellular context and the type of cancer [180]. Tissue-specific alterations of CaR expression level, functional activity, or coupling to signaling pathways might explain such opposing behavior.

In tissues such as parathyroid or colon, CaR inhibits cell proliferation and induces terminal differentiation and apoptosis. In tumors of these organs (and more recently also of the gastric mucosa), the observed and progressive reduction in CaR expression (e.g., from normal tissue to carcinomas) grants malignant potential, indicating for the receptor and the physiological significance of its activation a putative role as tumor suppressor [181–185]. Accordingly, when taking also CaR functional activity into consideration, inactivating mutations of the CaR or calcilytics might induce hyperplasia, while calcimimetics have the opposite outcome [186,187]. This scenario is extremely important assuming that the CaR is a novel target for the prevention/treatment of these tumors. However, as reported throughout this review, the incredible number of CaR interactors (both extracellular and intracellular) increases the complexity associated with this kind of pharmacological approach.

Indeed, some of the first evidence connecting extracellular Ca^{2+} and tumors was the inverse correlation between dietary Ca^{2+} level and risk of colorectal cancer [188]. This effect seemed to be mediated by the CaR and potentiated by the active vitamin D metabolite calcitriol [189]. Since CaR transcription is strongly upregulated by vitamin D [190], a Ca^{2+}-containing diet could protect against colon cancer partly via upregulation of CaR expression. The work by the group of Enikö Kallay provided clear evidence about the role of dietary Ca^{2+} and dissected the reciprocal interplay between extracellular Ca^{2+}/CaR and the Vitamin D system (for an exhaustive review refer to [189]). Significant reduction in CaR expression level has also been observed in unfavorable neuroblastomas when compared with benign, differentiated neuroblastic tumors. However, information about the mechanisms underlying the loss of CaR expression is still scattered and thus under intense investigation. Yet, in colon cancer and neuroblastomas, genetic or epigenetic mechanisms, such as hypermethylation and histone deacetylation of CaR promoter or monosomy of the chromosome 3, have been proposed [9,191].

On the other hand, in prostate and breast tumors, the expression level of the CaR is increased [192,193]. This experimental evidence is in line with an oncogene-like behavior of the CaR (e.g., increasing proliferation and inhibiting cell death), which seems strictly related to the Gαs/cAMP-dependent production of the parathyroid hormone-related protein (PTHrP). Strikingly, in normal mammary epithelial cells, during lactation, the CaR inhibits PTHrP synthesis and secretion through a Gαi, clearly indicating the ability of the receptor to bind and activate opposing G proteins in normal and breast cancer cells [157]. In addition, in breast cancer cells, CaR signalling through Gα12/13 affects cell migration via rho-dependent actin filament formation facilitating the metastatic spread of the tumor [194]. These tumors often lead to bone metastasis, since bone provides homing signals (such as PTHrP-induced high extracellular Ca^{2+}) to several cancers, such as breast, prostate, lung, and kidney cancer.

4.5.2. The CaR in the Heart

Another field of emerging interest for its implications on human health concerns CaR function in cardiovascular physiology and pathology.

The CaR has been demonstrated to exert a fundamental role in the vasculature and to affect blood pressure in different experimental models [160,195–198]. Nice reviews on this topic have been recently published by experts in the field [199–202].

Hence, the focus of this section is on the specific role of the CaR in the heart, a topic which is still quite controversial and, in our opinion, needs further study in the near future.

The first paper describing the functional expression of the CaR in the heart was published 15 years ago [177]. In that first report, the authors identified CaR mRNA and protein both in atrial and ventricular tissues from adult rats. Also, a tentative imaging approach was performed by using high extracellular Ca^{2+} concentrations, spermine (1–10 mM), and gadolinium on isolated adult ventricular cardiomyocytes. Although the interpretation of data collected with high extracellular Ca^{2+} on Fura-2-loaded cells could be hampered by a protocol in which the cardiomyocytes were first bathed in 0 Ca^{2+}, thus activating a serious Ca^{2+} overload upon Ca^{2+} re-addition, the authors provided evidence of effects of extracellular calcium, spermine, and gadolinium on changes of Inositol Phosphate (IP) levels, thus implicating the PLC/InsP3 signalling pathway in CaR activation.

The existence of a functional CaR was next clearly demonstrated in neonatal rat ventricular cardiomyocytes by Tfelt-Hansen and collaborators [178]. Importantly, in this study the calcimimetic AMG 073 was used to activate the CaR and proven to be effective both in IPs accumulation and ERK1/2 activation. Also, the involvement of CaR in IPs modulation was straightforwardly demonstrated by the significant inhibition of Ca^{2+}-induced IP accumulation upon infection of neonatal cardiac myocytes with adeno-associated viruses containing the dominant negative CaR R185Q.

Starting from 2006, a conspicuous number of papers has been published by researchers from the Harbin University, where the first paper on the CaR in cardiac tissues was published [177]. The results collected during the last decade by this group depict a general scenario in which CaR activation is mostly causative of cardiac diseases.

On the contrary, other groups have provided evidence of a cardioprotective role of the CaR. Various experimental models and approaches have been used to inquire into the CaR function in cardiac apoptosis induced by different drugs or by ischemia/reperfusion, hypertrophy and fibrosis, and diabetic cardiomyopathy. The role of the CaR in pre- and post-conditioning has also been investigated.

Role of the CaR in Cardiac Apoptosis

A pro-apototic effect of GdCl3, used as the only CaR agonist (in the presence of NiCl2 and CdCl2 to block L-type Ca^{2+} channels and Na+/Ca^{2+} exchangers, respectively) was reported by Sun and colleagues in neonatal cardiac myocytes [203]. CaR pro-apoptotic action was suggested to be mediated by intracellular Ca^{2+}-induced activation of extracellular signal-regulated protein kinase (ERK), c-Jun NH2-terminal protein kinases (JNK), and p38. GdCl3 was also shown to activate caspase 9 under these experimental conditions.

In parallel, another work was published on isolated rat hearts subjected to an ischemia/reperfusion protocol [204]. By using GdCl3 as CaR agonist, the authors suggested that an increased CaR expression is involved in the induction of cardiomyocytes apoptosis via Ca^{2+} overload-induced cytochrome C/caspase 3 axis activation. Later, Sun and colleagues suggested, by using the same experimental protocol, that one of the molecular players involved in CaR-mediated apoptosis in neonatal cardiomyocytes is the Ca^{2+} channel TRPC6 [205], whose expression was found, like the CaR, to be increased after Gd^{3+} stimulation. Similarly, an increase of TRPC3 expression was described in neonatal rat cardiomyocytes in a following paper, although no evaluation of apoptosis was performed in this study [206].

The effects of known proapoptotic substances were also investigated. Ciclosporin A (CsA) was described to exert its adverse effects (also) by increasing CaR expression both in vitro [207] and in vivo [208].

In neonatal rat cardiomyocytes [207], the proapoptotic role of the CaR in CsA-induced apoptosis was investigated by using Gd^{3+} and NPS-2390 as CaR agonist and inhibitor, respectively. While in the neonatal cell model the CsA-induced CaR-mediated cell death was attributed to Ca^{2+} overload [207], in a similar work performed two years later on H9c2 cardiomyoblasts the involvement of the MAP kinases pathway was suggested [209]. In the same year, the CaR was also implicated in lipopolysaccaride (LPS)-induced apoptosis via its increased expression and consequent Ca^{2+} overload and TNF alpha and IL-6 release [210].

Role of the CaR in Apoptotic Pathways Activated by Ischemia/Reperfusion

During the years 2006–2011, some of the same researchers from Harbin University made a considerable effort to disclose the involvement of the CaR in the apoptotic pathway activated by ischemia/reperfusion (I/R) or hypoxia/reoxigenation protocols both on in vivo and in vitro models, i.e., neonatal rat cardiomyocytes [211–214], isolated adult rat cardiomyocytes [204,215], and adult rat hearts [212,215]. Altogether, the collected results suggest that I/R induces a significant increase of CaR expression. Subsequent CaR-mediated signalling pathways activate apoptosis. The involvement of mitochondria in such an effect was suggested to be mediated by Ca^{2+} overload via inter-organelle signalling [213], ER stress [212], and PKC delta [216]. Activation of the C-Jun NH2 terminal protein kinase signalling pathway was also described as one of the mechanisms relevant for CaR-mediated injury during ischemia/reperfusion in adult rats [215].

Later, another group at the Nanjing University confirmed the work of Jiang et al. [211,217] and implicated a downregulation of CaR expression and signalling in the anti-apoptotic effect of hepatocyte growth factor in neonatal cardiomyocytes subjected to simulated ischemia/reperfusion [218].

Role of CaR Inhibition in Post-Conditioning

Almost in the same years, Zhang et al. [219] suggested on isolated rat hearts subjected to an ischemia/reperfusion protocol followed by post-conditioning (PC) that PKCε is involved in the protective effect of the PC maneuver via its negative feedback on the CaR and the subsequent reduction of intracellular Ca^{2+}. Also, in these studies $GdCl_3 + NiCl_2 + CdCl_2$ was the only experimental maneuver used to activate the CaR.

Similar results were obtained some years later on the neonatal rat cardiomyocytes model, this time using also the calcilytic NPS 2390 and implicating the attenuation of CaR-induced sarcoplasmic reticulum–mitochondria crosstalk in the protective effects of PKCε during post-conditioning [220].

In 2012, in a work performed in vivo and on isolated adult rat cardiomyocytes subjected to a hypoxia post-conditioning protocol, the authors suggested that the pro-apoptotic action of CaR exerted via ER stress activation was counteracted during post-conditioning by a reduction of CaR protein expression [221]. Very recently, Zhang and colleagues confirmed that ischemic post-conditioning and KATP channel agonists diazoxide and pinacidil exert a protective effect against ischemia/reperfusion injury through a mitigation of Ca^{2+} overload induced by downregulation of the CaR [222].

Given the relevance of CaR expression levels for the rationale of these studies, it is worth mentioning a recent short communication from the group of Schluter [223] focused on the assessment of CaR expression levels after ischemia reperfusion and cardiac pre- and post-conditioning. This report pointed out that different CaR regulation patterns seem to exist between left and right ventricular tissues and that also the size of infarcts could be a determinant in regulating CaR expression [223].

Role of the CaR in Pre-Conditioning

In opposition to the theory of a deleterious action of the CaR in the heart, there was already in 2010 the work by Sun and Murphy in which the authors suggested a cardioprotective role for the CaR, strategically localized in caveolae, during ischemic pre-conditioning (IPC) [224].

In this study, the perfusion of adult rat hearts with the CaR antagonist NPS 2143 during the application of the ischemic pre-conditioning protocol was indeed able to attenuate cardioprotection of IPC, as assessed by measuring the levels of the cardioprotective phospho ERK1/2, AKT, and GSK3β kinases, infarct size, and functional recovery. The suggestive speculation of the authors depicts a scenario in which the changes in pH and ionic strength imaginable at the particular microdomains of caveolae may activate the CaR, which could thus "sense" and "mediate" IPC.

Another controversial topic is the role of CaR activation in the progress of diabetic cardiomyopathy [225,226].

On the one hand, Bai et al. suggested that CaR expression is reduced in diabetic rat hearts and could contribute to the progress of diabetic cardiomyopathy via impaired intracellular Ca^{2+} signalling in adult cardiac cells [225]. On the other hand, Qi and colleagues showed that CaR expression is rather increased in diabetic rat hearts and by using Gd^{3+}, NPS 2390, and a siRNA to silence CaR protein, suggested that CaR signalling contributes to apoptosis of neonatal cardiomyocytes exposed to high concentrations of glucose in vitro via Ca^{2+} overload, reduction of the Bcl2/Bax ratio protein expression, and modulation of the MAP kinase pathway [226]. The apparent discrepancy of results on CaR expression was explained by invoking a difference in the experimental models which, according to Qi et al., were representative of type 2 diabetes in the first work [225] and type 1 diabetes in the latter [226].

Role of the CaR in Hypertrophy and Heart Failure

A role for the CaR in cardiac hypertrophy has been postulated since 2006, when Tfelt-Hansen and collaborators [178] observed a CaR-induced decrease in DNA synthesis in neonatal cardiomyocytes. Considering that this parameter increases in cells undergoing hypertrophy [227], the authors supposed a protective role for the CaR against cardiac hypertrophy.

In opposition to this theory, in a study performed by Wang and colleagues [228] on an in vitro model of hypertrophy, an increased expression of the CaR was found and exacerbating effects of Gd^{3+} on Ca^{2+} signalling and hypertrophic markers were argued to be mediated by the CaR.

Later, Zhong et al. suggested a role for the CaR in the Gd^{3+}-induced increase of nuclear Ca^{2+} spikes' frequency and consequent activation of the Calcineurin/NFAT pathway [229], a well-known player in cardiac hypertrophy [230].

The role of the CaR in cardiac hypertrophy and heart failure was further investigated in a successive work [231], in which Calindol and Calhex 231, as agonist and inhibitor of the CaR, respectively, were subcutaneously injected in isoproterenol-treated Wistar rats to assess the role of the CaR in ER stress in vivo. CaR expression was found to significantly increase during heart failure induced by isoproterenol and Chalex 231 significantly reduced the cross-sectional diameter of hypertrophic cardiomyocytes. In the same study, the CaR was suggested to play a role in cardiac hypertrophy and heart failure also in mice subjected to thoracic aorta constriction (TAC) [231]. In this model, Calindol was found to induce significant cardiac hypertrophy.

The involvement of the CaR in ER stress-mediated apoptosis in failing hearts was also suggested by the observation that Calindol and Chalex 231 were able to increase and decrease, respectively, ER stress protein expression and apoptosis in both the rat and mouse models. Finally, a CaR-induced IP3-dependent Ca+ overload of mitochondria and subsequent initiation of apoptosis were suggested in rat hearts and neonatal rat cardiomyocytes incubated for 48 h with isoproterenol. On these bases, the authors proposed that the CaR could represent a putative target for pharmacological intervention in heart failure.

Very recently, Liu et al. suggested the possibility to ameliorate isoproterenol-induced cardiac hypertrophy through suppression of CaR signalling with the CaR inhibitor Chalex 231 both in vivo and in vitro [232]. In this paper, the authors suggested that the anti-hypertrophic effect and amelioration of cardiac function by Chalex 231 may be mediated through inhibition of autophagy and the CaMKK-AMPK-mTOR signalling pathway.

A recent in vitro and in vivo study also suggested a role for the CaR in promoting cardiac fibrosis [233].

In opposition to a maladaptive pro-hypertrophic effect of the CaR, which is the common denominator of the above-described studies, there are recent contributions from the group of Schluter [223] which sustain the theory of a cardioprotective effect of the CaR in a pharmacological model of hypertension in adult rats. Interestingly, these authors showed that upon NO-synthase inhibition an increased expression of endothelin A and/or B1 receptors and CaR was measurable. This upregulation of CaR, most probably mediated by a paracrine action of endothelin 1, was suggestive of a CaR-induced adaptive hypertrophy. Indeed, CaR inhibition or destabilization by NPS 2390 and siRNA against RAMP, respectively, decreased load-free cell shortening, which was used as a read-out of cardiac function.

Role of the CaR on the Electrical Properties of Cardiac Cells

Notwithstanding the abrupt proliferation of literature on the role of the CaR in the heart in the last 15 years, very few data were available, until recently, on the effect of CaR agonists and modulators on the electrical properties of cardiac cells.

In very recent times, the group of Schluter [234] and Shieh [235] have published interesting data on this topic on adult rat hearts and guinea pig cardiomyocytes, respectively, by using different physiologically relevant experimental approaches.

First, Schreckenberg and colleagues examined the effects of CaR acute stimulation on the cardiac physiology of adult male rats [234]. The experimental approaches were diverse: the contractile response of ventricular cardiomyocytes was recorded with a cell-edge-detection system in parallel with Fura-2 measurements of cytosolic Ca^{2+}, while cardiac performance in response to acute CaR stimulation or inhibition was assessed on left ventricle muscle stripes and whole hearts by the Langendorff system.

Putrescine and Gd^{3+} were used as CaR agonists, while NPS-2390 (10/100 μM) and downregulation of CaR by siRNA were used to assess the specificity of the responses.

The data collected, while confirming the expression of CaR by adult rat ventricular cardiomyocytes, demonstrated that CaR action is relevant for basal cell shortening at physiological extracellular Ca^{2+} concentrations and that additional activation of the CaR speeds up cardiomyocytes relaxation and augments cell shortening, supposedly via increased Ca^{2+} transients induced by activation of the Gq/PLC/IP3 pathway.

Very interestingly, Liu et al. clearly demonstrated that activation of the CaR in guinea pig cardiomyocytes by the allosteric agonist NPS R 568 activates both the PLC and phsphatidylinositol-4 kinase (PI4) Kinase pathways and that the PI4 Kinase pathway predominates [235]. The net effect is an increase in PIP2 at the plasma membrane, which in turn determines a significant increase in currents through inward rectifier K+ channels. This action could thus exert a stabilization of the resting membrane potential of cardiomyocytes and contribute to a re-evaluation of old data on the stabilizing and cardioprotective effects of polyamines [236].

Finally, interesting hints on the CaR's role in the heart come from a paper focused on the function of the CaR in regulating blood vessel tone and blood pressure [237].

Since the experimental model of CaR KO mice (SM22αCaSRΔflox/ΔfloxKO mice) used in this work also display a deletion of the CaR in cardiac cells, the authors tested the hearts of KO mice by perfusion in the Langendorff mode. Given the significantly reduced basal heart rate found in KO mice versus wild-type (WT) animals, the authors suggested that CaR deletion from heart directly affects chronotropy. Also, ex vivo lung slice preparations from KO mice showed a reduced rate of spontaneous activity, suggesting that the decreased chronotropy in KO mice could be due to decreased pacemaker activity.

In summary, notwithstanding the numerous results collected on physio-pathological roles of the CaR in the heart, the current findings are far from providing a coherent picture.

Considering CaR promiscuity and pleiotropicity and the emerging concept of biased signalling at the CaR (see below), an interesting field of future investigation would be the accurate pharmacological assessment of the effects exerted by panels of CaR agonists and modulators on different signalling pathways in physiologically relevant experimental models.

Given the actual and potential use of calcimimetics and calcilytics in the clinic, these studies would be highly significant for their implication for human health.

4.6. Novel Trafficking/Signaling Modes of the CaR

Although GPCR have often been seen as cell surface proteins able to go through a simple on-off switch, research of the past few years has drawn a much more complex scenario, in which receptors exist in multiple conformations and, in dependence of the different ligands, can be stabilized in different conformational states, in turn activating multiple, yet specifically selected, downstream pathways whose identity depends on the combination of ligands and the specific signalling toolkit of the cell where they are expressed [238–242].

The development of new technologies, such as microscopy techniques and probes to follow receptor trafficking and subcellular signal dynamics (such as FRET and single molecule microscopy) as well as BRET-based high throughput assays for biased signalling [243], have been essential to demonstrate that these proteins are capable of trafficking among a variety of subcellular compartments, initiating specific signalling pathways at different locations: nuclear [244], mitochondrial [245], and endosomal–Golgi membrane [246,247].

As stated in the paragraph regarding the structural features of the CaR, functional receptors localized at the plasma membrane are dimers of the glycosilated CaR which are formed intracellularly [248] and are stabilized by ligand binding at the cell surface [249]. As for other GPCRs, the expression level of CaR is the result of a fine regulation of the dynamic equilibrium between mechanisms that precede its biosynthesis (transcription/mRNA level, translation efficiency, epigenetic control [250,251]) and regulate its anterograde trafficking and those controlling its retrograde trafficking, including degradation. In fact, it has been known since a long time that the fine balance between maturation, internalization, recycling, and degradation can influence the net level of cell surface receptor and thus represent a mechanism for the cell to regulate receptor desensitization and modulate strength of signal transduction [252]. Thus, the strength of signaling depends upon the level of GPCRs expression at the plasma membrane and the consequent receptor availability for agonists and modulators. This is also the case for CaR signalling as recently established by Conigrave's group [253].

A significant contribution to the field of CaR trafficking has come in the last ten years from different groups who have focused on the elucidation of the key steps involved in CaR biosynthesis and trafficking from the ER membrane to the Golgi membrane and then to the plasma membrane (see [254] for a recent review). A fundamental role in this field is also played by research focused on the CaR-interacting proteins, which finely modulate anterograde and retrograde trafficking of the CaR (for comprehensive reviews see [10,133,254–256]).

The Agonist-Driven Insertional Signalling (ADIS)

Both early and recent studies have revealed two peculiar characteristics of the CaR: a minimal functional desensitization (both by phosphorylation and/or β-arrestins) and a significant "reservoir" of CaR proteins in intracellular membrane compartments. Western blotting or immunohistochemistry analyses [11,76,257] have clearly showed, since early studies, that CaR immunoreactivity is predominantly represented by an intracellular, core-glycosilated form. Recent data are making increasingly clear that such an observation is not an artifact but is strictly related to the complex interaction between CaR trafficking and signalling. In fact, both the hallmarks of the CaR, minimal desensitization and high levels of intracellular proteins, can be explained by a novel signalling mode described by the group of Dr. Breitwieser and termed Agonist-driven insertional signalling (ADIS) [250,258].

In this model, CaR signalling is regulated by an agonist-dependent modulation of the net amount of receptor proteins at the plasma membrane. In particular, the level of CaR cell-surface expression results from the balance between the rate of trafficking from a large, pre-plasma membrane pool located at the Golgi/trans-Golgi/post-Golgi vesicles (which depends upon the concentration of CaR agonists and/or allosteric modulators) and the process of constitutive endocytosis of the receptor already at the plasma membrane. The sensitivity of parathyroid and renal tubular cells to extracellular Ca^{2+}, for example, is likely influenced by the balance between ADIS, which enhances anterograde trafficking of newly synthesized CaR proteins to the plasma membrane [258] and clathrin-mediated endocytosis and retrograde trafficking of cell-surface CaR receptors [10].

4.7. *Synthetic Allosteric Modulators of the CaR Can Function as Pharmacoperones*

Strictly related to the research in the field of CaR trafficking is the observation that CaR allosteric modulators (see Paragraph 4.3) can work as pharmacoperones. Pharmacoperones are permeant substances (which could be agonists, antagonists, or allosteric modulators) that can reach a misfolded protein at its intracellular location (most commonly the endoplasmic reticulum), stabilize it, and rescue the receptor to the cell membrane surface.

A relevant effort on the capacity of CaR allosteric modulators to work as pharmacoperones comes from Breitwieser's group [259–261]. These authors firstly demonstrated that the positive allosteric modulator NPS-R-568 was able to rescue almost fifty percent of the loss-of-function mutant CaRs analyzed [259,260]. In parallel, another group demonstrated the same ability of NPS-R-568 to rescue intracellular signalling in four mutants (out of seven) although the plasma membrane localization of the receptors was not assessed [262].

In the last five years, relevant findings in the field of pharmacoperone action of type II calcimimetics and calcilytics have been provided by studies from the group of Dr. Leach. These authors, while confirming data on NPS-R-568, showed the ability of cinacalcet, the only CaR modulator approved in the clinic, to effectively rescue trafficking and signalling of CaR mutants with loss of cell surface expression. [128,263]. They also showed that NPS-2143, a negative allosteric modulator, is able to positively modulate CaR trafficking to the cell membrane while negatively modulating CaR signalling in the presence of the orthosteric substrate [144,263], underlining a striking example of positive bias (see also above) toward trafficking by a CaR allosteric modulator.

Nonetheless, both early and recent studies suggest that the pharmacoperone action of NPS R-568 [264] and NPS-2143 [259,264] is mutant-specific.

Importantly, the effectiveness of old and new calcilytics on gain-of-function mutations of the CaR was further explored in vitro [265,266], demonstrating that a novel treatment option for patients with activating CaR mutations could be represented by the new quinazolinone-derived calcilytics, which have been shown to be effective in attenuating enhanced cytosolic Ca^{2+} signalling in four known mutations causing Bartter syndrome (BS) type 5 and autosomal dominant hypocalcemia (ADH). Very interestingly, recent in vivo works involving a mouse model harboring an ADH gain-of-function CaR mutation demonstrated the effectiveness of both old (i.e., NPS 2143) [267] and new (i.e., JTT-305/MK-5442) [268] calcilytics in normalizing the defect and improving the associated hypocalcemia.

4.8. Biased Signalling

Biased signalling (also termed biased agonism or ligand-directed signalling or stimulus bias) represents a recently described, yet quite general, modality of the signalling characteristic of diverse GPCRs [240].

It concerns the capability of different ligands to stabilize diverse conformations of the receptor and thus preferentially head GPCR signalling towards a specific transduction cascade.

This new concept, which substitutes the idea of one receptor/one GPCR conformation/one-uniform activation of signaling pathway(s), greatly complicates the scenario of GPCRs signalling, but opens new perspectives in the design of smart and tissue-specific drugs that avoid off-target actions or on-target effects on multiple tissues [269].

Given the promiscuity and pleiotropicity of the CaR and the wide expression of this receptor in tissues involved or not in the regulation of systemic Ca^{2+}, it is not surprising that fundamental insights into this new ligand- and tissue-specific pharmacology come from the study of biased signalling at the CaR.

In fact, the existence of ligand- and tissue-specific effects in the signalling pathways activated by the CaR, although not precisely quantified/measured, are traceable in a vast number of papers published throughout the years.

It could also be hypothesized that the experimental approaches used in the recent past could have contributed to an underestimation of the phenomenon of biased signalling at the CaR. Most of the work on CaR signalling is in fact performed by testing only a small number of CaR agonists and modulators on a limited array of signal transduction cascades (most frequently cytosolic Ca^{2+} dynamics). A striking example of biased signalling at the CaR comes from an acquired hypocalciuric hypocalcemia autoantibody, which was shown to increase the effects of external Ca^{2+} on Gq signalling to the detriment of the Gi-induced signalling pathway [270]. As far as the differential effects of CaR agonists and modulators are concerned, both early studies by Riccardi's group on rat pancreas [271] and more recent work by Ziegelstein and colleagues on human aortic endothelial cells showed that only some CaR ligands were able to induce intracellular Ca^{2+} release, while others were ineffective [160]. Also, at the level of physiological functions, Tfelt-Hansen's group demonstrated that while cinacalcet was able to induce vasodilatation of rat aorta, the neomycin and Gd^{3+} were ineffective [161].

In the last years, a large number of findings has highlighted the physiopathological significance of biased signalling.

A major contribution in this area has come from Bräuner-Osborne's group [145,272,273]. Specifically, these authors explored the action of a panel of well-known orthosteric CaR agonists on the Gq/11 protein, Gi/o protein, and extracellular signal-regulated kinases 1 and 2 (ERK1/2) pathways. Thomsen and colleagues [145] revealed that in HEK-293 cells stably transfected with rat CaR, the natural agonist Ca^{2+} is biased toward cAMP inhibition and IP3 accumulation over ERK1/2 phosphorylation. Differently, a significant bias toward ERK1/2 phosphorylation was shown for the polyamine spermine.

Interestingly, while two different groups showed that strontium ranelate, a drug used as treatment for osteoporosis, is a biased agonist of the CaR toward the ERK1/2 pathway and another cascade independent of Gq/11 signalling [273,274], different results were reported by others [275]. In another study, while both Sr^{2+} and Ca^{2+} were shown to stimulate PLC and NF-kB, only strontium ranelate induced apoptosis via PKC activation [276].

In recent years, the possibility of exploiting and "inducing" stimulus bias at the CaR has been deeply explored also by the group of Christopoulos and Leach [95,263,277,278] (reviewed in [144,269]). The first quantitative evidence that an allosteric modulator used in clinical practice exhibits stimulus bias was provided a few years ago by Davey and colleagues [277], which straightforwardly proved that allosteric modulators are biased toward intracellular Ca^{2+} signalling in a HEK293-TREx c-myc-CaR cell line.

Next, the same group analyzed in detail the effect of CaR allosteric modulators on naturally occurring CaR mutations [278], demonstrating that mutations can alter signalling bias by stabilizing receptor conformations that differentially couple to downstream signalling pathways. Importantly, and as stated above, in a subsequent paper, both Cinacalcet and the calcylitic NPS-2143 were demonstrated to have a bias toward Ca^{2+} signalling and rescue CaR mutants to the cell membrane [263]. The therapeutic implications of such data are enormous.

Just as an example, the calcimimetic cinacalcet has been used in clinic for the treatment of patients carrying loss-of-function CaR mutations only in the cases of end-stage renal disease. This is because of its hypocalcemic properties, most probably due to the Ca^{2+}-mediated increase of calcitonin secretion [272]. It is thus clear that designing a novel drug able to act on parathyroid glands without affecting calcitonin secretion from parafollicular cells would be the best therapeutic option.

Remarkably, in a very recent paper, the search for a calcimimetic able to suppress PTH levels at concentrations with low/no effect on serum calcitonin seems to have found good candidates among the third generation of calcimimetics [128]. Altogether, these findings highlight the importance of a deeper comprehension of the mechanisms of action of different orthosteric agonists and modulators in the different tissues for the better design of drugs that are devoid of side effects and are able to activate specific pathways in a cell-selective fashion [144,269].

It is worthwhile noting that an even higher level of complexity in the scenario of biased agonism and pharmacoperones action of CaR ligands is the observation that GPCRs have been recently shown to sustain G (Gs or Gi) protein signalling at different subcellular sites even after internalization of ligand–receptor complexes in endosomes (for a recent review see [242]).

5. Old and New Extracellular Ca^{2+} Sensors

Even though the CaR is so far the best-studied extracellular Ca^{2+} sensor, other receptors (either membrane proteins or channels) able to change their function following significant fluctuations in extracellular Ca^{2+} level have been identified [279]. Of note, those of the same family of the CaR have been the focus of intense studies based on a high degree of structural homologies, conservation of specific domains, and a common functional role as nutrient/salinity sensor. This family includes eight metabotropic glutamate receptors (mGluR1–8), two gamma-aminobutyric acid (GABA) receptors, three taste receptors (T1-T3), the recently "deorphanized" GPRC6A, and seven orphan receptors [280].

5.1. Metabotropic Glutamate Receptors

Metabotropic glutamate receptors (mGluRs) are neurotransmitters expressed mainly in the central nervous system, where they modulate key functions in a variety of different neurological processes (e.g., memory, learning, pain, and synaptic plasticity) [281]. mGluRs (especially mGluR1 and -R5) possess significant sensitivity to extracellular Ca^{2+} [282,283]; however, it is still unclear whether they exist in nature as proper Ca^{2+} sensors [282] or whether extracellular Ca^{2+} just sensitizes the receptor to its principal ligand, the excitatory amino acid L-glutamate [284,285]. For example, Ca^{2+} depletion in the synaptic cleft during burst firing induces significant inhibition of postsynaptic mGluR1 function [286]. In addition, Gerber et al. showed that increasing concentrations of extracellular Ca^{2+} might enhance the potency/efficacy of glutamate action on mGluRs1 [287]. Recently, a novel potential extracellular Ca^{2+}-binding site was identified within the extracellular domain of the mGluR1 adjacent to the known L-Glu-binding site [288,289]. When both L-Glu and Ca^{2+} are bound to their partially overlapping binding sites, they synergistically modulate mGluR1 activation. Strikingly, a significant reduction in the sensitivity of mGluR1 to extracellular Ca^{2+} and, in some cases, to L-Glu is induced by a mutation in the predicted Ca^{2+}-binding residue [290]. Furthermore, this Ca^{2+}-binding site is also adjacent to the interaction sites of orthosteric agonists and antagonists; thus, extracellular Ca^{2+} might modulate the sensitivity of mGluR1 not only to orthosteric agonists and antagonists but also to allosteric modulators [291].

5.2. The $GABA_B$ Receptor

The gamma-aminobutyric acid subtype B ($GABA_B$) receptor plays a key role in regulating membrane excitability and synaptic transmission in the brain. The functional receptor consists of a heterodimer of $GABA_{B\text{-}R1}$ and $GABA_{B\text{-}R2}$ [292]. As for the CaR and mGluRs, the $GABA_B$ receptor has the ability to sense extracellular Ca^{2+} (0.001–1 mM) but not other divalent and trivalent cations. Wise et al. showed that an addition of 1 mM extracellular Ca^{2+} potentiated GABA stimulation (EC50 was reduced from 72 to 7.7 μM) in membrane preparations from both CHO cells (that stably express the $GABA_{B\text{-}R1/R2}$ heterodimer) and rat brain cortex. Of note, even though Ca^{2+} does not act as a direct agonist of the receptor, extracellular Ca^{2+} may allosterically enhance the GABA-induced responses of the $GABA_B$ receptor [293].

5.3. Taste Receptors

T1R1 and T1R3 belong to the T1R family and form an heterodimer that is considered a broadly tuned L-amino acid sensor. Silve et al. reported a Ca^{2+}-binding pocket for T1R3. Of note, this receptor subunit is characterized by Ca^{2+}-binding residues that are conserved in the Venus fly trap region of the CaR [294]. Recently, in vivo taste studies have suggested the involvement of T1R3 in Ca^{2+}-Mg^{2+} taste in mice and humans [295,296]. These studies have also supposed that T1R3 might form a dimer with another receptor able to sense extracellular Ca^{2+}, such as the CaR.

5.4. The G Protein-Coupled Receptor Family C Group 6 Member A (GPRC6A)

The G protein-coupled receptor family C group 6 member A (GPRC6A) displays the closest mammalian homology to the CaR. This receptor, cloned in 2004, is extensively expressed at the tissue level [297] while limited information is available at the sub-tissue level. So far, experiments with KO mice for the receptor have shown GPCR6 involvement in the regulation of biological processes such as inflammation, metabolism, and hormonal functions. GPCR6 is activated by a large number of basic and small aliphatic L-α-amino acids (being L-arginine, L-lysine, and L-ornithine the most potent compounds). In addition, high extracellular Ca^{2+} levels (5–10 mM) either directly activate or positively modulate the receptor. This range of extracellular Ca^{2+} concentration can be physiological only in a tissue, such as the bone, where the receptor seems to be involved in bone remodelling processes [298]. However, it remains unclear which of the proposed agonists are the most important physiologically.

Furthermore, CaR allosteric modulators, such as NPS568, Calindol, and NPS2143 (IC50~10 μM), were shown to be effective also on GPRC6A [67,298,299].

5.5. Notch Receptor

The epidermal growth factor (EGF)-like domain contains a Ca^{2+}-binding domain with a wide range of Ca^{2+} affinities [300]. This type of module is probably the most prevalent extracellular Ca^{2+}-binding site. Among the proteins containing Ca^{2+}-binding EGF-like domains, there are proteins involved in blood coagulation, fibrinolysis and its complement system, matrix proteins, and the developmentally important cell surface receptors Notch [300]. Left–right asymmetry in all vertebrates is established during embryo development by several different molecular events. Raya et al., using a combination of modelling and experimental procedures, showed that, in chick embryos, asymmetric patterns of gene expression were the result of an extracellular Ca^{2+} gradient toward the left margin of the developing embryo. Here, extracellular Ca^{2+}-induced activation of Notch results in the asymmetric secretion of Nodal, a specific signalling molecule [301]. Of note is that the perturbed establishment of the left–right patterning in embryo maturation was induced by the disruption of the extracellular Ca^{2+} gradient.

5.6. Cadherins

Cadherins are a family of Ca^{2+}-dependent adherent receptors involved in development, morphogenesis, synaptogenesis, differentiation, and carcinogenesis. Most of the members are single membrane-spanning proteins possessing Ca^{2+}-binding extracellular domains with a wide range of Ca^{2+} affinities (Kd from μM to mM) [302]. The cadherin extracellular domains act as a Ca^{2+}-activated "mechanical switch" through which extracellular Ca^{2+} fluctuations are detected and integrated to control their mechanical integrity and mechano-transduction capability.

5.7. Stromal Interaction Molecule 1 (STIM1)

Ca^{2+}-binding proteins containing EF-hand domains are present in the extracellular environment or matrix. These proteins do also localize at the plasma membrane, with the EF-hand domain facing the extracellular medium [303,304]. An important example of this class of proteins is the single transmembrane-spanning Ca^{2+}-binding protein, STIM1, generally known as the endoplasmic reticulum Ca^{2+} sensor, which provides the trigger for store-operated channels (SOCs) activation. STIM1 at the cell surface may represent a fundamental pharmacological target to finely tune Ca^{2+}-regulated functions that are mediated by SOCs.

5.8. Hemichannels

Gap junctions are constituted by the juxtaposition of two membrane-spanning connexons (hemichannels) positioned end-to-end. However, functionally active hemichannels exist at the plasma membrane where they allow the rapid bi-directional movement of ions, second messengers, and metabolites between the cytosol and the extracellular space and vice-versa [305]. Opening of most hemichannels depends critically on decreases in extracellular Ca^{2+} levels (of about ~200 μM) around the physiological Ca^{2+} levels [306,307]. The vestibule of Connexin 32 has a Ca^{2+}-binding site involved in the regulatory effect of external Ca^{2+}. Of note is that it has been shown that the complete Ca^{2+} deregulation of connexin 32 hemichannel activity is induced by a mutation causing an inherited peripheral neuropathy. This evidence led to the assumption that the pathogenesis of the neuropathy may be strictly related to this dysfunction [308]. The pore of connexin 43 hemichannels might be in an open or closed conformation, with the open probability significantly dependent on external Ca^{2+} levels (from 1.8 to 1.0 mM) [309].

5.9. Ion Channels

Dynamic changes in extracellular Ca^{2+} are known to impact on a number of ion channels, especially those located in the brain.

Ca^{2+} homeostasis modulator 1 (CALHM1) is an ion channel expressed at the cell membrane of cortical neurons and type II taste bud cells of the tongue [310]. This peculiar channel, closed at resting membrane potential and modulated by significant membrane depolarizations, is also able to sense a decrease in extracellular Ca^{2+} that thus works, in synergy with voltage, to allosterically increase the CALHM channel gating open probability.

Interestingly, Ca^{2+} and other cations, anions, and ATP are able to permeate through the non-specific pore of CALHM1 [310].

Acid-sensing ion channels (ASICs) are trimeric cationic channels activated by extracellular protons and primarily expressed in the nervous system. These channels are permeable to Na+ and Ca^{2+} and "sense" extracellular Ca^{2+}, which modulates their open probability by competing with protons [311–313]. In addition, two negatively charged residues near the entrance of the channel pore have been shown to be critical for the direct Ca^{2+} blockade of ASIC1a currents [313]. Of note is that these channels seem to be also modulated by small fluctuations in extracellular Zn^{2+}, Mg^{2+}, and spermine [311,314].

Another class of ion channels modulated by dynamic changes of extracellular [Ca^{2+}] are the voltage-gated Human ether-à-go-go-related gene (HERG) encoded K+ channels, which have been shown to be implicated in a variety of physiopathological mechanisms, such as neuronal spike frequency adaptation and tumor cell growth [315]. Johnson et al. showed that small changes in extracellular Ca^{2+} under physiological conditions shift the voltage dependence of channel activation to more positive membrane potentials, allowing for fine-tuned cell repolarization [316]. Under this scenario, HERG channels seem to be responsible for a delayed rectifier K+ current functional to cellular repolarization of the cardiac contractile cycle in the heart [317].

Finally, the Transient receptor potential cation channel, subfamily M, member 7 (TRPM7) was shown to sense a reduction of extracellular Ca^{2+} in hippocampal and pyramidal neurons and aid an inward cation current [318–320]. These 36-pS channels (originally named Ca^{2+}-sensing nonselective cation channels, IcsNSC), which are almost silent at the 2.0 mM extracellular Ca^{2+} level, become robustly activated by small decreases in extracellular Ca^{2+} [321].

The history of research on the CaR's role in physiology and pathological states is a perfect example of the relevance of basic research for ameliorating human health. It is in fact clear that none of the clinically relevant discoveries in the pharmacological field today could have ever taken place without a deep knowledge of the specific mechanisms underlying the functionality of the tissues and organs on whom CaR mutations have their more striking effects.

It is also clear that the novel methodological tools available to researchers in the field of GPCRs and signal transduction will be of paramount importance to further improve the knowledge of basic signalling modes, a step that is indispensable for the design of smart drugs able to rescue GPCR defects in human diseases.

Acknowledgments: The authors gratefully acknowledge the contribution and support of all the colleagues and students who participated in the studies described in the present article.

Author Contributions: Matilde Colella and Andrea Gerbino contributed equally to this work.

Abbreviations

AA	arachidonic acid
AC	adenylate cyclase
ATP	adenosine triphosphate

CaM	Calmodulin
CaMK	Ca^{2+}/calmodulin-dependent protein kinase
cAMP	cyclic AMP
DAG	Diacylglycerol
eNOS	endothelial nitric oxide synthase
ER	endoplasmic reticulum
ERK 1/2	extracellullar-signal regulated kinase
iNOS	inducible nitric oxide synthase
InsP3	inositol-1,4,5-trisphosphate
JNK	Jun amino-terminal kinase
MAPK	mitogen-activated protein kinase
MEK	MAPK kinase
NO	nitric oxide
p38	p38 mitogen-activated protein kinase
PA	phosphatidic acid
PHP	Pharmacoperones
PI3K	phosphatidylinositol 3-kinase
PI4K	phosphatidylinositol 4-kinase
PIP2	phosphatidylinositol 4,5-bisphosphate
PKC	protein kinase C
PLA2	phospholipase A2
PLC	phospholipase C
PLD	phospholipase D
RhoA	Ras homolog gene family, member A
SOC	store-operated Ca^{2+} channel

References

1. Berridge, M.J.; Bootman, M.D.; Roderick, H.L. Calcium signalling: Dynamics, homeostasis and remodelling. *Nat. Rev. Mol. Cell Biol.* **2003**, *4*, 517–529. [CrossRef] [PubMed]
2. Rizzuto, R.; Bernardi, P.; Pozzan, T. Mitochondria as all-round players of the calcium game. *J. Physiol.* **2000**, *529 Pt 1*, 37–47. [CrossRef] [PubMed]
3. Bootman, M.D.; Fearnley, C.; Smyrnias, I.; MacDonald, F.; Roderick, H.L. An update on nuclear calcium signalling. *J. Cell Sci.* **2009**, *122*, 2337–2350. [CrossRef] [PubMed]
4. Medina, D.L.; Di Paola, S.; Peluso, I.; Armani, A.; De Stefani, D.; Venditti, R.; Montefusco, S.; Scotto-Rosato, A.; Prezioso, C.; Forrester, A.; et al. Lysosomal calcium signalling regulates autophagy through calcineurin and tfeb. *Nat. Cell Biol.* **2015**, *17*, 288–299. [CrossRef] [PubMed]
5. Corbett, E.F.; Michalak, M. Calcium, a signaling molecule in the endoplasmic reticulum? *Trends Biochem. Sci.* **2000**, *25*, 307–311. [CrossRef]
6. Brown, E.M.; Gamba, G.; Riccardi, D.; Lombardi, M.; Butters, R.; Kifor, O.; Sun, A.; Hediger, M.A.; Lytton, J.; Hebert, S.C. Cloning and characterization of an extracellular Ca^{2+}-sensing receptor from bovine parathyroid. *Nature* **1993**, *366*, 575–580. [CrossRef] [PubMed]
7. Brown, E.M.; Conigrave, A.D. Preface. *Best Pract. Res. Clin. Endocrinol. Metab.* **2013**, *27*, 283–284. [CrossRef] [PubMed]
8. Zhang, C.; Miller, C.L.; Brown, E.M.; Yang, J.J. The calcium sensing receptor: From calcium sensing to signaling. *Sci. China Life Sci.* **2015**, *58*, 14–27. [CrossRef] [PubMed]
9. Tennakoon, S.; Aggarwal, A.; Kallay, E. The calcium-sensing receptor and the hallmarks of cancer. *Biochim. Biophys. Acta* **2016**, *1863*, 1398–1407. [CrossRef] [PubMed]
10. Hannan, F.M.; Olesen, M.K.; Thakker, R.V. Calcimimetic and calcilytic therapies for inherited disorders of the calcium-sensing receptor signalling pathway. *Br. J. Pharmacol.* **2017**. [CrossRef] [PubMed]

11. Brown, E.M.; MacLeod, R.J. Extracellular calcium sensing and extracellular calcium signaling. *Physiol. Rev.* **2001**, *81*, 239–297. [CrossRef] [PubMed]
12. Hofer, A.M. Another dimension to calcium signaling: A look at extracellular calcium. *J. Cell Sci.* **2005**, *118*, 855–862. [CrossRef] [PubMed]
13. Ashby, M.C.; Tepikin, A.V. Polarized calcium and calmodulin signaling in secretory epithelia. *Physiol. Rev.* **2002**, *82*, 701–734. [CrossRef] [PubMed]
14. Belan, P.; Gerasimenko, O.; Petersen, O.H.; Tepikin, A.V. Distribution of Ca^{2+} extrusion sites on the mouse pancreatic acinar cell surface. *Cell Calcium* **1997**, *22*, 5–10. [CrossRef]
15. Peng, J.B.; Brown, E.M.; Hediger, M.A. Apical entry channels in calcium-transporting epithelia. *News Physiol. Sci.* **2003**, *18*, 158–163. [CrossRef] [PubMed]
16. Petersen, O.H. Localization and regulation of Ca^{2+} entry and exit pathways in exocrine gland cells. *Cell Calcium* **2003**, *33*, 337–344. [CrossRef]
17. Caroppo, R.; Gerbino, A.; Debellis, L.; Kifor, O.; Soybel, D.I.; Brown, E.M.; Hofer, A.M.; Curci, S. Asymmetrical, agonist-induced fluctuations in local extracellular $[Ca^{2+}]$ in intact polarized epithelia. *EMBO J.* **2001**, *20*, 6316–6326. [CrossRef] [PubMed]
18. Caroppo, R.; Gerbino, A.; Fistetto, G.; Colella, M.; Debellis, L.; Hofer, A.M.; Curci, S. Extracellular calcium acts as a "Third messenger" To regulate enzyme and alkaline secretion. *J. Cell Biol.* **2004**, *166*, 111–119. [CrossRef] [PubMed]
19. Gerbino, A.; Fistetto, G.; Colella, M.; Hofer, A.M.; Debellis, L.; Caroppo, R.; Curci, S. Real time measurements of water flow in amphibian gastric glands: Modulation via the extracellular Ca^{2+}-sensing receptor. *J. Biol. Chem.* **2007**, *282*, 13477–13486. [CrossRef] [PubMed]
20. Andersson, T.; Berggren, P.O.; Gylfe, E.; Hellman, B. Amounts and distribution of intracellular magnesium and calcium in pancreatic beta-cells. *Acta Physiol. Scand.* **1982**, *114*, 235–241. [CrossRef] [PubMed]
21. Nicaise, G.; Maggio, K.; Thirion, S.; Horoyan, M.; Keicher, E. The calcium loading of secretory granules. A possible key event in stimulus-secretion coupling. *Biol. Cell* **1992**, *75*, 89–99. [CrossRef]
22. Gillot, I.; Ciapa, B.; Payan, P.; Sardet, C. The calcium content of cortical granules and the loss of calcium from sea urchin eggs at fertilization. *Dev. Biol.* **1991**, *146*, 396–405. [CrossRef]
23. Hutton, J.C.; Penn, E.J.; Peshavaria, M. Low-molecular-weight constituents of isolated insulin-secretory granules. Bivalent cations, adenine nucleotides and inorganic phosphate. *Biochem. J.* **1983**, *210*, 297–305. [CrossRef] [PubMed]
24. Gerbino, A.; Maiellaro, I.; Carmone, C.; Caroppo, R.; Debellis, L.; Barile, M.; Busco, G.; Colella, M. Glucose increases extracellular $[Ca^{2+}]$ in rat insulinoma (ins-1e) pseudoislets as measured with Ca^{2+}-sensitive microelectrodes. *Cell Calcium* **2012**, *51*, 393–401. [CrossRef] [PubMed]
25. Belan, P.; Gardner, J.; Gerasimenko, O.; Gerasimenko, J.; Mills, C.L.; Petersen, O.H.; Tepikin, A.V. Isoproterenol evokes extracellular Ca^{2+} spikes due to secretory events in salivary gland cells. *J. Biol. Chem.* **1998**, *273*, 4106–4111. [CrossRef] [PubMed]
26. Von Grafenstein, H.R.; Powis, D.A. Calcium is released by exocytosis together with catecholamines from bovine adrenal medullary cells. *J. Neurochem.* **1989**, *53*, 428–435. [CrossRef] [PubMed]
27. Thirion, S.; Stuenkel, E.L.; Nicaise, G. Calcium loading of secretory granules in stimulated neurohypophysial nerve endings. *Neuroscience* **1995**, *64*, 125–137. [CrossRef]
28. Kuhtreiber, W.M.; Gillot, I.; Sardet, C.; Jaffe, L.F. Net calcium and acid release at fertilization in eggs of sea urchins and ascidians. *Cell Calcium* **1993**, *14*, 73–86. [CrossRef]
29. Hofer, A.M.; Gerbino, A.; Caroppo, R.; Curci, S. The extracellular calcium-sensing receptor and cell-cell signaling in epithelia. *Cell Calcium* **2004**, *35*, 297–306. [CrossRef] [PubMed]
30. Cohen, J.E.; Fields, R.D. Extracellular calcium depletion in synaptic transmission. *Neuroscientist* **2004**, *10*, 12–17. [CrossRef] [PubMed]
31. Egelman, D.M.; Montague, P.R. Calcium dynamics in the extracellular space of mammalian neural tissue. *Biophys. J.* **1999**, *76*, 1856–1867. [CrossRef]
32. Vassilev, P.M.; Mitchel, J.; Vassilev, M.; Kanazirska, M.; Brown, E.M. Assessment of frequency-dependent alterations in the level of extracellular Ca^{2+} in the synaptic cleft. *Biophys. J.* **1997**, *72*, 2103–2116. [CrossRef]
33. Pumain, R.; Heinemann, U. Stimulus- and amino acid-induced calcium and potassium changes in rat neocortex. *J. Neurophysiol.* **1985**, *53*, 1–16. [CrossRef] [PubMed]

34. Rusakov, D.A.; Fine, A. Extracellular Ca^{2+} depletion contributes to fast activity-dependent modulation of synaptic transmission in the brain. *Neuron* **2003**, *37*, 287–297. [CrossRef]

35. Moura, A.S. Membrane potential and intercellular calcium during glucose challenge in mouse islet of langerhans. *Biochem. Biophys. Res. Commun.* **1995**, *214*, 798–802. [CrossRef] [PubMed]

36. Perez-Armendariz, E.; Atwater, I. Glucose-evoked changes in $[K^+]$ and $[Ca^{2+}]$ in the intercellular spaces of the mouse islet of langerhans. *Adv. Exp. Med. Biol.* **1986**, *211*, 31–51. [PubMed]

37. Cleemann, L.; Pizarro, G.; Morad, M. Optical measurements of extracellular calcium depletion during a single heartbeat. *Science* **1984**, *226*, 174–177. [CrossRef] [PubMed]

38. Nicholson, C.; Bruggencate, G.T.; Steinberg, R.; Stöckle, H. Calcium modulation in brain extracellular microenvironment demonstrated with ion-selective micropipette. *Proc. Natl. Acad. Sci. USA* **1977**, *74*, 1287–1290. [CrossRef] [PubMed]

39. Jaffe, L.F.; Nuccitelli, R. An ultrasensitive vibrating probe for measuring steady extracellular currents. *J. Cell Biol.* **1974**, *63*, 614–628. [CrossRef] [PubMed]

40. Kuhtreiber, W.M.; Jaffe, L.F. Detection of extracellular calcium gradients with a calcium-specific vibrating electrode. *J. Cell Biol.* **1990**, *110*, 1565–1573. [CrossRef] [PubMed]

41. Yamoah, E.N.; Lumpkin, E.A.; Dumont, R.A.; Smith, P.J.; Hudspeth, A.J.; Gillespie, P.G. Plasma membrane Ca^{2+}-atpase extrudes Ca^{2+} from hair cell stereocilia. *J. Neurosci.* **1998**, *18*, 610–624. [PubMed]

42. Pepperell, J.R.; Kommineni, K.; Buradagunta, S.; Smith, P.J.; Keefe, D.L. Transmembrane regulation of intracellular calcium by a plasma membrane sodium/calcium exchanger in mouse ova. *Biol. Reprod.* **1999**, *60*, 1137–1143. [CrossRef] [PubMed]

43. Knox, R.J.; Jonas, E.A.; Kao, L.S.; Smith, P.J.; Connor, J.A.; Kaczmarek, L.K. Ca^{2+} influx and activation of a cation current are coupled to intracellular Ca^{2+} release in peptidergic neurons of aplysia californica. *J. Physiol.* **1996**, *494 Pt 3*, 627–639. [CrossRef] [PubMed]

44. Smith, P.J.; Hammar, K.; Porterfield, D.M.; Sanger, R.H.; Trimarchi, J.R. Self-referencing, non-invasive, ion selective electrode for single cell detection of trans-plasma membrane calcium flux. *Microsc. Res. Tech.* **1999**, *46*, 398–417. [CrossRef]

45. Mupanomunda, M.M.; Ishioka, N.; Bukoski, R.D. Interstitial Ca^{2+} undergoes dynamic changes sufficient to stimulate nerve-dependent Ca^{2+}-induced relaxation. *Am. J. Physiol.* **1999**, *276*, H1035–H1042. [CrossRef] [PubMed]

46. Mupanomunda, M.M.; Tian, B.; Ishioka, N.; Bukoski, R.D. Renal interstitial Ca^{2+}. *Am. J. Physiol. Renal Physiol.* **2000**, *278*, F644–F649. [CrossRef] [PubMed]

47. Hilgemann, D.W. Extracellular calcium transients at single excitations in rabbit atrium measured with tetramethylmurexide. *J. Gen. Physiol.* **1986**, *87*, 707–735. [CrossRef] [PubMed]

48. Hilgemann, D.W.; Langer, G.A. Transsarcolemmal calcium movements in arterially perfused rabbit right ventricle measured with extracellular calcium-sensitive dyes. *Circ. Res.* **1984**, *54*, 461–467. [CrossRef] [PubMed]

49. Tepikin, A.V.; Llopis, J.; Snitsarev, V.A.; Gallacher, D.V.; Petersen, O.H. The droplet technique: Measurement of calcium extrusion from single isolated mammalian cells. *Pflugers Arch.* **1994**, *428*, 664–670. [CrossRef] [PubMed]

50. Tepikin, A.V.; Voronina, S.G.; Gallacher, D.V.; Petersen, O.H. Pulsatile Ca^{2+} extrusion from single pancreatic acinar cells during receptor-activated cytosolic Ca^{2+} spiking. *J. Biol. Chem.* **1992**, *267*, 14073–14076. [PubMed]

51. Tepikin, A.V.; Voronina, S.G.; Gallacher, D.V.; Petersen, O.H. Acetylcholine-evoked increase in the cytoplasmic Ca^{2+} concentration and Ca^{2+} extrusion measured simultaneously in single mouse pancreatic acinar cells. *J. Biol. Chem.* **1992**, *267*, 3569–3572. [PubMed]

52. Belan, P.V.; Gerasimenko, O.V.; Tepikin, A.V.; Petersen, O.H. Localization of Ca^{2+} extrusion sites in pancreatic acinar cells. *J. Biol. Chem.* **1996**, *271*, 7615–7619. [CrossRef] [PubMed]

53. Blatter, L.A.; Niggli, E. Confocal near-membrane detection of calcium in cardiac myocytes. *Cell Calcium* **1998**, *23*, 269–279. [CrossRef]

54. Etter, E.F.; Kuhn, M.A.; Fay, F.S. Detection of changes in near-membrane Ca^{2+} concentration using a novel membrane-associated Ca^{2+} indicator. *J. Biol. Chem.* **1994**, *269*, 10141–10149. [PubMed]

55. Etter, E.F.; Minta, A.; Poenie, M.; Fay, F.S. Near-membrane $[Ca^{2+}]$ transients resolved using the Ca^{2+} indicator ffp18. *Proc. Natl. Acad. Sci. USA* **1996**, *93*, 5368–5373. [CrossRef]

56. De Luisi, A.; Hofer, A.M. Evidence that Ca^{2+} cycling by the plasma membrane Ca^{2+}-atpase increases the 'excitability' of the extracellular Ca^{2+}-sensing receptor. *J. Cell Sci.* **2003**, *116*, 1527–1538. [CrossRef] [PubMed]
57. Care, A.D.; Sherwood, L.M.; Potts, J.T., Jr.; Aurbach, G.D. Perfusion of the isolated parathyroid gland of the goat and sheep. *Nature* **1966**, *209*, 55–57. [CrossRef] [PubMed]
58. Sherwood, L.M.; Potts, J.T., Jr.; Care, A.D.; Mayer, G.P.; Aurbach, G.D. Evaluation by radioimmunoassay of factors controlling the secretion of parathyroid hormone. *Nature* **1966**, *209*, 52–55. [CrossRef] [PubMed]
59. Lopez-Barneo, J.; Armstrong, C.M. Depolarizing response of rat parathyroid cells to divalent cations. *J. Gen. Physiol.* **1983**, *82*, 269–294. [CrossRef] [PubMed]
60. Shoback, D.; Thatcher, J.; Leombruno, R.; Brown, E. Effects of extracellular Ca^{2+} and Mg^{2+} on cytosolic Ca^{2+} and pth release in dispersed bovine parathyroid cells. *Endocrinology* **1983**, *113*, 424–426. [CrossRef] [PubMed]
61. Brown, E.; Enyedi, P.; LeBoff, M.; Rotberg, J.; Preston, J.; Chen, C. High extracellular Ca^{2+} and Mg^{2+} stimulate accumulation of inositol phosphates in bovine parathyroid cells. *FEBS Lett.* **1987**, *218*, 113–118. [CrossRef]
62. Kifor, O.; Brown, E.M. Relationship between diacylglycerol levels and extracellular Ca^{2+} in dispersed bovine parathyroid cells. *Endocrinology* **1988**, *123*, 2723–2729. [CrossRef] [PubMed]
63. Shoback, D.M.; McGhee, J.M. Fluoride stimulates the accumulation of inositol phosphates, increases intracellular free calcium, and inhibits parathyroid hormone release in dispersed bovine parathyroid cells. *Endocrinology* **1988**, *122*, 2833–2839. [CrossRef] [PubMed]
64. Shoback, D.M.; Membreno, L.A.; McGhee, J.G. High calcium and other divalent cations increase inositol trisphosphate in bovine parathyroid cells. *Endocrinology* **1988**, *123*, 382–389. [CrossRef] [PubMed]
65. Nemeth, E.F.; Scarpa, A. Are changes in intracellular free calcium necessary for regulating secretion in parathyroid cells? *Ann. N. Y. Acad. Sci.* **1987**, *493*, 542–551. [CrossRef] [PubMed]
66. Foord, S.M.; Bonner, T.I.; Neubig, R.R.; Rosser, E.M.; Pin, J.P.; Davenport, A.P.; Spedding, M.; Harmar, A.J. International union of pharmacology. Xlvi. G protein-coupled receptor list. *Pharmacol. Rev.* **2005**, *57*, 279–288. [CrossRef] [PubMed]
67. Clemmensen, C.; Smajilovic, S.; Wellendorph, P.; Brauner-Osborne, H. The GPCR, class c, group 6, subtype a (GPRC6A) receptor: From cloning to physiological function. *Br. J. Pharmacol.* **2014**, *171*, 1129–1141. [CrossRef] [PubMed]
68. Norskov-Lauritsen, L.; Brauner-Osborne, H. Role of post-translational modifications on structure, function and pharmacology of class c g protein-coupled receptors. *Eur. J. Pharmacol.* **2016**, *763*, 233–240. [CrossRef] [PubMed]
69. Garrett, J.E.; Capuano, I.V.; Hammerland, L.G.; Hung, B.C.; Brown, E.M.; Hebert, S.C.; Nemeth, E.F.; Fuller, F. Molecular cloning and functional expression of human parathyroid calcium receptor cdnas. *J. Biol. Chem.* **1995**, *270*, 12919–12925. [CrossRef] [PubMed]
70. Hu, J.; Spiegel, A.M. Structure and function of the human calcium-sensing receptor: Insights from natural and engineered mutations and allosteric modulators. *J. Cell. Mol. Med.* **2007**, *11*, 908–922. [CrossRef] [PubMed]
71. Kunishima, N.; Shimada, Y.; Tsuji, Y.; Sato, T.; Yamamoto, M.; Kumasaka, T.; Nakanishi, S.; Jingami, H.; Morikawa, K. Structural basis of glutamate recognition by a dimeric metabotropic glutamate receptor. *Nature* **2000**, *407*, 971–977. [PubMed]
72. Muto, T.; Tsuchiya, D.; Morikawa, K.; Jingami, H. Structures of the extracellular regions of the group ii/iii metabotropic glutamate receptors. *Proc. Natl. Acad. Sci. USA* **2007**, *104*, 3759–3764. [CrossRef] [PubMed]
73. Geng, Y.; Xiong, D.; Mosyak, L.; Malito, D.L.; Kniazeff, J.; Chen, Y.; Burmakina, S.; Quick, M.; Bush, M.; Javitch, J.A.; et al. Structure and functional interaction of the extracellular domain of human $GABA_B$ receptor GBR2. *Nat. Neurosci.* **2012**, *15*, 970–978. [CrossRef] [PubMed]
74. Chakravarti, B.; Chattopadhyay, N.; Brown, E.M. Signaling through the extracellular calcium-sensing receptor (CASR). *Adv. Exp. Med. Biol.* **2012**, *740*, 103–142. [PubMed]
75. Hu, J.; Reyes-Cruz, G.; Goldsmith, P.K.; Gantt, N.M.; Miller, J.L.; Spiegel, A.M. Functional effects of monoclonal antibodies to the purified amino-terminal extracellular domain of the human Ca^{2+} receptor. *J. Bone Miner Res.* **2007**, *22*, 601–608. [CrossRef] [PubMed]
76. Bai, M.; Trivedi, S.; Brown, E.M. Dimerization of the extracellular calcium-sensing receptor (CAR) on the cell surface of car-transfected hek293 cells. *J. Biol. Chem.* **1998**, *273*, 23605–23610. [CrossRef] [PubMed]

77. Bai, M.; Trivedi, S.; Kifor, O.; Quinn, S.J.; Brown, E.M. Intermolecular interactions between dimeric calcium-sensing receptor monomers are important for its normal function. *Proc. Natl. Acad. Sci. USA* **1999**, *96*, 2834–2839. [CrossRef] [PubMed]
78. Gama, L.; Wilt, S.G.; Breitwieser, G.E. Heterodimerization of calcium sensing receptors with metabotropic glutamate receptors in neurons. *J. Biol. Chem.* **2001**, *276*, 39053–39059. [CrossRef] [PubMed]
79. Chang, W.; Tu, C.; Cheng, Z.; Rodriguez, L.; Chen, T.H.; Gassmann, M.; Bettler, B.; Margeta, M.; Jan, L.Y.; Shoback, D. Complex formation with the type b gamma-aminobutyric acid receptor affects the expression and signal transduction of the extracellular calcium-sensing receptor. Studies with HEK-293 cells and neurons. *J. Biol. Chem.* **2007**, *282*, 25030–25040. [CrossRef] [PubMed]
80. Jiang, Y.; Minet, E.; Zhang, Z.; Silver, P.A.; Bai, M. Modulation of interprotomer relationships is important for activation of dimeric calcium-sensing receptor. *J. Biol. Chem.* **2004**, *279*, 14147–14156. [CrossRef] [PubMed]
81. Ray, K.; Hauschild, B.C.; Steinbach, P.J.; Goldsmith, P.K.; Hauache, O.; Spiegel, A.M. Identification of the cysteine residues in the amino-terminal extracellular domain of the human Ca^{2+} receptor critical for dimerization. Implications for function of monomeric Ca^{2+} receptor. *J. Biol. Chem.* **1999**, *274*, 27642–27650. [CrossRef] [PubMed]
82. Reyes-Cruz, G.; Hu, J.; Goldsmith, P.K.; Steinbach, P.J.; Spiegel, A.M. Human Ca^{2+} receptor extracellular domain. Analysis of function of lobe i loop deletion mutants. *J. Biol. Chem.* **2001**, *276*, 32145–32151. [CrossRef] [PubMed]
83. Pace, A.J.; Gama, L.; Breitwieser, G.E. Dimerization of the calcium-sensing receptor occurs within the extracellular domain and is eliminated by cys → ser mutations at cys101 and cys236. *J. Biol. Chem.* **1999**, *274*, 11629–11634. [CrossRef] [PubMed]
84. Zhang, Z.; Sun, S.; Quinn, S.J.; Brown, E.M.; Bai, M. The extracellular calcium-sensing receptor dimerizes through multiple types of intermolecular interactions. *J. Biol. Chem.* **2001**, *276*, 5316–5322. [CrossRef] [PubMed]
85. Fan, G.F.; Ray, K.; Zhao, X.M.; Goldsmith, P.K.; Spiegel, A.M. Mutational analysis of the cysteines in the extracellular domain of the human Ca^{2+} receptor: Effects on cell surface expression, dimerization and signal transduction. *FEBS Lett.* **1998**, *436*, 353–356. [CrossRef]
86. Jacobsen, S.E.; Gether, U.; Brauner-Osborne, H. Investigating the molecular mechanism of positive and negative allosteric modulators in the calcium-sensing receptor dimer. *Sci. Rep.* **2018**, *7*, 46355. [CrossRef] [PubMed]
87. Hu, J.; Hauache, O.; Spiegel, A.M. Human Ca^{2+} receptor cysteine-rich domain. Analysis of function of mutant and chimeric receptors. *J. Biol. Chem.* **2000**, *275*, 16382–16389. [CrossRef] [PubMed]
88. Pin, J.P.; Galvez, T.; Prezeau, L. Evolution, structure, and activation mechanism of family 3/c G-protein-coupled receptors. *Pharmacol. Ther.* **2003**, *98*, 325–354. [CrossRef]
89. Ray, J.M.; Squires, P.E.; Curtis, S.B.; Meloche, M.R.; Buchan, A.M. Expression of the calcium-sensing receptor on human antral gastrin cells in culture. *J. Clin. Invest.* **1997**, *99*, 2328–2333. [CrossRef] [PubMed]
90. Awata, H.; Huang, C.; Handlogten, M.E.; Miller, R.T. Interaction of the calcium-sensing receptor and filamin, a potential scaffolding protein. *J. Biol. Chem.* **2001**, *276*, 34871–34879. [CrossRef] [PubMed]
91. Hjalm, G.; MacLeod, R.J.; Kifor, O.; Chattopadhyay, N.; Brown, E.M. Filamin-a binds to the carboxyl-terminal tail of the calcium-sensing receptor, an interaction that participates in car-mediated activation of mitogen-activated protein kinase. *J. Biol. Chem.* **2001**, *276*, 34880–34887. [CrossRef] [PubMed]
92. Bosel, J.; John, M.; Freichel, M.; Blind, E. Signaling of the human calcium-sensing receptor expressed in HEK293-cells is modulated by protein kinases a and c. *Exp. Clin. Endocrinol. Diabetes* **2003**, *111*, 21–26. [CrossRef] [PubMed]
93. Stepanchick, A.; McKenna, J.; McGovern, O.; Huang, Y.; Breitwieser, G.E. Calcium sensing receptor mutations implicated in pancreatitis and idiopathic epilepsy syndrome disrupt an arginine-rich retention motif. *Cell. Physiol. Biochem.* **2010**, *26*, 363–374. [CrossRef] [PubMed]
94. Hendy, G.N.; Canaff, L.; Cole, D.E. The casr gene: Alternative splicing and transcriptional control, and calcium-sensing receptor (CASR) protein: Structure and ligand binding sites. *Best Pract. Res. Clin. Endocrinol. Metab.* **2013**, *27*, 285–301. [CrossRef] [PubMed]
95. Leach, K.; Gregory, K.J.; Kufareva, I.; Khajehali, E.; Cook, A.E.; Abagyan, R.; Conigrave, A.D.; Sexton, P.M.; Christopoulos, A. Towards a structural understanding of allosteric drugs at the human calcium-sensing receptor. *Cell Res.* **2016**, *26*, 574–592. [CrossRef] [PubMed]

96. Huang, Y.; Zhou, Y.; Yang, W.; Butters, R.; Lee, H.W.; Li, S.; Castiblanco, A.; Brown, E.M.; Yang, J.J. Identification and dissection of Ca^{2+}-binding sites in the extracellular domain of Ca^{2+}-sensing receptor. *J. Biol. Chem.* **2007**, *282*, 19000–19010. [CrossRef] [PubMed]
97. Huang, Y.; Zhou, Y.; Castiblanco, A.; Yang, W.; Brown, E.M.; Yang, J.J. Multiple Ca^{2+}-binding sites in the extracellular domain of the Ca^{2+}-sensing receptor corresponding to cooperative Ca^{2+} response. *Biochemistry* **2009**, *48*, 388–398. [CrossRef] [PubMed]
98. Huang, Y.; Zhou, Y.; Wong, H.C.; Castiblanco, A.; Chen, Y.; Brown, E.M.; Yang, J.J. Calmodulin regulates Ca^{2+}-sensing receptor-mediated Ca^{2+} signaling and its cell surface expression. *J. Biol. Chem.* **2010**, *285*, 35919–35931. [CrossRef] [PubMed]
99. Zhang, C.; Zhuo, Y.; Moniz, H.A.; Wang, S.; Moremen, K.W.; Prestegard, J.H.; Brown, E.M.; Yang, J.J. Direct determination of multiple ligand interactions with the extracellular domain of the calcium-sensing receptor. *J. Biol. Chem.* **2014**, *289*, 33529–33542. [CrossRef] [PubMed]
100. Zhang, C.; Huang, Y.; Jiang, Y.; Mulpuri, N.; Wei, L.; Hamelberg, D.; Brown, E.M.; Yang, J.J. Identification of an L-phenylalanine binding site enhancing the cooperative responses of the calcium-sensing receptor to calcium. *J. Biol. Chem.* **2014**, *289*, 5296–5309. [CrossRef] [PubMed]
101. Zhang, C.; Mulpuri, N.; Hannan, F.M.; Nesbit, M.A.; Thakker, R.V.; Hamelberg, D.; Brown, E.M.; Yang, J.J. Role of Ca^{2+} and l-phe in regulating functional cooperativity of disease-associated "Toggle" Calcium-sensing receptor mutations. *PLoS ONE* **2014**, *9*, e113622. [CrossRef] [PubMed]
102. Zhang, C.; Zhang, T.; Zou, J.; Miller, C.L.; Gorkhali, R.; Yang, J.Y.; Schilmiller, A.; Wang, S.; Huang, K.; Brown, E.M.; et al. Structural basis for regulation of human calcium-sensing receptor by magnesium ions and an unexpected tryptophan derivative co-agonist. *Sci. Adv.* **2016**, *2*, e1600241. [CrossRef] [PubMed]
103. Zhang, C.; Miller, C.L.; Gorkhali, R.; Zou, J.; Huang, K.; Brown, E.M.; Yang, J.J. Molecular basis of the extracellular ligands mediated signaling by the calcium sensing receptor. *Front. Physiol.* **2016**, *7*, 441. [CrossRef] [PubMed]
104. Brown, E.M.; Fuleihan, G.E.-H.; Chen, C.J.; Kifor, O. A comparison of the effects of divalent and trivalent cations on parathyroid hormone release, 3′,5′-cyclic-adenosine monophosphate accumulation, and the levels of inositol phosphates in bovine parathyroid cells. *Endocrinology* **1990**, *127*, 1064–1071. [CrossRef] [PubMed]
105. Quinn, S.J.; Ye, C.P.; Diaz, R.; Kifor, O.; Bai, M.; Vassilev, P.; Brown, E. The Ca^{2+}-sensing receptor: A target for polyamines. *Am. J. Physiol.* **1997**, *273*, C1315–C1323. [CrossRef] [PubMed]
106. Brown, E.M.; Katz, C.; Butters, R.; Kifor, O. Polyarginine, polylysine, and protamine mimic the effects of high extracellular calcium concentrations on dispersed bovine parathyroid cells. *J. Bone Miner Res.* **1991**, *6*, 1217–1225. [CrossRef] [PubMed]
107. Ye, C.; Ho-Pao, C.L.; Kanazirska, M.; Quinn, S.; Rogers, K.; Seidman, C.E.; Seidman, J.G.; Brown, E.M.; Vassilev, P.M. Amyloid-beta proteins activate Ca^{2+}-permeable channels through calcium-sensing receptors. *J. Neurosci. Res.* **1997**, *47*, 547–554. [CrossRef]
108. Brown, E.M.; Butters, R.; Katz, C.; Kifor, O. Neomycin mimics the effects of high extracellular calcium concentrations on parathyroid function in dispersed bovine parathyroid cells. *Endocrinology* **1991**, *128*, 3047–3054. [CrossRef] [PubMed]
109. Nemeth, E.F. Regulation of cytosolic calcium by extracellular divalent cations in c-cells and parathyroid cells. *Cell Calcium* **1990**, *11*, 323–327. [CrossRef]
110. Colella, M.; Gerbino, A.; Hofer, A.M.; Curci, S. Recent advances in understanding the extracellular calcium-sensing receptor. *F1000Res* **2016**, *5*. [CrossRef] [PubMed]
111. Conigrave, A.D.; Quinn, S.J.; Brown, E.M. L-amino acid sensing by the extracellular Ca^{2+}-sensing receptor. *Proc. Natl. Acad. Sci. USA* **2000**, *97*, 4814–4819. [CrossRef] [PubMed]
112. Conigrave, A.D.; Mun, H.C.; Lok, H.C. Aromatic l-amino acids activate the calcium-sensing receptor. *J. Nutr.* **2007**, *137*, 1524S–1527S; discussion 1548S. [CrossRef] [PubMed]
113. Conigrave, A.D.; Hampson, D.R. Broad-spectrum amino acid-sensing class c G-protein coupled receptors: Molecular mechanisms, physiological significance and options for drug development. *Pharmacol. Ther.* **2010**, *127*, 252–260. [CrossRef] [PubMed]
114. Wang, M.; Yao, Y.; Kuang, D.; Hampson, D.R. Activation of family c G-protein-coupled receptors by the tripeptide glutathione. *J. Biol. Chem.* **2006**, *281*, 8864–8870. [CrossRef] [PubMed]

115. Broadhead, G.K.; Mun, H.C.; Avlani, V.A.; Jourdon, O.; Church, W.B.; Christopoulos, A.; Delbridge, L.; Conigrave, A.D. Allosteric modulation of the calcium-sensing receptor by gamma-glutamyl peptides: Inhibition of pth secretion, suppression of intracellular camp levels, and a common mechanism of action with l-amino acids. *J. Biol. Chem.* **2011**, *286*, 8786–8797. [CrossRef] [PubMed]
116. Quinn, S.J.; Bai, M.; Brown, E.M. Ph sensing by the calcium-sensing receptor. *J. Biol. Chem.* **2004**, *279*, 37241–37249. [CrossRef] [PubMed]
117. Quinn, S.J.; Kifor, O.; Trivedi, S.; Diaz, R.; Vassilev, P.; Brown, E. Sodium and ionic strength sensing by the calcium receptor. *J. Biol. Chem.* **1998**, *273*, 19579–19586. [CrossRef] [PubMed]
118. Nemeth, E.F.; Steffey, M.E.; Hammerland, L.G.; Hung, B.C.; van Wagenen, B.C.; DelMar, E.G.; Balandrin, M.F. Calcimimetics with potent and selective activity on the parathyroid calcium receptor. *Proc. Natl. Acad. Sci. USA* **1998**, *95*, 4040–4045. [CrossRef] [PubMed]
119. Kiefer, L.; Leiris, S.; Dodd, R.H. Novel calcium sensing receptor ligands: A patent survey. *Expert Opin. Ther. Pat.* **2011**, *21*, 681–698. [CrossRef] [PubMed]
120. Kiefer, L.; Beaumard, F.; Gorojankina, T.; Faure, H.; Ruat, M.; Dodd, R.H. Design and synthesis of calindol derivatives as potent and selective calcium sensing receptor agonists. *Bioorg. Med. Chem.* **2015**, *24*, 554–569. [CrossRef] [PubMed]
121. Nemeth, E.F.; Goodman, W.G. Calcimimetic and calcilytic drugs: Feats, flops, and futures. *Calcif. Tissue Int.* **2016**, *98*, 341–358. [CrossRef] [PubMed]
122. Nemeth, E.F. Calcimimetic and calcilytic drugs: Just for parathyroid cells? *Cell Calcium* **2004**, *35*, 283–289. [CrossRef] [PubMed]
123. Lindberg, J.S.; Culleton, B.; Wong, G.; Borah, M.F.; Clark, R.V.; Shapiro, W.B.; Roger, S.D.; Husserl, F.E.; Klassen, P.S.; Guo, M.D.; et al. Cinacalcet hcl, an oral calcimimetic agent for the treatment of secondary hyperparathyroidism in hemodialysis and peritoneal dialysis: A randomized, double-blind, multicenter study. *J. Am. Soc. Nephrol.* **2005**, *16*, 800–807. [CrossRef] [PubMed]
124. Hebert, S.C. Therapeutic use of calcimimetics. *Annu. Rev. Med.* **2006**, *57*, 349–364. [CrossRef] [PubMed]
125. Brown, E.M. Clinical utility of calcimimetics targeting the extracellular calcium-sensing receptor (casr). *Biochem. Pharmacol.* **2010**, *80*, 297–307. [CrossRef] [PubMed]
126. Henley, C., 3rd; Yang, Y.; Davis, J.; Lu, J.Y.; Morony, S.; Fan, W.; Florio, M.; Sun, B.; Shatzen, E.; Pretorius, J.K.; et al. Discovery of a calcimimetic with differential effects on parathyroid hormone and calcitonin secretion. *J. Pharmacol. Exp. Ther.* **2011**, *337*, 681–691. [CrossRef] [PubMed]
127. Ma, J.N.; Owens, M.; Gustafsson, M.; Jensen, J.; Tabatabaei, A.; Schmelzer, K.; Olsson, R.; Burstein, E.S. Characterization of highly efficacious allosteric agonists of the human calcium-sensing receptor. *J. Pharmacol. Exp. Ther.* **2011**, *337*, 275–284. [CrossRef] [PubMed]
128. Cook, A.E.; Mistry, S.N.; Gregory, K.J.; Furness, S.G.; Sexton, P.M.; Scammells, P.J.; Conigrave, A.D.; Christopoulos, A.; Leach, K. Biased allosteric modulation at the cas receptor engendered by structurally diverse calcimimetics. *Br. J. Pharmacol.* **2015**, *172*, 185–200. [CrossRef] [PubMed]
129. Nemeth, E.F. The search for calcium receptor antagonists (calcilytics). *J. Mol. Endocrinol.* **2002**, *29*, 15–21. [CrossRef] [PubMed]
130. Magno, A.L.; Ward, B.K.; Ratajczak, T. The calcium-sensing receptor: A molecular perspective. *Endocr. Rev.* **2011**, *32*, 3–30. [CrossRef] [PubMed]
131. Conigrave, A.D.; Ward, D.T. Calcium-sensing receptor (CASR): Pharmacological properties and signaling pathways. *Best Pract. Res. Clin. Endocrinol. Metab.* **2013**, *27*, 315–331. [CrossRef] [PubMed]
132. Nemeth, E.F.; Scarpa, A. Rapid mobilization of cellular Ca^{2+} in bovine parathyroid cells evoked by extracellular divalent cations. Evidence for a cell surface calcium receptor. *J. Biol. Chem.* **1987**, *262*, 5188–5196. [PubMed]
133. Hannan, F.M.; Babinsky, V.N.; Thakker, R.V. Disorders of the calcium-sensing receptor and partner proteins: Insights into the molecular basis of calcium homeostasis. *J. Mol. Endocrinol.* **2016**, *57*, R127–R142. [CrossRef] [PubMed]
134. Kifor, O.; MacLeod, R.J.; Diaz, R.; Bai, M.; Yamaguchi, T.; Yao, T.; Kifor, I.; Brown, E.M. Regulation of map kinase by calcium-sensing receptor in bovine parathyroid and car-transfected hek293 cells. *Am. J. Physiol. Renal Physiol.* **2001**, *280*, F291–F302. [CrossRef] [PubMed]

135. Kifor, O.; Diaz, R.; Butters, R.; Brown, E.M. The Ca^{2+}-sensing receptor (CAR) activates phospholipases c, a2, and d in bovine parathyroid and car-transfected, human embryonic kidney (HEK293) cells. *J. Bone Miner. Res.* **1997**, *12*, 715–725. [CrossRef] [PubMed]
136. Boudot, C.; Saidak, Z.; Boulanouar, A.K.; Petit, L.; Gouilleux, F.; Massy, Z.; Brazier, M.; Mentaverri, R.; Kamel, S. Implication of the calcium sensing receptor and the phosphoinositide 3-kinase/AKT pathway in the extracellular calcium-mediated migration of raw 264.7 osteoclast precursor cells. *Bone* **2010**, *46*, 1416–1423. [CrossRef] [PubMed]
137. Li, H.X.; Kong, F.J.; Bai, S.Z.; He, W.; Xing, W.J.; Xi, Y.H.; Li, G.W.; Guo, J.; Li, H.Z.; Wu, L.Y.; et al. Involvement of calcium-sensing receptor in oxldl-induced MMP-2 production in vascular smooth muscle cells via PI3K/AKT pathway. *Mol. Cell. Biochem.* **2012**, *362*, 115–122. [CrossRef] [PubMed]
138. de Jesus Ferreira, M.C.; Helies-Toussaint, C.; Imbert-Teboul, M.; Bailly, C.; Verbavatz, J.M.; Bellanger, A.C.; Chabardes, D. Co-expression of a Ca^{2+}-inhibitable adenylyl cyclase and of a Ca^{2+}-sensing receptor in the cortical thick ascending limb cell of the rat kidney. Inhibition of hormone-dependent camp accumulation by extracellular Ca^{2+}. *J. Biol. Chem.* **1998**, *273*, 15192–15202. [CrossRef] [PubMed]
139. Gerbino, A.; Ruder, W.C.; Curci, S.; Pozzan, T.; Zaccolo, M.; Hofer, A.M. Termination of camp signals by Ca^{2+} and Gαi via extracellular Ca^{2+} sensors: A link to intracellular Ca^{2+} oscillations. *J. Cell Biol.* **2005**, *171*, 303–312. [CrossRef] [PubMed]
140. Beierwaltes, W.H. The role of calcium in the regulation of renin secretion. *Am. J. Physiol. Renal Physiol.* **2010**, *298*, F1–F11. [CrossRef] [PubMed]
141. Ward, B.K.; Magno, A.L.; Walsh, J.P.; Ratajczak, T. The role of the calcium-sensing receptor in human disease. *Clin. Biochem.* **2012**, *45*, 943–953. [CrossRef] [PubMed]
142. Fitzpatrick, L.A.; Brandi, M.L.; Aurbach, G.D. Prostaglandin F2 alpha and alpha-adrenergic agonists regulate parathyroid cell function via the inhibitory guanine nucleotide regulatory protein. *Endocrinology* **1986**, *118*, 2115–2119. [CrossRef] [PubMed]
143. Chen, C.J.; Barnett, J.V.; Congo, D.A.; Brown, E.M. Divalent cations suppress 3′,5′-adenosine monophosphate accumulation by stimulating a pertussis toxin-sensitive guanine nucleotide-binding protein in cultured bovine parathyroid cells. *Endocrinology* **1989**, *124*, 233–239. [CrossRef] [PubMed]
144. Leach, K.; Sexton, P.M.; Christopoulos, A.; Conigrave, A.D. Engendering biased signalling from the calcium-sensing receptor for the pharmacotherapy of diverse disorders. *Br. J. Pharmacol.* **2014**, *171*, 1142–1155. [CrossRef] [PubMed]
145. Thomsen, A.R.B.; Hvidtfeldt, M.; Bräuner-Osborne, H. Biased agonism of the calcium-sensing receptor. *Cell Calcium* **2012**, *51*, 107–116. [CrossRef] [PubMed]
146. Huang, C.; Handlogten, M.E.; Miller, R.T. Parallel activation of phosphatidylinositol 4-kinase and phospholipase c by the extracellular calcium-sensing receptor. *J. Biol. Chem.* **2002**, *277*, 20293–20300. [CrossRef] [PubMed]
147. Avlani, V.A.; Ma, W.; Mun, H.C.; Leach, K.; Delbridge, L.; Christopoulos, A.; Conigrave, A.D. Calcium-sensing receptor-dependent activation of CREB phosphorylation in HEK293 cells and human parathyroid cells. *Am. J. Physiol. Endocrinol. Metab.* **2013**, *304*, E1097–E1104. [CrossRef] [PubMed]
148. Rey, O.; Young, S.H.; Yuan, J.; Slice, L.; Rozengurt, E. Amino acid-stimulated Ca^{2+} oscillations produced by the Ca^{2+}-sensing receptor are mediated by a phospholipase c/inositol 1,4,5-trisphosphate-independent pathway that requires g12, rho, filamin-a, and the actin cytoskeleton. *J. Biol. Chem.* **2005**, *280*, 22875–22882. [CrossRef] [PubMed]
149. Rey, O.; Young, S.H.; Papazyan, R.; Shapiro, M.S.; Rozengurt, E. Requirement of the trpc1 cation channel in the generation of transient Ca^{2+} oscillations by the calcium-sensing receptor. *J. Biol. Chem.* **2006**, *281*, 38730–38737. [CrossRef] [PubMed]
150. Rey, O.; Young, S.H.; Jacamo, R.; Moyer, M.P.; Rozengurt, E. Extracellular calcium sensing receptor stimulation in human colonic epithelial cells induces intracellular calcium oscillations and proliferation inhibition. *J. Cell. Physiol.* **2010**, *225*, 73–83. [CrossRef] [PubMed]
151. Zhu, D.; Kosik, K.S.; Meigs, T.E.; Yanamadala, V.; Denker, B.M. Galpha12 directly interacts with pp2a: Evidence for Gα12-stimulated PP2A phosphatase activity and dephosphorylation of microtubule-associated protein, tau. *J. Biol. Chem.* **2004**, *279*, 54983–54986. [CrossRef] [PubMed]

152. Zhu, D.; Tate, R.I.; Ruediger, R.; Meigs, T.E.; Denker, B.M. Domains necessary for galpha12 binding and stimulation of protein phosphatase-2a (PP2A): Is galpha12 a novel regulatory subunit of PP2A? *Mol. Pharmacol.* **2007**, *71*, 1268–1276. [CrossRef] [PubMed]
153. McCormick, W.D.; Atkinson-Dell, R.; Campion, K.L.; Mun, H.C.; Conigrave, A.D.; Ward, D.T. Increased receptor stimulation elicits differential calcium-sensing receptor(T888) dephosphorylation. *J. Biol. Chem.* **2010**, *285*, 14170–14177. [CrossRef] [PubMed]
154. Rybchyn, M.S.; Slater, M.; Conigrave, A.D.; Mason, R.S. An AKT-dependent increase in canonical wnt signaling and a decrease in sclerostin protein levels are involved in strontium ranelate-induced osteogenic effects in human osteoblasts. *J. Biol. Chem.* **2011**, *286*, 23771–23779. [CrossRef] [PubMed]
155. Brennan, T.C.; Rybchyn, M.S.; Green, W.; Atwa, S.; Conigrave, A.D.; Mason, R.S. Osteoblasts play key roles in the mechanisms of action of strontium ranelate. *Br. J. Pharmacol.* **2009**, *157*, 1291–1300. [CrossRef] [PubMed]
156. Brennan, S.C.; Thiem, U.; Roth, S.; Aggarwal, A.; Fetahu, I.; Tennakoon, S.; Gomes, A.R.; Brandi, M.L.; Bruggeman, F.; Mentaverri, R.; et al. Calcium sensing receptor signalling in physiology and cancer. *Biochim. Biophys. Acta* **2013**, *1833*, 1732–1744. [CrossRef] [PubMed]
157. Mamillapalli, R.; VanHouten, J.; Zawalich, W.; Wysolmerski, J. Switching of G-protein usage by the calcium-sensing receptor reverses its effect on parathyroid hormone-related protein secretion in normal versus malignant breast cells. *J. Biol. Chem.* **2008**, *283*, 24435–24447. [CrossRef] [PubMed]
158. Mamillapalli, R.; Wysolmerski, J. The calcium-sensing receptor couples to Galpha(s) and regulates pthrp and acth secretion in pituitary cells. *J. Endocrinol.* **2010**, *204*, 287–297. [CrossRef] [PubMed]
159. Brennan, S.C.; Wilkinson, W.J.; Tseng, H.E.; Finney, B.; Monk, B.; Dibble, H.; Quilliam, S.; Warburton, D.; Galietta, L.J.; Kemp, P.J.; et al. The extracellular calcium-sensing receptor regulates human fetal lung development via cftr. *Sci. Rep.* **2016**, *6*, 21975. [CrossRef] [PubMed]
160. Ziegelstein, R.C.; Xiong, Y.; He, C.; Hu, Q. Expression of a functional extracellular calcium-sensing receptor in human aortic endothelial cells. *Biochem. Biophys. Res. Commun.* **2006**, *342*, 153–163. [CrossRef] [PubMed]
161. Smajilovic, S.; Sheykhzade, M.; Holmegard, H.N.; Haunso, S.; Tfelt-Hansen, J. Calcimimetic, AMG 073, induces relaxation on isolated rat aorta. *Vascul. Pharmacol.* **2007**, *47*, 222–228. [CrossRef] [PubMed]
162. Greenberg, H.Z.E.; Shi, J.; Jahan, K.S.; Martinucci, M.C.; Gilbert, S.J.; Vanessa Ho, W.S.; Albert, A.P. Stimulation of calcium-sensing receptors induces endothelium-dependent vasorelaxations via nitric oxide production and activation of ikca channels. *Vascul. Pharmacol.* **2016**, *80*, 75–84. [CrossRef] [PubMed]
163. Qu, Y.Y.; Wang, L.M.; Zhong, H.; Liu, Y.M.; Tang, N.; Zhu, L.P.; He, F.; Hu, Q.H. TRPC1 stimulates calciumsensing receptorinduced storeoperated Ca^{2+} entry and nitric oxide production in endothelial cells. *Mol. Med. Rep.* **2017**, *16*, 4613–4619. [CrossRef] [PubMed]
164. Greenberg, H.Z.E.; Carlton-Carew, S.R.E.; Khan, D.M.; Zargaran, A.K.; Jahan, K.S.; Vanessa Ho, W.S.; Albert, A.P. Heteromeric TRPV4/TRPC1 channels mediate calcium-sensing receptor-induced nitric oxide production and vasorelaxation in rabbit mesenteric arteries. *Vascul. Pharmacol.* **2017**, *96–98*, 53–62. [CrossRef] [PubMed]
165. Garrett, J.E.; Tamir, H.; Kifor, O.; Simin, R.T.; Rogers, K.V.; Mithal, A.; Gagel, R.F.; Brown, E.M. Calcitonin-secreting cells of the thyroid express an extracellular calcium receptor gene. *Endocrinology* **1995**, *136*, 5202–5211. [CrossRef] [PubMed]
166. Hofer, A.M.; Brown, E.M. Extracellular calcium sensing and signalling. *Nat. Rev. Mol. Cell Biol.* **2003**, *4*, 530–538. [CrossRef] [PubMed]
167. Brown, E.M. Mutations in the calcium-sensing receptor and their clinical implications. *Horm. Res.* **1997**, *48*, 199–208. [CrossRef] [PubMed]
168. Pollak, M.R.; Brown, E.M.; Chou, Y.H.; Hebert, S.C.; Marx, S.J.; Steinmann, B.; Levi, T.; Seidman, C.E.; Seidman, J.G. Mutations in the human Ca^{2+}-sensing receptor gene cause familial hypocalciuric hypercalcemia and neonatal severe hyperparathyroidism. *Cell* **1993**, *75*, 1297–1303. [CrossRef]
169. Watanabe, S.; Fukumoto, S.; Chang, H.; Takeuchi, Y.; Hasegawa, Y.; Okazaki, R.; Chikatsu, N.; Fujita, T. Association between activating mutations of calcium-sensing receptor and Bartter's syndrome. *Lancet* **2002**, *360*, 692–694. [CrossRef]
170. Pollak, M.R.; Brown, E.M.; Estep, H.L.; McLaine, P.N.; Kifor, O.; Park, J.; Hebert, S.C.; Seidman, C.E.; Seidman, J.G. Autosomal dominant hypocalcaemia caused by a Ca^{2+}-sensing receptor gene mutation. *Nat. Genet.* **1994**, *8*, 303–307. [CrossRef] [PubMed]

171. Zhao, X.M.; Hauache, O.; Goldsmith, P.K.; Collins, R.; Spiegel, A.M. A missense mutation in the seventh transmembrane domain constitutively activates the human Ca^{2+} receptor. *FEBS Lett.* **1999**, *448*, 180–184. [CrossRef]
172. Carmosino, M.; Gerbino, A.; Hendy, G.N.; Torretta, S.; Rizzo, F.; Debellis, L.; Procino, G.; Svelto, M. NKCC2 activity is inhibited by the Bartter's syndrome type 5 gain-of-function CAR-A843E mutant in renal cells. *Biol. Cell* **2015**, *107*, 98–110. [CrossRef] [PubMed]
173. Tfelt-Hansen, J.; Brown, E.M. The calcium-sensing receptor in normal physiology and pathophysiology: A review. *Crit. Rev. Clin. Lab. Sci.* **2005**, *42*, 35–70. [CrossRef] [PubMed]
174. Ruat, M.; Molliver, M.E.; Snowman, A.M.; Snyder, S.H. Calcium sensing receptor: Molecular cloning in rat and localization to nerve terminals. *Proc. Natl. Acad. Sci. USA* **1995**, *92*, 3161–3165. [CrossRef] [PubMed]
175. Squires, P.E. Non-Ca^{2+}-homeostatic functions of the extracellular Ca^{2+}-sensing receptor (CAR) in endocrine tissues. *J. Endocrinol.* **2000**, *165*, 173–177. [CrossRef] [PubMed]
176. Canaff, L.; Petit, J.L.; Kisiel, M.; Watson, P.H.; Gascon-Barre, M.; Hendy, G.N. Extracellular calcium-sensing receptor is expressed in rat hepatocytes. Coupling to intracellular calcium mobilization and stimulation of bile flow. *J. Biol. Chem.* **2001**, *276*, 4070–4079. [CrossRef] [PubMed]
177. Wang, R.; Xu, C.; Zhao, W.; Zhang, J.; Cao, K.; Yang, B.; Wu, L. Calcium and polyamine regulated calcium-sensing receptors in cardiac tissues. *Eur. J. Biochem.* **2003**, *270*, 2680–2688. [CrossRef] [PubMed]
178. Tfelt-Hansen, J.; Hansen, J.L.; Smajilovic, S.; Terwilliger, E.F.; Haunso, S.; Sheikh, S.P. Calcium receptor is functionally expressed in rat neonatal ventricular cardiomyocytes. *Am. J. Physiol. Heart Circ. Physiol.* **2006**, *290*, H1165–H1171. [CrossRef] [PubMed]
179. Alfadda, T.I.; Saleh, A.M.; Houillier, P.; Geibel, J.P. Calcium-sensing receptor 20 years later. *Am. J. Physiol. Cell Physiol.* **2014**, *307*, C221–C231. [CrossRef] [PubMed]
180. Saidak, Z.; Mentaverri, R.; Brown, E.M. The role of the calcium-sensing receptor in the development and progression of cancer. *Endocr. Rev.* **2009**, *30*, 178–195. [CrossRef] [PubMed]
181. Corbetta, S.; Mantovani, G.; Lania, A.; Borgato, S.; Vicentini, L.; Beretta, E.; Faglia, G.; di Blasio, A.M.; Spada, A. Calcium-sensing receptor expression and signalling in human parathyroid adenomas and primary hyperplasia. *Clin. Endocrinol.* **2000**, *52*, 339–348. [CrossRef]
182. Gogusev, J.; Duchambon, P.; Hory, B.; Giovannini, M.; Goureau, Y.; Sarfati, E.; Drueke, T.B. Depressed expression of calcium receptor in parathyroid gland tissue of patients with hyperparathyroidism. *Kidney Int.* **1997**, *51*, 328–336. [CrossRef] [PubMed]
183. Farnebo, F.; Enberg, U.; Grimelius, L.; Backdahl, M.; Schalling, M.; Larsson, C.; Farnebo, L.O. Tumor-specific decreased expression of calcium sensing receptor messenger ribonucleic acid in sporadic primary hyperparathyroidism. *J. Clin. Endocrinol. Metab.* **1997**, *82*, 3481–3486. [CrossRef] [PubMed]
184. Sheinin, Y.; Kallay, E.; Wrba, F.; Kriwanek, S.; Peterlik, M.; Cross, H.S. Immunocytochemical localization of the extracellular calcium-sensing receptor in normal and malignant human large intestinal mucosa. *J. Histochem. Cytochem.* **2000**, *48*, 595–602. [CrossRef] [PubMed]
185. Chakrabarty, S.; Wang, H.; Canaff, L.; Hendy, G.N.; Appelman, H.; Varani, J. Calcium sensing receptor in human colon carcinoma: Interaction with Ca^{2+} and 1,25-dihydroxyvitamin D_3. *Cancer Res.* **2005**, *65*, 493–498. [PubMed]
186. Miller, G.; Davis, J.; Shatzen, E.; Colloton, M.; Martin, D.; Henley, C.M. Cinacalcet hcl prevents development of parathyroid gland hyperplasia and reverses established parathyroid gland hyperplasia in a rodent model of ckd. *Nephrol. Dial. Transplant.* **2012**, *27*, 2198–2205. [CrossRef] [PubMed]
187. Mizobuchi, M.; Hatamura, I.; Ogata, H.; Saji, F.; Uda, S.; Shiizaki, K.; Sakaguchi, T.; Negi, S.; Kinugasa, E.; Koshikawa, S.; et al. Calcimimetic compound upregulates decreased calcium-sensing receptor expression level in parathyroid glands of rats with chronic renal insufficiency. *J. Am. Soc. Nephrol.* **2004**, *15*, 2579–2587. [CrossRef] [PubMed]
188. Garland, C.; Shekelle, R.B.; Barrett-Connor, E.; Criqui, M.H.; Rossof, A.H.; Paul, O. Dietary vitamin d and calcium and risk of colorectal cancer: A 19-year prospective study in men. *Lancet* **1985**, *1*, 307–309. [CrossRef]
189. Aggarwal, A.; Hobaus, J.; Tennakoon, S.; Prinz-Wohlgenannt, M.; Graca, J.; Price, S.A.; Heffeter, P.; Berger, W.; Baumgartner-Parzer, S.; Kallay, E. Active vitamin d potentiates the anti-neoplastic effects of calcium in the colon: A cross talk through the calcium-sensing receptor. *J. Steroid Biochem. Mol. Biol.* **2016**, *155*, 231–238. [CrossRef] [PubMed]

190. Canaff, L.; Hendy, G.N. Human calcium-sensing receptor gene. Vitamin d response elements in promoters P1 and P2 confer transcriptional responsiveness to 1,25-dihydroxyvitamin D. *J. Biol. Chem.* **2002**, *277*, 30337–30350. [CrossRef] [PubMed]
191. Hendy, G.N.; Canaff, L. Calcium-sensing receptor gene: Regulation of expression. *Front. Physiol.* **2016**, *7*, 394. [CrossRef] [PubMed]
192. Mihai, R.; Stevens, J.; McKinney, C.; Ibrahim, N.B. Expression of the calcium receptor in human breast cancer—A potential new marker predicting the risk of bone metastases. *Eur. J. Surg. Oncol.* **2006**, *32*, 511–515. [CrossRef] [PubMed]
193. Sanders, J.L.; Chattopadhyay, N.; Kifor, O.; Yamaguchi, T.; Brown, E.M. Ca^{2+}-sensing receptor expression and pthrp secretion in pc-3 human prostate cancer cells. *Am. J. Physiol. Endocrinol. Metab.* **2001**, *281*, E1267–E1274. [CrossRef] [PubMed]
194. Ward, D.T.; Riccardi, D. New concepts in calcium-sensing receptor pharmacology and signalling. *Br. J. Pharmacol.* **2012**, *165*, 35–48. [CrossRef] [PubMed]
195. Weston, A.H.; Absi, M.; Ward, D.T.; Ohanian, J.; Dodd, R.H.; Dauban, P.; Petrel, C.; Ruat, M.; Edwards, G. Evidence in favor of a calcium-sensing receptor in arterial endothelial cells: Studies with calindol and calhex 231. *Circ. Res.* **2005**, *97*, 391–398. [CrossRef] [PubMed]
196. Smajilovic, S.; Hansen, J.L.; Christoffersen, T.E.; Lewin, E.; Sheikh, S.P.; Terwilliger, E.F.; Brown, E.M.; Haunso, S.; Tfelt-Hansen, J. Extracellular calcium sensing in rat aortic vascular smooth muscle cells. *Biochem. Biophys. Res. Commun.* **2006**, *348*, 1215–1223. [CrossRef] [PubMed]
197. Berra Romani, R.; Raqeeb, A.; Laforenza, U.; Scaffino, M.F.; Moccia, F.; Avelino-Cruz, J.E.; Oldani, A.; Coltrini, D.; Milesi, V.; Taglietti, V.; et al. Cardiac microvascular endothelial cells express a functional Ca^{2+}-sensing receptor. *J. Vascul. Res.* **2009**, *46*, 73–82. [CrossRef] [PubMed]
198. Weston, A.H.; Absi, M.; Harno, E.; Geraghty, A.R.; Ward, D.T.; Ruat, M.; Dodd, R.H.; Dauban, P.; Edwards, G. The expression and function of Ca^{2+}-sensing receptors in rat mesenteric artery; comparative studies using a model of type ii diabetes. *Br. J. Pharmacol.* **2008**, *154*, 652–662. [CrossRef] [PubMed]
199. Smajilovic, S.; Tfelt-Hansen, J. Calcium acts as a first messenger through the calcium-sensing receptor in the cardiovascular system. *Cardiovasc. Res.* **2007**, *75*, 457–467. [CrossRef] [PubMed]
200. Weston, A.H.; Geraghty, A.; Egner, I.; Edwards, G. The vascular extracellular calcium-sensing receptor: An update. *Acta Physiol.* **2011**, *203*, 127–137. [CrossRef] [PubMed]
201. Smajilovic, S.; Yano, S.; Jabbari, R.; Tfelt-Hansen, J. The calcium-sensing receptor and calcimimetics in blood pressure modulation. *Br. J. Pharmacol.* **2011**, *164*, 884–893. [CrossRef] [PubMed]
202. Schreckenberg, R.; Schluter, K.D. Calcium sensing receptor expression and signalling in cardiovascular physiology and disease. *Vascul. Pharmacol.* **2018**. [CrossRef] [PubMed]
203. Sun, Y.H.; Liu, M.N.; Li, H.; Shi, S.; Zhao, Y.J.; Wang, R.; Xu, C.Q. Calcium-sensing receptor induces rat neonatal ventricular cardiomyocyte apoptosis. *Biochem. Biophys. Res. Commun.* **2006**, *350*, 942–948. [CrossRef] [PubMed]
204. Zhang, W.H.; Fu, S.B.; Lu, F.H.; Wu, B.; Gong, D.M.; Pan, Z.W.; Lv, Y.J.; Zhao, Y.J.; Li, Q.F.; Wang, R.; et al. Involvement of calcium-sensing receptor in ischemia/reperfusion-induced apoptosis in rat cardiomyocytes. *Biochem. Biophys. Res. Commun.* **2006**, *347*, 872–881. [CrossRef] [PubMed]
205. Sun, Y.H.; Li, Y.Q.; Feng, S.L.; Li, B.X.; Pan, Z.W.; Xu, C.Q.; Li, T.T.; Yang, B.F. Calcium-sensing receptor activation contributed to apoptosis stimulates trpc6 channel in rat neonatal ventricular myocytes. *Biochem. Biophys. Res. Commun.* **2010**, *394*, 955–961. [CrossRef] [PubMed]
206. Feng, S.L.; Sun, M.R.; Li, T.T.; Yin, X.; Xu, C.Q.; Sun, Y.H. Activation of calcium-sensing receptor increases TRPC3 expression in rat cardiomyocytes. *Biochem. Biophys. Res. Commun.* **2011**, *406*, 278–284. [CrossRef] [PubMed]
207. Tang, J.; Wang, G.; Liu, Y.; Fu, Y.; Chi, J.; Zhu, Y.; Zhao, Y.; Yin, X. Cyclosporin a induces cardiomyocyte injury through calcium-sensing receptor-mediated calcium overload. *Pharmazie* **2011**, *66*, 52–57. [PubMed]
208. Zhao, Y.; Hou, G.; Zhang, Y.; Chi, J.; Zhang, L.; Zou, X.; Tang, J.; Liu, Y.; Fu, Y.; Yin, X. Involvement of the calcium-sensing receptor in cyclosporin a-induced cardiomyocyte apoptosis in rats. *Pharmazie* **2011**, *66*, 968–974. [PubMed]
209. Chi, J.; Zhu, Y.; Fu, Y.; Liu, Y.; Zhang, X.; Han, L.; Yin, X.; Zhao, D. Cyclosporin a induces apoptosis in H9C2 cardiomyoblast cells through calcium-sensing receptor-mediated activation of the erk MAPK and p38 MAPK pathways. *Mol. Cell. Biochem.* **2013**, *367*, 227–236. [CrossRef] [PubMed]

210. Wang, H.-Y.; Liu, X.-Y.; Han, G.; Wang, Z.-Y.; Li, X.-X.; Jiang, Z.-M.; Jiang, C.-M. LPS induces cardiomyocyte injury through calcium-sensing receptor. *Mol. Cell. Biochem.* **2013**, *379*, 153–159. [CrossRef] [PubMed]
211. Jiang, C.M.; Han, L.P.; Li, H.Z.; Qu, Y.B.; Zhang, Z.R.; Wang, R.; Xu, C.Q.; Li, W.M. Calcium-sensing receptors induce apoptosis in cultured neonatal rat ventricular cardiomyocytes during simulated ischemia/reperfusion. *Cell Biol. Int.* **2008**, *32*, 792–800. [CrossRef] [PubMed]
212. Lu, F.; Tian, Z.; Zhang, W.; Zhao, Y.; Bai, S.; Ren, H.; Chen, H.; Yu, X.; Wang, J.; Wang, L.; et al. Calcium-sensing receptors induce apoptosis in rat cardiomyocytes via the Endo (sarco) plasmic reticulum pathway during hypoxia/reoxygenation. *Basic Clin. Pharmacol. Toxicol.* **2010**, *106*, 396–405. [CrossRef] [PubMed]
213. Lu, F.H.; Tian, Z.; Zhang, W.H.; Zhao, Y.J.; Li, H.L.; Ren, H.; Zheng, H.S.; Liu, C.; Hu, G.X.; Tian, Y.; et al. Calcium-sensing receptors regulate cardiomyocyte Ca^{2+} signaling via the sarcoplasmic reticulum-mitochondrion interface during hypoxia/reoxygenation. *J. Biomed. Sci.* **2010**, *17*, 50. [CrossRef] [PubMed]
214. Guo, J.; Li, H.Z.; Zhang, W.H.; Wang, L.C.; Wang, L.N.; Zhang, L.; Li, G.W.; Li, H.X.; Yang, B.F.; Wu, L.; et al. Increased expression of calcium-sensing receptors induced by ox-LDL amplifies apoptosis of cardiomyocytes during simulated ischaemia-reperfusion. *Clin. Exp. Pharmacol. Physiol.* **2010**, *37*, e128–e135. [CrossRef] [PubMed]
215. Jiang, C.M.; Xu, C.Q.; Mi, Y.; Li, H.Z.; Wang, R.; Li., W.-M. Calcium-sensing receptor induced myocardial ischemia/reperfusion injury via the c-jun nh2-terminal protein kinase pathway. *Acta Cardiol. Sin.* **2010**, *26*, 102–110.
216. Zheng, H.S.; Liu, J.; Liu, C.; Lu, F.H.; Zhao, Y.J.; Jin, Z.F.; Ren, H.; Leng, X.N.; Jia, J.; Hu, G.X.; et al. Calcium-sensing receptor activating phosphorylation of PKC delta translocation on mitochondria to induce cardiomyocyte apoptosis during ischemia/reperfusion. *Mol. Cell. Biochem.* **2011**, *358*, 335–343. [CrossRef] [PubMed]
217. Yan, L.; Zhu, T.; Sun, T.; Wang, L.; Pan, S.; Tao, Z.; Yang, Z.; Cao, K. Activation of calcium-sensing receptors is associated with apoptosis in a model of simulated cardiomyocytes ischemia/reperfusion. *J. Biomed. Res.* **2010**, *24*, 301–307. [CrossRef]
218. Yan, L.; Zhu, T.B.; Wang, L.S.; Pan, S.Y.; Tao, Z.X.; Yang, Z.; Cao, K.; Huang, J. Inhibitory effect of hepatocyte growth factor on cardiomyocytes apoptosis is partly related to reduced calcium sensing receptor expression during a model of simulated ischemia/reperfusion. *Mol. Biol. Rep.* **2011**, *38*, 2695–2701. [CrossRef] [PubMed]
219. Zhang, W.H.; Lu, F.H.; Zhao, Y.J.; Wang, L.N.; Tian, Y.; Pan, Z.W.; Lv, Y.J.; Wang, Y.L.; Du, L.J.; Sun, Z.R.; et al. Post-conditioning protects rat cardiomyocytes via pkcepsilon-mediated calcium-sensing receptors. *Biochem. Biophys. Res. Commun.* **2007**, *361*, 659–664. [CrossRef] [PubMed]
220. Dong, S.; Teng, Z.; Lu, F.H.; Zhao, Y.J.; Li, H.; Ren, H.; Chen, H.; Pan, Z.W.; Lv, Y.J.; Yang, B.F.; et al. Post-conditioning protects cardiomyocytes from apoptosis via PKC (epsilon)-interacting with calcium-sensing receptors to inhibit Endo (sarco) plasmic reticulum-mitochondria crosstalk. *Mol. Cell. Biochem.* **2010**, *341*, 195–206. [CrossRef] [PubMed]
221. Gan, R.T.; Hu, G.X.; Zhao, Y.J.; Li, H.L.; Jin, Z.F.; Ren, H.; Dong, S.Y.; Zhong, X.; Li, H.Z.; Yang, B.F.; et al. Post-conditioning protecting rat cardiomyocytes from apoptosis via attenuating calcium-sensing receptor-induced Endo (sarco) plasmic reticulum stress. *Mol. Cell. Biochem.* **2012**, *361*, 123–134. [CrossRef] [PubMed]
222. Zhang, L.; Cao, S.; Deng, S.; Yao, G.; Yu, T. Ischemic postconditioning and pinacidil suppress calcium overload in anoxia-reoxygenation cardiomyocytes via down-regulation of the calcium-sensing receptor. *PeerJ* **2016**, *4*, e2612. [CrossRef] [PubMed]
223. Dyukova, E.; Schreckenberg, R.; Sitdikova, G.; Schlüter, K.-D. Influence of ischemic pre- and post-conditioning on cardiac expression of calcium-sensing receptor. *BioNanoScience* **2017**, *7*, 112–114. [CrossRef]
224. Sun, J.; Murphy, E. Calcium sensing receptor: A sensor and mediator of ischemic preconditioning in the hearts. *Am. J. Physiol. Heart Circ. Physiol.* **2010**, *299*. [CrossRef] [PubMed]
225. Bai, S.Z.; Sun, J.; Wu, H.; Zhang, N.; Li, H.X.; Li, G.W.; Li, H.Z.; He, W.; Zhang, W.H.; Zhao, Y.J.; et al. Decrease in calcium-sensing receptor in the progress of diabetic cardiomyopathy. *Diabetes Res. Clin. Pract.* **2012**, *95*, 378–385. [CrossRef] [PubMed]
226. Qi, H.; Cao, Y.; Huang, W.; Liu, Y.; Wang, Y.; Li, L.; Liu, L.; Ji, Z.; Sun, H. Crucial role of calcium-sensing receptor activation in cardiac injury of diabetic rats. *PLoS ONE* **2013**, *8*, e65147. [CrossRef] [PubMed]

227. Matturri, L.; Milei, J.; Grana, D.R.; Lavezzi, A.M. Characterization of myocardial hypertrophy by DNA content, pcna expression and apoptotic index. *Int. J. Cardiol.* **2002**, *82*, 33–39. [CrossRef]
228. Wang, L.N.; Wang, C.; Lin, Y.; Xi, Y.H.; Zhang, W.H.; Zhao, Y.J.; Li, H.Z.; Tian, Y.; Lv, Y.J.; Yang, B.F.; et al. Involvement of calcium-sensing receptor in cardiac hypertrophy-induced by angiotensinii through calcineurin pathway in cultured neonatal rat cardiomyocytes. *Biochem. Biophys. Res. Commun.* **2008**, *369*, 584–589. [CrossRef] [PubMed]
229. Zhong, X.L.J.; Lu, F.; Wang, Y.; Zhao, Y.; Dong, S.; Leng, X.; Jia, J.; Ren, H.; Xu, C.; Zhang, W. Calcium sensing receptor regulates cardiomyocyte function through nuclear calcium. *Cell Biol. Int.* **2012**, *36*, 937–943. [CrossRef] [PubMed]
230. Colella, M.; Pozzan, T. Cardiac cell hypertrophy in vitro: Role of calcineurin/nfat as Ca^{2+} signal integrators. *Ann. N. Y. Acad. Sci.* **2008**, *1123*, 64–68. [CrossRef] [PubMed]
231. Lu, F.H.; Fu, S.B.; Leng, X.; Zhang, X.; Dong, S.; Zhao, Y.J.; Ren, H.; Li, H.; Zhong, X.; Xu, C.Q.; et al. Role of the calcium-sensing receptor in cardiomyocyte apoptosis via the sarcoplasmic reticulum and mitochondrial death pathway in cardiac hypertrophy and heart failure. *Cell. Physiol. Biochem.* **2013**, *31*, 728–743. [CrossRef] [PubMed]
232. Liu, L.; Wang, C.; Lin, Y.; Xi, Y.; Li, H.; Shi, S.; Li, H.; Zhang, W.; Zhao, Y.; Tian, Y.; et al. Suppression of calcium-sensing receptor ameliorates cardiac hypertrophy through inhibition of autophagy. *Mol. Med. Rep.* **2016**, *14*, 111–120. [CrossRef] [PubMed]
233. Zhang, X.; Zhang, T.; Wu, J.; Yu, X.; Zheng, D.; Yang, F.; Li, T.; Wang, L.; Zhao, Y.; Dong, S.; et al. Calcium sensing receptor promotes cardiac fibroblast proliferation and extracellular matrix secretion. *Cell. Physiol. Biochem.* **2014**, *33*, 557–568. [CrossRef] [PubMed]
234. Schreckenberg, R.; Dyukova, E.; Sitdikova, G.; Abdallah, Y.; Schluter, K.D. Mechanisms by which calcium receptor stimulation modifies electromechanical coupling in isolated ventricular cardiomyocytes. *Pflugers Arch.* **2015**, *467*, 379–388. [CrossRef] [PubMed]
235. Liu, C.-H.; Chang, H.-K.; Lee, S.-P.; Shieh, R.-C. Activation of the Ca^{2+}-sensing receptors increases currents through inward rectifier k+ channels via activation of phosphatidylinositol 4-kinase. *Pflugers Arch.* **2017**, *468*, 1931–1943. [CrossRef] [PubMed]
236. Tagliavini, S.; Genedani, S.; Bertolini, A.; Bazzani, C. Ischemia- and reperfusion-induced arrhythmias are prevented by putrescine. *Eur. J. Pharmacol.* **1991**, *194*, 7–10. [CrossRef]
237. Schepelmann, M.; Yarova, P.L.; Lopez-Fernandez, I.; Davies, T.S.; Brennan, S.C.; Edwards, P.J.; Aggarwal, A.; Graca, J.; Rietdorf, K.; Matchkov, V.; et al. The vascular Ca^{2+}-sensing receptor regulates blood vessel tone and blood pressure. *Am. J. Physiol. Cell Physiol.* **2016**, *310*, C193–C204. [CrossRef] [PubMed]
238. Wess, J. G-protein-coupled receptors: Molecular mechanisms involved in receptor activation and selectivity of g-protein recognition. *FASEB J.* **1997**, *11*, 346–354. [CrossRef] [PubMed]
239. Lefkowitz, R.J. Historical review: A brief history and personal retrospective of seven-transmembrane receptors. *Trends Pharmacol. Sci.* **2004**, *25*, 413–422. [CrossRef] [PubMed]
240. Rajagopal, S.; Ahn, S.; Rominger, D.H.; Gowen-MacDonald, W.; Lam, C.M.; Dewire, S.M.; Violin, J.D.; Lefkowitz, R.J. Quantifying ligand bias at seven-transmembrane receptors. *Mol. Pharmacol.* **2010**, *80*, 367–377. [CrossRef] [PubMed]
241. Lohse, M.J.; Hofmann, K.P. Spatial and temporal aspects of signaling by G-protein-coupled receptors. *Mol. Pharmacol.* **2015**, *88*, 572–578. [CrossRef] [PubMed]
242. Pupo, A.S.; Duarte, D.A.; Lima, V.; Teixeira, L.B.; Parreiras, E.S.L.T.; Costa-Neto, C.M. Recent updates on gpcr biased agonism. *Pharmacol. Res.* **2016**, *112*, 49–57. [CrossRef] [PubMed]
243. Calebiro, D.; Sungkaworn, T.; Maiellaro, I. Real-time monitoring of GPCR/CAMP signalling by fret and single-molecule microscopy. *Horm. Metab. Res.* **2014**, *46*, 827–832. [CrossRef] [PubMed]
244. Ibarra, C.; Vicencio, J.M.; Estrada, M.; Lin, Y.; Rocco, P.; Rebellato, P.; Munoz, J.P.; Garcia-Prieto, J.; Quest, A.F.; Chiong, M.; et al. Local control of nuclear calcium signaling in cardiac myocytes by perinuclear microdomains of sarcolemmal insulin-like growth factor 1 receptors. *Circ. Res.* **2013**, *112*, 236–245. [CrossRef] [PubMed]
245. Benard, G.; Massa, F.; Puente, N.; Lourenco, J.; Bellocchio, L.; Soria-Gomez, E.; Matias, I.; Delamarre, A.; Metna-Laurent, M.; Cannich, A.; et al. Mitochondrial cb1 receptors regulate neuronal energy metabolism. *Nat. Neurosci.* **2012**, *15*, 558–564. [CrossRef] [PubMed]

246. Calebiro, D.; Nikolaev, V.O.; Gagliani, M.C.; de Filippis, T.; Dees, C.; Tacchetti, C.; Persani, L.; Lohse, M.J. Persistent camp-signals triggered by internalized g-protein coupled receptors. *PLoS Biol.* **2009**, *7*, e1000172. [CrossRef] [PubMed]
247. Irannejad, R.; Tomshine, J.C.; Tomshine, J.R.; Chevalier, M.; Mahoney, J.P.; Steyaert, J.; Rasmussen, S.G.; Sunahara, R.K.; El-Samad, H.; Huang, B.; et al. Conformational biosensors reveal GPCR signalling from endosomes. *Nature* **2013**, *495*, 534–538. [CrossRef] [PubMed]
248. Pidasheva, S.; Grant, M.; Canaff, L.; Ercan, O.; Kumar, U.; Hendy, G.N. Calcium-sensing receptor dimerizes in the endoplasmic reticulum: Biochemical and biophysical characterization of casr mutants retained intracellularly. *Hum. Mol. Genet.* **2006**, *15*, 2200–2209. [CrossRef] [PubMed]
249. Ward, D.T.; Brown, E.M.; Harris, H.W. Disulfide bonds in the extracellular calcium-polyvalent cation-sensing receptor correlate with dimer formation and its response to divalent cations in vitro. *J. Biol. Chem.* **1998**, *273*, 14476–14483. [CrossRef] [PubMed]
250. Breitwieser, G.E. Minireview: The intimate link between calcium sensing receptor trafficking and signaling: Implications for disorders of calcium homeostasis. *Mol. Endocrinol.* **2012**, *26*, 1482–1495. [CrossRef] [PubMed]
251. Breitwieser, G.E. Pharmacoperones and the calcium sensing receptor: Exogenous and endogenous regulators. *Pharmacol. Res.* **2014**, *83*, 30–37. [CrossRef] [PubMed]
252. Holtback, U.; Brismar, H.; DiBona, G.F.; Fu, M.; Greengard, P.; Aperia, A. Receptor recruitment: A mechanism for interactions between g protein-coupled receptors. *Proc. Natl. Acad. Sci. USA* **1999**, *96*, 7271–7275. [CrossRef] [PubMed]
253. Brennan, S.C.; Mun, H.C.; Leach, K.; Kuchel, P.W.; Christopoulos, A.; Conigrave, A.D. Receptor expression modulates calcium-sensing receptor mediated intracellular Ca^{2+} mobilization. *Endocrinology* **2015**, *156*, 1330–1342. [CrossRef] [PubMed]
254. Breitwieser, G.E. The calcium sensing receptor life cycle: Trafficking, cell surface expression, and degradation. *Best Pract. Res. Clin. Endocrinol. Metab.* **2013**, *27*, 303–313. [CrossRef] [PubMed]
255. Ray, K. Calcium-sensing receptor: Trafficking, endocytosis, recycling, and importance of interacting proteins. *Prog. Mol. Biol. Transl. Sci.* **2015**, *132*, 127–150. [PubMed]
256. Huang, C.; Miller, R.T. The calcium-sensing receptor and its interacting proteins. *J. Cell Mol. Med.* **2007**, *11*, 923–934. [CrossRef] [PubMed]
257. Bai, M.; Quinn, S.; Trivedi, S.; Kifor, O.; Pearce, S.H.; Pollak, M.R.; Krapcho, K.; Hebert, S.C.; Brown, E.M. Expression and characterization of inactivating and activating mutations in the human Ca^{2+} o-sensing receptor. *J. Biol. Chem.* **1996**, *271*, 19537–19545. [CrossRef] [PubMed]
258. Grant, M.P.; Stepanchick, A.; Cavanaugh, A.; Breitwieser, G.E. Agonist-driven maturation and plasma membrane insertion of calcium-sensing receptors dynamically control signal amplitude. *Sci. Signal.* **2011**, *4*, ra78. [CrossRef] [PubMed]
259. Huang, Y.; Breitwieser, G.E. Rescue of calcium-sensing receptor mutants by allosteric modulators reveals a conformational checkpoint in receptor biogenesis. *J. Biol. Chem.* **2007**, *282*, 9517–9525. [CrossRef] [PubMed]
260. White, E.; McKenna, J.; Cavanaugh, A.; Breitwieser, G.E. Pharmacochaperone-mediated rescue of calcium-sensing receptor loss-of-function mutants. *Mol. Endocrinol.* **2009**, *23*, 1115–1123. [CrossRef] [PubMed]
261. Cavanaugh, A.; McKenna, J.; Stepanchick, A.; Breitwieser, G.E. Calcium-sensing receptor biosynthesis includes a cotranslational conformational checkpoint and endoplasmic reticulum retention. *J. Biol. Chem.* **2010**, *285*, 19854–19864. [CrossRef] [PubMed]
262. Rus, R.; Haag, C.; Bumke-Vogt, C.; Bahr, V.; Mayr, B.; Mohlig, M.; Schulze, E.; Frank-Raue, K.; Raue, F.; Schofl, C. Novel inactivating mutations of the calcium-sensing receptor: The calcimimetic nps r-568 improves signal transduction of mutant receptors. *J. Clin. Endocrinol. Metab.* **2008**, *93*, 4797–4803. [CrossRef] [PubMed]
263. Leach, K.; Wen, A.; Cook, A.E.; Sexton, P.M.; Conigrave, A.D.; Christopoulos, A. Impact of clinically relevant mutations on the pharmacoregulation and signaling bias of the calcium-sensing receptor by positive and negative allosteric modulators. *Endocrinology* **2013**, *154*, 1105–1116. [CrossRef] [PubMed]
264. Nakamura, A.; Hotsubo, T.; Kobayashi, K.; Mochizuki, H.; Ishizu, K.; Tajima, T. Loss-of-function and gain-of-function mutations of calcium-sensing receptor: Functional analysis and the effect of allosteric modulators NPS R-568 and NPS 2143. *J. Clin. Endocrinol. Metab.* **2013**, *98*, E1692–E1701. [CrossRef] [PubMed]

265. Letz, S.; Rus, R.; Haag, C.; Dorr, H.G.; Schnabel, D.; Mohlig, M.; Schulze, E.; Frank-Raue, K.; Raue, F.; Mayr, B.; et al. Novel activating mutations of the calcium-sensing receptor: The calcilytic NPS-2143 mitigates excessive signal transduction of mutant receptors. *J. Clin. Endocrinol. Metab.* **2010**, *95*, E229–E233. [CrossRef] [PubMed]
266. Letz, S.; Haag, C.; Schulze, E.; Frank-Raue, K.; Raue, F.; Hofner, B.; Mayr, B.; Schofl, C. Amino alcohol-(NPS-2143) and quinazolinone-derived calcilytics (ATF936 and AXT914) differentially mitigate excessive signalling of calcium-sensing receptor mutants causing bartter syndrome type 5 and autosomal dominant hypocalcemia. *PLoS ONE* **2014**, *9*, e115178. [CrossRef] [PubMed]
267. Hannan, F.M.; Walls, G.V.; Babinsky, V.N.; Nesbit, M.A.; Kallay, E.; Hough, T.A.; Fraser, W.D.; Cox, R.D.; Hu, J.; Spiegel, A.M.; et al. The calcilytic agent NPS 2143 rectifies hypocalcemia in a mouse model with an activating calcium-sensing receptor (CASR) mutation: Relevance to autosomal dominant hypocalcemia type 1 (ADH1). *Endocrinology* **2015**, *156*, 3114–3121. [CrossRef] [PubMed]
268. Dong, B.; Endo, I.; Ohnishi, Y.; Kondo, T.; Hasegawa, T.; Amizuka, N.; Kiyonari, H.; Shioi, G.; Abe, M.; Fukumoto, S.; et al. Calcilytic ameliorates abnormalities of mutant calcium-sensing receptor (CASR) knock-in mice mimicking autosomal dominant hypocalcemia (ADH). *J. Bone Miner. Res.* **2015**, *30*, 1980–1993. [CrossRef] [PubMed]
269. Leach, K.; Conigrave, A.D.; Sexton, P.M.; Christopoulos, A. Towards tissue-specific pharmacology: Insights from the calcium-sensing receptor as a paradigm for GPCR (patho)physiological bias. *Trends Pharmacol. Sci.* **2015**, *36*, 215–225. [CrossRef] [PubMed]
270. Makita, N.; Sato, J.; Manaka, K.; Shoji, Y.; Oishi, A.; Hashimoto, M.; Fujita, T.; Iiri, T. An acquired hypocalciuric hypercalcemia autoantibody induces allosteric transition among active human ca-sensing receptor conformations. *Proc. Natl. Acad. Sci. USA* **2007**, *104*, 5443–5448. [CrossRef] [PubMed]
271. Bruce, J.I.; Yang, X.; Ferguson, C.J.; Elliott, A.C.; Steward, M.C.; Case, R.M.; Riccardi, D. Molecular and functional identification of a Ca^{2+} (polyvalent cation)-sensing receptor in rat pancreas. *J. Biol. Chem.* **1999**, *274*, 20561–20568. [CrossRef] [PubMed]
272. Thomsen, A.R.; Smajilovic, S.; Brauner-Osborn, H. Novel strategies in drug discovery of the calcium-sensing receptor based on biased signaling. *Curr. Drug. Targets* **2012**, *13*, 1324–1335. [CrossRef] [PubMed]
273. Thomsen, A.R.; Worm, J.; Jacobsen, S.E.; Stahlhut, M.; Latta, M.; Brauner-Osborne, H. Strontium is a biased agonist of the calcium-sensing receptor in rat medullary thyroid carcinoma 6-23 cells. *J. Pharmacol. Exp. Ther.* **2012**, *343*, 638–649. [CrossRef] [PubMed]
274. Chattopadhyay, N.; Quinn, S.J.; Kifor, O.; Ye, C.; Brown, E.M. The calcium-sensing receptor (CAR) is involved in strontium ranelate-induced osteoblast proliferation. *Biochem. Pharmacol.* **2007**, *74*, 438–447. [CrossRef] [PubMed]
275. Coulombe, J.; Faure, H.; Robin, B.; Ruat, M. In vitro effects of strontium ranelate on the extracellular calcium-sensing receptor. *Biochem. Biophys. Res. Commun.* **2004**, *323*, 1184–1190. [CrossRef] [PubMed]
276. Hurtel-Lemaire, A.S.; Mentaverri, R.; Caudrillier, A.; Cournarie, F.; Wattel, A.; Kamel, S.; Terwilliger, E.F.; Brown, E.M.; Brazier, M. The calcium-sensing receptor is involved in strontium ranelate-induced osteoclast apoptosis: New insights into the associated signaling pathways. *J. Biol. Chem.* **2009**, *284*, 575–584. [CrossRef] [PubMed]
277. Davey, A.E.; Leach, K.; Valant, C.; Conigrave, A.D.; Sexton, P.M.; Christopoulos, A. Positive and negative allosteric modulators promote biased signaling at the calcium-sensing receptor. *Endocrinology* **2012**, *153*, 1232–1241. [CrossRef] [PubMed]
278. Leach, K.; Wen, A.; Davey, A.E.; Sexton, P.M.; Conigrave, A.D.; Christopoulos, A. Identification of molecular phenotypes and biased signaling induced by naturally occurring mutations of the human calcium-sensing receptor. *Endocrinology* **2012**, *153*, 4304–4316. [CrossRef] [PubMed]
279. Hofer, A.M.; Lefkimmiatis, K. Extracellular calcium and cAMP: Second messengers as "Third messengers"? *Physiology* **2007**, *22*, 320–327. [CrossRef] [PubMed]
280. Wellendorph, P.; Johansen, L.D.; Brauner-Osborne, H. Molecular pharmacology of promiscuous seven transmembrane receptors sensing organic nutrients. *Mol. Pharmacol.* **2009**, *76*, 453–465. [CrossRef] [PubMed]
281. Kano, M.; Kato, M. Quisqualate receptors are specifically involved in cerebellar synaptic plasticity. *Nature* **1987**, *325*, 276–279. [CrossRef] [PubMed]
282. Francesconi, A.; Duvoisin, R.M. Divalent cations modulate the activity of metabotropic glutamate receptors. *J. Neurosci. Res.* **2004**, *75*, 472–479. [CrossRef] [PubMed]

283. Kubo, Y.; Miyashita, T.; Murata, Y. Structural basis for a Ca^{2+}-sensing function of the metabotropic glutamate receptors. *Science* **1998**, *279*, 1722–1725. [CrossRef] [PubMed]
284. Nash, M.S.; Saunders, R.; Young, K.W.; Challiss, R.A.; Nahorski, S.R. Reassessment of the Ca^{2+} sensing property of a type i metabotropic glutamate receptor by simultaneous measurement of inositol 1,4,5-trisphosphate and Ca^{2+} in single cells. *J. Biol. Chem.* **2001**, *276*, 19286–19293. [CrossRef] [PubMed]
285. Tabata, T.; Kano, M. Calcium dependence of native metabotropic glutamate receptor signaling in central neurons. *Mol. Neurobiol.* **2004**, *29*, 261–270. [CrossRef]
286. Hardingham, N.R.; Bannister, N.J.; Read, J.C.; Fox, K.D.; Hardingham, G.E.; Jack, J.J. Extracellular calcium regulates postsynaptic efficacy through group 1 metabotropic glutamate receptors. *J. Neurosci.* **2006**, *26*, 6337–6345. [CrossRef] [PubMed]
287. Gerber, U.; Gee, C.E.; Benquet, P. Metabotropic glutamate receptors: Intracellular signaling pathways. *Curr. Opin. Pharmacol.* **2007**, *7*, 56–61. [CrossRef] [PubMed]
288. Wang, X.; Kirberger, M.; Qiu, F.; Chen, G.; Yang, J.J. Towards predicting Ca^{2+}-binding sites with different coordination numbers in proteins with atomic resolution. *Proteins* **2009**, *75*, 787–798. [CrossRef] [PubMed]
289. Wang, X.; Zhao, K.; Kirberger, M.; Wong, H.; Chen, G.; Yang, J.J. Analysis and prediction of calcium-binding pockets from apo-protein structures exhibiting calcium-induced localized conformational changes. *Protein Sci.* **2010**, *19*, 1180–1190. [CrossRef] [PubMed]
290. Jiang, Y.; Huang, Y.; Wong, H.C.; Zhou, Y.; Wang, X.; Yang, J.; Hall, R.A.; Brown, E.M.; Yang, J.J. Elucidation of a novel extracellular calcium-binding site on metabotropic glutamate receptor 1α (mGluR1α) that controls receptor activation. *J. Biol. Chem.* **2010**, *285*, 33463–33474. [CrossRef] [PubMed]
291. Jiang, J.Y.; Nagaraju, M.; Meyer, R.C.; Zhang, L.; Hamelberg, D.; Hall, R.A.; Brown, E.M.; Conn, P.J.; Yang, J.J. Extracellular calcium modulates actions of orthosteric and allosteric ligands on metabotropic glutamate receptor 1α. *J. Biol. Chem.* **2014**, *289*, 1649–1661. [CrossRef] [PubMed]
292. Jones, K.A.; Borowsky, B.; Tamm, J.A.; Craig, D.A.; Durkin, M.M.; Dai, M.; Yao, W.J.; Johnson, M.; Gunwaldsen, C.; Huang, L.Y.; et al. Gaba(b) receptors function as a heteromeric assembly of the subunits $GABA_bR1$ and $GABA_bR2$. *Nature* **1998**, *396*, 674–679. [CrossRef] [PubMed]
293. Wise, A.; Green, A.; Main, M.J.; Wilson, R.; Fraser, N.; Marshall, F.H. Calcium sensing properties of the $GABA_b$ receptor. *Neuropharmacology* **1999**, *38*, 1647–1656. [CrossRef]
294. Silve, C.; Petrel, C.; Leroy, C.; Bruel, H.; Mallet, E.; Rognan, D.; Ruat, M. Delineating a Ca^{2+} binding pocket within the venus flytrap module of the human calcium-sensing receptor. *J. Biol. Chem.* **2005**, *280*, 37917–37923. [CrossRef] [PubMed]
295. Tordoff, M.G.; Shao, H.; Alarcón, L.K.; Margolskee, R.F.; Mosinger, B.; Bachmanov, A.A.; Reed, D.R.; McCaughey, S. Involvement of t1r3 in calcium-magnesium taste. *Physiol. Genomics* **2008**, *34*, 338–348. [CrossRef] [PubMed]
296. Tordoff, M.G.; Alarcón, L.K.; Valmeki, S.; Jiang, P. T1R3: A human calcium taste receptor. *Sci. Rep.* **2012**, *2*, 496. [CrossRef] [PubMed]
297. Wellendorph, P.; Bräuner-Osborne, H. Molecular cloning, expression, and sequence analysis of gprc6a, a novel family c G-protein-coupled receptor. *Gene* **2004**, *335*, 37–46. [CrossRef] [PubMed]
298. Pi, M.; Quarles, L.D. Osteoblast calcium-sensing receptor has characteristics of ANF/7TM receptors. *J. Cell Biochem.* **2005**, *95*, 1081–1092. [CrossRef] [PubMed]
299. Faure, H.; Gorojankina, T.; Rice, N.; Dauban, P.; Dodd, R.H.; Bräuner-Osborne, H.; Rognan, D.; Ruat, M. Molecular determinants of non-competitive antagonist binding to the mouse gprc6a receptor. *Cell Calcium* **2009**, *46*, 323–332. [CrossRef] [PubMed]
300. Stenflo, J.; Stenberg, Y.; Muranyi, A. Calcium-binding EGF-like modules in coagulation proteinases: Function of the calcium ion in module interactions. *Biochim. Biophys. Acta* **2000**, *1477*, 51–63. [CrossRef]
301. Raya, A.; Kawakami, Y.; Rodríguez-Esteban, C.; Ibañes, M.; Rasskin-Gutman, D.; Rodríguez-León, J.; Büscher, D.; Feijó, J.A.; Izpisúa Belmonte, J.C. Notch activity acts as a sensor for extracellular calcium during vertebrate left-right determination. *Nature* **2004**, *427*, 121–128. [CrossRef] [PubMed]
302. Halbleib, J.M.; Nelson, W.J. Cadherins in development: Cell adhesion, sorting, and tissue morphogenesis. *Genes Dev.* **2006**, *20*, 3199–3214. [CrossRef] [PubMed]
303. Spassova, M.A.; Soboloff, J.; He, L.P.; Xu, W.; Dziadek, M.A.; Gill, D.L. Stim1 has a plasma membrane role in the activation of store-operated Ca^{2+} channels. *Proc. Natl. Acad. Sci. USA* **2006**, *103*, 4040–4045. [CrossRef] [PubMed]

304. Jardin, I.; López, J.J.; Redondo, P.C.; Salido, G.M.; Rosado, J.A. Store-operated Ca^{2+} entry is sensitive to the extracellular Ca^{2+} concentration through plasma membrane stim1. *Biochim. Biophys. Acta* **2009**, *1793*, 1614–1622. [CrossRef] [PubMed]
305. Fasciani, I.; Temperán, A.; Pérez-Atencio, L.F.; Escudero, A.; Martínez-Montero, P.; Molano, J.; Gómez-Hernández, J.M.; Paino, C.L.; González-Nieto, D.; Barrio, L.C. Regulation of connexin hemichannel activity by membrane potential and the extracellular calcium in health and disease. *Neuropharmacology* **2013**, *75*, 479–490. [CrossRef] [PubMed]
306. Ebihara, L.; Liu, X.; Pal, J.D. Effect of external magnesium and calcium on human connexin46 hemichannels. *Biophys. J.* **2003**, *84*, 277–286. [CrossRef]
307. Quist, A.P.; Rhee, S.K.; Lin, H.; Lal, R. Physiological role of Gap-junctional hemichannels. Extracellular calcium-dependent isosmotic volume regulation. *J. Cell Biol.* **2000**, *148*, 1063–1074. [CrossRef] [PubMed]
308. Gomez-Hernandez, J.M.; de Miguel, M.; Larrosa, B.; Gonzalez, D.; Barrio, L.C. Molecular basis of calcium regulation in connexin-32 hemichannels. *Proc. Natl. Acad. Sci. USA* **2003**, *100*, 16030–16035. [CrossRef] [PubMed]
309. Thimm, J.; Mechler, A.; Lin, H.; Rhee, S.; Lal, R. Calcium-dependent open/closed conformations and interfacial energy maps of reconstituted hemichannels. *J. Biol. Chem.* **2005**, *280*, 10646–10654. [CrossRef] [PubMed]
310. Ma, Z.; Tanis, J.E.; Taruno, A.; Foskett, J.K. Calcium homeostasis modulator (CALHM) ion channels. *Pflugers Arch.* **2016**, *468*, 395–403. [CrossRef] [PubMed]
311. Babini, E.; Paukert, M.; Geisler, H.S.; Grunder, S. Alternative splicing and interaction with di- and polyvalent cations control the dynamic range of acid-sensing ion channel 1 (ASIC1). *J. Biol. Chem.* **2002**, *277*, 41597–41603. [CrossRef] [PubMed]
312. Immke, D.C.; McCleskey, E.W. Protons open acid-sensing ion channels by catalyzing relief of Ca^{2+} blockade. *Neuron* **2003**, *37*, 75–84. [CrossRef]
313. Paukert, M.; Babini, E.; Pusch, M.; Gründer, S. Identification of the Ca^{2+} blocking site of acid-sensing ion channel (ASIC) 1: Implications for channel gating. *J. Gen. Physiol.* **2004**, *124*, 383–394. [CrossRef] [PubMed]
314. Jiang, Q.; Inoue, K.; Wu, X.; Papasian, C.J.; Wang, J.Q.; Xiong, Z.G.; Chu, X.P. Cysteine 149 in the extracellular finger domain of acid-sensing ion channel 1b subunit is critical for zinc-mediated inhibition. *Neuroscience* **2011**, *193*, 89–99. [CrossRef] [PubMed]
315. Johnson, J.P.; Mullins, F.M.; Bennett, P.B. Human ether-à-go-go-related gene k+ channel gating probed with extracellular Ca^{2+}. Evidence for two distinct voltage sensors. *J. Gen. Physiol.* **1999**, *113*, 565–580. [CrossRef] [PubMed]
316. Johnson, J.P.; Balser, J.R.; Bennett, P.B. A novel extracellular calcium sensing mechanism in voltage-gated potassium ion channels. *J. Neurosci.* **2001**, *21*, 4143–4153. [PubMed]
317. Sanguinetti, M.C.; Jiang, C.; Curran, M.E.; Keating, M.T. A mechanistic link between an inherited and an acquired cardiac arrhythmia: Herg encodes the ikr potassium channel. *Cell* **1995**, *81*, 299–307. [CrossRef]
318. Xiong, Z.G.; Chu, X.P.; MacDonald, J.F. Effect of lamotrigine on the Ca^{2+}-sensing cation current in cultured hippocampal neurons. *J. Neurophysiol.* **2001**, *86*, 2520–2526. [CrossRef] [PubMed]
319. Xiong, Z.; Lu, W.; MacDonald, J.F. Extracellular calcium sensed by a novel cation channel in hippocampal neurons. *Proc. Natl. Acad. Sci. USA* **1997**, *94*, 7012–7017. [CrossRef] [PubMed]
320. Burgo, A.; Carmignoto, G.; Pizzo, P.; Pozzan, T.; Fasolato, C. Paradoxical Ca^{2+} rises induced by low external Ca^{2+} in rat hippocampal neurones. *J. Physiol.* **2003**, *549*, 537–552. [CrossRef] [PubMed]
321. Wei, W.L.; Sun, H.S.; Olah, M.E.; Sun, X.; Czerwinska, E.; Czerwinski, W.; Mori, Y.; Orser, B.A.; Xiong, Z.G.; Jackson, M.F.; et al. Trpm7 channels in hippocampal neurons detect levels of extracellular divalent cations. *Proc. Natl. Acad. Sci. USA* **2007**, *104*, 16323–16328. [CrossRef] [PubMed]

8

T Cell Calcium Signaling Regulation by the Co-Receptor CD5

Claudia M. Tellez Freitas, Deborah K. Johnson and K. Scott Weber *

Department of Microbiology and Molecular Biology, Brigham Young University, Provo, UT 84604, USA; claudiamsmicrobiology@gmail.com (C.M.T.F.); deborahkj@gmail.com (D.K.J.)

* Correspondence: scott_weber@byu.edu

Abstract: Calcium influx is critical for T cell effector function and fate. T cells are activated when T cell receptors (TCRs) engage peptides presented by antigen-presenting cells (APC), causing an increase of intracellular calcium (Ca^{2+}) concentration. Co-receptors stabilize interactions between the TCR and its ligand, the peptide-major histocompatibility complex (pMHC), and enhance Ca^{2+} signaling and T cell activation. Conversely, some co-receptors can dampen Ca^{2+} signaling and inhibit T cell activation. Immune checkpoint therapies block inhibitory co-receptors, such as cytotoxic T-lymphocyte associated antigen 4 (CTLA-4) and programmed death 1 (PD-1), to increase T cell Ca^{2+} signaling and promote T cell survival. Similar to CTLA-4 and PD-1, the co-receptor CD5 has been known to act as a negative regulator of T cell activation and to alter Ca^{2+} signaling and T cell function. Though much is known about the role of CD5 in B cells, recent research has expanded our understanding of CD5 function in T cells. Here we review these recent findings and discuss how our improved understanding of CD5 Ca^{2+} signaling regulation could be useful for basic and clinical research.

Keywords: calcium signaling; T cell receptor (TCR); co-receptors; CD-5; PD-1; CTL-4

---

1. Introduction

T cells are a critical component of the adaptive immune system. T cell responses are influenced by signals that modulate the effects of the T cell receptor (TCR) and peptide-major histocompatibility complex (pMHC) interaction and initiate the transcription of genes involved in cytokine production, proliferation, and differentiation [1–3]. T cell activation requires multiple signals. First, the TCR engages the pMHC leading to tyrosine phosphorylation of CD3 and initiation of the Ca^{2+}/Calcineurin/Nuclear factor of activated T cells (NFAT) or Protein kinase C-theta (PKCθ)/Nuclear factor-κ-light chain enhancer of activated B cells (NF-κB) or Mitogen-activated protein kinase (MAP kinase)/AP-1 pathways [4–6]. Second, cell surface costimulatory molecules, such as co-receptor CD28, amplify TCR-pMHC complex signals and promote stronger intracellular interactions to prevent T cell anergy [7,8]. Finally, cytokines such as interleukin-12 (IL-12), interferon α (INFα), and interleukin-1 (IL-1) promote T cell proliferation, differentiation, and effector functions [6].

Co-receptors such as CD4 and CD8 interact with MHC molecules and additional co-receptors interact with surface ligands present on antigen-presenting cells (APCs) to regulate T cell homeostasis, survival, and effector functions with stimulatory or inhibitory signals [9]. Altering co-receptor levels, balance, or function dramatically affects immune responses and their dysfunction is implicated in autoimmune diseases [10]. Stimulatory co-receptors such as CD28, inducible T cell co-stimulator (ICOS), Tumor necrosis factor receptor superfamily member 9 (TNFRSF9 or 4-1BB), member of the TNR-superfamily receptor (CD134 or OX40), glucocorticoid-induced tumor necrosis factor (TNF) receptor (GITR), CD137, and CD77 promote T cell activation and protective responses [11]. Co-receptor signaling is initiated by the phosphorylation of tyrosine residues located in immunoreceptor tyrosine-based activation motifs (ITAMs) or immunoreceptor tyrosine-based inhibitory motifs (ITIMs) [7,12]. The phosphorylated tyrosines serve as docking sites for spleen tyrosine kinase (Syk) family members such as zeta-chain-associated protein kinase 10 (ZAP-70) and Syk which activate the phospholipase C γ (PLCγ), RAS, and extracellular signal-regulated kinase (ERK) pathways in addition to mobilizing intracellular Ca^{2+} stores [13].

One of the best described T cell co-receptors, CD28, is a stimulatory T cell surface receptor from the Ig superfamily with a single Ig variable-like domain which binds to B7-1 (CD80) and B7-2 (CD86) [2]. Ligand binding phosphorylates CD28 cytoplasmic domain tyrosine motifs such as YMNM and PYAP and initiates binding and activation of phosphatidylinositide 3 kinase (PI3K) which interacts with protein kinase B (Akt) and promotes T cell proliferation and survival [1]. CD28 also activates the NFAT pathway and mobilizes intracellular Ca^{2+} stores through association with growth factor receptor-bound protein 2 (GRB2) and the production of phosphatidylinositol 4,5-bisphosphate (PIP2), the substrate of PLCγ1, respectively [2,14]. Blocking stimulatory co-receptors suppresses T cell effector function. For example, blocking stimulatory CD28 with anti-CD28 antibodies promotes regulatory T cell function and represses activation of auto- and allo-reactive T effector cells after organ transplantation [8,15].

T cells also have inhibitory co-receptors which regulate T cell responses [8]. The best characterized are immunoglobulin (Ig) superfamily members cytotoxic T-lymphocyte-associated protein 4 (CTLA-4) and programmed cell death protein 1 (PD-1) [8,16]. CTLA-4 binds CD80 and CD86 with greater avidity than CD28, and its inhibitory role refines early phase activation signals for proliferation and cytokine production [16–19]. PD-1, another CD28/B7 family member, regulates late phase effector and memory response [20]. Inhibitory co-receptors such as CTLA-4 and PD-1, known as "immune checkpoints", block the interaction between CD28 and its ligands altering downstream secondary T cell activation signals [19]. Therefore, blocking CTLA-4 or PD-1 promotes effector T cell function in immunosuppressive environments [19,21].

There are also a number of co-receptors that have differential modulatory properties. For example, CD5, a lymphocyte glycoprotein expressed on thymocytes and all mature T cells, has contradictory roles at different time points. CD5 expression is set during thymocyte development and decreases the perceived strength of TCR-pMHC signaling in naïve T cells by clustering at the TCR-pMHC complex and reducing TCR downstream signals such as the Ca^{2+} response when its cytoplasmic pseudo-ITAM domain is phosphorylated [22–25]. The CD5 cytoplasmic domain has four tyrosine residues (Y378, Y429, Y411, and Y463), and residues Y429 and Y441 are found in a YSQP-(x8)-YPAL pseudo ITAM motif while other tyrosine residues make up a pseudo-ITIM domain [23]. Phosphorylated tyrosines recruit several effector molecules and may sequester activation kinases away from the TCR complex, effectively reducing activation signaling strength [23]. Recruited proteins include Src homology-2 protein phosphatase-1 (SHP-1), Ras GTPase protein (rasGAP), CBL, casein kinase II (CK2), zeta-chain-associated protein kinase 70 (ZAP70), and PI3K which are involved in regulating both positive and negative TCR-induced responses [26–28]. For example, ZAP-70 phosphorylates other substrates and eventually recruits effector molecules such as PLC gamma and promotes Ca^{2+} signaling and Ras activation which stimulates the ERK pathway and leads to cellular activation [13,29]. Conversely, SHP1 inhibits Ca^{2+} signaling and PKC activation via decreased tyrosine phosphorylation of PLCγ [13,26,30,31]. Further, Y463 serves as a docking site for c-Cb1, a ubiquitin ligase, which is phosphorylated upon CD3–CD5 ligation and leads to increased ubiquitylation and lysosomal/proteasomal degradation of TCR downstream signaling effectors and CD5 itself [32]. Thus, CD5 has a mix of downstream effects that both promote and inhibit T cell activation. Curiously, recent work suggests that in contrast to its initial inhibitory nature, CD5 also co-stimulates resting and mature T cells by augmenting CD3-mediated signaling [25,33–35].

Ca^{2+} is an important second messenger in many cells types, including lymphocytes, and plays a key role in shaping immune responses. In naïve T cells, intracellular Ca^{2+} is maintained at low levels, but when TCR-pMHC complexes are formed, inositol triphosphate (IP3) initiates Ca^{2+} release from intracellular stores of the endoplasmic reticulum (ER) which opens the Ca^{2+} release-activated Ca^{2+} channels (CRAC) and initiates influx of extracellular Ca^{2+} through store-operated Ca^{2+} entry (SOCE) [36–41]. The resulting elevation of intracellular Ca^{2+} levels activates transcription factors involved in T cell proliferation, differentiation, and cytokine production (e.g., nuclear factor of activated cells (NFAT)) [36,37]. Thus, impaired Ca^{2+} mobilization affects T cell development, activation, differentiation, and function [42,43]. Examples of diseases with impaired Ca^{2+} signaling in T cells include systemic lupus erythematosus, type 1 diabetes mellitus, and others [44,45].

In this review, we will focus on CD5 co-receptor signaling and its functional effects on T cell activation. First, we will discuss how the inhibitory co-receptors CTLA-4 and PD-1 modulate T cell function. Then we will compare CTLA-4 and PD-1 function to CD5 function, examine recent findings that expand our understanding of the role of CD5, and assess how these findings apply to T cell Ca^{2+} signaling. Finally, we will consider CD5 Ca^{2+} signaling regulation in T cells and its potential physiological impact on immunometabolism, cell differentiation, homeostasis, and behavior.

2. Roles of Negative Regulatory T Cell Co-Receptors

2.1. *Cytotoxic T-Lymphocyte Antigen-4 (CTLA-4)*

Cytotoxic T-lymphocyte antigen-4 (CTLA-4, CD152) inhibits early stages of T cell activation by recruiting inhibitory proteins such as SHP-2 and type II serine/threonine phosphatase PP2A that interfere with T cell synapse signaling [21,46–48]. CTLA-4 binds B7, a protein on activated APCs, with higher affinity than the stimulatory co-receptor CD28; the resulting balance between inhibitory and stimulatory signals controls T cell activation or anergy [19,49]. In naïve T cells, CTLA-4 is located in intracellular vesicles which localize at TCR binding sites following antigen recognition and intracellular Ca^{2+} mobilization [19,50]. Like CD28, CTLA-4 aggregates to the central supramolecular activation complex (cSMAC) where it then extrinsically controls activation by decreasing immunological synapse contact time [51–53]. This suppresses proactivation signals by activating ligands (B7-1 and B7-2) and induces the enzyme Inoleamine 2,3-dioxygenase (IDO) which impairs Ca^{2+} mobilization and suppresses T cell activation, ultimately altering IL-2 production and other effector functions in T cells [51,54,55]. CTLA-4 also stimulates production of regulatory cytokines, such as transforming growth factor beta (TGF-β), which inhibit APC presentation and T cell effector function [47,52,53]. Compared to effector T cells (T_{eff}), CTLA-4 is highly expressed in regulatory T cells (T_{reg}) and plays a role in maintaining T_{reg} homeostasis, proliferation, and immune responses [16,56,57]. Total or partial CTLA-4 deficiency inhibits T_{reg}'s ability to control cytokine production and can cause immune dysregulation [58–61]. Thus, CTLA-4 has an important role in the T_{reg} suppressive response [60]. Additionally, CTLA-4 mutations are associated with autoimmune diseases as thoroughly reviewed by Kristiansen et al. [62].

The loss of CTLA-4 results in removal of CTLA-4 competition with CD28 for B7-1 and B7-2 and is implicated in autoimmunity and cancer [15,63]. Because CTLA-4 inhibits TCR signaling, CTLA-4 deficiency leads to T cell overactivation as measured by increased CD3ζ phosphorylation and Ca^{2+} mobilization [64]. Thus, modulating CTLA-4 signaling is an attractive target for immunotherapies that seek to boost or impair early TCR signaling for cancer and autoinflammatory diseases [65,66]. For example, Ipilimunab, an IgG1 antibody-based melanoma treatment, is a T cell potentiator that blocks CTLA-4 to stimulate T cell proliferation and stem malignant disease progression by delaying tumor progression and has been shown to significantly increase life expectancy [19,67,68]. Additionally, Tremelimumab, a noncomplement fixing IgG2 antibody, has been tested alone or in combination with other antibodies such as Durvalumab (a PD-1 inhibitor) and improves antitumor activity in patients with non-small cell lung cancer (NSCLC), melanoma, colon cancer, gastric cancer, and mesothelioma treatment [69–74].

2.2. *Programmed Death 1 (PD-1)*

Programmed cell death protein-1 (PD-1, CD279) is a 288-amino acid (50–55 KDa) type I transmembrane protein and a member of the B7/CD28 immunoglobulin superfamily expressed on activated T cells, B cells, and myeloid cells [19,75,76]. PD-1 has two known ligands, PD-L1 and PD-L2, which inhibit T cell activation signals [77]. Like CTLA-4, PD-1 also inhibits T cell proliferation and cytokine production (INF-γ, TNF and IL-2) but is expressed at a later phase of T cell activation [19]. PD-1 has an extracellular single immunoglobulin (Ig) superfamily domain and a cytoplasmic domain containing an ITIM and an immunoreceptor tyrosine-based switch motif (ITSM) subunit critical for PD-1 inhibitory function [78]. Upon T cell activation, PD-1 is upregulated and initiates ITIM and ITSM tyrosine interaction with SHP-2 which mediates TCR signaling inhibition by decreasing ERK phosphorylation and intracellular Ca^{2+} mobilization [79,80]. PD-1 can block the activation signaling pathways PI3K-Akt and Ras-Mek-ERK, which inhibit or regulate T cell activation [79,81]. Thus, engagement of PD-1 by its ligand affects intracellular Ca^{2+} mobilization, IL-2 and TNF-α production, supporting PD-1's inhibitory role in TCR strength-mediated signals [82].

PD-1 signaling also affects regulatory T cell (T_{reg}) homeostasis, expansion, and function [83]. T_{reg} activation and proliferation are impacted by PD-1 expression which enhances their development and function while inhibiting T effector cells [75,84]. PD-1, PD-L, and T_{regs} help terminate immune responses [85]. Thus, PD-1 deficiency results not only in increased T cell activation, but in the breakdown of tolerance and the development of autoimmunity in diseases such as multiple sclerosis and systemic lupus erythematosus [85–89]. PD-1 and its ligands protect tissues from autoimmune attacks by regulating T cell activation and inducing and maintaining peripheral tolerance [90,91]. Studies done in PD-1-deficient mice observed the development of lupus-like glomerulonephritis and arthritis, cardiomyopathy, autoimmune hydronephrosis, and Type I diabetes, among other ailments [92–94]. PD-1 protects against autoimmunity and promotes T_{reg} function. [85]. Enhancing T_{reg} response with a PD-L1 agonist shows therapeutic potential for asthma and other autoimmune disorders [85,95]. Because PD-1 specifically modulates lymphocyte function, effective FDA-approved monoclonal antibodies targeting PD-1 are clinically available (i.e., Pembrolizumab and Nivolumab) to treat advanced malignancies [20]. Not only does blocking PD-1 decrease immunotolerance of tumor cells, it also increases cytotoxic T lymphocyte antitumor activity [20].

3. CD5: A Contradictory Co-Receptor

3.1. *Overview of CD5 Signaling and Ca^{2+} Mobilization in T Cells*

CD5, known as Ly-1 antigen in mice or as Leu-1 in humans, is a type I transmembrane glycoprotein (67 kDa) expressed on the surface of thymocytes, mature T cells, and a subset of B cells (B-1a) [96,97]. Although CD5 was discovered over 30 years ago, it was only in the last decade that CD5 gained attention as a key T cell activation regulator [98,99]. CD5 expression is set in the thymus during positive selection and correlates with how tightly the thymocyte TCR binds to self-peptide-MHC (self-pMHC); greater TCR affinity for self-peptide leads to increased CD5 expression in double positive (DP) thymocytes [100]. In other words, DP thymocytes that receive strong activation signals through their TCR express more CD5 than those DP thymocytes that receive weak TCR signals [100]. CD5 knockout mice ($CD5^{-/-}$) have a defective negative and positive selection process, and therefore their thymocytes are hyper-responsive to TCR stimulation with increased Ca^{2+} mobilization, proliferation, and cytokine production [23,98]. On the other hand, because of the increased TCR avidity for self-pMHC, mature T cells with high CD5 expression ($CD5^{hi}$) (peripheral or postpositive selection T cells) respond to foreign peptide with increased survival and activation compared to mature T cells with low CD5 expression ($CD5^{lo}$) [34,101]. Therefore, CD5 is a negative regulator of TCR signaling in the thymus and modulates mature T cell response in the periphery [23,34,100,102].

While CTLA-4 and PD-1 belong to the immunoglobulin (Ig) family, CD5 belongs to group B of the scavenger receptor cysteine-rich (SRCR) superfamily and contains three extracellular SRCR domains [30,96,103]. The cytoplasmic tail of CD5 contains several tyrosine residues which mediate the negative regulatory activity independent of extracellular engagement [100,104,105]. As CD5 physically associates with TCRζ/CD3 complex upon TCR and pMHC interaction, the tyrosine residues in both

TCRζ and CD5 are phosphorylated by tyrosine kinases associated with the complex [30,106–110]. This interaction is so intrinsic to T cell signaling that CD5 expression levels are proportional to the degree of TCRζ phosphorylation, IL-2 production capacity, and ERK phosphorylation which are critical for CD3-mediated signaling [33,111]. It is unknown whether posttranslational modifications, such as conserved domain 1 and domain 2 glycosylations, impact CD5 signaling [112,113]. CD5 is present in membrane lipids rafts of mature T cells where, upon activation, it helps augment TCR signaling, increases Ca^{2+} mobilization, and upregulates ZAP-70/LAT (linker for activation of T cells) activation [114–116]. This suggests that CD5 is not only a negative regulator in thymocytes, but also appears to positively influence T cell immune response to foreign antigens [117,118]. See Figure 1.

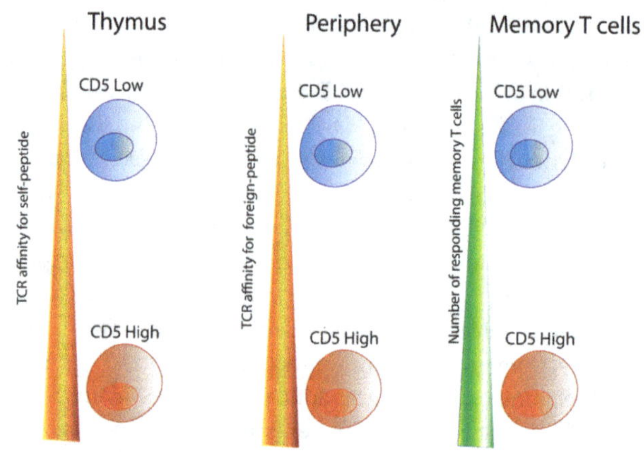

Figure 1. Effects of CD5 on different stages of T cell development. CD5 expression on thymocytes is directly proportional to the signaling intensity of the TCR:self-pMHC interaction. In the periphery, T cells with higher CD5 levels ($CD5^{hi}$) are better responders to foreign-peptide. Long-lived memory cells populations are enriched for $CD5^{hi}$ T cells [34,102,119].

CD5 has three known ligands: CD72, a glycoprotein expressed by B cells, CD5 ligand or CD5L, an activation antigen expressed on splenocytes, and CD5 itself [120–122]. Crosslinking CD5L to CD5 increases intracellular Ca^{2+} concentrations [30,120,121,123,124]. Early studies with anti-CD5 monoclonal antibodies also demonstrated enhanced Ca^{2+} mobilization and proliferation, suggesting that CD5 co-stimulates and increases the T cell activation signal [125,126]. Following TCR:pMHC interaction, CD5 cytoplasmic ITAM and ITIM like-domains are phosphorylated by p56lck and bound by Src homology 2 (SH2) domain-containing protein tyrosine phosphatase (SHP-1) [108,127,128]. However, while SHP-1 affects Ca^{2+} mobilization and is a purported down-regulator of thymocyte activation, recent findings suggest that SHP-1 is not necessary for CD5 signaling as T cells deficient in SHP-1 have normal CD5 expression and continue to signal normally [26,129]. Thus, while CD5 is not a SHP-1 substrate and SHP-1 is likely unnecessary for CD5 signaling, CD5 signaling results in increased Ca^{2+} mobilization. It has yet to be resolved how CD5 can act as an inhibiting co-receptor in the thymus and as an activating co-receptor in the periphery.

3.2. CD5 as a Ca^{2+} Signaling Modulator

As previously mentioned, CD5 expression levels are set in the thymus during T cell development and are maintained on peripheral lymphocytes [117]. CD5 expression in T cells plays an important role during development and primes naïve T cells for responsiveness in the periphery [35,111,130]. $CD5^{hi}$ T cells have the highest affinity for self-peptides and respond with increased cytokine production and proliferation to infection [101,131,132].

Our laboratory works with two TCR transgenic mouse lines with different levels of CD5 expression: LLO56 ($CD5^{hi}$) and LLO118 ($CD5^{lo}$) [111,117,130]. While LLO56 ($CD5^{hi}$) and LLO118

(CD5^{lo}) have similar affinity for the same immunodominant epitope (listeriolysin O amino acids 190–205 or LLO$_{190-205}$) from *Listeria monocytogenes*, on day 7 of primary response, LLO118 (CD5^{lo}) has approximately three times the number of responding cells compared to LLO56 (CD5^{hi}), and conversely, on day 4 during secondary infection, LLO56 (CD5^{hi}) has approximately fifteen times more cells than LLO118 (CD5^{lo}) [130]. This difference is not due to differential proliferative capacity, rather LLO56 (CD5^{hi}) has higher levels of apoptosis during the primary response [130]. Thus, LLO56 CD5^{hi} and LLO118 CD5^{lo}'s capacity to respond to infection appears to be regulated by their CD5 expression levels [117]. LLO56 (CD5^{hi}) thymocytes have greater affinity for self-peptide, which primes them to be highly apoptotic [130].

Recently we reported that in response to foreign peptide, LLO56 (CD5^{hi}) naïve T cells have higher intracellular Ca^{2+} mobilization than LLO118 (CD5^{lo}), which correlates with increased rate of apoptosis of LLO56 (CD5^{hi}), as Ca^{2+} overloaded mitochondria release cytochrome c which activates caspase and nuclease enzymes, thus initiating the apoptotic pathways [35,133,134]. LLO56 (CD5^{hi}) naïve T cell increased Ca^{2+} mobilization also provides additional support to the idea that CD5^{hi} T cells have an enhanced response to foreign peptide [35,134]. This supports previous research that found that upon T cell activation, increased CD5 expression is correlated with greater basal TCRζ phosphorylation, increased ERK phosphorylation, and more IL-2 production [101,111].

Thus, unlike CTLA-4 and PD-1 which are expressed only on activated T cells in the periphery during early and late phases of immune response, respectively, CD5 is set during T cell development, and influences T cells both during thymic development and during postthymic immune responses [19, 101,111] (see Figure 2). CD5 not only has an important inhibitory role in the thymus, but also appears to positively influence the T cell population response; for example, more CD5^{hi} T cells populate the memory T cell repertoire because CD5^{hi} naïve T cells have a stronger primary response [34,135]. CD5 finetunes the sensitivity of TCR signaling to pMHC, altering intracellular Ca^{2+} mobilization and NFAT transcription, key players in T cell effector function [19,64,126]. As Ca^{2+} signaling plays a key role in T cell activation and function, controlling Ca^{2+} mobilization in T cells through CD5 expression could influence diverse areas of clinical research including metabolism, cancer treatments, and even cognitive behavior.

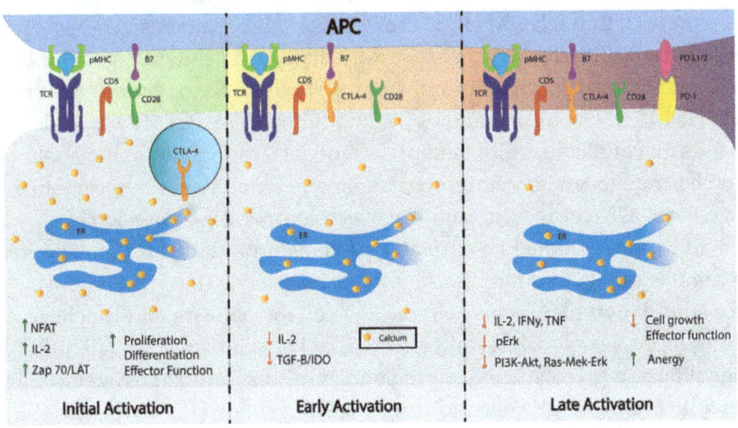

Figure 2. Inhibiting co-receptors modulate T cell activation by increasing (green arrows) or decreasing activity (red arrows). CD5 is present in naïve T cells and localizes to the TCR:pMHC complex during activation. Initial activation cascades signal for the release of CTLA-4 from vesicles to the cell surface while the transcription factor NFAT transcribes PD-1. CTLA-4 provides inhibitory signals during early activation while PD-1 is expressed later and inhibits later stages of T cell activation. The initial Ca^{2+} mobilization is decreased by CTLA-4 and PD-1 downstream signals. A more detailed illustration of the calcium signaling pathway (i.e., IP3, STIM 1/2, CRAC channel, calmodulin, etc.) is outlined in Figure 3.

4. Physiological Impact of CD5 Expression in T Cells

4.1. Metabolism

Naive T cells are in a quiescent state and rely on oxidative phosphorylation (OXPHOS) to generate ATP for survival [136,137]. Upon TCR-pMHC interaction, T cells undergo metabolic reprograming to meet energetic demands by switching from OXPHOS to glycolysis [138]. Glycolysis is a rapid source of ATP and regulates posttranscriptional production of INF-γ, a critical effector cytokine [139]. Following the immune response, most effector T cells undergo apoptosis while a subset become quiescent memory T cells. Memory T cells have lower energetic requirements and rely on OXPHOS and Fatty Acid Oxidation (FAO) to enhance mitochondrial capacity for maintenance and survival [140].

Ca^{2+} signaling is a key second messenger in T cell activation and Ca^{2+} ions also modulate T cell metabolism through CRAC channel activity and NFAT activation [3,141]. During TCR-pMHC binding Ca^{2+} is released from the endoplasmic reticulum (ER) where it is absorbed by the mitochondria and initiates an influx of extracellular Ca^{2+} [3]. First, the rise of cytoplasmic Ca^{2+} activates stromal interaction molecule 1 (STIM1) located on the ER membrane to interact with the CRAC channel located on the cell membrane [142]. The release of the ER store and resulting extracellular Ca^{2+} influx increases the intracellular Ca^{2+} concentration and promotes AMPK (adenosine monophosphates activated protein kinase) expression and CaMKK (calmodulin-dependent protein kinase kinase) activity [3,142,143]. AMPK senses cellular energy levels through the ratio of AMP to ATP and generates ATP by inhibiting ATP-dependent pathways and stimulating catabolic pathways [144]. This indirectly controls T cell fate as AMPK indirectly inhibits mTOR (mammalian target of rapamycin complex) [145]. Because mTOR coordinates the metabolic cues that control T cell homeostasis, it plays a critical role in T cell fate [146]. T cells that are TSC1 (Tuberous sclerosis complex 1)-deficient show metabolic alterations through increased glucose uptake and glycolytic flux [147].

The rise of cytoplasmic Ca^{2+} also encourages mitochondria to uptake cytoplasmic Ca^{2+} through the mitochondrial Ca^{2+} uniporter (MCU) [148]. This MCU uptake increases Ca^{2+} influx by depleting Ca^{2+} near the ER which further activates the CRAC channels and promotes STIM1 oligomerization [3,149–151]. Ca^{2+} uptake in the mitochondria also enhances the function of the tricarboxylic acid cycle (TAC), which generates more ATP through OXPHOS [152,153]. OXPHOS is maintained by a glycolysis product, phosphoenolpyruvate (PEP), which sustains TCR-mediated Ca^{2+}-NFAT signaling by inhibiting the sarcoendoplasmic reticulum (SR) calcium transport ATPase (SERCA) pump, thus promoting T cell effector function [154,155]. Downregulation of calmodulin kinase, CaMKK2, which controls NFAT signaling, decreases glycolytic flux, glucose uptake, and lactate and citrate metabolic processes [156]. Ca^{2+} may also orchestrate the metabolic reprogramming of naïve T cells by promoting glycolysis and OXPHOS through the SOCE/calcineurin pathway which controls the expression of glucose transporters GLUT1/GLUT3 and transcriptional co-regulator proteins important for the expression of electron transport chain complexes required for mitochondria respiration [141].

Co-receptor stimulation plays a pivotal role in T cell metabolism and function. A decrease in T cell Ca^{2+} signaling represses glycolysis and affects T cell effector function [152]. PD-1 and CTLA-4 depress Ca^{2+} signaling and glycolysis while promoting FAO and antibodies against CTLA-4 and PD-1 increase Ca^{2+} mobilization and glycolysis during T cell activation [157,158]. Like CTLA-4 and PD-1, CD5 modulatory function has the potential to influence T cell metabolism. Analysis of gene families modulated by CD5 in B cells found that CD5 upregulates metabolic-related genes including VEFG, Wnt signaling pathways genes, MAPK cascade genes, I-kB/NF-kB cascade genes, TGF β signaling genes, and adipogenesis process genes [159]. Therefore, proliferation differences correlated with CD5 expression in T cells may be caused by improved metabolic function as $CD5^{lo}$ T cells seem to be more quiescent than $CD5^{hi}$ T cells [160]. Although not much is known about how CD5 alters metabolic function in T cells, signaling strength differences of $CD5^{hi}$ and $CD5^{lo}$ T cell populations correlate with intracellular Ca^{2+} mobilization during activation and influence their immune response [35,111,130]. This implies that different metabolic processes may be initiated which would influence proliferation, memory cell generation, and cytokine production. Figure 3 summarizes how Ca^{2+} may be mobilized in $CD5^{hi}$ and $CD5^{lo}$ naïve T cells and the role Ca^{2+} may play on metabolism.

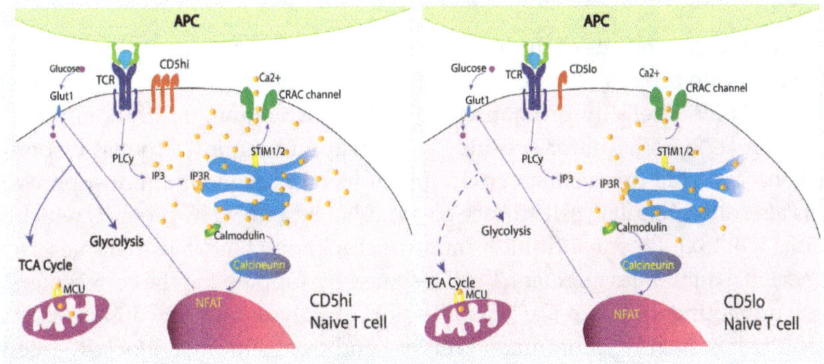

Figure 3. CD5 expression levels in naïve T cells may influence T cell metabolism and function. Differential levels of CD5 result in differences in Ca^{2+} mobilization in naïve T cells. $CD5^{hi}$ naïve T cells have higher Ca^{2+} influx than $CD5^{lo}$ naïve T cells upon TCR:pMHC interaction [35]. Ca^{2+} signaling plays a significant role in T cell activation and influences metabolism and T cell function. Differential Ca^{2+} mobilization and expression of calcineurin and NFAT affect glycolysis and mitochondrial respiration (hypothetical levels of metabolic activation are shown with dashed (low) or solid (high) arrows), suggesting CD5 expression may affect metabolic reprograming during T cell activation [141].

4.2. Neuroimmunology

The field of neuroimmunology examines the interplay between the immune system and the central nervous system (CNS) [161]. The adaptive immune system does influence the CNS as cognition is impaired by the absence of mature T cells [162]. In wild type mice, there is an increase in the number of T cells present in the meninges during the learning process, in stark contrast to mice with T helper 2 cytokine deficiencies (such as IL-4 and IL-13) who have decreased T cell recruitment and impaired learning [163]. Furthermore, regulation of T cell activation and cytokine production critically assists neuronal function and behavior, suggesting that manipulation of T cells could be a potential therapeutic target in treating neuroimmunological diseases [164,165].

T cells go through several microenvironments before reaching the CNS [166]. Many of the signal interactions present in these microenvironments affect T cell function and involve changes in intracellular Ca^{2+} levels [166,167]. In experimental autoimmune encephalitis (EAE), a model for human multiple sclerosis, autoreactive T cells have Ca^{2+} fluctuations throughout their journey to the CNS [166]. Prior to reaching the CNS, T cells interact with splenic stroma cells that do not display the cognate auto-antigen and this interaction produces short-lived low Ca^{2+} mobilization spikes [166]. Following entrance into the CNS, T cells encounter autoantigen-presenting cells and have sustained Ca^{2+} mobilization which results in NFAT translocation and T cell activation [166,168]. EAE mice display reduced social interaction and cognition demonstrating that autoimmune response impairs neuronal function and organismal behavior [169].

Inhibitory T cell co-receptors are implicated in CNS dysregulation and disease. Varicella zoster virus (VZV) infection is characterized by lifelong persistence in neurons. VZV increases the expression of CTLA-4 and PD-1 in infected T cells which reduces IL-2 production and increases T cell anergy [170,171]. PD-1-deficient mice ($Pdcd1^{-/-}$) have increased T cell activation, leading to greater intracellular Ca^{2+} mobilization, and as previously discussed, increased glycolysis [86]. PD-1 deficiency causes elevated concentration of aromatic amino acids in the serum, specifically tryptophan and tyrosine, which decreases their availability in the brain where they are important for the synthesis of neurotransmitters such as dopamine and serotonin; consequently, there is an increase in anxiety-like behavior and fear in Pdcd1$^{-/-}$ mice [86]. Therefore, increased T cell activation caused by PD-1 deficiency can affect brain function and thus, affects cognitive behavior [86].

4.3. Cancer

T cells are critical components of the immune response to cancer. Helper T cells directly activate killer T cells to eradicate tumors and are essential in generating a strong antitumor response alone or in concert with killer T cells by promoting killer T cell activation, infiltration, persistence, and memory formation [172–177]. Tumor-specific T cells may not mount a robust response towards cancerous cells because the tumor microenvironment has numerous immunosuppressive factors; cancerous cells also downregulate cell surface co-stimulatory and MHC proteins which suppresses T cell activation [178–182]. Potent antitumor immune checkpoint blockade therapies using CTLA-4 and PD-1 monoclonal antibodies augment T cell response by suppressing the co-receptors' inhibitory signals, thereby promoting increased Ca^{2+} mobilization, glycolysis, and activation [183,184]. CTLA-4 monoclonal antibodies such as ipilimumab (Yervoy) and tremelimumab block B7-interaction and have been used to treat melanoma [47,185,186]. The monoclonal antibody pembrolizumab is highly selective for PD-1 and prevents PD-1 from engaging PD-L1 and PD-L2, thus enhancing T cell immune response [19,187,188]. Further research will address whether combining anti-CTLA-4 and anti-PD-1 antibodies will improve cancer treatments [19].

As previously mentioned, Ca^{2+} is critical for T cell activation and immune response. Manipulating Ca^{2+} signaling to enhance T cell-directed immune response against cancer is an intriguing notion, yet the means to target the Ca^{2+} response of specific cells without tampering with the metabolic processes of other cells remains elusive [189]. Antitumor activity of tumor-infiltrating lymphocytes (TIL) is inversely related to CD5 expression [99]. CD5 levels in naïve T cells are constantly tuned in the periphery by interactions with self pMHC complexes to maintain homeostasis; therefore, CD5 expression on TILs can be downregulated in response to low affinity for cancer antigens [190–192]. Thus, the majority of TILs are CD5^{lo} which increase their reactivity while CD5^{hi} TILs do not elicit a Ca^{2+} response and become anergic and are unable to eliminate malignant cells [99,192]. While downregulation of CD5 on TILs enhances antitumor T cell activity, CD5^{lo} T cells are also more likely to experience activation-induced cell death (AICD) as CD5 protects T cells from overstimulation [23]. To maximize TIL effectiveness, the inhibitory effects of CD5 could be blocked by neutralizing monoclonal antibodies or soluble CD5-Fc molecules combined with soluble FAS-Fc molecules to reduce the inherent AICD [23,193,194]. Soluble human CD5 (shCD5) may have a similar effect but avoids targeting issues by blocking CD5-mediated interaction via a "decoy receptor" effect. Mice constitutively expressing shCD5 had reduced melanoma and thyoma tumor cell growth and increased numbers of CD4^{+} and CD8^{+} T cells [195]. Wild type mice treated with an injection of recombinant shCD5 also had reduced tumor growth [195]. Finally, CD5-deficient mice engrafted with B16-F10 melanoma cells had slower tumor growth compared to wild type C57BL/6 mice [196]. This evidence suggests that CD5, along with PD-1 and CTLA-4, may be a potential target to specifically modulate T cell Ca^{2+} mobilization in an immunosuppressive tumor setting.

4.4. Microbiome

The gut microbiome, including the bacteria and their products, forms a dynamic beneficial symbiosis with the immune system influencing host genes and cellular response. The gut microbiome shapes and directs immune responses while the immune system dictates the bacterial composition of the gut microbiome [197]. As the gut is the major symbiotic system intersecting the immune system and microbiota, understanding their connection has implications for immune system development and function as the gut microbiome is involved in protecting against pathogens, influencing states of inflammation, and even affecting cancer patient outcomes [198,199].

The gut microbiome primes immune responses [200]. Alteration in the microbial composition can induce changes in T cell function in infectious disease, autoimmunity, and cancer [201]. For example, mice treated with antibiotics which restrict or reduce the microbial environment exhibit impaired immune response because their T cells have altered TCR signaling and compromised intracellular Ca^{2+} mobilization in infectious disease and cystic fibrosis models [202–204]. In contrast, administering oral antibiotics to mice with EAE increases the frequency of $CD5^{+}$ B cell subpopulations in distal lymphoid sites and confers disease protection [205]. In cancer, the microbiome also influences patient response to immune checkpoint inhibitors such as CTLA-4 and PD-1 [206,207]. Mice and melanoma patients immunized or populated with *Bacteriodes fragilis* respond better to treatment with Ipilimumab, a monoclonal antibody against CTLA-4 [198]. Similarly, tumor-specific immunity improved when anti-PD-1/PD-L1 monoclonal antibodies where used in the presence of *Bifidobacterium* [208].

Though little is known about how CD5 influences T cell interaction with the microbiome, some tantalizing details are available. As specific bacterium promotes cancer regression during CTLA-4 and PD-1 checkpoint blockades, a CD5 blockade in conjunction with bacterial selection may also improve immune response. Such studies would lead to novel immunotherapeutic treatments for cancer and autoimmune diseases.

5. Conclusions

CD5, widely known as an inhibitory co-receptor in the thymus, appears to modulate the signaling intensity of peripheral T cells by increasing Ca^{2+} signaling activity and efficacy of $CD5^{hi}$ T cells. CD5 expression levels in the periphery correlates with intracellular Ca^{2+} mobilization, suggesting that CD5 promotes peripheral T cell activation and immune response. As such, CD5 may be a novel checkpoint therapy to regulate T cell activation and metabolism through altering Ca^{2+} mobilization, and could be used to affect neurological behavior, alter microbiome interactions, and treat cancer and autoinflammatory diseases. While this paper focuses on the role of co-receptor CD5 effects on calcium signaling and activation of T cells, CD5 itself may be regulated through posttranslational modifications, such as *N*-glycosylation, which may affect Ca^{2+} mobilization, T cell metabolism, activation, and function. In the future it would be interesting to determine the role of other posttranslational modifications (e.g., *N*-glycosylation, *S*-glutathionylation, lipidation) in CD5 signaling.

Author Contributions: C.M.T.F. is the first author and wrote the manuscript, D.K.J. contributed additional material and editing help, K.S.W. helped with the plan for the manuscript and editing and is the corresponding author.

Funding: This work was supported by a National Institute of Allergy and Infectious Diseases grant (R0102063) to K.S.W. The funder had no role in preparation of the manuscript.

Acknowledgments: We thank Kiara Vaden Whitley, Jeralyn Jones Fransen, Tyler Cox and Josie Tueller for their critical reviews of this manuscript.

Abbreviations

CTLA-4	Cytotoxic T-lymphocyte antigen 4
CD	Cluster of differenciation
PD-1	Programmed cell death protein 1
AMP	Adenosine monophosphate
ATP	Adenosine triphosphate
CaMKK	Calmodulin-dependent protein kinase kinase
AMPK	AMP-activated protein kinase
SOCE	Store-operated calcium channels
CRAC	$Calcium^{+}$-release-activated channel
STIM	Stromal interaction molecule
SERCA	Sarcoendoplasmic reticulum calcium transport ATPase
ER	Endoplasmic reticulum
NFAT	Nuclear factor of activated T cells
INF-γ	Interferon gamma

Abbreviations

TNF	Tumor necrosis factor
IL-2	Interleukin 2
GLUT1	Glucose transporter 1
GLUT3	Glucose transporter 3
TIL	Tumor infiltrating lymphocytes
ERK	Extracellular signal-regulated kinases

References

1. Chen, L.; Flies, D.B. Molecular mechanisms of T cell co-stimulation and co-inhibition. *Nat. Rev. Immunol.* **2013**, *13*, 227–242. [CrossRef] [PubMed]
2. Beyersdorf, N.; Kerkau, T.; Hünig, T. CD28 co-stimulation in T cell homeostasis: A recent perspective. *Immunotargets Ther.* **2015**, *4*, 111–122. [PubMed]
3. Fracchia, K.M.; Pai, C.Y.; Walsh, C.M. Modulation of T cell metabolism and function through calcium signaling. *Front. Immunol.* **2013**, *4*, 324. [CrossRef] [PubMed]
4. Cunningham, A.J.; Lafferty, K.J. Letter: Cellular proliferation can be an unreliable index of immune competence. *J. Immunol.* **1974**, *112*, 436–437. [CrossRef] [PubMed]
5. Nakayama, T.; Yamashita, M. The TCR-mediated signaling pathways that control the direction of helper T cell differentiation. *Semin. Immunol.* **2010**, *22*, 303–309. [CrossRef] [PubMed]
6. Goral, S. The three-signal hypothesis of lymphocyte activation/targets for immunosuppression. *Dial. Transplant.* **2011**, *40*, 14–16. [CrossRef]
7. Pennock, N.D.; White, J.T.; Cross, E.W.; Cheney, E.E.; Tamburini, B.A.; Kedl, R.M. T cell responses: Naïve to memory and everything in between. *Adv. Physiol. Educ.* **2013**, *37*, 273–283. [CrossRef] [PubMed]
8. Sharpe, A.H.; Abbas, A.K. T cell costimulation—Biology, therapeutic potential, and challenges. *N. Engl. J. Med.* **2006**, *355*, 973–975. [CrossRef] [PubMed]
9. Artyomov, M.N.; Lis, M.; Devadas, S.; Davis, M.M.; Chakraborty, A.K. CD4 and CD8 binding to MHC molecules primarily acts to enhance LCK delivery. *Proc. Natl. Acad. Sci. USA* **2010**, *107*, 16916–16921. [CrossRef] [PubMed]
10. Ravetch, J.V.; Lanier, L.L. Immune inhibitory receptors. *Science* **2000**, *290*, 84–89. [CrossRef] [PubMed]
11. Mellman, I.; Coukos, G.; Dranoff, G. Cancer immunotherapy comes of age. *Nature* **2011**, *480*, 480. [CrossRef] [PubMed]
12. Fuertes Marraco, S.A.; Neubert, N.J.; Verdeil, G.; Speiser, D.E. Inhibitory receptors beyond T cell exhaustion. *Front. Immunol.* **2015**, *6*, 310. [CrossRef] [PubMed]
13. Barrow, A.D.; Trowsdale, J. You say ITAM and I say ITIM, let's call the whole thing off: The ambiguity of immunoreceptor signalling. *Eur. J. Immunol.* **2006**, *36*, 1646–1653. [CrossRef] [PubMed]
14. Esensten, J.H.; Helou, Y.A.; Chopra, G.; Weiss, A.; Bluestone, J.A. CD28 costimulation: From mechanism to therapy. *Immunity* **2016**, *44*, 973–988. [CrossRef] [PubMed]
15. Dilek, N.; Poirier, N.; Hulin, P.; Coulon, F.; Mary, C.; Ville, S.; Vie, H.; Clémenceau, B.; Blancho, G.; Vanhove, B. Targeting CD28, CTLA-4 and PD-L1 costimulation differentially controls immune synapses and function of human regulatory and conventional t cells. *PLoS ONE* **2013**, *8*, e83139. [CrossRef] [PubMed]
16. Chambers, C.A.; Sullivan, T.J.; Allison, J.P. Lymphoproliferation in CTLA-4-deficient mice is mediated by costimulation-dependent activation of $CD4^+$ T cells. *Immunity* **1997**, *7*, 885–895. [CrossRef]
17. Lindsten, T.; Lee, K.P.; Harris, E.S.; Petryniak, B.; Craighead, N.; Reynolds, P.J.; Lombard, D.B.; Freeman, G.J.; Nadler, L.M.; Gray, G.S.; et al. Characterization of CTLA-4 structure and expression on human T cells. *J. Immunol.* **1993**, *151*, 3489–3499. [PubMed]
18. Boise, L.H.; Minn, A.J.; Noel, P.J.; June, C.H.; Accavitti, M.A.; Lindsten, T.; Thompson, C.B. CD28 costimulation can promote T cell survival by enhancing the expression of Bcl-XL. *Immunity* **1995**, *3*, 87–98. [CrossRef]
19. Buchbinder, E.I.; Desai, A. CTLA-4 and PD-1 pathways: Similarities, differences, and implications of their inhibition. *Am. J. Clin. Oncol.* **2016**, *39*, 98–106. [CrossRef] [PubMed]
20. Iwai, Y.; Hamanishi, J.; Chamoto, K.; Honjo, T. Cancer immunotherapies targeting the PD-1 signaling pathway. *J. Biomed. Sci.* **2017**, *24*, 26. [CrossRef] [PubMed]

21. Chambers, C.A.; Kuhns, M.S.; Egen, J.G.; Allison, J.P. CTLA-4-mediated inhibition in regulation of T cell responses: Mechanisms and manipulation in tumor immunotherapy. *Annu. Rev. Immunol.* **2001**, *19*, 565–594. [CrossRef] [PubMed]
22. Brossard, C.; Semichon, M.; Trautmann, A.; Bismuth, G. CD5 inhibits signaling at the immunological synapse without impairing its formation. *J. Immunol.* **2003**, *170*, 4623–4629. [CrossRef] [PubMed]
23. Tabbekh, M.; Mokrani-Hammani, M.B.; Bismuth, G.; Mami-Chouaib, F. T cell modulatory properties of CD5 and its role in antitumor immune responses. *Oncoimmunology* **2013**, *2*, e22841. [CrossRef] [PubMed]
24. Mahoney, K.M.; Freeman, G.J.; McDermott, D.F. The next immune-checkpoint inhibitors: PD-1/PD-L1 blockade in melanoma. *Clin. Ther.* **2015**, *37*, 764–782. [CrossRef] [PubMed]
25. De Wit, J.; Souwer, Y.; van Beelen, A.J.; de Groot, R.; Muller, F.J.; Klaasse Bos, H.; Jorritsma, T.; Kapsenberg, M.L.; de Jong, E.C.; van Ham, S.M. CD5 costimulation induces stable Th17 development by promoting IL-23R expression and sustained STAT3 activation. *Blood* **2011**, *118*, 6107–6114. [CrossRef] [PubMed]
26. Perez-Villar, J.J.; Whitney, G.S.; Bowen, M.A.; Hewgill, D.H.; Aruffo, A.A.; Kanner, S.B. CD5 negatively regulates the T cell antigen receptor signal transduction pathway: Involvement of SH2-containing phosphotyrosine phosphatase SHP-1. *Mol. Cell. Biol.* **1999**, *19*, 2903–2912. [CrossRef] [PubMed]
27. Gary-Gouy, H.; Harriague, J.; Dalloul, A.; Donnadieu, E.; Bismuth, G. CD5-negative regulation of B cell receptor signaling pathways originates from tyrosine residue Y429 outside an immunoreceptor tyrosine-based inhibitory motif. *J. Immunol.* **2002**, *168*, 232–239. [CrossRef] [PubMed]
28. Dennehy, K.M.; Broszeit, R.; Garnett, D.; Durrheim, G.A.; Spruyt, L.L.; Beyers, A.D. Thymocyte activation induces the association of phosphatidylinositol 3-kinase and pp120 with CD5. *Eur. J. Immunol.* **1997**, *27*, 679–686. [CrossRef] [PubMed]
29. Samelson, L.E. Signal transduction mediated by the T cell antigen receptor: The role of adapter proteins. *Annu. Rev. Immunol.* **2002**, *20*, 371–394. [CrossRef] [PubMed]
30. Burgess, K.E.; Yamamoto, M.; Prasad, K.V.S.; Rudd, C.E. CD5 acts as a tyrosine kinase substrate within a receptor complex comprising T cell receptor ζ-chain CD3 and protein-tyrosine kinases P56LCK and P59FYN. *Proc. Natl. Acad. Sci. USA* **1992**, *89*, 9311–9315. [CrossRef] [PubMed]
31. Consuegra-Fernandez, M.; Aranda, F.; Simoes, I.; Orta, M.; Sarukhan, A.; Lozano, F. CD5 as a Target for Immune-Based Therapies. *Crit. Rev. Immunol.* **2015**, *35*, 85–115. [CrossRef] [PubMed]
32. Roa, N.S.; Ordonez-Rueda, D.; Chavez-Rios, J.R.; Raman, C.; Garcia-Zepeda, E.A.; Lozano, F.; Soldevila, G. The carboxy-terminal region of CD5 is required for c-CBL mediated TCR signaling downmodulation in thymocytes. *Biochem. Biophys. Res. Commun.* **2013**, *432*, 52–59. [CrossRef] [PubMed]
33. Berney, S.M.; Schaan, T.; Wolf, R.E.; Kimpel, D.L.; van der Heyde, H.; Atkinson, T.P. CD5 (OKT1) augments CD3-mediated intracellular signaling events in human T lymphocytes. *Inflammation* **2001**, *25*, 215–221. [CrossRef] [PubMed]
34. Azzam, H.S.; DeJarnette, J.B.; Huang, K.; Emmons, R.; Park, C.-S.; Sommers, C.L.; El-Khoury, D.; Shores, E.W.; Love, P.E. Fine tuning of TCR signaling by CD5. *J. Immunol.* **2001**, *166*, 5464–5472. [CrossRef] [PubMed]
35. Freitas, C.M.T.; Hamblin, G.J.; Raymond, C.M.; Weber, K.S. Naive helper T cells with high CD5 expression have increased calcium signaling. *PLoS ONE* **2017**, *12*, e0178799. [CrossRef] [PubMed]
36. Feske, S. Calcium signalling in lymphocyte activation and disease. *Nat. Rev. Immunol.* **2007**, *7*, 690–702. [CrossRef] [PubMed]
37. Joseph, N.; Reicher, B.; Barda-Saad, M. The calcium feedback loop and T cell activation: How cytoskeleton networks control intracellular calcium flux. *Biochim. Biophys. Acta Biomembr.* **2014**, *1838*, 557–568. [CrossRef] [PubMed]
38. Vig, M.; Kinet, J.-P. Calcium signaling in immune cells. *Nat. Immunol.* **2009**, *10*, 21–27. [CrossRef] [PubMed]
39. Wolf, I.M.A.; Guse, A.H. Ca^{2+} microdomains in T-lymphocytes. *Front. Oncol.* **2017**, *7*, 73. [CrossRef] [PubMed]
40. Hogan, P.G.; Lewis, R.S.; Rao, A. Molecular basis of calcium signaling in lymphocytes: STIM and ORAI. *Annu. Rev. Immunol.* **2010**, *28*, 491–533. [CrossRef] [PubMed]
41. Oh-hora, M.; Rao, A. Calcium signaling in lymphocytes. *Curr. Opin. Immunol.* **2008**, *20*, 250–258. [CrossRef] [PubMed]
42. Janeway, C.A., Jr. The co-receptor function of CD4. *Semin. Immunol.* **1991**, *3*, 153–160. [PubMed]

43. Moran, A.E.; Hogquist, K.A. T cell receptor affinity in thymic development. *Immunology* **2012**, *135*, 261–267. [CrossRef] [PubMed]
44. Kyttaris, V.C.; Zhang, Z.; Kampagianni, O.; Tsokos, G.C. Calcium signaling in systemic lupus erythematosus T cells: A treatment target. *Arthritis Rheum.* **2011**, *63*, 2058–2066. [CrossRef] [PubMed]
45. Demkow, U.; Winklewski, P.; Ciepiela, O.; Popko, K.; Lipinska, A.; Kucharska, A.; Michalska, B.; Wasik, M. Modulatory effect of insulin on T cell receptor mediated calcium signaling is blunted in long lasting type 1 diabetes mellitus. *Pharmacol. Rep.* **2012**, *64*, 150–156. [CrossRef]
46. Parry, R.V.; Chemnitz, J.M.; Frauwirth, K.A.; Lanfranco, A.R.; Braunstein, I.; Kobayashi, S.V.; Linsley, P.S.; Thompson, C.B.; Riley, J.L. CTLA-4 and PD-1 receptors inhibit T cell activation by distinct mechanisms. *Mol. Cell. Biol.* **2005**, *25*, 9543–9553. [CrossRef] [PubMed]
47. Grosso, J.F.; Jure-Kunkel, M.N. CTLA-4 blockade in tumor models: An overview of preclinical and translational research. *Cancer Immun.* **2013**, *13*, 5. [PubMed]
48. Rudd, C.E.; Taylor, A.; Schneider, H. CD28 and CTLA-4 coreceptor expression and signal transduction. *Immunol. Rev.* **2009**, *229*, 12–26. [CrossRef] [PubMed]
49. Jago, C.B.; Yates, J.; Olsen Saraiva CÂMara, N.; Lechler, R.I.; Lombardi, G. Differential expression of CTLA-4 among T cell subsets. *Clin. Exp. Immunol.* **2004**, *136*, 463–471. [CrossRef] [PubMed]
50. Linsley, P.S.; Bradshaw, J.; Greene, J.; Peach, R.; Bennett, K.L.; Mittler, R.S. Intracellular trafficking of CTLA-4 and focal localization towards sites of TCR engagement. *Immunity* **1996**, *4*, 535–543. [CrossRef]
51. Schneider, H.; Smith, X.; Liu, H.; Bismuth, G.; Rudd, C.E. CTLA-4 disrupts ZAP70 microcluster formation with reduced T cell/APC dwell times and calcium mobilization. *Eur. J. Immunol.* **2008**, *38*, 40–47. [CrossRef] [PubMed]
52. Grohmann, U.; Orabona, C.; Fallarino, F.; Vacca, C.; Calcinaro, F.; Falorni, A.; Candeloro, P.; Belladonna, M.L.; Bianchi, R.; Fioretti, M.C.; et al. CTLA-4-Ig regulates tryptophan catabolism in vivo. *Nat. Immunol.* **2002**, *3*, 1097–1101. [CrossRef] [PubMed]
53. Chen, W.; Jin, W.; Wahl, S.M. Engagement of cytotoxic T lymphocyte-associated antigen 4 (CTLA-4) induces transforming growth factor beta (TGF-β) production by murine $CD4^+$ T cells. *J. Exp. Med.* **1998**, *188*, 1849–1857. [CrossRef] [PubMed]
54. Hryniewicz, A.; Boasso, A.; Edghill-Smith, Y.; Vaccari, M.; Fuchs, D.; Venzon, D.; Nacsa, J.; Betts, M.R.; Tsai, W.-P.; Heraud, J.-M.; et al. CTLA-4 blockade decreases TGF-β, IDO, and viral RNA expression in tissues of SIVmac251-infected macaques. *Blood* **2006**, *108*, 3834–3842. [CrossRef] [PubMed]
55. Iken, K.; Liu, K.; Liu, H.; Bizargity, P.; Wang, L.; Hancock, W.W.; Visner, G.A. Indoleamine 2,3-dioxygenase and metabolites protect murine lung allografts and impair the calcium mobilization of T cells. *Am. J. Respir. Cell Mol. Biol.* **2012**, *47*, 405–416. [CrossRef] [PubMed]
56. Walker, L.S.K.; Sansom, D.M. Confusing signals: Recent progress in CTLA-4 biology. *Trends Immunol.* **2015**, *36*, 63–70. [CrossRef] [PubMed]
57. Cederbom, L.; Hall, H.; Ivars, F. $CD4^+CD25^+$ regulatory T cells down-regulate co-stimulatory molecules on antigen-presenting cells. *Eur. J. Immunol.* **2000**, *30*, 1538–1543. [CrossRef]
58. Burnett, D.L.; Parish, I.A.; Masle-Farquhar, E.; Brink, R.; Goodnow, C.C. Murine LRBA deficiency causes CTLA-4 deficiency in Tregs without progression to immune dysregulation. *Immunol. Cell Biol.* **2017**, *95*, 775–778. [CrossRef] [PubMed]
59. Verma, N.; Burns, S.O.; Walker, L.S.K.; Sansom, D.M. Immune deficiency and autoimmunity in patients with CTLA-4 (CD152) mutations. *Clin. Exp. Immunol.* **2017**, *190*, 1–7. [CrossRef] [PubMed]
60. Wing, K.; Onishi, Y.; Prieto-Martin, P.; Yamaguchi, T.; Miyara, M.; Fehervari, Z.; Nomura, T.; Sakaguchi, S. CTLA-4 control over $Foxp3^+$ regulatory T cell function. *Science* **2008**, *322*, 271–275. [CrossRef] [PubMed]
61. Sojka, D.K.; Hughson, A.; Fowell, D.J. CTLA-4 is Required by $CD4^+CD25^+$ treg to control $CD4^+$ T cell lymphopenia-induced proliferation. *Eur. J. Immunol.* **2009**, *39*, 1544–1551. [CrossRef] [PubMed]
62. Kristiansen, O.P.; Larsen, Z.M.; Pociot, F. CTLA-4 in autoimmune diseases–a general susceptibility gene to autoimmunity? *Genes Immun.* **2000**, *1*, 170–184. [CrossRef] [PubMed]
63. Chikuma, S. CTLA-4, an essential immune-checkpoint for T cell activation. *Curr. Top. Microbiol. Immunol.* **2017**, *410*, 99–126. [PubMed]
64. Tai, X.; Van Laethem, F.; Pobezinsky, L.; Guinter, T.; Sharrow, S.O.; Adams, A.; Granger, L.; Kruhlak, M.; Lindsten, T.; Thompson, C.B.; et al. Basis of CTLA-4 function in regulatory and conventional $CD4^+$ T cells. *Blood* **2012**, *119*, 5155–5163. [CrossRef] [PubMed]

65. Lo, B.; Abdel-Motal, U.M. Lessons from CTLA-4 deficiency and checkpoint inhibition. *Curr. Opin. Immunol.* **2017**, *49*, 14–19. [CrossRef] [PubMed]
66. Avogadri, F.; Yuan, J.; Yang, A.; Schaer, D.; Wolchok, J.D. Modulation of CTLA-4 and GITR for cancer immunotherapy. *Curr. Top. Microbiol. Immunol.* **2011**, *344*, 211–244. [PubMed]
67. Royal, R.E.; Levy, C.; Turner, K.; Mathur, A.; Hughes, M.; Kammula, U.S.; Sherry, R.M.; Topalian, S.L.; Yang, J.C.; Lowy, I.; et al. Phase 2 trial of single agent Ipilimumab (anti-CTLA-4) for locally advanced or metastatic pancreatic adenocarcinoma. *J. Immunother.* **2010**, *33*, 828–833. [CrossRef] [PubMed]
68. Le, D.T.; Lutz, E.; Uram, J.N.; Sugar, E.A.; Onners, B.; Solt, S.; Zheng, L.; Diaz, L.A., Jr.; Donehower, R.C.; Jaffee, E.M.; et al. Evaluation of ipilimumab in combination with allogeneic pancreatic tumor cells transfected with a GM-CSF gene in previously treated pancreatic cancer. *J. Immunother.* **2013**, *36*, 382–389. [CrossRef] [PubMed]
69. Chung, K.Y.; Gore, I.; Fong, L.; Venook, A.; Beck, S.B.; Dorazio, P.; Criscitiello, P.J.; Healey, D.I.; Huang, B.; Gomez-Navarro, J.; et al. Phase II study of the anti-cytotoxic T-lymphocyte-associated antigen 4 monoclonal antibody, tremelimumab, in patients with refractory metastatic colorectal cancer. *J. Clin. Oncol.* **2010**, *28*, 3485–3490. [CrossRef] [PubMed]
70. Ribas, A.; Camacho, L.H.; Lopez-Berestein, G.; Pavlov, D.; Bulanhagui, C.A.; Millham, R.; Comin-Anduix, B.; Reuben, J.M.; Seja, E.; Parker, C.A.; et al. Antitumor activity in melanoma and anti-self responses in a phase I trial with the anti-cytotoxic T lymphocyte-associated antigen 4 monoclonal antibody CP-675,206. *J. Clin. Oncol.* **2005**, *23*, 8968–8977. [CrossRef] [PubMed]
71. Calabro, L.; Morra, A.; Fonsatti, E.; Cutaia, O.; Fazio, C.; Annesi, D.; Lenoci, M.; Amato, G.; Danielli, R.; Altomonte, M.; et al. Efficacy and safety of an intensified schedule of tremelimumab for chemotherapy-resistant malignant mesothelioma: An open-label, single-arm, phase 2 study. *Lancet Respir. Med.* **2015**, *3*, 301–309. [CrossRef]
72. Comin-Anduix, B.; Escuin-Ordinas, H.; Ibarrondo, F.J. Tremelimumab: Research and clinical development. *OncoTargets Ther.* **2016**, *9*, 1767–1776.
73. Ribas, A.; Comin-Anduix, B.; Chmielowski, B.; Jalil, J.; de la Rocha, P.; McCannel, T.A.; Ochoa, M.T.; Seja, E.; Villanueva, A.; Oseguera, D.K.; et al. Dendritic cell vaccination combined with CTLA4 blockade in patients with metastatic melanoma. *Clin. Cancer Res.* **2009**, *15*, 6267–6276. [CrossRef] [PubMed]
74. Antonia, S.; Goldberg, S.B.; Balmanoukian, A.; Chaft, J.E.; Sanborn, R.E.; Gupta, A.; Narwal, R.; Steele, K.; Gu, Y.; Karakunnel, J.J.; et al. Safety and antitumour activity of durvalumab plus tremelimumab in non-small cell lung cancer: A multicentre, phase 1b study. *Lancet Oncol.* **2016**, *17*, 299–308. [CrossRef]
75. Dong, Y.; Sun, Q.; Zhang, X. PD-1 and its ligands are important immune checkpoints in cancer. *Oncotarget* **2017**, *8*, 2171–2186. [CrossRef] [PubMed]
76. Shi, L.; Chen, S.; Yang, L.; Li, Y. The role of PD-1 and PD-L1 in T cell immune suppression in patients with hematological malignancies. *J. Hematol. Oncol.* **2013**, *6*, 74. [CrossRef] [PubMed]
77. Keir, M.E.; Butte, M.J.; Freeman, G.J.; Sharpe, A.H. PD-1 and its ligands in tolerance and immunity. *Annu. Rev. Immunol.* **2008**, *26*, 677–704. [CrossRef] [PubMed]
78. Okazaki, T.; Wang, J. PD-1/PD-L pathway and autoimmunity. *Autoimmunity* **2005**, *38*, 353–357. [CrossRef] [PubMed]
79. Boussiotis, V.A. Molecular and biochemical aspects of the PD-1 checkpoint pathway. *N. Engl. J. Med.* **2016**, *375*, 1767–1778. [CrossRef] [PubMed]
80. Wang, S.-F.; Fouquet, S.; Chapon, M.; Salmon, H.; Regnier, F.; Labroquère, K.; Badoual, C.; Damotte, D.; Validire, P.; Maubec, E.; et al. Early T cell signalling is reversibly altered in PD-1^+ T lymphocytes infiltrating human tumors. *PLoS ONE* **2011**, *6*, e17621. [CrossRef] [PubMed]
81. Gorentla, B.K.; Zhong, X.-P. T cell receptor signal transduction in T lymphocytes. *J. Clin. Cell. Immunol.* **2012**, *2012*, 005.
82. Wei, F.; Zhong, S.; Ma, Z.; Kong, H.; Medvec, A.; Ahmed, R.; Freeman, G.J.; Krogsgaard, M.; Riley, J.L. Strength of PD-1 signaling differentially affects T cell effector functions. *Proc. Natl. Acad. Sci. USA* **2013**, *110*, E2480–E2489. [CrossRef] [PubMed]
83. Cochain, C.; Chaudhari, S.M.; Koch, M.; Wiendl, H.; Eckstein, H.-H.; Zernecke, A. Programmed cell death-1 deficiency exacerbates T cell activation and atherogenesis despite expansion of regulatory T cells in atherosclerosis-prone mice. *PLoS ONE* **2014**, *9*, e93280. [CrossRef] [PubMed]

84. Asano, T.; Kishi, Y.; Meguri, Y.; Yoshioka, T.; Iwamoto, M.; Maeda, Y.; Yagita, H.; Tanimoto, M.; Koreth, J.; Ritz, J.; et al. PD-1 signaling has a critical role in maintaining regulatory T cell homeostasis; implication for treg depletion therapy by PD-1 blockade. *Blood* **2015**, *126*, 848.
85. Francisco, L.M.; Sage, P.T.; Sharpe, A.H. The PD-1 pathway in tolerance and autoimmunity. *Immunol. Rev.* **2010**, *236*, 219–242. [CrossRef] [PubMed]
86. Miyajima, M.; Zhang, B.; Sugiura, Y.; Sonomura, K.; Guerrini, M.M.; Tsutsui, Y.; Maruya, M.; Vogelzang, A.; Chamoto, K.; Honda, K.; et al. Metabolic shift induced by systemic activation of T cells in PD-1-deficient mice perturbs brain monoamines and emotional behavior. *Nat. Immunol.* **2017**, *18*, 1342–1352. [CrossRef] [PubMed]
87. Riella, L.V.; Paterson, A.M.; Sharpe, A.H.; Chandraker, A. Role of the PD-1 pathway in the immune response. *Am. J. Transplant.* **2012**, *12*, 2575–2587. [CrossRef] [PubMed]
88. Kroner, A.; Mehling, M.; Hemmer, B.; Rieckmann, P.; Toyka, K.V.; Maurer, M.; Wiendl, H. A PD-1 polymorphism is associated with disease progression in multiple sclerosis. *Ann. Neurol* **2005**, *58*, 50–57. [CrossRef] [PubMed]
89. Pawlak-Adamska, E.; Nowak, O.; Karabon, L.; Pokryszko-Dragan, A.; Partyka, A.; Tomkiewicz, A.; Ptaszkowski, J.; Frydecka, I.; Podemski, R.; Dybko, J.; et al. *PD-1* gene polymorphic variation is linked with first symptom of disease and severity of relapsing-remitting form of MS. *J. Neuroimmunol.* **2017**, *305*, 115–127. [CrossRef] [PubMed]
90. Dai, S.; Jia, R.; Zhang, X.; Fang, Q.; Huang, L. The PD-1/PD-Ls pathway and autoimmune diseases. *Cell. Immunol.* **2014**, *290*, 72–79. [CrossRef] [PubMed]
91. Gianchecchi, E.; Delfino, D.V.; Fierabracci, A. Recent insights into the role of the PD-1/PD-L1 pathway in immunological tolerance and autoimmunity. *Autoimmun. Rev.* **2013**, *12*, 1091–1100. [CrossRef] [PubMed]
92. Wang, J.; Yoshida, T.; Nakaki, F.; Hiai, H.; Okazaki, T.; Honjo, T. Establishment of NOD-Pdcd1$^{-/-}$ mice as an efficient animal model of type I diabetes. *Proc. Natl. Acad. Sci. USA* **2005**, *102*, 11823–11828. [CrossRef] [PubMed]
93. Okazaki, T.; Otaka, Y.; Wang, J.; Hiai, H.; Takai, T.; Ravetch, J.V.; Honjo, T. Hydronephrosis associated with antiurothelial and antinuclear autoantibodies in BALB/c-Fcgr2b$^{-/-}$Pdcd1$^{-/-}$ mice. *J. Exp. Med.* **2005**, *202*, 1643–1648. [CrossRef] [PubMed]
94. Nishimura, H.; Okazaki, T.; Tanaka, Y.; Nakatani, K.; Hara, M.; Matsumori, A.; Sasayama, S.; Mizoguchi, A.; Hiai, H.; Minato, N.; et al. Autoimmune dilated cardiomyopathy in PD-1 receptor-deficient mice. *Science* **2001**, *291*, 319–322. [CrossRef] [PubMed]
95. Xiao, Y.; Yu, S.; Zhu, B.; Bedoret, D.; Bu, X.; Francisco, L.M.; Hua, P.; Duke-Cohan, J.S.; Umetsu, D.T.; Sharpe, A.H.; et al. RGMb is a novel binding partner for PD-L2 and its engagement with PD-L2 promotes respiratory tolerance. *J. Exp. Med.* **2014**, *211*, 943–959. [CrossRef] [PubMed]
96. Masuda, K.; Kishimoto, T. CD5: A new partner for IL-6. *Immunity* **2016**, *44*, 720–722. [CrossRef] [PubMed]
97. Huang, H.J.; Jones, N.H.; Strominger, J.L.; Herzenberg, L.A. Molecular cloning of Ly-1, a membrane glycoprotein of mouse T lymphocytes and a subset of B cells: Molecular homology to its human counterpart Leu-1/T1 (CD5). *Proc. Natl. Acad. Sci. USA* **1987**, *84*, 204–208. [CrossRef] [PubMed]
98. Tarakhovsky, A.; Kanner, S.B.; Hombach, J.; Ledbetter, J.A.; Muller, W.; Killeen, N.; Rajewsky, K. A role for CD5 in TCR-mediated signal transduction and thymocyte selection. *Science* **1995**, *269*, 535–537. [CrossRef] [PubMed]
99. Dalloul, A. CD5: A safeguard against autoimmunity and a shield for cancer cells. *Autoimmun. Rev.* **2009**, *8*, 349–353. [CrossRef] [PubMed]
100. Bhandoola, A.; Bosselut, R.; Yu, Q.; Cowan, M.L.; Feigenbaum, L.; Love, P.E.; Singer, A. CD5-mediated inhibition of TCR signaling during intrathymic selection and development does not require the CD5 extracellular domain. *Eur. J. Immunol.* **2002**, *32*, 1811–1817. [CrossRef]
101. Mandl, J.N.; Monteiro, J.P.; Vrisekoop, N.; Germain, R.N. T cell positive selection uses self-ligand binding strength to optimize repertoire recognition of foreign antigens. *Immunity* **2013**, *38*, 263–274. [CrossRef] [PubMed]
102. Henderson, J.G.; Opejin, A.; Jones, A.; Gross, C.; Hawiger, D. CD5 instructs extrathymic regulatory T cell development in response to self and tolerizing antigens. *Immunity* **2015**, *42*, 471–483. [CrossRef] [PubMed]

103. Gringhuis, S.I.; de Leij, L.F.; Wayman, G.A.; Tokumitsu, H.; Vellenga, E. The Ca^{2+}/calmodulin-dependent kinase type IV is involved in the CD5-mediated signaling pathway in human T lymphocytes. *J. Biol. Chem.* **1997**, *272*, 31809–31820. [CrossRef] [PubMed]
104. Hassan, N.J.; Simmonds, S.J.; Clarkson, N.G.; Hanrahan, S.; Puklavec, M.J.; Bomb, M.; Barclay, A.N.; Brown, M.H. CD6 regulates T cell responses through activation-dependent recruitment of the positive regulator SLP-76. *Mol. Cell. Biol.* **2006**, *26*, 6727–6738. [CrossRef] [PubMed]
105. Pena-Rossi, C.; Zuckerman, L.A.; Strong, J.; Kwan, J.; Ferris, W.; Chan, S.; Tarakhovsky, A.; Beyers, A.D.; Killeen, N. Negative regulation of CD4 lineage development and responses by CD5. *J. Immunol.* **1999**, *163*, 6494–6501. [PubMed]
106. Davies, A.A.; Ley, S.C.; Crumpton, M.J. CD5 is phosphorylated on tyrosine after stimulation of the T cell antigen receptor complex. *Proc. Natl. Acad. Sci. USA* **1992**, *89*, 6368–6372. [CrossRef] [PubMed]
107. Samelson, L.E.; Phillips, A.F.; Luong, E.T.; Klausner, R.D. Association of the fyn protein-tyrosine kinase with the T cell antigen receptor. *Proc. Natl. Acad. Sci. USA* **1990**, *87*, 4358–4362. [CrossRef] [PubMed]
108. Raab, M.; Yamamoto, M.; Rudd, C.E. The T cell antigen CD5 acts as a receptor and substrate for the protein-tyrosine kinase p56lck. *Mol.Cell. Biol.* **1994**, *14*, 2862–2870. [CrossRef] [PubMed]
109. Beyers, A.D.; Spruyt, L.L.; Williams, A.F. Molecular associations between the T-lymphocyte antigen receptor complex and the surface antigens CD2, CD4, or CD8 and CD5. *Proc. Natl. Acad. Sci. USA* **1992**, *89*, 2945–2949. [CrossRef] [PubMed]
110. Spertini, F.; Stohl, W.; Ramesh, N.; Moody, C.; Geha, R.S. Induction of human T cell proliferation by a monoclonal antibody to CD5. *J. Immunol.* **1991**, *146*, 47–52. [PubMed]
111. Persaud, S.P.; Parker, C.R.; Lo, W.-L.; Weber, K.S.; Allen, P.M. Intrinsic $CD4^+$ T cell sensitivity and response to pathogen are set and sustained by avidity for thymic and peripheral self-pMHC. *Nat. Immunol.* **2014**, *15*, 266–274. [CrossRef] [PubMed]
112. Calvo, J.; Padilla, O.; Places, L.; Vigorito, E.; Vila, J.M.; Vilella, R.; Mila, J.; Vives, J.; Bowen, M.A.; Lozano, F. Relevance of individual CD5 extracellular domains on antibody recognition, glycosylation and co-mitogenic signalling. *Tissue Antigen.* **1999**, *54*, 16–26. [CrossRef]
113. McAlister, M.S.; Brown, M.H.; Willis, A.C.; Rudd, P.M.; Harvey, D.J.; Aplin, R.; Shotton, D.M.; Dwek, R.A.; Barclay, A.N.; Driscoll, P.C. Structural analysis of the CD5 antigen—Expression, disulphide bond analysis and physical characterisation of CD5 scavenger receptor superfamily domain 1. *Eur J. Biochem.* **1998**, *257*, 131–141. [CrossRef] [PubMed]
114. Cho, J.-H.; Kim, H.-O.; Surh, C.D.; Sprent, J. T cell receptor-dependent regulation of lipid rafts controls naive $CD8^+$ T cell homeostasis. *Immunity* **2010**, *32*, 214–226. [CrossRef] [PubMed]
115. Yashiro-Ohtani, Y.; Zhou, X.-Y.; Toyo-oka, K.; Tai, X.-G.; Park, C.-S.; Hamaoka, T.; Abe, R.; Miyake, K.; Fujiwara, H. Non-CD28 costimulatory molecules present in T cell rafts induce T cell costimulation by enhancing the association of TCR with rafts. *J. Immunol.* **2000**, *164*, 1251–1259. [CrossRef] [PubMed]
116. König, R. Chapter 315—Signal Transduction in T Lymphocytes A2—Bradshaw, Ralph A. In *Handbook of Cell Signaling*, 2nd ed.; Dennis, E.A., Ed.; Academic Press: San Diego, CA, USA, 2010; pp. 2679–2688.
117. Milam, A.V.; Allen, P.M. Functional heterogeneity in $CD4^+$ T cell responses against a bacterial pathogen. *Front. Immunol* **2015**, *6*, 621. [CrossRef] [PubMed]
118. Lozano, F.; Simarro, M.; Calvo, J.; Vila, J.M.; Padilla, O.; Bowen, M.A.; Campbell, K.S. CD5 signal transduction: Positive or negative modulation of antigen receptor signaling. *Crit. Rev. Immunol.* **2000**, *20*, 347–358. [CrossRef] [PubMed]
119. Hogquist, K.A.; Jameson, S.C. The self-obsession of T cells: How TCR signaling thresholds affect fate decisions in the thymus and effector function in the periphery. *Nat. Immunol.* **2014**, *15*, 815–823. [CrossRef] [PubMed]
120. Van de Velde, H.; von Hoegen, I.; Luo, W.; Parnes, J.R.; Thielemans, K. The B-cell surface protein CD72/Lyb-2 is the ligand for CD5. *Nature* **1991**, *351*, 662–665. [CrossRef] [PubMed]
121. Biancone, L.; Bowen, M.A.; Lim, A.; Aruffo, A.; Andres, G.; Stamenkovic, I. Identification of a novel inducible cell-surface ligand of CD5 on activated lymphocytes. *J. Exp. Med.* **1996**, *184*, 811–819. [CrossRef] [PubMed]
122. Brown, M.H.; Lacey, E. A ligand for CD5 is CD5. *J. Immunol.* **2010**, *185*, 6068–6074. [CrossRef] [PubMed]
123. Luo, W.; Van de Velde, H.; von Hoegen, I.; Parnes, J.R.; Thielemans, K. Ly-1 (CD5), a membrane glycoprotein of mouse T lymphocytes and a subset of B cells, is a natural ligand of the B cell surface protein Lyb-2 (CD72). *J. Immunol.* **1992**, *148*, 1630–1634. [PubMed]

124. Vandenberghe, P.; Verwilghen, J.; Van Vaeck, F.; Ceuppens, J.L. Ligation of the CD5 or CD28 molecules on resting human T cells induces expression of the early activation antigen CD69 by a calcium- and tyrosine kinase-dependent mechanism. *Immunology* **1993**, *78*, 210–217. [PubMed]
125. Ceuppens, J.L.; Baroja, M.L. Monoclonal antibodies to the CD5 antigen can provide the necessary second signal for activation of isolated resting T cells by solid-phase-bound OKT3. *J. Immunol.* **1986**, *137*, 1816–1821. [PubMed]
126. June, C.H.; Rabinovitch, P.S.; Ledbetter, J.A. CD5 antibodies increase intracellular ionized calcium concentration in T cells. *J. Immunol.* **1987**, *138*, 2782–2792. [PubMed]
127. Reth, M. Antigen receptor tail clue. *Nature* **1989**, *338*, 383–384. [CrossRef] [PubMed]
128. Unkeless, J.C.; Jin, J. Inhibitory receptors, ITIM sequences and phosphatases. *Curr. Opin. Immunol* **1997**, *9*, 338–343. [CrossRef]
129. Dong, B.; Somani, A.K.; Love, P.E.; Zheng, X.; Chen, X.; Zhang, J. CD5-mediated inhibition of TCR signaling proceeds normally in the absence of SHP-1. *Int. J. Mol. Med.* **2016**, *38*, 45–56. [CrossRef] [PubMed]
130. Weber, K.S.; Li, Q.J.; Persaud, S.P.; Campbell, J.D.; Davis, M.M.; Allen, P.M. Distinct $CD4^+$ helper T cells involved in primary and secondary responses to infection. *Proc. Natl. Acad. Sci. USA* **2012**, *109*, 9511–9516. [CrossRef] [PubMed]
131. Fulton, R.B.; Hamilton, S.E.; Xing, Y.; Best, J.A.; Goldrath, A.W.; Hogquist, K.A.; Jameson, S.C. The TCR's sensitivity to self peptide–MHC dictates the ability of naive $CD8^+$ T cells to respond to foreign antigens. *Nat. Immunol.* **2014**, *16*, 107–117. [CrossRef] [PubMed]
132. Palin, A.C.; Love, P.E. CD5 helps aspiring regulatory T cells ward off unwelcome cytokine advances. *Immunity* **2015**, *42*, 395–396. [CrossRef] [PubMed]
133. Mattson, M.P.; Chan, S.L. Calcium orchestrates apoptosis. *Nat. Cell Biol.* **2003**, *5*, 1041–1043. [CrossRef] [PubMed]
134. Orrenius, S.; Nicotera, P. The calcium ion and cell death. *J. Neural Transm. Suppl.* **1994**, *43*, 1–11. [PubMed]
135. Zhao, C.; Davies, J.D. A peripheral $CD4^+$ T cell precursor for naive, memory, and regulatory T cells. *J. Exp. Med.* **2010**, *207*, 2883–2894. [CrossRef] [PubMed]
136. Wahl, D.R.; Byersdorfer, C.A.; Ferrara, J.L.M.; Opipari, A.W.; Glick, G.D. Distinct metabolic programs in activated T cells: Opportunities for selective immunomodulation. *Immunol. Rev.* **2012**, *249*, 104–115. [CrossRef] [PubMed]
137. Pearce, E.L.; Pearce, E.J. Metabolic pathways in immune cell activation and quiescence. *Immunity* **2013**, *38*, 633–643. [CrossRef] [PubMed]
138. Van der Windt, G.J.; Pearce, E.L. Metabolic switching and fuel choice during T cell differentiation and memory development. *Immunol. Rev.* **2012**, *249*, 27–42. [CrossRef] [PubMed]
139. Chang, C.-H.; Curtis, J.D.; Maggi, L.B., Jr.; Faubert, B.; Villarino, A.V.; O'Sullivan, D.; Huang, S.C.-C.; van der Windt, G.J.W.; Blagih, J.; Qiu, J.; et al. Posttranscriptional control of T cell effector function by aerobic glycolysis. *Cell* **2013**, *153*, 1239–1251. [CrossRef] [PubMed]
140. Almeida, L.; Lochner, M.; Berod, L.; Sparwasser, T. Metabolic pathways in T cell activation and lineage differentiation. *Semin. Immunol.* **2016**, *28*, 514–524. [CrossRef] [PubMed]
141. Vaeth, M.; Maus, M.; Klein-Hessling, S.; Freinkman, E.; Yang, J.; Eckstein, M.; Cameron, S.; Turvey, S.E.; Serfling, E.; Berberich-Siebelt, F.; et al. Store-operated Ca^{2+} entry controls clonal expansion of T cells through metabolic reprogramming. *Immunity* **2017**, *47*, 664–679. [CrossRef] [PubMed]
142. Feske, S.; Skolnik, E.Y.; Prakriya, M. Ion channels and transporters in lymphocyte function and immunity. *Nat. Rev. Immunol.* **2012**, *12*, 532–547. [CrossRef] [PubMed]
143. Tamás, P.; Hawley, S.A.; Clarke, R.G.; Mustard, K.J.; Green, K.; Hardie, D.G.; Cantrell, D.A. Regulation of the energy sensor AMP-activated protein kinase by antigen receptor and Ca^{2+} in T lymphocytes. *J. Exp. Med.* **2006**, *203*, 1665–1670. [CrossRef] [PubMed]
144. Ma, E.H.; Poffenberger, M.C.; Wong, A.H.; Jones, R.G. The role of AMPK in T cell metabolism and function. *Curr. Opin. Immunol.* **2017**, *46*, 45–52. [CrossRef] [PubMed]
145. Huang, J.; Manning, B.D. The TSC1–TSC2 complex: A molecular switchboard controlling cell growth. *Biochem. J.* **2008**, *412*, 179–190. [CrossRef] [PubMed]
146. Chi, H. Regulation and function of mTOR signalling in T cell fate decision. *Nat. Rev. Immunol.* **2012**, *12*, 325–338. [CrossRef] [PubMed]

147. MacIver, N.J.; Blagih, J.; Saucillo, D.C.; Tonelli, L.; Griss, T.; Rathmell, J.C.; Jones, R.G. The liver kinase B1 is a central regulator of T cell development, activation, and metabolism. *J. Immunol.* **2011**, *187*, 4187–4198. [CrossRef] [PubMed]
148. Kirichok, Y.; Krapivinsky, G.; Clapham, D.E. The mitochondrial calcium uniporter is a highly selective ion channel. *Nature* **2004**, *427*, 360–364. [CrossRef] [PubMed]
149. Gilabert, J.A.; Bakowski, D.; Parekh, A.B. Energized mitochondria increase the dynamic range over which inositol 1,4,5-trisphosphate activates store-operated calcium influx. *EMBO J.* **2001**, *20*, 2672–2679. [CrossRef] [PubMed]
150. Gilabert, J.A.; Parekh, A.B. Respiring mitochondria determine the pattern of activation and inactivation of the store-operated Ca^{2+} current I (CRAC). *EMBO J.* **2000**, *19*, 6401–6407. [CrossRef] [PubMed]
151. Singaravelu, K.; Nelson, C.; Bakowski, D.; de Brito, O.M.; Ng, S.W.; di Capite, J.; Powell, T.; Scorrano, L.; Parekh, A.B. Mitofusin 2 regulates STIM1 migration from the Ca^{2+} store to the plasma membrane in cells with depolarized mitochondria. *J. Biol. Chem.* **2011**, *286*, 12189–12201. [CrossRef] [PubMed]
152. Dimeloe, S.; Burgener, A.V.; Grahlert, J.; Hess, C. T cell metabolism governing activation, proliferation and differentiation; a modular view. *Immunology* **2017**, *150*, 35–44. [CrossRef] [PubMed]
153. Jouaville, L.S.; Pinton, P.; Bastianutto, C.; Rutter, G.A.; Rizzuto, R. Regulation of mitochondrial ATP synthesis by calcium: Evidence for a long-term metabolic priming. *Proc. Natl. Acad. Sci. USA* **1999**, *96*, 13807–13812. [CrossRef] [PubMed]
154. Ho, P.C.; Bihuniak, J.D.; Macintyre, A.N.; Staron, M.; Liu, X.; Amezquita, R.; Tsui, Y.C.; Cui, G.; Micevic, G.; Perales, J.C.; et al. Phosphoenolpyruvate is a metabolic checkpoint of anti-tumor T cell responses. *Cell* **2015**, *162*, 1217–1228. [CrossRef] [PubMed]
155. Rumi-Masante, J.; Rusinga, F.I.; Lester, T.E.; Dunlap, T.B.; Williams, T.D.; Dunker, A.K.; Weis, D.D.; Creamer, T.P. Structural basis for activation of calcineurin by calmodulin. *J. Mol. Biol.* **2012**, *415*, 307–317. [CrossRef] [PubMed]
156. Racioppi, L.; Means, A.R. Calcium/calmodulin-dependent protein kinase kinase 2: Roles in signaling and pathophysiology. *J. Biol. Chem.* **2012**, *287*, 31658–31665. [CrossRef] [PubMed]
157. Chang, C.-H.; Qiu, J.; O'Sullivan, D.; Buck, M.D.; Noguchi, T.; Curtis, J.D.; Chen, Q.; Gindin, M.; Gubin, M.M.; van der Windt, G.J.W.; et al. Metabolic competition in the tumor microenvironment is a driver of cancer progression. *Cell* **2015**, *162*, 1229–1241. [CrossRef] [PubMed]
158. Patsoukis, N.; Bardhan, K.; Chatterjee, P.; Sari, D.; Liu, B.; Bell, L.N.; Karoly, E.D.; Freeman, G.J.; Petkova, V.; Seth, P.; et al. PD-1 alters T cell metabolic reprogramming by inhibiting glycolysis and promoting lipolysis and fatty acid oxidation. *Nat. Commun.* **2015**, *6*, 6692. [CrossRef] [PubMed]
159. Gary-Gouy, H.; Sainz-Perez, A.; Marteau, J.-B.; Marfaing-Koka, A.; Delic, J.; Merle-Beral, H.; Galanaud, P.; Dalloul, A. Natural phosphorylation of CD5 in chronic lymphocytic leukemia B cells and analysis of CD5-regulated genes in a B cell line suggest a role for CD5 in malignant phenotype. *J. Immunol.* **2007**, *179*, 4335–4344. [CrossRef] [PubMed]
160. Palmer, M.J.; Mahajan, V.S.; Chen, J.; Irvine, D.J.; Lauffenburger, D.A. Signaling thresholds govern heterogeneity in IL-7-receptor-mediated responses of naive $CD8^+$ T cells. *Immunol Cell. Biol.* **2011**, *89*, 581–594. [CrossRef] [PubMed]
161. Kipnis, J.; Gadani, S.; Derecki, N.C. Pro-cognitive properties of T cells. *Nat. Rev. Immunol.* **2012**, *12*, 663–669. [CrossRef] [PubMed]
162. Kipnis, J.; Cohen, H.; Cardon, M.; Ziv, Y.; Schwartz, M. T cell deficiency leads to cognitive dysfunction: Implications for therapeutic vaccination for schizophrenia and other psychiatric conditions. *Proc. Natl. Acad. Sci. USA* **2004**, *101*, 8180–8185. [CrossRef] [PubMed]
163. Brombacher, T.M.; Nono, J.K.; De Gouveia, K.S.; Makena, N.; Darby, M.; Womersley, J.; Tamgue, O.; Brombacher, F. IL-13–mediated regulation of learning and memory. *J. Immunol.* **2017**, *198*, 2681–2688. [CrossRef] [PubMed]
164. Oliveira-dos-Santos, A.J.; Matsumoto, G.; Snow, B.E.; Bai, D.; Houston, F.P.; Whishaw, I.Q.; Mariathasan, S.; Sasaki, T.; Wakeham, A.; Ohashi, P.S.; et al. Regulation of T cell activation, anxiety, and male aggression by RGS2. *Proc. Natl. Acad. Sci. USA* **2000**, *97*, 12272–12277. [CrossRef] [PubMed]
165. Filiano, A.J.; Gadani, S.P.; Kipnis, J. How and why do T cells and their derived cytokines affect the injured and healthy brain? *Nat. Rev. Neurosci.* **2017**, *18*, 375. [CrossRef] [PubMed]

166. Kyratsous, N.I.; Bauer, I.J.; Zhang, G.; Pesic, M.; Bartholomäus, I.; Mues, M.; Fang, P.; Wörner, M.; Everts, S.; Ellwart, J.W.; et al. Visualizing context-dependent calcium signaling in encephalitogenic T cells in vivo by two-photon microscopy. *Proc. Natl. Acad. Sci. USA* **2017**, *114*, E6381–E6389. [CrossRef] [PubMed]
167. Smedler, E.; Uhlén, P. Frequency decoding of calcium oscillations. *Biochim. Biophys. Acta Gen. Subj.* **2014**, *1840*, 964–969. [CrossRef] [PubMed]
168. Pesic, M.; Bartholomaus, I.; Kyratsous, N.I.; Heissmeyer, V.; Wekerle, H.; Kawakami, N. 2-photon imaging of phagocyte-mediated T cell activation in the CNS. *J. Clin. Investig.* **2013**, *123*, 1192–1201. [CrossRef] [PubMed]
169. de Bruin, N.M.W.J.; Schmitz, K.; Schiffmann, S.; Tafferner, N.; Schmidt, M.; Jordan, H.; Häußler, A.; Tegeder, I.; Geisslinger, G.; Parnham, M.J. Multiple rodent models and behavioral measures reveal unexpected responses to FTY720 and DMF in experimental autoimmune encephalomyelitis. *Behav. Brain Res.* **2016**, *300*, 160–174. [CrossRef] [PubMed]
170. Schub, D.; Janssen, E.; Leyking, S.; Sester, U.; Assmann, G.; Hennes, P.; Smola, S.; Vogt, T.; Rohrer, T.; Sester, M.; et al. Altered phenotype and functionality of varicella zoster virus–specific cellular immunity in individuals with active infection. *J. Infect. Dis.* **2015**, *211*, 600–612. [CrossRef] [PubMed]
171. Schub, D.; Fousse, M.; Faßbender, K.; Gärtner, B.C.; Sester, U.; Sester, M.; Schmidt, T. CTLA-4-expression on VZV-specific T cells in CSF and blood is specifically increased in patients with VZV related central nervous system infections. *Eur. J. Immunol.* **2018**, *48*, 151–160. [CrossRef] [PubMed]
172. Koebel, C.M.; Vermi, W.; Swann, J.B.; Zerafa, N.; Rodig, S.J.; Old, L.J.; Smyth, M.J.; Schreiber, R.D. Adaptive immunity maintains occult cancer in an equilibrium state. *Nature* **2007**, *450*, 903–907. [CrossRef] [PubMed]
173. Mattes, J.; Hulett, M.; Xie, W.; Hogan, S.; Rothenberg, M.E.; Foster, P.; Parish, C. Immunotherapy of cytotoxic T cell-resistant tumors by T helper 2 cells: An eotaxin and STAT6-dependent process. *J. Exp. Med.* **2003**, *197*, 387–393. [CrossRef] [PubMed]
174. Hung, K.; Hayashi, R.; Lafond-Walker, A.; Lowenstein, C.; Pardoll, D.; Levitsky, H. The central role of $CD4^+$ T cells in the antitumor immune response. *J. Exp. Med.* **1998**, *188*, 2357–2368. [CrossRef] [PubMed]
175. Scholler, J.; Brady, T.L.; Binder-Scholl, G.; Hwang, W.T.; Plesa, G.; Hege, K.M.; Vogel, A.N.; Kalos, M.; Riley, J.L.; Deeks, S.G.; et al. Decade-long safety and function of retroviral-modified chimeric antigen receptor T cells. *Sci. Transl. Med.* **2012**, *4*, 132ra153. [CrossRef] [PubMed]
176. Ho, Y.C.; Shan, L.; Hosmane, N.N.; Wang, J.; Laskey, S.B.; Rosenbloom, D.I.S.; Lai, J.; Blankson, J.N.; Siliciano, J.D.; Siliciano, R.F. Replication-competent noninduced proviruses in the latent reservoir increase barrier to HIV-1 cure. *Cell* **2013**, *155*, 540–551. [CrossRef] [PubMed]
177. Huetter, G.; Nowak, D.; Mossner, M.; Ganepola, S.; Muessig, A.; Allers, K.; Schneider, T.; Hofmann, J.; Kuecherer, C.; Blau, O.; et al. Long-Term Control of HIV by CCR5 Δ32/Δ32 Stem-Cell Transplantaion. *N. Engl. J. Med.* **2009**, *360*, 692–698. [CrossRef] [PubMed]
178. Ahmadzadeh, M.; Johnson, L.A.; Heemskerk, B.; Wunderlich, J.R.; Dudley, M.E.; White, D.E.; Rosenberg, S.A. Tumor antigen-specific CD8 T cells infiltrating the tumor express high levels of PD-1 and are functionally impaired. *Blood* **2009**, *114*, 1537–1544. [CrossRef] [PubMed]
179. Baitsch, L.; Baumgaertner, P.; Devevre, E.; Raghav, S.K.; Legat, A.; Barba, L.; Wieckowski, S.; Bouzourene, H.; Deplancke, B.; Romero, P.; et al. Exhaustion of tumor-specific $CD8^+$ T cells in metastases from melanoma patients. *J. Clin. Investig.* **2011**, *121*, 2350–2360. [CrossRef] [PubMed]
180. Staveley-O'Carroll, K.; Sotomayor, E.; Montgomery, J.; Borrello, I.; Hwang, L.; Fein, S.; Pardoll, D.; Levitsky, H. Induction of antigen-specific T cell anergy: An early event in the course of tumor progression. *Proc. Natl. Acad. Sci. USA* **1998**, *95*, 1178–1183. [CrossRef] [PubMed]
181. Rosenberg, S.A.; Yang, J.C.; Sherry, R.M.; Kammula, U.S.; Hughes, M.S.; Phan, G.Q.; Citrin, D.E.; Restifo, N.P.; Robbins, P.F.; Wunderlich, J.R.; et al. Durable complete responses in heavily pretreated patients with metastatic melanoma using T cell transfer immunotherapy. *Clin. Cancer Res.* **2011**, *17*, 4550–4557. [CrossRef] [PubMed]
182. Dudley, M.E.; Wunderlich, J.R.; Robbins, P.F.; Yang, J.C.; Hwu, P.; Schwartzentruber, D.J.; Topalian, S.L.; Sherry, R.; Restifo, N.P.; Hubicki, A.M.; et al. Cancer regression and autoimmunity in patients after clonal repopulation with antitumor lymphocytes. *Science* **2002**, *298*, 850–854. [CrossRef] [PubMed]
183. Postow, M.A.; Callahan, M.K.; Wolchok, J.D. Immune checkpoint blockade in cancer therapy. *J. Clin. Oncol.* **2015**, *33*, 1974–1982. [CrossRef] [PubMed]

184. Wei, S.C.; Levine, J.H.; Cogdill, A.P.; Zhao, Y.; Anang, N.-A.A.S.; Andrews, M.C.; Sharma, P.; Wang, J.; Wargo, J.A.; Pe'er, D.; et al. Distinct cellular mechanisms underlie anti-CTLA-4 and anti-PD-1 checkpoint blockade. *Cell* **2017**, *170*, 1120–1133.e17. [CrossRef] [PubMed]
185. Barbee, M.S.; Ogunniyi, A.; Horvat, T.Z.; Dang, T.O. Current status and future directions of the immune checkpoint inhibitors ipilimumab, pembrolizumab, and nivolumab in oncology. *Ann. Pharmacother.* **2015**, *49*, 907–937. [CrossRef] [PubMed]
186. Sangro, B.; Gomez-Martin, C.; de la Mata, M.; Inarrairaegui, M.; Garralda, E.; Barrera, P.; Riezu-Boj, J.I.; Larrea, E.; Alfaro, C.; Sarobe, P.; et al. A clinical trial of CTLA-4 blockade with tremelimumab in patients with hepatocellular carcinoma and chronic hepatitis C. *J. Hepatol.* **2013**, *59*, 81–88. [CrossRef] [PubMed]
187. Reck, M.; Rodríguez-Abreu, D.; Robinson, A.G.; Hui, R.; Csőszi, T.; Fülöp, A.; Gottfried, M.; Peled, N.; Tafreshi, A.; Cuffe, S.; et al. Pembrolizumab versus chemotherapy for PD-L1–positive non–small-cell lung cancer. *N. Engl. J. Med.* **2016**, *375*, 1823–1833. [CrossRef] [PubMed]
188. Hersey, P.; Gowrishankar, K. Pembrolizumab joins the anti-PD-1 armamentarium in the treatment of melanoma. *Future Oncol.* **2015**, *11*, 133–140. [CrossRef] [PubMed]
189. Rooke, R. Can calcium signaling be harnessed for cancer immunotherapy? *Biochim. Biophys. Acta Mol. Cell Res.* **2014**, *1843*, 2334–2340. [CrossRef] [PubMed]
190. Ernst, B.; Lee, D.S.; Chang, J.M.; Sprent, J.; Surh, C.D. The peptide ligands mediating positive selection in the thymus control T cell survival and homeostatic proliferation in the periphery. *Immunity* **1999**, *11*, 173–181. [CrossRef]
191. Smith, K.; Seddon, B.; Purbhoo, M.A.; Zamoyska, R.; Fisher, A.G.; Merkenschlager, M. Sensory adaptation in naive peripheral CD4 T cells. *J. Exp. Med.* **2001**, *194*, 1253–1261. [CrossRef] [PubMed]
192. Dorothée, G.; Vergnon, I.; El Hage, F.; Chansac, B.L.M.; Ferrand, V.; Lécluse, Y.; Opolon, P.; Chouaib, S.; Bismuth, G.; Mami-Chouaib, F. In situ sensory adaptation of tumor-infiltrating T lymphocytes to peptide-MHC levels elicits strong antitumor reactivity. *J. Immunol.* **2005**, *174*, 6888–6897. [CrossRef] [PubMed]
193. Friedlein, G.; El Hage, F.; Vergnon, I.; Richon, C.; Saulnier, P.; Lecluse, Y.; Caignard, A.; Boumsell, L.; Bismuth, G.; Chouaib, S.; et al. Human CD5 protects circulating tumor antigen-specific CTL from tumor-mediated activation-induced cell death. *J. Immunol.* **2007**, *178*, 6821–6827. [CrossRef] [PubMed]
194. Axtell, R.C.; Webb, M.S.; Barnum, S.R.; Raman, C. Cutting edge: Critical role for CD5 in experimental autoimmune encephalomyelitis: Inhibition of engagement reverses disease in mice. *J. Immunol.* **2004**, *173*, 2928–2932. [CrossRef] [PubMed]
195. Simoes, I.T.; Aranda, F.; Carreras, E.; Andres, M.V.; Casado-Llombart, S.; Martinez, V.G.; Lozano, F. Immunomodulatory effects of soluble CD5 on experimental tumor models. *Oncotarget* **2017**, *8*, 108156–108169. [CrossRef] [PubMed]
196. Tabbekh, M.; Franciszkiewicz, K.; Haouas, H.; Lecluse, Y.; Benihoud, K.; Raman, C.; Mami-Chouaib, F. Rescue of tumor-infiltrating lymphocytes from activation-induced cell death enhances the antitumor CTL response in CD5-deficient mice. *J. Immunol.* **2011**, *187*, 102–109. [CrossRef] [PubMed]
197. Round, J.L.; Mazmanian, S.K. The gut microbiome shapes intestinal immune responses during health and disease. *Nat. Rev. Immunol.* **2009**, *9*, 313–323. [CrossRef] [PubMed]
198. Vétizou, M.; Pitt, J.M.; Daillère, R.; Lepage, P.; Waldschmitt, N.; Flament, C.; Rusakiewicz, S.; Routy, B.; Roberti, M.P.; Duong, C.P.M.; et al. Anticancer immunotherapy by CTLA-4 blockade relies on the gut microbiota. *Science* **2015**, *350*, 1079–1084. [CrossRef] [PubMed]
199. Botticelli, A.; Zizzari, I.; Mazzuca, F.; Ascierto, P.A.; Putignani, L.; Marchetti, L.; Napoletano, C.; Nuti, M.; Marchetti, P. Cross-talk between microbiota and immune fitness to steer and control response to anti PD-1/PDL-1 treatment. *Oncotarget* **2017**, *8*, 8890–8899. [CrossRef] [PubMed]
200. Kosiewicz, M.M.; Dryden, G.W.; Chhabra, A.; Alard, P. Relationship between gut microbiota and development of T cell associated disease. *FEBS Lett.* **2014**, *588*, 4195–4206. [CrossRef] [PubMed]
201. Lathrop, S.K.; Bloom, S.M.; Rao, S.M.; Nutsch, K.; Lio, C.-W.; Santacruz, N.; Peterson, D.A.; Stappenbeck, T.S.; Hsieh, C.-S. Peripheral education of the immune system by colonic commensal microbiota. *Nature* **2011**, *478*, 250. [CrossRef] [PubMed]
202. Gonzalez-Perez, G.; Lamousé-Smith, E.S.N. Gastrointestinal microbiome dysbiosis in infant mice alters peripheral $CD8^+$ T cell receptor signaling. *Front. Immunol.* **2017**, *8*, 265. [CrossRef] [PubMed]

203. Huang, T.; Wei, B.; Velazquez, P.; Borneman, J.; Braun, J. Commensal microbiota alter the abundance and TCR responsiveness of splenic naïve CD4$^+$ T lymphocytes. *Clin. Immunol.* **2005**, *117*, 221–230. [CrossRef] [PubMed]
204. Bazett, M.; Bergeron, M.-E.; Haston, C.K. Streptomycin treatment alters the intestinal microbiome, pulmonary T cell profile and airway hyperresponsiveness in a cystic fibrosis mouse model. *Sci. Rep.* **2016**, *6*, 19189. [CrossRef] [PubMed]
205. Ochoa-Repáraz, J.; Mielcarz, D.W.; Haque-Begum, S.; Kasper, L.H. Induction of a regulatory B cell population in experimental allergic encephalomyelitis by alteration of the gut commensal microflora. *Gut Microbes* **2010**, *1*, 103–108. [CrossRef] [PubMed]
206. Allison, J.P.; Krummel, M.F. The Yin and Yang of T cell costimulation. *Science* **1995**, *270*, 932–933. [CrossRef] [PubMed]
207. Allison, J.P. Checkpoints. *Cell* **2015**, *162*, 1202–1205. [CrossRef] [PubMed]
208. Sivan, A.; Corrales, L.; Hubert, N.; Williams, J.B.; Aquino-Michaels, K.; Earley, Z.M.; Benyamin, F.W.; Lei, Y.M.; Jabri, B.; Alegre, M.-L.; et al. Commensal Bifidobacterium promotes antitumor immunity and facilitates anti–PD-L1 efficacy. *Science* **2015**, *350*, 1084–1089. [CrossRef] [PubMed]

9

Calcium and Nuclear Signaling in Prostate Cancer

Ivan V. Maly and Wilma A. Hofmann *

Department of Physiology and Biophysics, Jacobs School of Medicine and Biomedical Sciences, University at Buffalo, 955 Main Street, Buffalo, NY 14203, USA; ivanmaly@buffalo.edu

* Correspondence: whofmann@buffalo.edu

Abstract: Recently, there have been a number of developments in the fields of calcium and nuclear signaling that point to new avenues for a more effective diagnosis and treatment of prostate cancer. An example is the discovery of new classes of molecules involved in calcium-regulated nuclear import and nuclear calcium signaling, from the G protein-coupled receptor (GPCR) and myosin families. This review surveys the new state of the calcium and nuclear signaling fields with the aim of identifying the unifying themes that hold out promise in the context of the problems presented by prostate cancer. Genomic perturbations, kinase cascades, developmental pathways, and channels and transporters are covered, with an emphasis on nuclear transport and functions. Special attention is paid to the molecular mechanisms behind prostate cancer progression to the malignant forms and the unfavorable response to anti-androgen treatment. The survey leads to some new hypotheses that connect heretofore disparate results and may present a translational interest.

Keywords: metastasis; nuclear import; myosin IC; calcium; prostate cancer

---

1. Introduction

Despite the recent progress in diagnosis and treatment, prostate cancer remains one of the most common and lethal malignancies [1,2], with approximately 3.3 million men living with the condition only in the United States. The disease is characterized by the comparative ease of the initial diagnosis, a long period of indolence relative to other common cancers, and an abrupt and less predictable progression to the lethal stages. When sharpened up, the questions facing the practitioners and researchers have been formulated as concerning the possibility of treatment for the patients that need it, and the actual need of those for whom treatment is possible [3]. In many cases, the critical events in disease progression from the stage that is believed to merit only watchful waiting to one that is lethal are the emergence of castration resistance and metastasis [4]. Mechanistically grounded prognostic biomarkers and potential drug targets associated with these steps toward poor survival are of particular interest as outcomes of the molecular biological research.

From the perspective of molecular sciences, the puzzle of prostate cancer is encapsulated in a complex perturbation of multiple regulatory pathways and mechanisms. In the face of this complexity, the biological signal integration by the calcium ion (Ca^{2+}), in addition to the central importance of gene expression regulation, presents a particular interest. The intracellular propagation of a calcium-mediated signal [5] and the import of downstream molecules into the nucleus [6] can be seen as two common physicochemical events in the dysregulation involving the cell and its environment. Moreover, calcium channels and GPCRs are among the favored "druggable" targets in therapeutics development [7–9], and metastasis is manifested primarily by the molecular-motor driven cell motility [10,11]. The unexpected results of the recent experiments implicate the GPCR and myosin molecular motor molecules in the intranuclear calcium signaling and calcium-regulated nuclear transport [12,13]. Motivated by the emergent role of these mechanisms in multiple aspects of prostate cancer cell regulation, below we review the recent advances in molecular biology of prostate cancer with a special focus on calcium signaling and nuclear transport.

2. Genomic Background

2.1. *NKX3.1 Insufficiency*

85% of high-grade prostate intraepithelial neoplasia (PIN) and prostate adenocarcinomas display a loss of heterozygosity in the 8p21.2 locus that includes the *NKX3.1* homeobox gene [14]. The incidence correlates with disease grade [15], and there is evidence of an epigenetic downregulation at the locus as well [16]. *NKX3.1* is an early marker of developing prostate epithelium during budding from the urogenital sinus, and its expression throughout the development of the gland is important for ductal branching and expression of secreted proteins [17–19]. It is also important in adulthood, when homo- and heterozygous mutants display hyperplasia and often PIN [20,21]. As shown by Lei et al. [22], NKX3.1 stabilizes the tumor suppressor p53, inhibits the activation of protein kinase B (PKB/AKT), and blocks prostate cancer initiation caused by the phosphatase and tensin homolog deleted on chromosome 10 (PTEN) loss in xenografts. At the same time, *NKX3.1* inactivation in mice leads to a deficiency in the response to oxidative stress [23]. In human prostate cell lines, it has been shown that NKX3.1 promotes the response to DNA damage [24] but is ubiquitinated under the action of inflammatory cytokines [25]. Recent experiments have demonstrated, in addition, that the proteolytic control of NKX3.1 expression levels in prostate cancer cells is regulated by mitogens such as epithelial growth factor (EGF) in an intracellular calcium- and protein kinase C (PKC)-sensitive manner [26]. Thus, on the protein level, the effect of most common genomic abnormality that has been identified in prostate cancer converges with the effect of the posttranslational expression regulation by calcium signaling. Additional experiments are needed to establish the relative impact of these two factors at different stages of the disease development.

2.2. *Amplification and Susceptibility in MYC*

The *MYC* oncogene is somatically amplified (as part of the 8q24 region) in a subset of advanced prostate tumors [27,28]. Specifically nuclear MYC was found to be upregulated in many PIN lesions and most carcinomas in the absence of gene amplification [29]. Numerous single nucleotide polymorphisms have been identified in the gene-poor region next to *MYC* that contains *MYC*'s long-range regulatory elements [30–33]. It has also been established that overexpression of *MYC* in transgenic mice drives formation of PIN as well as progression to invasive adenocarcinoma [34]. The tumors formed under these conditions are characterized by downregulation of *NKX3.1* and a simultaneous upregulation of *Pim1*, *MYC*'s partner in lymphomas. Coexpression of *MYC* and *Pim1*, on the other hand, leads to formation of carcinomas that exhibit the neuroendocrine phenotype [35]. Similarly to the recent results concerning NKX3.1, it has been known for some time that *MYC* in the prostate cancer cells is under the control of calcium-mediated signaling. Specifically, it has been shown that *MYC* is controlled by Notch, which involves the calcium/calmodulin-dependent kinase II (CAMKII)-regulated production of the nuclear import-competent cleaved form of Notch-1 [36]. The crosstalk of this calcium-regulated nuclear signaling pathway, which connects with one of the key regulators of prostate cancer progression, with other factors influencing the intracellular calcium homeostasis appears to be a promising direction for future research.

2.3. *TMPRSS2-ERG Fusion*

Most prostate carcinomas exhibit deletions (less commonly translocations) on chromosome 21q that activate ETS-family transcription factors, usually via an N-terminal fusion with the androgen receptor (AR)-activated gene *TMPRSS2* [37–41]. *TMPRSS2-ERG* is found in approximately 15% of high-grade PINs and 50% of localized prostate cancer [42,43]. In agreement with these clinical sample analyses, AR binding in the lymph node cancer of prostate (LNCaP) cell line brings the *TMPRSS2* and *ERG* loci in physical proximity [44,45]. This finding suggests a role for androgen signaling in the induction of fusions under conditions leading to DNA damage. In this connection, it is interesting that double-stranded breaks also occur under the action of topoisomerase II, which is recruited to the AR-responsive elements by androgen signaling [46]. At the same time, whole-genome chromatin immunoprecipitation analysis has shown that ERG binding can silence AR-responsive genes [47]. Conceivably, this effect may reinforce the physiological hormonal changes in the aging prostate, as well as those accompanying therapeutic castration. Furthermore, in cell culture and transgenic mice, activation of *ETS* expression contributes quantitatively to promotion of invasivity and epithelial-mesenchymal transition (EMT) [38,48,49]. By itself, expression of truncated human *ERG* in mice leads only to a weak PIN [48,49]. It however synergizes with a loss of *PTEN*, leading to a high-grade PIN and carcinoma [50,51]. Collectively, these results put the *TMPRSS2-ERG* fusion at the center of a web of mechanisms responsible for the prostate cancer progression.

Recent work has begun to clarify the relationship of this common genomic condition with perturbations of calcium signaling and nuclear import. Epigenomic profiling has identified the genes that are differentially methylated in the *TMPRSS2-ERG* fusion vs. non-fusion prostate tumors [52]. Among the top-ranked genes is the calcium-channel gene *CACNA1D*, a target of ERG. The most recent extension of this work employed retrospective analysis to demonstrate a negative correlation of the calcium channel blocker use by the prostate cancer patients with both high-grade disease and occurrence of fusion tumors [53]. The latter result can be interpreted as a selective disadvantage of a fusion in the absence of a functional reinforcement by means of the perturbation of calcium signaling. Prostate cancer and benign epithelial prostate cells overexpressing ERG display a characteristic invasivity in vitro [48], an effect that was most recently reversed using peptidomimetic inhibitors of ERG that were fused with a nuclear localization sequence (NLS) peptides to colocalize with ERG in the nucleus [54]. It is remarkable that of the two forms of the fusion protein found in prostate cancer cells, one lacks the NLS and does not enter the nucleus [55]. It binds the fully functional form of the fusion and results in its downregulation on the protein level as well as a reduced expression of MYC. The evidence points to the importance of further study of prostate oncoprotein isoforms that may be characterized by different nucleocytoplasmic partitioning.

The *TMPRSS2-ERG* fusion activity is also associated with Y-box binding protein 1 (YB-1), a genomic-instability response protein that enters the nucleus in response to genotoxic stress [56]. YB-1 contributes to the upregulation of AR during the development of castration resistance [57]. The nuclear localization of this protein has been most recently associated statistically with a poor prognosis for prostate cancer patients [58]. Its accumulation was at the same time shown to be linked to the *TMPRSS2-ERG* fusion, as well as *PTEN* deletion. It appears worth investigating whether the conditions for nuclear import, for example, the calcium-mediated depolarization of the nuclear envelope that is reviewed below, impact the YB-1 regulated downstream effects of the *TMPRSS2-ERG* fusion.

Additionally, the fusion drives expression of Toll-like receptors (TLR), including TLR4 [59]. This results in an increased phosphorylation of nuclear factor κB (NF-κB) on the serine residue that prevents binding of inhibitor of κB (IκB) that masks the NLS. The active fusion isoform and the phosphorylated factor both localize exclusively in the nucleus, as seen also in the orthotopic tumor samples. In vitro, the same experiments demonstrate, the cells' proliferation is controlled by the transcriptional activity of NF-κB, and the latter is increased with the expression of the fusion product. These findings have led to the speculation that the activation of TLR4 by bacterial lipopolysaccharides, as well as tumor microenvironment proteins, may impact the progression of prostate cancer. Adding to the hypothetical effects that may be induced by the *TMPRSS2-ERG* fusion, it should be noted that NF-κB is known to be able to drive an aberrant ligand-independent nuclear accumulation of AR by direct binding [60]. Conceivably, this may mean that the fusion can aid the cells in bypassing the critical androgen-regulated step during the establishment of androgen independence in the progressing tumor. One more gene activated downstream from *TMPRSS2-ERG* is the chemokine receptor *CXCR4* [61], whose role in intranuclear calcium signaling will be reviewed in Section 8.1.

2.4. PTEN Loss

The phosphatase gene *PTEN* undergoes homozygous deletion early in prostate carcinogenesis [62]. The loss has been found to be correlated with the development of aggressive and castration-resistant disease [63–65]. At the same time, it has been shown that conditional deletion of *PTEN* in prostate epithelium leads to PIN and adenocarcinoma [66]. Inactivation of *PTEN* in genetically engineered mice was found to cooperate with the loss of function of *NKX3.1* [67], upregulation of *c-MYC* [68], and expression of the human *TMPRSS2-ERG* fusion [50,51]. Among the molecular partners required for tumor formation on the *PTEN* loss background, the studies have identified mammalian target of rapamycin complex 2 (mTORC2) and phosphoinositide 3-kinase (PI3K) isoform p110β [69,70]. It is remarkable that prostate epithelial cells from *NKX3.1; PTEN* mutant mice display androgen independence even before development of the cancerous phenotype [71]. If this finding is translatable to the conditions in the human prostate, it indicates sufficiency of the combination of these common early and advanced genomic abnormalities for the critical step in the disease progression. This possibility raises the importance of further investigation, as suggested in Section 2.1, of the calcium-dependent regulation of NKX3.1, which appears to parallel the background genomic effect on the protein level.

In LNCaP cells, PTEN was found to interact with AR directly, resulting in an inhibition of the latter's nuclear localization and promotion of its degradation [72]. PTEN's own localization in heterozygous (+/−) prostate cancer cells was found to be predominantly nuclear as assessed by immunofluorescence in vitro [73]. The nuclear retention of PTEN was investigated also on model non-prostate-cancer cell lines [74] and found to be regulated by reactive oxygen species (ROS). The functional importance of the nuclear localization has been demonstrated in the same study in mouse xenograft models. Using human prostate cancer cells expressing mutant forms of PTEN, it was possible to demonstrate that nuclear PTEN acts in concert with p53 to suppress tumor formation. Most recently, the nuclear transport receptor importin 11 has been identified as a tumor suppressor that protects PTEN from proteolytic degradation [75]. This suggest the existence of a secondary function of the molecular interaction that presumably underpins the phosphatase's nuclear import. Generally, it appears to be an example of the nuclear import serving also as a regulatory mechanism impacting the protein-level expression of a critical tumor suppressor, akin to the chaperone-mediated sequestration mechanisms that will be reviewed in Section 7.

3. Developmental Pathways

Gene expression studies suggest that developmental signaling pathways may be reactivated during neoplastic processes in the adult prostate [76,77]. In the mouse, the wingless-integration adenomatous-polyposis-coli (WNT-APC)-β-catenin pathway is seen as contributing to carcinogenesis positively [78–80]. Consistent with this is the observation of an enhanced nuclear localization of β-catenin specifically in castration-resistant tumors [81]. In man, on the other hand, the nuclear localization of β-catenin has been seen as negatively correlated with tumor progression [82,83]. Further pointing to an explanation whereby the WNT-β-catenin signaling in prostate cancer development may be stage-specific is the finding that the WNT5A protein is upregulated specifically in the metastatic human samples [84]. The same work also demonstrated the existence of calcium waves that are induced by WNT5A in cultured metastatic prostate cancer cells, as well as this protein's stimulation of the actin cytoskeleton activity and cell motility (the latter being also under the control of CAMKII). Subsequent experiments showed that the calcium wave induced by this and other WNT proteins results in an intranuclear calcium mobilization and import of β-catenin into the nucleus, which is dependent on a thapsigargin-sensitive depolarization of the nuclear envelope [85]. The results support an interpretation whereby WNT signaling has a dual role as an anti-dedifferentiation regulator and a promoter of cell motility. It will be interesting to determine whether the intranuclear calcium mobilization by WNT impacts the functions of nuclear myosin (reviewed in Section 8.2), and what mechanisms or quantitative properties distinguish it from the one that has for some time been viewed as associated with apoptosis [86,87].

Among other developmental pathways, Hedgehog and fibroblast growth factor (FGF) signaling should be mentioned. Hedgehog signaling plays a significant role in prostate cancer progression [88]. Although experiments on tumor-derived primary cultures were seen initially as pointing toward an autocrine mechanism [89], targeted experimentation identified the mechanism as paracrine signaling to the epithelium from the prostate stroma [90], similarly to other epithelial cancers [91]. Nuclear accumulation of the downstream effector of Hedgehog signaling, transcription factor Gli1, as seen in clinical samples, is associated with metastasis and poor survival [92]. A breaking report [93] identifies the mode of the intracellular Hedgehog signal propagation in the prostate cancer cells as a novel non-canonical pathway involving AR binding-driven nuclear entry of the Gli transcription factors. The contribution of FGF signaling is analogous, in that both epithelial activation of FGF receptor 1 and stromal overexpression of FGF10 play a role [94,95], potentially leading downstream to the activation of extracellular signal-regulated kinase (ERK). Kwabi-Addo et al. [96] call attention to the alternative translation initiation isoforms of FGF2, which display an intranuclear accumulation and are expressed in transgenic adenocarcinoma of the mouse prostate (TRAMP) models, and posit that these isoforms may present a special interest in the context of paracrine signaling, because of the possibility of direct translation of the extracellular signal to the gene expression effects in the target cell. Overall, the ongoing work on the developmental pathways' dysregulation in prostate cancer emphasizes the need for the elucidation of the pivotal molecular mechanisms behind the calcium-regulated nuclear import, which may involve such novel biophysical principles as gating by the nuclear envelope depolarization.

4. Kinase Cascades

The *PTEN* loss of function, which may result, in particular, from the reviewed genomic abnormality, has been shown to cause an upregulation of the AKT/mTOR pathway, primarily via activation of the kinase AKT1 [97–99]. An alternative activating mechanism for this signaling pathway is the activating mutation in AKT1 itself, which has been found in human prostate cancer [100]. Experimental expression of the constitutively activated isoform p110β of PI3K is also sufficient for induction of neoplasia in mice [101]. In addition, activation of either AKT or ERK enables androgen-independent growth both in vitro and in vivo [102]. Collectively, these findings establish a broad molecular background among the intracellular signaling kinase classes, whose perturbation may support the development of prostate cancer.

Underscoring the functional importance of the activated kinases' localization in the cell, it has been shown that tumor suppressor PML recruits both phospho-AKT (pAKT) and its phosphatase PP2 to nuclear bodies [103]. Complicating the picture, loss of *PML* leads to an impairment of the PP2 activity toward pAKT and an accumulation of the latter in the nucleus. The kinase acts in the nucleus by inhibiting FOXO transcription factors, in particular the FOXO3A-mediated transcription of Bim and p27 (Kip1), suggesting a two-pronged mechanism that involves both the suppression of apoptosis and an acceleration of the cell cycle. Promisingly for the translation of this insight into the practice of prostate cancer drug development, the AKT inhibitor ML-9 has recently been found effective at inducing prostate cancer cell death in vitro in a calcium-dependent manner [104].

Adding to the in vitro and in vivo evidence, ERK is frequently found to be activated in the clinical prostate cancer, often at advanced stages and in combination with AKT. As reviewed above, the simultaneous activation promotes progression and androgen independence [102,105]. Moreover, simultaneous targeting of the two pathways was sufficient to inhibit hormone-refractory prostate cancer in the mouse model [106]. Mechanistically, the activation of mTOR and IκB kinase (IKK) by AKT leads to stimulation of NF-κB [107]. Supporting the significance of this regulation, the abnormal

NF-κB signaling correlates with androgen sensitivity, metastasis, and outcome in human prostate cancer and the mouse model. The studies implicate both the expression and the nuclear localization of NF-κB [108–111], and identify among the downstream mechanisms its ability to regulate expression of AR [112].

Phosphorylation of mitogen-activated protein kinases (MAPK) by expression of constitutively active RAF or RAS in the mouse prostate epithelium also leads to a nuclear accumulation of pERK, and this action promotes induction of tumorigenesis [79,113]. However, such mutations are rare in human prostate cancer. Notwithstanding, the collective perturbation of the RAS/RAF pathway via small but collectively significant alterations in the expression of the individual pathway members is prevalent in advanced human prostate cancer [114]. Phospho-MAPK, whether induced by EGF in vitro or prevalent in certain subsets of prostate cancer epithelial cells as assessed by immunofluorescence in vivo, preferentially partitions into the nucleus [115], which can be seen as indicative of the predominantly nuclear downstream functions. These functions remain to be investigated. Taken together, the data reviewed in the last sections demonstrate how a large number of fundamental molecular perturbations behind the development of prostate cancer are interlinked by calcium signaling and the calcium-regulated nuclear import mechanisms (Table 1).

Table 1. General regulatory factors involved in the calcium and nuclear signaling in prostate cancer. The molecular factors and their cell-physiological effects are categorized, and the corresponding calcium and nuclear events sketched out in the approximate order of their treatment in the text.

Category	Factor	Physiology	Partners	Calcium Regulation/Nuclear Signal
genomic	*NKX3.1* insufficiency	development	p53, AKT, PTEN	Ca^{2+}-regulated proteolysis
	MYC amplification	cell cycle	NKX3.1, PIM1	Ca^{2+}-regulated Notch nuclear entry
	TMPRSS2-ERG fusion	EMT	PTEN, *CACNA1D*	NF-κB nuclear entry, Ca^{2+} channel expression
	PTEN loss	hormone independence	mTORC2, PI3K	AR nuclear localization
developmental	WNT	cell motility	APC	Ca^{2+} wave, β-catenin nuclear entry
	Hedgehog	signaling from stroma	AR	Gli1 nuclear entry
	FGF	signaling from stroma	ERK	nuclear entry of alternative translation isoforms
kinases and phosphatases	AKT-mTOR	apoptosis	FOXO3A, Bim, p27	pAKT accumulation in nuclear bodies, Ca^{2+}-dependent inhibition
	ERK, MAPK	hormone independence	RAS, RAF, AKT	pERK/pMAPK accumulation in nucleus
	calcineurin	anti-senescence	NFATc1, MYC, IL6, STAT3	NFATc1 nuclear import

5. Anti-Senescence Signaling

A limitless proliferation potential achieved through anti-senescence factors has been proposed to be one of the biological hallmarks of cancer cells [116,117]. The recent characterization of store-operated calcium entry (SOCE) channel components STIM1 and ORAI1 in prostate cancer sheds light on the involvement of calcium in these processes. It was found [118] that the expression of both STIM1 and ORAI1 correlated negatively with the Gleason score (the commonly used histopathological assessment of prostate cancer progression) and was lower in prostate carcinomas relative to the normal tissue. STIM1 was also significantly reduced in hyperplasia. In agreement with these results obtained by immunohistochemistry on patient samples, STIM1 expression was reduced in carcinoma cell lines relative to the benign prostate hyperplasia (BPH) cell line BPH-1, and additionally reduced in the malignant PC3 line relative to the less metastatic DU145. The endogenous expression level differences in these experiments had the expected functional consequences for SOCE activity. Overexpression of the two factors, on the other hand, induced growth inhibition via a cell cycle arrest and senescence, as detected morphologically and by β-galactosidase staining. The apoptosis markers were similarly affected. The complex phenotype was additionally characterized by an increased migration of the cultured cells and a decreased ability of their conditioned medium to induce recruitment of macrophages in vitro. These effects were correlated with a signature of downstream TGF-β signaling activation, manifested in an overexpression of Snail and WNT1 and a rise of p-Smads and nuclear β-catenin, as well as a perturbed secretion of cytokines. On the other hand, inoculation of the STIM1- and ORAI1-overespressing cells into non-obese diabetic, severe combined immunodeficiency (NOD/SCID) mice showed a comparative reduction of tumor growth and a relative loss of BrdU-positive cells and E-cadherin immunostaining. The latter can be seen as indicative of EMT, in agreement with the complex phenotype observed in vitro. An additional complicating finding both in vivo and in vitro has been an induction, in STIM1- and ORAI1-overexpressing cells and tumors, of a relative overexpression of the apoptosis-inhibiting [119,120] form of the tumor necrosis factor (TNF) receptor, decoy receptor 2 (DcR2), even though the anti-apoptotic Bcl-2 and XIAP in these experiments were reduced. A consistent interpretation may be possible, stemming from the fact that DcR2 is, at the same time, an established senescence marker [121,122].

Additional recent findings linking senescence in prostate cancer to calcium signaling pertain to the nuclear factor of activated T-cells c1 (NFATc1), a protein previously implicated in the regulation of expression of prostate specific membrane antigen in an ionomycin-responsive manner in LNCaP cells [123], as well as in the expression of the osteomimicry markers in PC3 and C4-2B cells [124]. NFATc subfamily proteins (the majority of the NFAT types) are activated via dephosphorylation by the calcium/calmodulin-dependent serine/threonine phosphatase calcineurin, upon which the NLS is exposed and the protein undergoes import into the nucleus, where it performs the functions of a transcription activator [125]. In a general agreement with previous work on NFATc1, Manda et al. [126] find the protein expressed in human adenocarcinoma samples (both in the neoplastic epithelium and in the stroma) as well as in tumorigenic prostate cancer cell lines, but not in the epithelium of the normal prostate or benign RWPE-1 cells. Targeted expression of the activated nuclear form of NFATc1 in a mouse model leads to PIN and subsequently adenocarcinoma marked by an elevated expression of proinflammatory cytokines, apparent proliferative influence of NFATc1$^+$ cells on the neighboring cells, and castration resistance. Revealingly, MYC was upregulated in both NFATc1$^+$ and—apparently via the elevated IL6 and pSTAT3—in the neighboring cells. A double transgenic model with a *PTEN* deletion displayed a synergy of the factors and an accelerated tumorigenesis. Compared with the common *PTEN* deletion model, the prostates were marked by a suppressed expression and nuclear localization of the senescence marker p21 and an insignificant β-galactosidase staining. The mechanism of the p21 suppression may involve the STAT3-SKP2 pathway demonstrated in the gastric and cervical cancer cells [127,128]. Altogether, the data suggest that the calcium-regulated nuclear import step may be critical for the establishment of anti-senescence signaling in advanced prostate cancer.

6. Channels and Transporters

6.1. Transient Receptor Potential (TRP) Channels

Channels and transporters are among the best-studied molecular classes orchestrating the calcium signaling. In the recent years, considerable attention has been given to the role of TRP channels in the development of prostate cancer. The function of TRPV6, a vanilloid subfamily plasma membrane calcium channel, in particular, is now comparatively well characterized. Earlier termed CaT-like, it is a marker of locally advanced, metastatic, and androgen-insensitive prostate cancer [129]. On mRNA level, in situ hybridization uncovered a correlation of its expression with a number of cancer progression characteristics, including the Gleason score [130]. Remarkably, while its expression is detectable in most androgen-insensitive lesions, the level is reduced relative to untreated tumors. In-vitro studies have shown that this channel, as well as the related TRPC6, increases prostate cancer cell proliferation, acting through NFAT [131,132]. A subsequent detailed immunohistochemical analysis established also the correlation of the TRPV6 expression with prostate-specific antigen (PSA) and TRPC1, and with a lack of the apoptosis marker caspase-3 cleaved fragment [133]. Intriguingly, the presence of TRPV6 in the more advanced tumors, where it is found in luminal cells, was correlated with the appearance of ORAI1 in the same cells, which normally localizes to the basal cells. It has been possible to establish the involvement of TRPV6 in SOCE in prostate cancer cells, as well as its physical association with ORAI1 via STIM1 and TRPC1 under the conditions of SOCE. The latter conditions also induced a translocation of TRPV6 to the plasma membrane, as observed using a confocal microscope. The same study showed that overexpression of TPRV6 in prostate cancer cells in vitro increases their resistance to apoptosis induced, for example, by cisplatin, while in heterotopic xenografts in mice its experimentally manipulated level of expression correlates with the resulting tumor mass. Most recently, an inhibitor of this channel was evaluated in a first-in-human phase I clinical trial with promising results [134]. The work on TRPV6, to date, can serve as a model of calcium-mediated mechanism elucidation leading to translationally relevant results in the prostate cancer field.

Among the other TRP channels, as reviewed most recently by Cui et al. [8], the nonselective cationic channel vanilloid 2 (TRPV2) has been associated specifically with metastatic prostate cancer. Experiments in an earlier TRPV2 study [135] demonstrated the channel's involvement in the promotion of prostate cancer cells' migration and invasion both in vitro and in vivo. The effect, accompanied by elevated expression levels of the invasion enzymes MMP2 and MMP9, as well as cathepsin B, appeared to be mediated by an elevated basal cytosolic concentration of Ca^{2+} that was dependent on the channel's expression level. For progress in system-level understanding of the mechanism behind the emergence of prostate cancer invasivity, it will be of interest to characterize the functional interplay of these effects with the myosin-driven secretion of metalloproteases [136], a process that is likely open to calcium regulation in the light of the molecular dynamics investigations that will be reviewed in the last section.

Another TRP channel, melastatin family 8 (TRPM8, initially termed TRP-P8), was first found to be overexpressed in prostate cancer by Tsavaler et al. [137]. In androgen-responsive prostate cancer cells in vitro, it is expressed on the plasma membrane and in the endoplasmic reticulum (ER), and the level of expression is positively regulated by androgen [138]. The channel is capable of initiating an agonist-induced elevation of the intracellular calcium ion concentration. In keeping with the general theme of fine-tuned homeostasis, which is widely accepted to characterize the calcium regulation [139], both the agonist and antagonist action on TRPM8 in in-vitro experiments was able to induce apoptosis. In the same experiments, the channel was found to be expressed at a low and unregulated level in androgen-insensitive cells. This finding agrees with the loss of TRPM8 in transition to androgen independence that was detected in a genome-wide expression profiling study [140]. Subsequent work [141] found that experimental re-expression of this channel in androgen-insensitive cells reduced their migration in vitro. At the same time, in non-prostate-cancer model cell culture, PSA was able to potentiate the TRPM8 agonist-induced calcium current by enhancing trafficking of the channel to the plasma membrane via its phosphorylation downstream of the bradykinin 2 receptor signaling pathway. In resection samples and primary cultures, the plasma membrane localization and activity of this channel was found to be characteristic of the luminal cells of normal tissue, BPH, and especially in situ tumors, whereas the dedifferentiated transit amplifying and intermediate phenotype retained primarily the endoplasmic isoform [142]. As pointed out by the authors of this investigation, one functional consequence of the endoplasmic localization of the channel may be the apoptosis-resistant state characterized by a lowered intraendoplasmic calcium concentration, which had been seen in other studies of advanced prostate cancer [143,144]. Thus, the channel's role appears to be different, depending on the stage of prostate cancer, and change from its expression maintaining the pro-survival calcium homeostasis to its partial loss potentially contributing to the development of metastatic potential in parallel with the acquisition of androgen independence.

A recent study [145] has implicated a second melastatin family TRP channel, TRPM7, in the positive regulation of prostate cancer cells' migration, invasion, and expression of MMPs, as well as in downregulation of E-cadherin that accompanies EMT. Furthermore, the related TRPM4 has a function in the proliferation of prostate cancer cells, according to a breaking report [146]. In this instance, the downstream cascade includes a relative dephosphorylation of β-catenin and an elevation of this regulator's nuclear localization. Completing the present picture of a broad and mechanistically diverse involvement of the melastatin subfamily channels in the prostate cancer development, TRPM2 is one more known member that controls proliferation of prostate cancer cells by inhibiting the nuclear ADP-ribosylation [147]. The intracellular localization of TRPM2, which is associated with the plasmalemma of benign prostate epithelial cells, is altered in tumorigenic prostate cell lines, where it appears in intranuclear clusters. It will be interesting to investigate if the channels of this broad class are similar to the GPCRs to be reviewed in Section 8.1 in exhibiting nuclear translocation and influencing the trans-nuclear envelope potential and intranuclear calcium concentration in prostate cancer cells. Their downstream targets, hypothetically, could include the physical nuclear organization and transcription-related functions of nuclear myosins (see Section 8.2).

6.2. Other Channels and Transporters

The other types of channels and transporters that have been implicated in the development of prostate cancer include molecules responsible for a remarkably diverse set of calcium homeostasis mechanisms, such as voltage-gated and calcium-activated channels, as well as intracellular pumps. In particular, there is pharmacological evidence of a contribution from T-type calcium channels, including, specifically, $Ca_V3.2$, to the proliferative activity of prostate cancer cells in vitro and to cancer progression in xenograft experiments [148,149]. A similar involvement of this type of channels has been documented in other tumor types [150]. The $Ca_V3.2$ isoform may also be involved in neuroendocrine differentiation in prostate cancer, which is associated with castration resistance, distant metastasis, and poor prognosis [151]. Revealingly, induction of neuroendocrine differentiation of LNCaP epithelial prostate cancer cells by androgen deprivation in vitro was found to be accompanied by an upregulation of $Ca_V3.2$ and an increase in the associated calcium current [152,153]. These results suggest that a calcium-mediated reinforcement loop similar to the one already discussed in connection with the *CACNA1D* gene (which encodes $Ca_V1.3$, see Section 2.3) may be contributing to the neuroendocrine differentiation in the androgen-deprived prostate.

An intriguing connection to the cytoskeleton signaling has recently been uncovered that involves a plasma membrane channel that is calcium-activated but conducts potassium ions. The big potassium calcium-sensitive (BKCa) channels are expressed in excitable cells [154] and aberrantly in cancers of various origin [155]. They are overexpressed in a significant subset of advanced prostate cancers, and inhibiting their expression can suppress PC3 cell proliferation in vitro [156]. A recent study [157] has extended these results, using a xenograft model, and additionally demonstrated a stimulating effect of these channels' overexpression on cell migration and invasion. Interestingly, the effects depended only slightly on the channels' conductance, but were exerted instead in an integrin- and phosphorylated focal adhesion kinase (FAK)-dependent manner. The immunofluorescence colocalization and immunoprecipitation showed a formation of a BKCa-αVβ3 integrin complex, accompanied by recruitment and phosphorylation of FAK. The possible interplay of this mechanism with the tumor stroma's influence on the focal adhesion signaling, as well as the potential feedback from the extracellular matrix modification by the invasive epithelial cells themselves, present interest for future research.

The sarcoendoplasmic calcium pump SERCA's expression in epithelial prostate cancer cells is upregulated by the action of EGF, dihydrotestosterone (DHT), or serum, according to the in-vitro studies on LNCaP cells [158]. Stimulation of proliferation caused by these agents is suppressed by the SERCA inhibitor thapsigargin. Thus, the pump appears to be both an intermediary of the response to mitogens and subject to the downstream regulation caused by these factors, which may sensitize the tumor to the subsequent stimulation. Although the SERCA isoform implicated by the cited study was the non-tissue specific isoform (2B), a prostate-selective thapsigargin analog has been designed [159] and, most recently, reached phase II in the clinical trials for prostate cancer [8].

In addition to the reviewed roles of calcium-release activated calcium channel protein ORAI1 in senescence and SOCE, the channel appears to be involved also in the capacitative entry mediating apoptosis in androgen-sensitive prostate cancer cells in culture [160]. The androgen dependence of expression of ORAI1 itself, which was demonstrated in this study, raises the possibility of its downregulation being part of the mechanism of the emergence of aggressive cancer following the androgen deprivation therapy. Extending these results, the store-independent entry was recently associated with prostate cancer-specific enhanced expression of ORAI3 and formation of a heteromeric channel with ORAI1 [161]. Interestingly, the channel is gated by arachidonic acid, opening a new link between calcium signaling and the regulation by metabolism and inflammation.

The mitochondrial calcium uniporter, which contributes to the proapoptotic accumulation of calcium in mitochondria, is a comparatively new pump that has been studied in relation to cancer [8]. In-silico screens pointed to a negative regulation of its expression by miR-25, and this prediction was borne out by the experimental measurements [162]. Established prostate cancer cell lines express large quantities of this micro-RNA and are, at the same time, characterized by suppressed levels of the uniporter. Experimental restoration of its expression has been shown to sensitize prostate cancer cells in vitro to apoptosis. This is a promising direction for a further investigation in vivo.

An important apoptotic pathway is mediated by a sustained Ca^{2+} flux from the ER to mitochondria through the endoplasmic inositol 1,4,5-trisphosphate receptors, InsP3R [163]. The regulation of this mechanism has been linked to the deficiency of the *PTEN* gene, which, as reviewed above, is common in prostate cancer. The role of InsP3R has been elucidated in a series of experiments on established prostate cancer cell lines, which similarly exhibit the *PTEN* inactivation [164]. LNCaP cells, in particular, are characterized by a frameshift mutation in *PTEN*, and PKB/ACT in these cells was found to be constitutively phosphorylated and activated in the PI3K-dependent manner [165]. The phosphorylation in these experiments has proved determinative of the cells' resistance to apoptosis upon withdrawal of the serum factors. Subsequent work [166] has traced the mechanism of the resistance to phosphorylation of InsP3R by AKT. Most recently, a similar effect has been described in PC3 cells, where it is mediated by the lack of competition from the deficient *PTEN* product for binding to InsP3R3, which opens the latter to binding the F-box protein FBXL2 [167]. FBXL2 has the function of the receptor subunit of the ubiquitin ligase complexes, and can target InsP3R3 for degradation. The different mechanisms in the lymph node- and bone-metastatic cell lines may be indicative of diverging functions that can be discovered even for the comparatively well-characterized molecular classes at the different stages of prostate cancer progression. While the conditions-dependent multifunctionality may play the role of a complicating factor in drug development, it can also present an opportunity to more selectively target the tumor development at its specific crucial steps.

Finally, in the discussion of the channels involved in calcium signaling in prostate cancer, ryanodine receptors (RyR) have also been found to be expressed in prostate cancer cell lines [168]. The isoforms RyR1-3 vary in quantity depending on the cell line, a finding that should be further investigated, as it may be indicative of modification of the response mediated by these channels, depending on the stage in the tumor development or its environment. Calcium release is observed under the action of the RyR agonist caffeine and has been found to mildly promote apoptosis of the lymph-node metastatic LNCaP cells in vitro [169]. A synthetic agonist 4-chloro-*m*-cresol was also effective at eliciting this calcium response in these experiments, pointing to a possibility of targeting the receptor through a prostate-selective drug design in the future. On the whole, the calcium and calcium-operated channels and transporters remain one of the most systematically studied molecular classes in relation to their roles in prostate cancer (Table 2).

Table 2. Calcium and nuclear signaling in prostate cancer that is associated with specific channels and transporters.

Category	Channel/Transporter	Related Cell Physiology	Partners	Calcium Regulation/Nuclear Signal
TRP channels	TRPM2	proliferation	nuclear ADP-ribosylase	channel accumulation in nuclear clusters
	TRPM4	proliferation	β-catenin	β-catenin nuclear accumulation
	TRPM7	EMT	E-cadherin, MMPs	PM Ca^{2+} channel
	TRPM8	apoptosis	AR, PSA	lowered ER Ca^{2+}
	TRPV2	invasion	MMPs	elevated basal Ca^{2+}
	TRPV6	proliferation	NFAT, ORAI1, STIM1	SOCE
other channels, transporters	BKCa	invasion	integrin	Ca^{2+}-activated phosphorylation of FAK
	SERCA	proliferation	EGF, DHT	ER Ca^{2+}
	ORAI1	apoptosis	ORAI3, arachidonic acid	capacitative entry via the gated heteromer
	mitochondrial uniporter	apoptosis	miR-25	mitochondrial Ca^{2+}
	InsP3R	apoptosis	AKT, PTEN, FBXL2	ER-to-mitochondria current

7. Additional Topics

7.1. Calpain Proteolysis

Upon elevation of the intracellular calcium concentration in prostate cancer cells, as demonstrated in experiments in vitro [170], the calcium-regulated protease calpain cleaves β-catenin, producing a stable, transcriptionally active fragment that enters the nucleus. At the same time, microarray experiments demonstrated an elevation of the calpain expression in metastatic prostate cancer, which may suggest sensitization of this nuclear import pathway to calcium signaling in cells undergoing the transition to malignancy. Alongside the calcineurin-NFAT (Section 5) and calmodulin-myosin (Section 8.2) mechanisms, the calpain-β-catenin mechanism can be seen as part of an emerging paradigmatic triad representative of the mechanistically diverse calcium-regulated nuclear import pathways, many of which probably remain to be elucidated.

Another target for the calcium-induced cleavage by calpain in prostate cancer cells is E-cadherin, whose inactivation is characteristic of adenocarcinomas and is often seen as a marker for EMT [171]. The intracellular calcium-induced calpain proteolysis of FAK [172], at the same time, is a contributing factor to the modulation of cell adhesion in developing prostate cancer. The likely BKCa-mediated crosstalk with the mechanism reviewed in the last section is a promising avenue for future research. On the whole, the data on calpain demonstrate a remarkably multifaceted role of the calcium-regulated proteolysis in prostate cancer development and emphasize the simultaneous challenge and opportunity for the systems-based design of novel therapeutics that could reverse the dysregulation of this process.

7.2. 14-3-3 Mediated Nucleocytoplasmic Redistribution

As reviewed above (Section 4), AKT is one of the key signaling kinases involved in prostate cancer development, which exerts its effects, in particular, by inhibiting the FOXO3A transcription activator of the proapoptotic factor Bim through a mechanism that involves nucleocytoplasmic partitioning. Additional light has been shed on this issue by the work that has implicated the chaperone protein 14-3-3 [173]. The serine phosphorylation of FOXO3A by AKT creates a binding site for 14-3-3 and leads to retention of the complex in the cytoplasm, sequestering the transcription factor. The cytoplasmic localization of FOXO3A and 14-3-3 is correlated with the Gleason score in human samples and with disease progression in TRAMP mice [174]. In the reviewed work by Trotman et al. [103], it was shown, in particular, that pAKT in the mouse model prostate lesions accumulates in the nuclei. The mechanistic significance of this fact has been illuminated by the results that were obtained on non-prostate-cancer cells in vitro [175]. In the latter experiments, the nuclear pAKT was found to phosphorylate FOXO3A/FKHL1 immediately prior to the 14-3-3 binding and nuclear export of this transcription factor. The chaperone-mediated sequestration and nuclear export induced by the intranuclear activated kinase is an intriguing element of the nuclear transport-related mechanistic repertoire of prostate cancer, which merits further study.

7.3. *Autonomic-System Regulation*

In addition to the *PTEN* inactivation and constitutive activation of AKT, an acetylcholine-induced and calcium-dependent pathway leading to this kinase's activation has been uncovered recently [176]. Following the in-vivo indications that the autonomic, including cholinergic, nerve infiltration promoted prostate cancer development [177], the in-vitro work by Wang et al. [176] revealed the existence of autocrine acetylcholine signaling in epithelial prostate cancer cells, which proceeds through the muscarinic GPCR CHRM3 and a Ca^{2+} influx. Phosphorylation of AKT under these conditions was found to be mediated by calmodulin and CAMK kinase (CAMKK). CHRM3 is known to be upregulated in human prostate cancer samples, and experiments in vivo and in vitro have implicated the autocrine mechanism in the prostate cancer cells' proliferative and migratory potential. An antagonist treatment in vivo also increased the percentage of apoptotic cells and reduced the castration-resistant tumor growth. The most recent extension of this work has implicated the autocrine calcium-dependent mechanism in BPH and maintaining the prostate epithelial progenitor cells in the proliferative state [178]. It has also been demonstrated in LNCaP cells that carbachol-induced CAMKK-dependent blockage of apoptosis is mediated by AKT phosphorylation of the Bcl-2 associated death promoter protein BAD [179]. The effect on BAD is linked to its 14-3-3 binding and sequestration in the cytosol from the mitochondria, as shown in non-prostate-cancer cells [180]. Similar to the cytoplasmic sequestration of FOXO3A from the nucleus (last section), this mechanism adds complexity to the web of the spatially-restricted interactions of the molecules involved in the intracellular transport downstream of calcium signaling.

It is interesting to note that although the sympathetic, adrenergic effects in the Magnon et al. experiments [177] were found to be mediated by the stromal β1 and β2 receptors, a similar adrenergic effect on proliferation of primary prostate epithelial cells was demonstrated in the reviewed work of Thebault et al. [131], where it was mediated by α1 receptors. The downstream adrenergic signaling seen in the epithelial cell culture proceeded through a store-independent calcium entry via the TRPC6 channels, calcineurin, and NFAT-dependent transcription, thereby implicating the reviewed (Section 5) calcium-dependent nuclear import pathway.

8. New Molecular Classes in Calcium Signaling to the Nucleus

8.1. *GPCRs*

Alongside the adrenergic receptors, whose role was reviewed above, the chemokine cysteine (C)-X-C receptor 4 (CXCR4) is another GPCR that is remarkable for its involvement in the nuclear calcium signaling. It has been implicated in bone tropism of prostate cancer cells [181], in which it participates through its interaction with its bone-associated agonist stromal cell-derived factor 1 (SDF-1). It has also been found to promote migration and invasion of prostate cancer cells under the conditions when PTEN is inactivated by ROS [182]. It resides on the plasma membrane but also localizes specifically to the nuclei in human prostate cancer samples, as revealed by means of immunohistochemistry [13]. It similarly accumulates in the nuclear fractions of cultured prostate cancer cells, possesses an NLS, and associates with the nuclear import receptor transportin-β1. Exposure to SDF-1 results in dissociation of the α subunit from the CXCR4 in the nuclear fraction, leading to intranuclear calcium release. In addition to the identification of a new class of nuclear calcium signaling molecules associated with prostate cancer, this recent study points to a novel mechanism of escape from any therapeutic receptor antagonist action that is limited to the cell surface.

8.2. Myosins

Non-muscle myosins have recently been shown to have multiple functions in prostate cancer. The longest-known functions of the myosin family concern cell motility, and recent work on prostate cancer cells has finely delineated this class of functions among the myosin family members including IB, IXB, X, and XVIIIA [183]. The contributions of these molecules to the shaping of the actin cytoskeleton and motile morphology of prostate cancer cells underpin mechanically the acquisition of a metastatic phenotype by the originally non-motile epithelial cells [184]. Myosin IC too has long been known as a contributor to cell migration and to intracellular transport in various cell types [185]. The most recent results have implicated it in the prostate cancer cell migration and invasion [136]. The three isoforms of this protein, including the recently discovered one (A) that is associated with metastatic prostate cancer [186,187]; and the one (B) originally described as the nuclear myosin [188], differ kinetically, according to the latest work [189].

Following the discovery of the NLS [190] within the IQ domain of myosin IC that mediates the interaction with apo-calmodulin, it was demonstrated that in prostate cancer cells, elevation of the intracellular calcium concentration drives the nuclear import of this myosin [12]. This import depends on importin-β1 and is presumed to be mediated by the unmasking of the NLS when, as established for myosin IC in vitro [191], the calcium-calmodulin dissociates from the IQ domain. The most recent work [192] has demonstrated that binding to phosphoinositides, mediated in part by the NLS region, contributes to the nuclear import of myosin IC. It will be of interest to determine whether similar mechanisms regulate the nucleocytoplasmic partitioning of the other myosins known to localize to the nucleus, such as subfamilies II, V, VI, X, XVI, and XVIII [193].

The nuclear functions have to date been best characterized for myosin IC isoform B. They include promotion of transcription, which is effected through an association with RNA polymerases I and II [194–200]. This myosin isoform also associates with RNA transcripts and pre-ribosomal units, participating in the export of the ribosome subunits from the nucleus [201,202]. The movements of chromosomes inside the nucleus similarly depend on nuclear myosin and actin [203,204]. Myosin IC isoform C—long thought to be cytoplasmic but now recognized as the most responsive to the calcium-regulated nuclear import in prostate cancer cells [12]—displays a capacity for interaction with RNA polymerase II and for maintaining the level of polymerase I activity [205]. Since calmodulin, acting as the light chain, modulates the motor properties of myosin IC [206] and itself undergoes nuclear import via the facilitated pathway [207], it appears promising to investigate next the intranuclear calcium regulation of the myosin functions in prostate cancer.

The tumor-suppressor qualities of myosin IC have recently been characterized in non-prostate-cancer cells [208]. Partially similar results involving suppression of anchorage-independent growth had previously been obtained in relation to myosin XVIII in other cancers, as reviewed by Oudekirk and Krendel [10], and most recently in a clinical association study for myosin VC in colorectal cancer [209]. At the same time, in breast and prostate cancer *MYO18A* (which encodes the non-muscle myosin XVIIIA) appeared to be an oncogene [210]. Future work should be able to shed light on the possible reciprocal regulation by means of the calcium-controlled switch between the cytoplasmic and nuclear functions of the non-muscle myosins, leading to a unified picture of these phenomena.

9. Summary and Outlook

The intracellular calcium signal and the nuclear entry of the activated factors can be seen as dual, linked integrating events that undergird the molecular complexity behind the critical stages of prostate cancer development (Tables 1–3). As these multiple examples in the reviewed literature from the recent years show, the nuclear vs. cytoplasmic localization frequently presents a unique differentiating feature that is associated with a prognosis for progression of the disease. Although localized calcium signals have the capacity to channel the information transmitted by the specific pathways, the influences impinging on the intracellular calcium homeostasis establish a network for a physical integration of the conditions of the tumor microenvironment. Of special interest from the molecular biological viewpoint are the events linking the calcium signaling to the regulation of the nuclear entry, as well as the dynamic steady-state nucleocytoplasmic distribution of the transcription regulators and cytoskeletal effector molecules. The polyfunctionality and novel intracellular localizations of the previously known factors in prostate cancer development that have been uncovered in the most recent work point, on the one hand, at a greater complexity that lies ahead in the molecular exploration of the disease mechanisms. On the other hand, however, they hold out a promise for a mechanistically grounded validation of new prognostic markers and potential targets for drug development, which would be able to disrupt the critical nodes in the prostate cancer network.

Table 3. Additional molecular classes involved in the calcium and nuclear signaling in prostate cancer.

Category	Factor	Physiology	Partners	Calcium Regulation/ Nuclear Signal
proteases	calpain	metastasis, EMT	AR, β-catenin, E-cadherin	nuclear entry of cleaved β-catenin
chaperones	14-3-3	apoptosis	AKT, FOXO3A, Bim	nuclear export of FOXO3A
autonomic regulation	CHRM3	castration resistance	CAMKK, AKT, BAD	acetylcholine-induced Ca^{2+} influx
	adrenergic receptors	proliferation	TPRC6, calcineurin, NFAT	store-independent entry
GPCRs	CXCR4	bone tropism	SDF-1, PTEN, ROS, importin-β1	nuclear entry, intranuclear Ca^{2+} release
molecular motors	myosin IC	migration, invasion, metastasis	calmodulin, importin-β1, RNA polymerase I and II, nuclear actin	Ca^{2+}-induced nuclear entry

Acknowledgments: Supported in part by the NCI of the National Institutes of Health under award number R21CA220155 to Wilma A. Hofmann.

Author Contributions: Ivan V. Maly—conceptualization and writing; Wilma A. Hofmann—conceptualization, review and editing.

Abbreviations

AR	Androgen receptor
BKCa	Big potassium calcium-sensitive (channel)
BPH	Benign prostate hyperplasia
CAMK	Calcium/calmodulin-dependent kinase
CAMKK	Calcium/calmodulin-dependent kinase kinase
DcR2	Decoy receptor 2
DHT	Dihydrotestosterone
EGF	Epithelial growth factor
EMT	Epithelial-mesenchymal transition
ER	Endoplasmic reticulum
ERK	Extracellular signal-regulated kinase
FAK	Focal adhesion kinase
FGF	Fibroblast growth factor
GPCR	G protein-coupled receptor
IκB	Inhibitor of κB
IKK	Inhibitor of κB kinase
InsP3R	Inositol 1,4,5-trisphosphate receptor

LNCaP	Lymph node cancer of prostate (cell line)
MAPK	Mitogen-activated protein kinase
mTORC2	Mammalian target of rapamycin complex 2
NFAT	Nuclear factor of activated T-cells
NF-κB	Nuclear factor κB
NLS	Nuclear localization signal (sequence)
PI3K	Phosphoinositide 3-kinase
PIN	Prostate intraepithelial neoplasia
PKB	Protein kinase B
PKC	Protein kinase C
PSA	Prostate-specific antigen
PTEN	Phosphatase and tensin homolog deleted on chromosome 10
ROS	Reactive oxygen species
RyR	Ryanodine receptor
SDF-1	Stromal cell-derived factor 1
SOCE	Store-operated calcium entry
TLR	Toll-like receptor
TRAMP	Transgenic adenocarcinoma of the mouse prostate
TRP	Transient receptor potential (channel)
YB-1	Y-box binding protein 1

References

1. Miller, K.D.; Siegel, R.L.; Lin, C.C.; Mariotto, A.B.; Kramer, J.L.; Rowland, J.H.; Stein, K.D.; Alteri, R.; Jemal, A. Cancer treatment and survivorship statistics, 2016. *CA Cancer J. Clin.* **2016**, *66*, 271–289. [CrossRef] [PubMed]
2. Siegel, R.L.; Miller, K.D.; Jemal, A. Cancer statistics, 2016. *CA Cancer J. Clin.* **2016**, *66*, 7–30. [CrossRef] [PubMed]
3. Carlsson, S.; Vickers, A. Spotlight on prostate cancer: The latest evidence and current controversies. *BMC Med.* **2015**, *13*, 60. [CrossRef] [PubMed]
4. Shen, M.M.; Abate-Shen, C. Molecular genetics of prostate cancer: New prospects for old challenges. *Genes Dev.* **2010**, *24*, 1967–2000. [CrossRef] [PubMed]
5. Bootman, M.D. Calcium signaling. *Cold Spring Harb. Perspect. Biol.* **2012**, *4*, a011171. [CrossRef] [PubMed]
6. Cautain, B.; Hill, R.; de Pedro, N.; Link, W. Components and regulation of nuclear transport processes. *FEBS J.* **2015**, *282*, 445–462. [CrossRef] [PubMed]
7. Godfraind, T. Discovery and Development of Calcium Channel Blockers. *Front. Pharmacol.* **2017**, *8*, 286. [CrossRef] [PubMed]
8. Cui, C.; Merritt, R.; Fu, L.; Pan, Z. Targeting calcium signaling in cancer therapy. *Acta Pharm. Sin. B* **2017**, *7*, 3–17. [CrossRef] [PubMed]
9. Hauser, A.S.; Attwood, M.M.; Rask-Andersen, M.; Schioth, H.B.; Gloriam, D.E. Trends in GPCR drug discovery: New agents, targets and indications. *Nat. Rev. Drug Discov.* **2017**, *16*, 829–842. [CrossRef] [PubMed]
10. Ouderkirk, J.L.; Krendel, M. Non-muscle myosins in tumor progression, cancer cell invasion, and metastasis. *Cytoskeleton* **2014**, *71*, 447–463. [CrossRef] [PubMed]
11. Li, Y.R.; Yang, W.X. Myosins as fundamental components during tumorigenesis: Diverse and indispensable. *Oncotarget* **2016**, *7*, 46785–46812. [CrossRef] [PubMed]
12. Maly, I.V.; Hofmann, W.A. Calcium-regulated import of myosin IC into the nucleus. *Cytoskeleton* **2016**, *73*, 341–350. [CrossRef] [PubMed]
13. Don-Salu-Hewage, A.S.; Chan, S.Y.; McAndrews, K.M.; Chetram, M.A.; Dawson, M.R.; Bethea, D.A.; Hinton, C.V. Cysteine (C)-X-C receptor 4 undergoes transportin 1-dependent nuclear localization and remains functional at the nucleus of metastatic prostate cancer cells. *PLoS ONE* **2013**, *8*, e57194. [CrossRef] [PubMed]
14. Emmert-Buck, M.R.; Vocke, C.D.; Pozzatti, R.O.; Duray, P.H.; Jennings, S.B.; Florence, C.D.; Zhuang, Z.; Bostwick, D.G.; Liotta, L.A.; Linehan, W.M. Allelic loss on chromosome 8p12-21 in microdissected prostatic intraepithelial neoplasia. *Cancer Res.* **1995**, *55*, 2959–2962. [PubMed]

15. Bethel, C.R.; Faith, D.; Li, X.; Guan, B.; Hicks, J.L.; Lan, F.; Jenkins, R.B.; Bieberich, C.J.; De Marzo, A.M. Decreased NKX3.1 protein expression in focal prostatic atrophy, prostatic intraepithelial neoplasia, and adenocarcinoma: Association with gleason score and chromosome 8p deletion. *Cancer Res.* **2006**, *66*, 10683–10690. [CrossRef] [PubMed]
16. Asatiani, E.; Huang, W.X.; Wang, A.; Rodriguez Ortner, E.; Cavalli, L.R.; Haddad, B.R.; Gelmann, E.P. Deletion, methylation, and expression of the NKX3.1 suppressor gene in primary human prostate cancer. *Cancer Res.* **2005**, *65*, 1164–1173. [CrossRef] [PubMed]
17. Bhatia-Gaur, R.; Donjacour, A.A.; Sciavolino, P.J.; Kim, M.; Desai, N.; Young, P.; Norton, C.R.; Gridley, T.; Cardiff, R.D.; Cunha, G.R.; et al. Roles for Nkx3.1 in prostate development and cancer. *Genes Dev.* **1999**, *13*, 966–977. [CrossRef] [PubMed]
18. Schneider, A.; Brand, T.; Zweigerdt, R.; Arnold, H. Targeted disruption of the Nkx3.1 gene in mice results in morphogenetic defects of minor salivary glands: Parallels to glandular duct morphogenesis in prostate. *Mech. Dev.* **2000**, *95*, 163–174. [CrossRef]
19. Tanaka, M.; Komuro, I.; Inagaki, H.; Jenkins, N.A.; Copeland, N.G.; Izumo, S. Nkx3.1, a murine homolog of Ddrosophila bagpipe, regulates epithelial ductal branching and proliferation of the prostate and palatine glands. *Dev. Dyn.* **2000**, *219*, 248–260. [CrossRef]
20. Abdulkadir, S.A.; Magee, J.A.; Peters, T.J.; Kaleem, Z.; Naughton, C.K.; Humphrey, P.A.; Milbrandt, J. Conditional loss of Nkx3.1 in adult mice induces prostatic intraepithelial neoplasia. *Mol. Cell. Biol.* **2002**, *22*, 1495–1503. [CrossRef] [PubMed]
21. Kim, M.J.; Bhatia-Gaur, R.; Banach-Petrosky, W.A.; Desai, N.; Wang, Y.; Hayward, S.W.; Cunha, G.R.; Cardiff, R.D.; Shen, M.M.; Abate-Shen, C. Nkx3.1 mutant mice recapitulate early stages of prostate carcinogenesis. *Cancer Res.* **2002**, *62*, 2999–3004. [PubMed]
22. Lei, Q.; Jiao, J.; Xin, L.; Chang, C.J.; Wang, S.; Gao, J.; Gleave, M.E.; Witte, O.N.; Liu, X.; Wu, H. NKX3.1 stabilizes p53, inhibits AKT activation, and blocks prostate cancer initiation caused by PTEN loss. *Cancer Cell* **2006**, *9*, 367–378. [CrossRef] [PubMed]
23. Ouyang, X.; DeWeese, T.L.; Nelson, W.G.; Abate-Shen, C. Loss-of-function of Nkx3.1 promotes increased oxidative damage in prostate carcinogenesis. *Cancer Res.* **2005**, *65*, 6773–6779. [CrossRef] [PubMed]
24. Bowen, C.; Gelmann, E.P. NKX3.1 activates cellular response to DNA damage. *Cancer Res.* **2010**, *70*, 3089–3097. [CrossRef] [PubMed]
25. Markowski, M.C.; Bowen, C.; Gelmann, E.P. Inflammatory cytokines induce phosphorylation and ubiquitination of prostate suppressor protein NKX3.1. *Cancer Res.* **2008**, *68*, 6896–6901. [CrossRef] [PubMed]
26. Decker, J.; Jain, G.; Kießling, T.; Philip, S.; Rid, M.; Barth, T.T.; Möller, P.; Cronauer, M.V.; Marienfeld, R.B. Loss of the Tumor Suppressor NKX3.1 in Prostate Cancer Cells is Induced by Prostatitis Related Mitogens. *J. Clin. Exp. Oncol.* **2016**, *5*. [CrossRef]
27. Sato, K.; Qian, J.; Slezak, J.M.; Lieber, M.M.; Bostwick, D.G.; Bergstralh, E.J.; Jenkins, R.B. Clinical significance of alterations of chromosome 8 in high-grade, advanced, nonmetastatic prostate carcinoma. *J. Natl. Cancer Inst.* **1999**, *91*, 1574–1580. [CrossRef] [PubMed]
28. Jenkins, R.B.; Qian, J.; Lieber, M.M.; Bostwick, D.G. Detection of c-myc oncogene amplification and chromosomal anomalies in metastatic prostatic carcinoma by fluorescence in situ hybridization. *Cancer Res.* **1997**, *57*, 524–531. [PubMed]
29. Gurel, B.; Iwata, T.; Koh, C.M.; Jenkins, R.B.; Lan, F.; Van Dang, C.; Hicks, J.L.; Morgan, J.; Cornish, T.C.; Sutcliffe, S.; et al. Nuclear MYC protein overexpression is an early alteration in human prostate carcinogenesis. *Mod. Pathol.* **2008**, *21*, 1156–1167. [CrossRef] [PubMed]
30. Sotelo, J.; Esposito, D.; Duhagon, M.A.; Banfield, K.; Mehalko, J.; Liao, H.; Stephens, R.M.; Harris, T.J.; Munroe, D.J.; Wu, X. Long-range enhancers on 8q24 regulate c-Myc. *Proc. Natl. Acad. Sci. USA* **2010**, *107*, 3001–3005. [CrossRef] [PubMed]
31. Jia, L.; Landan, G.; Pomerantz, M.; Jaschek, R.; Herman, P.; Reich, D.; Yan, C.; Khalid, O.; Kantoff, P.; Oh, W.; et al. Functional enhancers at the gene-poor 8q24 cancer-linked locus. *PLoS Genet.* **2009**, *5*, e1000597. [CrossRef] [PubMed]
32. Al Olama, A.A.; Kote-Jarai, Z.; Giles, G.G.; Guy, M.; Morrison, J.; Severi, G.; Leongamornlert, D.A.; Tymrakiewicz, M.; Jhavar, S.; Saunders, E.; et al. Multiple loci on 8q24 associated with prostate cancer susceptibility. *Nat. Genet.* **2009**, *41*, 1058–1060. [CrossRef] [PubMed]

33. Gudmundsson, J.; Sulem, P.; Gudbjartsson, D.F.; Blondal, T.; Gylfason, A.; Agnarsson, B.A.; Benediktsdottir, K.R.; Magnusdottir, D.N.; Orlygsdottir, G.; Jakobsdottir, M.; et al. Genome-wide association and replication studies identify four variants associated with prostate cancer susceptibility. *Nat. Genet.* **2009**, *41*, 1122–1126. [CrossRef] [PubMed]
34. Ellwood-Yen, K.; Graeber, T.G.; Wongvipat, J.; Iruela-Arispe, M.L.; Zhang, J.; Matusik, R.; Thomas, G.V.; Sawyers, C.L. Myc-driven murine prostate cancer shares molecular features with human prostate tumors. *Cancer Cell* **2003**, *4*, 223–238. [CrossRef]
35. Wang, J.; Kim, J.; Roh, M.; Franco, O.E.; Hayward, S.W.; Wills, M.L.; Abdulkadir, S.A. Pim1 kinase synergizes with c-MYC to induce advanced prostate carcinoma. *Oncogene* **2010**, *29*, 2477–2487. [CrossRef] [PubMed]
36. Mamaeva, O.A.; Kim, J.; Feng, G.; McDonald, J.M. Calcium/calmodulin-dependent kinase II regulates notch-1 signaling in prostate cancer cells. *J. Cell. Biochem.* **2009**, *106*, 25–32. [CrossRef] [PubMed]
37. Mehra, R.; Tomlins, S.A.; Shen, R.; Nadeem, O.; Wang, L.; Wei, J.T.; Pienta, K.J.; Ghosh, D.; Rubin, M.A.; Chinnaiyan, A.M.; et al. Comprehensive assessment of TMPRSS2 and ETS family gene aberrations in clinically localized prostate cancer. *Mod. Pathol.* **2007**, *20*, 538–544. [CrossRef] [PubMed]
38. Tomlins, S.A.; Laxman, B.; Dhanasekaran, S.M.; Helgeson, B.E.; Cao, X.; Morris, D.S.; Menon, A.; Jing, X.; Cao, Q.; Han, B.; et al. Distinct classes of chromosomal rearrangements create oncogenic ETS gene fusions in prostate cancer. *Nature* **2007**, *448*, 595–599. [CrossRef] [PubMed]
39. Tomlins, S.A.; Rhodes, D.R.; Perner, S.; Dhanasekaran, S.M.; Mehra, R.; Sun, X.W.; Varambally, S.; Cao, X.; Tchinda, J.; Kuefer, R.; et al. Recurrent fusion of TMPRSS2 and ETS transcription factor genes in prostate cancer. *Science* **2005**, *310*, 644–648. [CrossRef] [PubMed]
40. Iljin, K.; Wolf, M.; Edgren, H.; Gupta, S.; Kilpinen, S.; Skotheim, R.I.; Peltola, M.; Smit, F.; Verhaegh, G.; Schalken, J.; et al. TMPRSS2 fusions with oncogenic ETS factors in prostate cancer involve unbalanced genomic rearrangements and are associated with HDAC1 and epigenetic reprogramming. *Cancer Res.* **2006**, *66*, 10242–10246. [CrossRef] [PubMed]
41. Rouzier, C.; Haudebourg, J.; Carpentier, X.; Valerio, L.; Amiel, J.; Michiels, J.F.; Pedeutour, F. Detection of the TMPRSS2-ETS fusion gene in prostate carcinomas: Retrospective analysis of 55 formalin-fixed and paraffin-embedded samples with clinical data. *Cancer Genet. Cytogenet.* **2008**, *183*, 21–27. [CrossRef] [PubMed]
42. Mosquera, J.M.; Perner, S.; Genega, E.M.; Sanda, M.; Hofer, M.D.; Mertz, K.D.; Paris, P.L.; Simko, J.; Bismar, T.A.; Ayala, G.; et al. Characterization of TMPRSS2-ERG fusion high-grade prostatic intraepithelial neoplasia and potential clinical implications. *Clin. Cancer Res.* **2008**, *14*, 3380–3385. [CrossRef] [PubMed]
43. Albadine, R.; Latour, M.; Toubaji, A.; Haffner, M.; Isaacs, W.B.; E, A.P.; Meeker, A.K.; Demarzo, A.M.; Epstein, J.I.; Netto, G.J. TMPRSS2-ERG gene fusion status in minute (minimal) prostatic adenocarcinoma. *Mod. Pathol.* **2009**, *22*, 1415–1422. [CrossRef] [PubMed]
44. Lin, C.; Yang, L.; Tanasa, B.; Hutt, K.; Ju, B.G.; Ohgi, K.; Zhang, J.; Rose, D.W.; Fu, X.D.; Glass, C.K.; et al. Nuclear receptor-induced chromosomal proximity and DNA breaks underlie specific translocations in cancer. *Cell* **2009**, *139*, 1069–1083. [CrossRef] [PubMed]
45. Mani, R.S.; Tomlins, S.A.; Callahan, K.; Ghosh, A.; Nyati, M.K.; Varambally, S.; Palanisamy, N.; Chinnaiyan, A.M. Induced chromosomal proximity and gene fusions in prostate cancer. *Science* **2009**, *326*, 1230. [CrossRef] [PubMed]
46. Haffner, M.C.; Aryee, M.J.; Toubaji, A.; Esopi, D.M.; Albadine, R.; Gurel, B.; Isaacs, W.B.; Bova, G.S.; Liu, W.; Xu, J.; et al. Androgen-induced TOP2B-mediated double-strand breaks and prostate cancer gene rearrangements. *Nat. Genet.* **2010**, *42*, 668–675. [CrossRef] [PubMed]
47. Yu, J.; Yu, J.; Mani, R.S.; Cao, Q.; Brenner, C.J.; Cao, X.; Wang, X.; Wu, L.; Li, J.; Hu, M.; et al. An integrated network of androgen receptor, polycomb, and TMPRSS2-ERG gene fusions in prostate cancer progression. *Cancer Cell.* **2010**, *17*, 443–454. [CrossRef] [PubMed]
48. Tomlins, S.A.; Laxman, B.; Varambally, S.; Cao, X.; Yu, J.; Helgeson, B.E.; Cao, Q.; Prensner, J.R.; Rubin, M.A.; Shah, R.B.; et al. Role of the TMPRSS2-ERG gene fusion in prostate cancer. *Neoplasia* **2008**, *10*, 177–188. [CrossRef] [PubMed]
49. Klezovitch, O.; Risk, M.; Coleman, I.; Lucas, J.M.; Null, M.; True, L.D.; Nelson, P.S.; Vasioukhin, V. A causal role for ERG in neoplastic transformation of prostate epithelium. *Proc. Natl. Acad. Sci. USA* **2008**, *105*, 2105–2110. [CrossRef] [PubMed]

50. Carver, B.S.; Tran, J.; Gopalan, A.; Chen, Z.; Shaikh, S.; Carracedo, A.; Alimonti, A.; Nardella, C.; Varmeh, S.; Scardino, P.T.; et al. Aberrant ERG expression cooperates with loss of PTEN to promote cancer progression in the prostate. *Nat. Genet.* **2009**, *41*, 619–624. [CrossRef] [PubMed]
51. King, J.C.; Xu, J.; Wongvipat, J.; Hieronymus, H.; Carver, B.S.; Leung, D.H.; Taylor, B.S.; Sander, C.; Cardiff, R.D.; Couto, S.S.; et al. Cooperativity of TMPRSS2-ERG with PI3-kinase pathway activation in prostate oncogenesis. *Nat. Genet.* **2009**, *41*, 524–526. [CrossRef] [PubMed]
52. Geybels, M.S.; Alumkal, J.J.; Luedeke, M.; Rinckleb, A.; Zhao, S.; Shui, I.M.; Bibikova, M.; Klotzle, B.; van den Brandt, P.A.; Ostrander, E.A.; et al. Epigenomic profiling of prostate cancer identifies differentially methylated genes in TMPRSS2:ERG fusion-positive versus fusion-negative tumors. *Clin. Epigenet.* **2015**, *7*, 128. [CrossRef] [PubMed]
53. Geybels, M.S.; McCloskey, K.D.; Mills, I.G.; Stanford, J.L. Calcium Channel Blocker Use and Risk of Prostate Cancer by TMPRSS2:ERG Gene Fusion Status. *Prostate* **2017**, *77*, 282–290. [CrossRef] [PubMed]
54. Wang, X.; Qiao, Y.; Asangani, I.A.; Ateeq, B.; Poliakov, A.; Cieslik, M.; Pitchiaya, S.; Chakravarthi, B.; Cao, X.; Jing, X.; et al. Development of Peptidomimetic Inhibitors of the ERG Gene Fusion Product in Prostate Cancer. *Cancer Cell* **2017**, *31*, 532–548. [CrossRef] [PubMed]
55. Rastogi, A.; Tan, S.H.; Mohamed, A.A.; Chen, Y.; Hu, Y.; Petrovics, G.; Sreenath, T.; Kagan, J.; Srivastava, S.; McLeod, D.G.; et al. Functional antagonism of TMPRSS2-ERG splice variants in prostate cancer. *Genes Cancer* **2014**, *5*, 273–284. [PubMed]
56. Gimenez-Bonafe, P.; Fedoruk, M.N.; Whitmore, T.G.; Akbari, M.; Ralph, J.L.; Ettinger, S.; Gleave, M.E.; Nelson, C.C. YB-1 is upregulated during prostate cancer tumor progression and increases P-glycoprotein activity. *Prostate* **2004**, *59*, 337–349. [CrossRef] [PubMed]
57. Shiota, M.; Takeuchi, A.; Song, Y.; Yokomizo, A.; Kashiwagi, E.; Uchiumi, T.; Kuroiwa, K.; Tatsugami, K.; Fujimoto, N.; Oda, Y.; et al. Y-box binding protein-1 promotes castration-resistant prostate cancer growth via androgen receptor expression. *Endocr. Relat. Cancer* **2011**, *18*, 505–517. [CrossRef] [PubMed]
58. Heumann, A.; Kaya, O.; Burdelski, C.; Hube-Magg, C.; Kluth, M.; Lang, D.S.; Simon, R.; Beyer, B.; Thederan, I.; Sauter, G.; et al. Up regulation and nuclear translocation of Y-box binding protein 1 (YB-1) is linked to poor prognosis in ERG-negative prostate cancer. *Sci. Rep.* **2017**, *7*, 2056. [CrossRef] [PubMed]
59. Wang, J.; Cai, Y.; Shao, L.J.; Siddiqui, J.; Palanisamy, N.; Li, R.; Ren, C.; Ayala, G.; Ittmann, M. Activation of NF-{kappa}B by TMPRSS2/ERG Fusion Isoforms through Toll-Like Receptor-4. *Cancer Res.* **2011**, *71*, 1325–1333. [CrossRef] [PubMed]
60. Nadiminty, N.; Lou, W.; Sun, M.; Chen, J.; Yue, J.; Kung, H.J.; Evans, C.P.; Zhou, Q.; Gao, A.C. Aberrant activation of the androgen receptor by NF-kappaB2/p52 in prostate cancer cells. *Cancer Res.* **2010**, *70*, 3309–3319. [CrossRef] [PubMed]
61. Singareddy, R.; Semaan, L.; Conley-Lacomb, M.K.; St John, J.; Powell, K.; Iyer, M.; Smith, D.; Heilbrun, L.K.; Shi, D.; Sakr, W.; et al. Transcriptional regulation of CXCR4 in prostate cancer: Significance of TMPRSS2-ERG fusions. *Mol. Cancer Res.* **2013**, *11*, 1349–1361. [CrossRef] [PubMed]
62. Wang, S.I.; Parsons, R.; Ittmann, M. Homozygous deletion of the PTEN tumor suppressor gene in a subset of prostate adenocarcinomas. *Clin. Cancer Res.* **1998**, *4*, 811–815. [PubMed]
63. Schmitz, M.; Grignard, G.; Margue, C.; Dippel, W.; Capesius, C.; Mossong, J.; Nathan, M.; Giacchi, S.; Scheiden, R.; Kieffer, N. Complete loss of PTEN expression as a possible early prognostic marker for prostate cancer metastasis. *Int. J. Cancer* **2007**, *120*, 1284–1292. [CrossRef] [PubMed]
64. Sircar, K.; Yoshimoto, M.; Monzon, F.A.; Koumakpayi, I.H.; Katz, R.L.; Khanna, A.; Alvarez, K.; Chen, G.; Darnel, A.D.; Aprikian, A.G.; et al. PTEN genomic deletion is associated with p-Akt and AR signalling in poorer outcome, hormone refractory prostate cancer. *J. Pathol.* **2009**, *218*, 505–513. [CrossRef] [PubMed]
65. McMenamin, M.E.; Soung, P.; Perera, S.; Kaplan, I.; Loda, M.; Sellers, W.R. Loss of PTEN expression in paraffin-embedded primary prostate cancer correlates with high Gleason score and advanced stage. *Cancer Res.* **1999**, *59*, 4291–4296. [PubMed]
66. Wang, S.; Gao, J.; Lei, Q.; Rozengurt, N.; Pritchard, C.; Jiao, J.; Thomas, G.V.; Li, G.; Roy-Burman, P.; Nelson, P.S.; et al. Prostate-specific deletion of the murine Pten tumor suppressor gene leads to metastatic prostate cancer. *Cancer Cell* **2003**, *4*, 209–221. [CrossRef]
67. Kim, M.J.; Cardiff, R.D.; Desai, N.; Banach-Petrosky, W.A.; Parsons, R.; Shen, M.M.; Abate-Shen, C. Cooperativity of Nkx3.1 and Pten loss of function in a mouse model of prostate carcinogenesis. *Proc. Natl. Acad. Sci. USA* **2002**, *99*, 2884–2889. [CrossRef] [PubMed]

68. Kim, J.; Eltoum, I.E.; Roh, M.; Wang, J.; Abdulkadir, S.A. Interactions between cells with distinct mutations in c-MYC and Pten in prostate cancer. *PLoS Genet.* **2009**, *5*, e1000542. [CrossRef] [PubMed]
69. Jia, S.; Liu, Z.; Zhang, S.; Liu, P.; Zhang, L.; Lee, S.H.; Zhang, J.; Signoretti, S.; Loda, M.; Roberts, T.M.; et al. Essential roles of PI(3)K-p110β in cell growth, metabolism and tumorigenesis. *Nature* **2008**, *454*, 776–779. [CrossRef] [PubMed]
70. Guertin, D.A.; Stevens, D.M.; Saitoh, M.; Kinkel, S.; Crosby, K.; Sheen, J.H.; Mullholland, D.J.; Magnuson, M.A.; Wu, H.; Sabatini, D.M. mTOR complex 2 is required for the development of prostate cancer induced by Pten loss in mice. *Cancer Cell* **2009**, *15*, 148–159. [CrossRef] [PubMed]
71. Gao, H.; Ouyang, X.; Banach-Petrosky, W.A.; Shen, M.M.; Abate-Shen, C. Emergence of androgen independence at early stages of prostate cancer progression in Nkx3.1; Pten mice. *Cancer Res.* **2006**, *66*, 7929–7933. [CrossRef] [PubMed]
72. Lin, H.K.; Hu, Y.C.; Lee, D.K.; Chang, C. Regulation of androgen receptor signaling by PTEN (phosphatase and tensin homolog deleted on chromosome 10) tumor suppressor through distinct mechanisms in prostate cancer cells. *Mol. Endocrinol.* **2004**, *18*, 2409–2423. [CrossRef] [PubMed]
73. Trotman, L.C.; Wang, X.; Alimonti, A.; Chen, Z.; Teruya-Feldstein, J.; Yang, H.; Pavletich, N.P.; Carver, B.S.; Cordon-Cardo, C.; Erdjument-Bromage, H.; et al. Ubiquitination regulates PTEN nuclear import and tumor suppression. *Cell* **2007**, *128*, 141–156. [CrossRef] [PubMed]
74. Chang, C.J.; Mulholland, D.J.; Valamehr, B.; Mosessian, S.; Sellers, W.R.; Wu, H. PTEN nuclear localization is regulated by oxidative stress and mediates p53-dependent tumor suppression. *Mol. Cell. Biol.* **2008**, *28*, 3281–3289. [CrossRef] [PubMed]
75. Chen, M.; Nowak, D.G.; Narula, N.; Robinson, B.; Watrud, K.; Ambrico, A.; Herzka, T.M.; Zeeman, M.E.; Minderer, M.; Zheng, W.; et al. The nuclear transport receptor Importin-11 is a tumor suppressor that maintains PTEN protein. *J. Cell Biol.* **2017**, *216*, 641–656. [CrossRef] [PubMed]
76. Schaeffer, E.M.; Marchionni, L.; Huang, Z.; Simons, B.; Blackman, A.; Yu, W.; Parmigiani, G.; Berman, D.M. Androgen-induced programs for prostate epithelial growth and invasion arise in embryogenesis and are reactivated in cancer. *Oncogene* **2008**, *27*, 7180–7191. [CrossRef] [PubMed]
77. Pritchard, C.; Mecham, B.; Dumpit, R.; Coleman, I.; Bhattacharjee, M.; Chen, Q.; Sikes, R.A.; Nelson, P.S. Conserved gene expression programs integrate mammalian prostate development and tumorigenesis. *Cancer Res.* **2009**, *69*, 1739–1747. [CrossRef] [PubMed]
78. Bruxvoort, K.J.; Charbonneau, H.M.; Giambernardi, T.A.; Goolsby, J.C.; Qian, C.N.; Zylstra, C.R.; Robinson, D.R.; Roy-Burman, P.; Shaw, A.K.; Buckner-Berghuis, B.D.; et al. Inactivation of Apc in the mouse prostate causes prostate carcinoma. *Cancer Res.* **2007**, *67*, 2490–2496. [CrossRef] [PubMed]
79. Pearson, H.B.; Phesse, T.J.; Clarke, A.R. K-ras and Wnt signaling synergize to accelerate prostate tumorigenesis in the mouse. *Cancer Res.* **2009**, *69*, 94–101. [CrossRef] [PubMed]
80. Yu, X.; Wang, Y.; Jiang, M.; Bierie, B.; Roy-Burman, P.; Shen, M.M.; Taketo, M.M.; Wills, M.; Matusik, R.J. Activation of β-Catenin in mouse prostate causes HGPIN and continuous prostate growth after castration. *Prostate* **2009**, *69*, 249–262. [CrossRef] [PubMed]
81. Wang, G.; Wang, J.; Sadar, M.D. Crosstalk between the androgen receptor and β-catenin in castrate-resistant prostate cancer. *Cancer Res.* **2008**, *68*, 9918–9927. [CrossRef] [PubMed]
82. Horvath, L.G.; Henshall, S.M.; Lee, C.S.; Kench, J.G.; Golovsky, D.; Brenner, P.C.; O'Neill, G.F.; Kooner, R.; Stricker, P.D.; Grygiel, J.J.; et al. Lower levels of nuclear β-catenin predict for a poorer prognosis in localized prostate cancer. *Int. J. Cancer* **2005**, *113*, 415–422. [CrossRef] [PubMed]
83. Whitaker, H.C.; Girling, J.; Warren, A.Y.; Leung, H.; Mills, I.G.; Neal, D.E. Alterations in β-catenin expression and localization in prostate cancer. *Prostate* **2008**, *68*, 1196–1205. [CrossRef] [PubMed]
84. Wang, Q.; Symes, A.J.; Kane, C.A.; Freeman, A.; Nariculam, J.; Munson, P.; Thrasivoulou, C.; Masters, J.R.; Ahmed, A. A novel role for Wnt/Ca^{2+} signaling in actin cytoskeleton remodeling and cell motility in prostate cancer. *PLoS ONE* **2010**, *5*, e10456. [CrossRef] [PubMed]
85. Thrasivoulou, C.; Millar, M.; Ahmed, A. Activation of intracellular calcium by multiple Wnt ligands and translocation of β-catenin into the nucleus: A convergent model of Wnt/Ca^{2+} and Wnt/β-catenin pathways. *J. Biol. Chem.* **2013**, *288*, 35651–35659. [CrossRef] [PubMed]
86. Nicotera, P.; Zhivotovsky, B.; Orrenius, S. Nuclear calcium transport and the role of calcium in apoptosis. *Cell Calcium* **1994**, *16*, 279–288. [CrossRef]

87. Hsu, J.L.; Pan, S.L.; Ho, Y.F.; Hwang, T.L.; Kung, F.L.; Guh, J.H. Costunolide induces apoptosis through nuclear Ca^{2+} overload and DNA damage response in human prostate cancer. *J. Urol.* **2011**, *185*, 1967–1974. [CrossRef] [PubMed]
88. Karhadkar, S.S.; Bova, G.S.; Abdallah, N.; Dhara, S.; Gardner, D.; Maitra, A.; Isaacs, J.T.; Berman, D.M.; Beachy, P.A. Hedgehog signalling in prostate regeneration, neoplasia and metastasis. *Nature* **2004**, *431*, 707–712. [CrossRef] [PubMed]
89. Sanchez, P.; Hernandez, A.M.; Stecca, B.; Kahler, A.J.; DeGueme, A.M.; Barrett, A.; Beyna, M.; Datta, M.W.; Datta, S.; Ruiz i Altaba, A. Inhibition of prostate cancer proliferation by interference with SONIC HEDGEHOG-GLI1 signaling. *Proc. Natl. Acad. Sci. USA* **2004**, *101*, 12561–12566. [CrossRef] [PubMed]
90. Shaw, A.; Gipp, J.; Bushman, W. The Sonic Hedgehog pathway stimulates prostate tumor growth by paracrine signaling and recapitulates embryonic gene expression in tumor myofibroblasts. *Oncogene* **2009**, *28*, 4480–4490. [CrossRef] [PubMed]
91. Yauch, R.L.; Gould, S.E.; Scales, S.J.; Tang, T.; Tian, H.; Ahn, C.P.; Marshall, D.; Fu, L.; Januario, T.; Kallop, D.; et al. A paracrine requirement for hedgehog signalling in cancer. *Nature* **2008**, *455*, 406–410. [CrossRef] [PubMed]
92. Gonnissen, A.; Isebaert, S.; Haustermans, K. Hedgehog signaling in prostate cancer and its therapeutic implication. *Int. J. Mol. Sci.* **2013**, *14*, 13979–14007. [CrossRef] [PubMed]
93. Li, N.; Truong, S.; Nouri, M.; Moore, J.; Al Nakouzi, N.; Lubik, A.A.; Buttyan, R. Non-canonical activation of hedgehog in prostate cancer cells mediated by the interaction of transcriptionally active androgen receptor proteins with Gli3. *Oncogene* **2018**. [CrossRef] [PubMed]
94. Acevedo, V.D.; Gangula, R.D.; Freeman, K.W.; Li, R.; Zhang, Y.; Wang, F.; Ayala, G.E.; Peterson, L.E.; Ittmann, M.; Spencer, D.M. Inducible FGFR-1 activation leads to irreversible prostate adenocarcinoma and an epithelial-to-mesenchymal transition. *Cancer Cell* **2007**, *12*, 559–571. [CrossRef] [PubMed]
95. Memarzadeh, S.; Xin, L.; Mulholland, D.J.; Mansukhani, A.; Wu, H.; Teitell, M.A.; Witte, O.N. Enhanced paracrine FGF10 expression promotes formation of multifocal prostate adenocarcinoma and an increase in epithelial androgen receptor. *Cancer Cell* **2007**, *12*, 572–585. [CrossRef] [PubMed]
96. Kwabi-Addo, B.; Ozen, M.; Ittmann, M. The role of fibroblast growth factors and their receptors in prostate cancer. *Endocr. Relat. Cancer* **2004**, *11*, 709–724. [CrossRef] [PubMed]
97. Chen, M.L.; Xu, P.Z.; Peng, X.D.; Chen, W.S.; Guzman, G.; Yang, X.; Di Cristofano, A.; Pandolfi, P.P.; Hay, N. The deficiency of *Akt1* is sufficient to suppress tumor development in *Pten+/−* mice. *Genes Dev.* **2006**, *20*, 1569–1574. [CrossRef] [PubMed]
98. Thomas, G.V.; Horvath, S.; Smith, B.L.; Crosby, K.; Lebel, L.A.; Schrage, M.; Said, J.; De Kernion, J.; Reiter, R.E.; Sawyers, C.L. Antibody-based profiling of the phosphoinositide 3-kinase pathway in clinical prostate cancer. *Clin. Cancer Res.* **2004**, *10*, 8351–8356. [CrossRef] [PubMed]
99. Shen, M.M.; Abate-Shen, C. Pten inactivation and the emergence of androgen-independent prostate cancer. *Cancer Res.* **2007**, *67*, 6535–6538. [CrossRef] [PubMed]
100. Boormans, J.L.; Hermans, K.G.; van Leenders, G.J.; Trapman, J.; Verhagen, P.C. An activating mutation in AKT1 in human prostate cancer. *Int. J. Cancer* **2008**, *123*, 2725–2726. [CrossRef] [PubMed]
101. Lee, S.H.; Poulogiannis, G.; Pyne, S.; Jia, S.; Zou, L.; Signoretti, S.; Loda, M.; Cantley, L.C.; Roberts, T.M. A constitutively activated form of the p110β isoform of PI3-kinase induces prostatic intraepithelial neoplasia in mice. *Proc. Natl. Acad. Sci. USA* **2010**, *107*, 11002–11007. [CrossRef] [PubMed]
102. Gao, H.; Ouyang, X.; Banach-Petrosky, W.A.; Gerald, W.L.; Shen, M.M.; Abate-Shen, C. Combinatorial activities of Akt and B-Raf/Erk signaling in a mouse model of androgen-independent prostate cancer. *Proc. Natl. Acad. Sci. USA* **2006**, *103*, 14477–14482. [CrossRef] [PubMed]
103. Trotman, L.C.; Alimonti, A.; Scaglioni, P.P.; Koutcher, J.A.; Cordon-Cardo, C.; Pandolfi, P.P. Identification of a tumour suppressor network opposing nuclear Akt function. *Nature* **2006**, *441*, 523–527. [CrossRef] [PubMed]
104. Kondratskyi, A.; Yassine, M.; Slomianny, C.; Kondratska, K.; Gordienko, D.; Dewailly, E.; Lehen'kyi, V.; Skryma, R.; Prevarskaya, N. Identification of ML-9 as a lysosomotropic agent targeting autophagy and cell death. *Cell Death Dis.* **2014**, *5*, e1193. [CrossRef] [PubMed]
105. Uzgare, A.R.; Isaacs, J.T. Enhanced redundancy in Akt and mitogen-activated protein kinase-induced survival of malignant versus normal prostate epithelial cells. *Cancer Res.* **2004**, *64*, 6190–6199. [CrossRef] [PubMed]

106. Kinkade, C.W.; Castillo-Martin, M.; Puzio-Kuter, A.; Yan, J.; Foster, T.H.; Gao, H.; Sun, Y.; Ouyang, X.; Gerald, W.L.; Cordon-Cardo, C.; et al. Targeting AKT/mTOR and ERK MAPK signaling inhibits hormone-refractory prostate cancer in a preclinical mouse model. *J. Clin. Investig.* **2008**, *118*, 3051–3064. [CrossRef] [PubMed]
107. Dan, H.C.; Cooper, M.J.; Cogswell, P.C.; Duncan, J.A.; Ting, J.P.; Baldwin, A.S. Akt-dependent regulation of NF-{kappa}B is controlled by mTOR and Raptor in association with IKK. *Genes Dev.* **2008**, *22*, 1490–1500. [CrossRef] [PubMed]
108. Fradet, V.; Lessard, L.; Begin, L.R.; Karakiewicz, P.; Masson, A.M.; Saad, F. Nuclear factor-kappaB nuclear localization is predictive of biochemical recurrence in patients with positive margin prostate cancer. *Clin. Cancer Res.* **2004**, *10*, 8460–8464. [CrossRef] [PubMed]
109. Ismail, H.A.; Lessard, L.; Mes-Masson, A.M.; Saad, F. Expression of NF-kappaB in prostate cancer lymph node metastases. *Prostate* **2004**, *58*, 308–313. [CrossRef] [PubMed]
110. Lessard, L.; Karakiewicz, P.I.; Bellon-Gagnon, P.; Alam-Fahmy, M.; Ismail, H.A.; Mes-Masson, A.M.; Saad, F. Nuclear localization of nuclear factor-kappaB p65 in primary prostate tumors is highly predictive of pelvic lymph node metastases. *Clin. Cancer Res.* **2006**, *12*, 5741–5745. [CrossRef] [PubMed]
111. Luo, J.L.; Tan, W.; Ricono, J.M.; Korchynskyi, O.; Zhang, M.; Gonias, S.L.; Cheresh, D.A.; Karin, M. Nuclear cytokine-activated IKKα controls prostate cancer metastasis by repressing Maspin. *Nature* **2007**, *446*, 690–694. [CrossRef] [PubMed]
112. Zhang, L.; Altuwaijri, S.; Deng, F.; Chen, L.; Lal, P.; Bhanot, U.K.; Korets, R.; Wenske, S.; Lilja, H.G.; Chang, C.; et al. NF-kappaB regulates androgen receptor expression and prostate cancer growth. *Am. J. Pathol.* **2009**, *175*, 489–499. [CrossRef] [PubMed]
113. Jeong, J.H.; Wang, Z.; Guimaraes, A.S.; Ouyang, X.; Figueiredo, J.L.; Ding, Z.; Jiang, S.; Guney, I.; Kang, G.H.; Shin, E.; et al. BRAF activation initiates but does not maintain invasive prostate adenocarcinoma. *PLoS ONE* **2008**, *3*, e3949. [CrossRef] [PubMed]
114. Taylor, B.S.; Schultz, N.; Hieronymus, H.; Gopalan, A.; Xiao, Y.; Carver, B.S.; Arora, V.K.; Kaushik, P.; Cerami, E.; Reva, B.; et al. Integrative genomic profiling of human prostate cancer. *Cancer Cell* **2010**, *18*, 11–22. [CrossRef] [PubMed]
115. Gioeli, D.; Mandell, J.W.; Petroni, G.R.; Frierson, H.F., Jr.; Weber, M.J. Activation of mitogen-activated protein kinase associated with prostate cancer progression. *Cancer Res.* **1999**, *59*, 279–284. [PubMed]
116. Roderick, H.L.; Cook, S.J. Ca^{2+} signalling checkpoints in cancer: Remodelling Ca^{2+} for cancer cell proliferation and survival. *Nat. Rev. Cancer* **2008**, *8*, 361–375. [CrossRef] [PubMed]
117. Hanahan, D.; Weinberg, R.A. Hallmarks of cancer: The next generation. *Cell* **2011**, *144*, 646–674. [CrossRef] [PubMed]
118. Xu, Y.; Zhang, S.; Niu, H.; Ye, Y.; Hu, F.; Chen, S.; Li, X.; Luo, X.; Jiang, S.; Liu, Y.; et al. STIM1 accelerates cell senescence in a remodeled microenvironment but enhances the epithelial-to-mesenchymal transition in prostate cancer. *Sci. Rep.* **2015**, *5*, 11754. [CrossRef] [PubMed]
119. van Noesel, M.M.; van Bezouw, S.; Salomons, G.S.; Voute, P.A.; Pieters, R.; Baylin, S.B.; Herman, J.G.; Versteeg, R. Tumor-specific down-regulation of the tumor necrosis factor-related apoptosis-inducing ligand decoy receptors DcR1 and DcR2 is associated with dense promoter hypermethylation. *Cancer Res.* **2002**, *62*, 2157–2161. [PubMed]
120. Sheikh, M.S.; Fornace, A.J., Jr. Death and decoy receptors and p53-mediated apoptosis. *Leukemia* **2000**, *14*, 1509–1513. [CrossRef] [PubMed]
121. Collado, M.; Gil, J.; Efeyan, A.; Guerra, C.; Schuhmacher, A.J.; Barradas, M.; Benguria, A.; Zaballos, A.; Flores, J.M.; Barbacid, M.; et al. Tumour biology: Senescence in premalignant tumours. *Nature* **2005**, *436*, 642. [CrossRef] [PubMed]
122. Althubiti, M.; Lezina, L.; Carrera, S.; Jukes-Jones, R.; Giblett, S.M.; Antonov, A.; Barlev, N.; Saldanha, G.S.; Pritchard, C.A.; Cain, K.; et al. Characterization of novel markers of senescence and their prognostic potential in cancer. *Cell Death Dis.* **2014**, *5*, e1528. [CrossRef] [PubMed]
123. Lee, S.J.; Lee, K.; Yang, X.; Jung, C.; Gardner, T.; Kim, H.S.; Jeng, M.H.; Kao, C. NFATc1 with AP-3 site binding specificity mediates gene expression of prostate-specific-membrane-antigen. *J. Mol. Biol.* **2003**, *330*, 749–760. [CrossRef]

124. Kavitha, C.V.; Deep, G.; Gangar, S.C.; Jain, A.K.; Agarwal, C.; Agarwal, R. Silibinin inhibits prostate cancer cells- and RANKL-induced osteoclastogenesis by targeting NFATc1, NF-kappaB, and AP-1 activation in RAW264.7 cells. *Mol. Carcinog.* **2014**, *53*, 169–180. [CrossRef] [PubMed]
125. Mognol, G.P.; Carneiro, F.R.; Robbs, B.K.; Faget, D.V.; Viola, J.P. Cell cycle and apoptosis regulation by NFAT transcription factors: New roles for an old player. *Cell Death Dis.* **2016**, *7*, e2199. [CrossRef] [PubMed]
126. Manda, K.R.; Tripathi, P.; Hsi, A.C.; Ning, J.; Ruzinova, M.B.; Liapis, H.; Bailey, M.; Zhang, H.; Maher, C.A.; Humphrey, P.A.; et al. NFATc1 promotes prostate tumorigenesis and overcomes PTEN loss-induced senescence. *Oncogene* **2016**, *35*, 3282–3292. [CrossRef] [PubMed]
127. Wei, Z.; Jiang, X.; Qiao, H.; Zhai, B.; Zhang, L.; Zhang, Q.; Wu, Y.; Jiang, H.; Sun, X. STAT3 interacts with Skp2/p27/p21 pathway to regulate the motility and invasion of gastric cancer cells. *Cell. Signal.* **2013**, *25*, 931–938. [CrossRef] [PubMed]
128. Huang, H.; Zhao, W.; Yang, D. Stat3 induces oncogenic Skp2 expression in human cervical carcinoma cells. *Biochem. Biophys. Res. Commun.* **2012**, *418*, 186–190. [CrossRef] [PubMed]
129. Wissenbach, U.; Niemeyer, B.A.; Fixemer, T.; Schneidewind, A.; Trost, C.; Cavalie, A.; Reus, K.; Meese, E.; Bonkhoff, H.; Flockerzi, V. Expression of CaT-like, a novel calcium-selective channel, correlates with the malignancy of prostate cancer. *J. Biol. Chem.* **2001**, *276*, 19461–19468. [CrossRef] [PubMed]
130. Fixemer, T.; Wissenbach, U.; Flockerzi, V.; Bonkhoff, H. Expression of the Ca^{2+}-selective cation channel TRPV6 in human prostate cancer: A novel prognostic marker for tumor progression. *Oncogene* **2003**, *22*, 7858–7861. [CrossRef] [PubMed]
131. Thebault, S.; Flourakis, M.; Vanoverberghe, K.; Vandermoere, F.; Roudbaraki, M.; Lehen'kyi, V.; Slomianny, C.; Beck, B.; Mariot, P.; Bonnal, J.L.; et al. Differential role of transient receptor potential channels in Ca^{2+} entry and proliferation of prostate cancer epithelial cells. *Cancer Res.* **2006**, *66*, 2038–2047. [CrossRef] [PubMed]
132. Lehen'kyi, V.; Flourakis, M.; Skryma, R.; Prevarskaya, N. TRPV6 channel controls prostate cancer cell proliferation via Ca^{2+}/NFAT-dependent pathways. *Oncogene* **2007**, *26*, 7380–7385. [CrossRef] [PubMed]
133. Raphael, M.; Lehen'kyi, V.; Vandenberghe, M.; Beck, B.; Khalimonchyk, S.; Vanden Abeele, F.; Farsetti, L.; Germain, E.; Bokhobza, A.; Mihalache, A.; et al. TRPV6 calcium channel translocates to the plasma membrane via Orai1-mediated mechanism and controls cancer cell survival. *Proc. Natl. Acad. Sci. USA* **2014**, *111*, E3870–E3879. [CrossRef] [PubMed]
134. Fu, S.; Hirte, H.; Welch, S.; Ilenchuk, T.T.; Lutes, T.; Rice, C.; Fields, N.; Nemet, A.; Dugourd, D.; Piha-Paul, S.; et al. First-in-human phase I study of SOR-C13, a TRPV6 calcium channel inhibitor, in patients with advanced solid tumors. *Investig. New Drugs* **2017**, *35*, 324–333. [CrossRef] [PubMed]
135. Monet, M.; Lehen'kyi, V.; Gackiere, F.; Firlej, V.; Vandenberghe, M.; Roudbaraki, M.; Gkika, D.; Pourtier, A.; Bidaux, G.; Slomianny, C.; et al. Role of cationic channel TRPV2 in promoting prostate cancer migration and progression to androgen resistance. *Cancer Res.* **2010**, *70*, 1225–1235. [CrossRef] [PubMed]
136. Maly, I.V.; Domaradzki, T.M.; Gosy, V.A.; Hofmann, W.A. Myosin isoform expressed in metastatic prostate cancer stimulates cell invasion. *Sci. Rep.* **2017**, *7*, 8476. [CrossRef] [PubMed]
137. Tsavaler, L.; Shapero, M.H.; Morkowski, S.; Laus, R. Trp-p8, a novel prostate-specific gene, is up-regulated in prostate cancer and other malignancies and shares high homology with transient receptor potential calcium channel proteins. *Cancer Res.* **2001**, *61*, 3760–3769. [PubMed]
138. Zhang, L.; Barritt, G.J. Evidence that TRPM8 is an androgen-dependent Ca^{2+} channel required for the survival of prostate cancer cells. *Cancer Res.* **2004**, *64*, 8365–8373. [CrossRef] [PubMed]
139. Monteith, G.R.; Davis, F.M.; Roberts-Thomson, S.J. Calcium channels and pumps in cancer: Changes and consequences. *J. Biol. Chem.* **2012**, *287*, 31666–31673. [CrossRef] [PubMed]
140. Henshall, S.M.; Afar, D.E.; Hiller, J.; Horvath, L.G.; Quinn, D.I.; Rasiah, K.K.; Gish, K.; Willhite, D.; Kench, J.G.; Gardiner-Garden, M.; et al. Survival analysis of genome-wide gene expression profiles of prostate cancers identifies new prognostic targets of disease relapse. *Cancer Res.* **2003**, *63*, 4196–4203. [PubMed]
141. Gkika, D.; Flourakis, M.; Lemonnier, L.; Prevarskaya, N. PSA reduces prostate cancer cell motility by stimulating TRPM8 activity and plasma membrane expression. *Oncogene* **2010**, *29*, 4611–4616. [CrossRef] [PubMed]
142. Bidaux, G.; Flourakis, M.; Thebault, S.; Zholos, A.; Beck, B.; Gkika, D.; Roudbaraki, M.; Bonnal, J.L.; Mauroy, B.; Shuba, Y.; et al. Prostate cell differentiation status determines transient receptor potential melastatin member 8 channel subcellular localization and function. *J. Clin. Investig.* **2007**, *117*, 1647–1657. [CrossRef] [PubMed]

143. Vanoverberghe, K.; Vanden Abeele, F.; Mariot, P.; Lepage, G.; Roudbaraki, M.; Bonnal, J.L.; Mauroy, B.; Shuba, Y.; Skryma, R.; Prevarskaya, N. Ca^{2+} homeostasis and apoptotic resistance of neuroendocrine-differentiated prostate cancer cells. *Cell Death Differ.* **2004**, *11*, 321–330. [CrossRef] [PubMed]
144. Vanden Abeele, F.; Skryma, R.; Shuba, Y.; Van Coppenolle, F.; Slomianny, C.; Roudbaraki, M.; Mauroy, B.; Wuytack, F.; Prevarskaya, N. Bcl-2-dependent modulation of Ca^{2+} homeostasis and store-operated channels in prostate cancer cells. *Cancer Cell* **2002**, *1*, 169–179. [CrossRef]
145. Chen, L.; Cao, R.; Wang, G.; Yuan, L.; Qian, G.; Guo, Z.; Wu, C.L.; Wang, X.; Xiao, Y. Downregulation of TRPM7 suppressed migration and invasion by regulating epithelial-mesenchymal transition in prostate cancer cells. *Med. Oncol.* **2017**, *34*, 127. [CrossRef] [PubMed]
146. Sagredo, A.I.; Sagredo, E.A.; Cappelli, C.; Baez, P.; Andaur, R.E.; Blanco, C.; Tapia, J.C.; Echeverria, C.; Cerda, O.; Stutzin, A.; et al. TRPM4 regulates Akt/GSK3-β activity and enhances β-catenin signaling and cell proliferation in prostate cancer cells. *Mol. Oncol.* **2018**, *12*, 151–165. [CrossRef] [PubMed]
147. Zeng, X.; Sikka, S.C.; Huang, L.; Sun, C.; Xu, C.; Jia, D.; Abdel-Mageed, A.B.; Pottle, J.E.; Taylor, J.T.; Li, M. Novel role for the transient receptor potential channel TRPM2 in prostate cancer cell proliferation. *Prostate Cancer Prostatic Dis.* **2010**, *13*, 195–201. [CrossRef] [PubMed]
148. Haverstick, D.M.; Heady, T.N.; Macdonald, T.L.; Gray, L.S. Inhibition of human prostate cancer proliferation in vitro and in a mouse model by a compound synthesized to block Ca^{2+} entry. *Cancer Res.* **2000**, *60*, 1002–1008. [PubMed]
149. Gray, L.S.; Perez-Reyes, E.; Gomora, J.C.; Haverstick, D.M.; Shattock, M.; McLatchie, L.; Harper, J.; Brooks, G.; Heady, T.; Macdonald, T.L. The role of voltage gated T-type Ca^{2+} channel isoforms in mediating "capacitative" Ca^{2+} entry in cancer cells. *Cell Calcium* **2004**, *36*, 489–497. [CrossRef] [PubMed]
150. Panner, A.; Wurster, R.D. T-type calcium channels and tumor proliferation. *Cell Calcium* **2006**, *40*, 253–259. [CrossRef] [PubMed]
151. Parimi, V.; Goyal, R.; Poropatich, K.; Yang, X.J. Neuroendocrine differentiation of prostate cancer: A review. *Am. J. Clin. Exp. Urol.* **2014**, *2*, 273–285. [PubMed]
152. Mariot, P.; Vanoverberghe, K.; Lalevee, N.; Rossier, M.F.; Prevarskaya, N. Overexpression of an α1H (Cav3.2) T-type calcium channel during neuroendocrine differentiation of human prostate cancer cells. *J. Biol. Chem.* **2002**, *277*, 10824–10833. [CrossRef] [PubMed]
153. Rossier, M.F.; Lesouhaitier, O.; Perrier, E.; Bockhorn, L.; Chiappe, A.; Lalevee, N. Aldosterone regulation of T-type calcium channels. *J. Steroid Biochem. Mol. Biol.* **2003**, *85*, 383–388. [CrossRef]
154. Contreras, G.F.; Castillo, K.; Enrique, N.; Carrasquel-Ursulaez, W.; Castillo, J.P.; Milesi, V.; Neely, A.; Alvarez, O.; Ferreira, G.; Gonzalez, C.; et al. A BK (Slo1) channel journey from molecule to physiology. *Channels* **2013**, *7*, 442–458. [CrossRef] [PubMed]
155. Ge, L.; Hoa, N.T.; Wilson, Z.; Arismendi-Morillo, G.; Kong, X.T.; Tajhya, R.B.; Beeton, C.; Jadus, M.R. Big Potassium (BK) ion channels in biology, disease and possible targets for cancer immunotherapy. *Int. Immunopharmacol.* **2014**, *22*, 427–443. [CrossRef] [PubMed]
156. Bloch, M.; Ousingsawat, J.; Simon, R.; Schraml, P.; Gasser, T.C.; Mihatsch, M.J.; Kunzelmann, K.; Bubendorf, L. KCNMA1 gene amplification promotes tumor cell proliferation in human prostate cancer. *Oncogene* **2007**, *26*, 2525–2534. [CrossRef] [PubMed]
157. Du, C.; Zheng, Z.; Li, D.; Chen, L.; Li, N.; Yi, X.; Yang, Y.; Guo, F.; Liu, W.; Xie, X.; et al. BKCa promotes growth and metastasis of prostate cancer through facilitating the coupling between αvβ3 integrin and FAK. *Oncotarget* **2016**, *7*, 40174–40188. [CrossRef] [PubMed]
158. Legrand, G.; Humez, S.; Slomianny, C.; Dewailly, E.; Vanden Abeele, F.; Mariot, P.; Wuytack, F.; Prevarskaya, N. Ca^{2+} pools and cell growth. Evidence for sarcoendoplasmic Ca^{2+}-ATPases 2B involvement in human prostate cancer cell growth control. *J. Biol. Chem.* **2001**, *276*, 47608–47614. [CrossRef] [PubMed]
159. Denmeade, S.R.; Mhaka, A.M.; Rosen, D.M.; Brennen, W.N.; Dalrymple, S.; Dach, I.; Olesen, C.; Gurel, B.; Demarzo, A.M.; Wilding, G.; et al. Engineering a prostate-specific membrane antigen-activated tumor endothelial cell prodrug for cancer therapy. *Sci. Transl. Med.* **2012**, *4*, 140ra186. [CrossRef] [PubMed]
160. Flourakis, M.; Lehen'kyi, V.; Beck, B.; Raphael, M.; Vandenberghe, M.; Abeele, F.V.; Roudbaraki, M.; Lepage, G.; Mauroy, B.; Romanin, C.; et al. Orai1 contributes to the establishment of an apoptosis-resistant phenotype in prostate cancer cells. *Cell Death Dis.* **2010**, *1*, e75. [CrossRef] [PubMed]

161. Dubois, C.; Vanden Abeele, F.; Lehen'kyi, V.; Gkika, D.; Guarmit, B.; Lepage, G.; Slomianny, C.; Borowiec, A.S.; Bidaux, G.; Benahmed, M.; et al. Remodeling of channel-forming ORAI proteins determines an oncogenic switch in prostate cancer. *Cancer Cell* **2014**, *26*, 19–32. [CrossRef] [PubMed]
162. Marchi, S.; Lupini, L.; Patergnani, S.; Rimessi, A.; Missiroli, S.; Bonora, M.; Bononi, A.; Corra, F.; Giorgi, C.; De Marchi, E.; et al. Downregulation of the mitochondrial calcium uniporter by cancer-related miR-25. *Curr. Biol.* **2013**, *23*, 58–63. [CrossRef] [PubMed]
163. Orrenius, S.; Zhivotovsky, B.; Nicotera, P. Regulation of cell death: The calcium-apoptosis link. *Nat. Rev. Mol. Cell Biol.* **2003**, *4*, 552–565. [CrossRef] [PubMed]
164. Vlietstra, R.J.; van Alewijk, D.C.; Hermans, K.G.; van Steenbrugge, G.J.; Trapman, J. Frequent inactivation of PTEN in prostate cancer cell lines and xenografts. *Cancer Res.* **1998**, *58*, 2720–2723. [PubMed]
165. Carson, J.P.; Kulik, G.; Weber, M.J. Antiapoptotic signaling in LNCaP prostate cancer cells: A survival signaling pathway independent of phosphatidylinositol 3′-kinase and Akt/protein kinase B. *Cancer Res.* **1999**, *59*, 1449–1453. [PubMed]
166. Khan, M.T.; Wagner, L., 2nd; Yule, D.I.; Bhanumathy, C.; Joseph, S.K. Akt kinase phosphorylation of inositol 1,4,5-trisphosphate receptors. *J. Biol. Chem.* **2006**, *281*, 3731–3737. [CrossRef] [PubMed]
167. Kuchay, S.; Giorgi, C.; Simoneschi, D.; Pagan, J.; Missiroli, S.; Saraf, A.; Florens, L.; Washburn, M.P.; Collazo-Lorduy, A.; Castillo-Martin, M.; et al. PTEN counteracts FBXL2 to promote IP3R3- and Ca^{2+}-mediated apoptosis limiting tumour growth. *Nature* **2017**, *546*, 554–558. [CrossRef] [PubMed]
168. Kobylewski, S.E.; Henderson, K.A.; Eckhert, C.D. Identification of ryanodine receptor isoforms in prostate DU-145, LNCaP, and PWR-1E cells. *Biochem. Biophys. Res. Commun.* **2012**, *425*, 431–435. [CrossRef] [PubMed]
169. Mariot, P.; Prevarskaya, N.; Roudbaraki, M.M.; Le Bourhis, X.; Van Coppenolle, F.; Vanoverberghe, K.; Skryma, R. Evidence of functional ryanodine receptor involved in apoptosis of prostate cancer (LNCaP) cells. *Prostate* **2000**, *43*, 205–214. [CrossRef]
170. Rios-Doria, J.; Kuefer, R.; Ethier, S.P.; Day, M.L. Cleavage of β-catenin by calpain in prostate and mammary tumor cells. *Cancer Res.* **2004**, *64*, 7237–7240. [CrossRef] [PubMed]
171. Rios-Doria, J.; Day, K.C.; Kuefer, R.; Rashid, M.G.; Chinnaiyan, A.M.; Rubin, M.A.; Day, M.L. The role of calpain in the proteolytic cleavage of E-cadherin in prostate and mammary epithelial cells. *J. Biol. Chem.* **2003**, *278*, 1372–1379. [CrossRef] [PubMed]
172. Park, J.J.; Rubio, M.V.; Zhang, Z.; Um, T.; Xie, Y.; Knoepp, S.M.; Snider, A.J.; Gibbs, T.C.; Meier, K.E. Effects of lysophosphatidic acid on calpain-mediated proteolysis of focal adhesion kinase in human prostate cancer cells. *Prostate* **2012**, *72*, 1595–1610. [CrossRef] [PubMed]
173. Shukla, S.; Shukla, M.; Maclennan, G.T.; Fu, P.; Gupta, S. Deregulation of FOXO3A during prostate cancer progression. *Int. J. Oncol.* **2009**, *34*, 1613–1620. [PubMed]
174. Shukla, S.; Bhaskaran, N.; Maclennan, G.T.; Gupta, S. Deregulation of FoxO3a accelerates prostate cancer progression in TRAMP mice. *Prostate* **2013**, *73*, 1507–1517. [CrossRef] [PubMed]
175. Brunet, A.; Kanai, F.; Stehn, J.; Xu, J.; Sarbassova, D.; Frangioni, J.V.; Dalal, S.N.; DeCaprio, J.A.; Greenberg, M.E.; Yaffe, M.B. 14-3-3 transits to the nucleus and participates in dynamic nucleocytoplasmic transport. *J. Cell Biol.* **2002**, *156*, 817–828. [CrossRef] [PubMed]
176. Wang, N.; Yao, M.; Xu, J.; Quan, Y.; Zhang, K.; Yang, R.; Gao, W.Q. Autocrine Activation of CHRM3 Promotes Prostate Cancer Growth and Castration Resistance via CaM/CaMKK-Mediated Phosphorylation of Akt. *Clin. Cancer Res.* **2015**, *21*, 4676–4685. [CrossRef] [PubMed]
177. Magnon, C.; Hall, S.J.; Lin, J.; Xue, X.; Gerber, L.; Freedland, S.J.; Frenette, P.S. Autonomic nerve development contributes to prostate cancer progression. *Science* **2013**, *341*, 1236361. [CrossRef] [PubMed]
178. Wang, N.; Dong, B.J.; Quan, Y.; Chen, Q.; Chu, M.; Xu, J.; Xue, W.; Huang, Y.R.; Yang, R.; Gao, W.Q. Regulation of Prostate Development and Benign Prostatic Hyperplasia by Autocrine Cholinergic Signaling via Maintaining the Epithelial Progenitor Cells in Proliferating Status. *Stem Cell Rep.* **2016**, *6*, 668–678. [CrossRef] [PubMed]
179. Schmitt, J.M.; Smith, S.; Hart, B.; Fletcher, L. CaM kinase control of AKT and LNCaP cell survival. *J. Cell. Biochem.* **2012**, *113*, 1514–1526. [CrossRef] [PubMed]
180. Hekman, M.; Albert, S.; Galmiche, A.; Rennefahrt, U.E.; Fueller, J.; Fischer, A.; Puehringer, D.; Wiese, S.; Rapp, U.R. Reversible membrane interaction of BAD requires two C-terminal lipid binding domains in conjunction with 14-3-3 protein binding. *J. Biol. Chem.* **2006**, *281*, 17321–17336. [CrossRef] [PubMed]

181. Taichman, R.S.; Cooper, C.; Keller, E.T.; Pienta, K.J.; Taichman, N.S.; McCauley, L.K. Use of the stromal cell-derived factor-1/CXCR4 pathway in prostate cancer metastasis to bone. *Cancer Res.* **2002**, *62*, 1832–1837. [PubMed]
182. Chetram, M.A.; Don-Salu-Hewage, A.S.; Hinton, C.V. ROS enhances CXCR4-mediated functions through inactivation of PTEN in prostate cancer cells. *Biochem. Biophys. Res. Commun.* **2011**, *410*, 195–200. [CrossRef] [PubMed]
183. Makowska, K.A.; Hughes, R.E.; White, K.J.; Wells, C.M.; Peckham, M. Specific Myosins Control Actin Organization, Cell Morphology, and Migration in Prostate Cancer Cells. *Cell Rep.* **2015**, *13*, 2118–2125. [CrossRef] [PubMed]
184. Peckham, M. How myosin organization of the actin cytoskeleton contributes to the cancer phenotype. *Biochem. Soc. Trans.* **2016**, *44*, 1026–1034. [CrossRef] [PubMed]
185. Barylko, B.; Jung, G.; Albanesi, J.P. Structure, function, and regulation of myosin 1C. *Acta Biochim. Pol.* **2005**, *52*, 373–380. [PubMed]
186. Ihnatovych, I.; Sielski, N.L.; Hofmann, W.A. Selective expression of myosin IC Isoform A in mouse and human cell lines and mouse prostate cancer tissues. *PLoS ONE* **2014**, *9*, e108609. [CrossRef] [PubMed]
187. Ihnatovych, I.; Migocka-Patrzalek, M.; Dukh, M.; Hofmann, W.A. Identification and characterization of a novel myosin Ic isoform that localizes to the nucleus. *Cytoskeleton* **2012**, *69*, 555–565. [CrossRef] [PubMed]
188. Nowak, G.; Pestic-Dragovich, L.; Hozak, P.; Philimonenko, A.; Simerly, C.; Schatten, G.; de Lanerolle, P. Evidence for the presence of myosin I in the nucleus. *J. Biol. Chem.* **1997**, *272*, 17176–17181. [CrossRef] [PubMed]
189. Zattelman, L.; Regev, R.; Usaj, M.; Reinke, P.Y.A.; Giese, S.; Samson, A.O.; Taft, M.H.; Manstein, D.J.; Henn, A. N-terminal splicing extensions of the human MYO1C gene fine-tune the kinetics of the three full-length myosin IC isoforms. *J. Biol. Chem.* **2017**, *292*, 17804–17818. [CrossRef] [PubMed]
190. Dzijak, R.; Yildirim, S.; Kahle, M.; Novak, P.; Hnilicova, J.; Venit, T.; Hozak, P. Specific nuclear localizing sequence directs two myosin isoforms to the cell nucleus in calmodulin-sensitive manner. *PLoS ONE* **2012**, *7*, e30529. [CrossRef] [PubMed]
191. Gillespie, P.G.; Cyr, J.L. Calmodulin binding to recombinant myosin-1c and myosin-1c IQ peptides. *BMC Biochem.* **2002**, *3*, 31. [CrossRef]
192. Nevzorov, I.; Sidorenko, E.; Wang, W.; Zhao, H.; Vartiainen, M.K. Myosin-1C uses a novel phosphoinositide-dependent pathway for nuclear localization. *EMBO Rep.* **2018**, *19*, 290–304. [CrossRef] [PubMed]
193. De Lanerolle, P. Nuclear actin and myosins at a glance. *J. Cell Sci.* **2012**, *125*, 4945–4949. [CrossRef] [PubMed]
194. Ye, J.; Zhao, J.; Hoffmann-Rohrer, U.; Grummt, I. Nuclear myosin I acts in concert with polymeric actin to drive RNA polymerase I transcription. *Genes Dev.* **2008**, *22*, 322–330. [CrossRef] [PubMed]
195. Percipalle, P.; Fomproix, N.; Cavellan, E.; Voit, R.; Reimer, G.; Kruger, T.; Thyberg, J.; Scheer, U.; Grummt, I.; Farrants, A.K. The chromatin remodelling complex WSTF-SNF2h interacts with nuclear myosin 1 and has a role in RNA polymerase I transcription. *EMBO Rep.* **2006**, *7*, 525–530. [CrossRef] [PubMed]
196. Hofmann, W.A.; Vargas, G.M.; Ramchandran, R.; Stojiljkovic, L.; Goodrich, J.A.; de Lanerolle, P. Nuclear myosin I is necessary for the formation of the first phosphodiester bond during transcription initiation by RNA polymerase II. *J. Cell. Biochem.* **2006**, *99*, 1001–1009. [CrossRef] [PubMed]
197. Kysela, K.; Philimonenko, A.A.; Philimonenko, V.V.; Janacek, J.; Kahle, M.; Hozak, P. Nuclear distribution of actin and myosin I depends on transcriptional activity of the cell. *Histochem. Cell Biol.* **2005**, *124*, 347–358. [CrossRef] [PubMed]
198. Philimonenko, V.V.; Zhao, J.; Iben, S.; Dingova, H.; Kysela, K.; Kahle, M.; Zentgraf, H.; Hofmann, W.A.; de Lanerolle, P.; Hozak, P.; et al. Nuclear actin and myosin I are required for RNA polymerase I transcription. *Nat. Cell Biol.* **2004**, *6*, 1165–1172. [CrossRef] [PubMed]
199. Fomproix, N.; Percipalle, P. An actin-myosin complex on actively transcribing genes. *Exp. Cell Res.* **2004**, *294*, 140–148. [CrossRef] [PubMed]
200. Pestic-Dragovich, L.; Stojiljkovic, L.; Philimonenko, A.A.; Nowak, G.; Ke, Y.; Settlage, R.E.; Shabanowitz, J.; Hunt, D.F.; Hozak, P.; de Lanerolle, P. A myosin I isoform in the nucleus. *Science* **2000**, *290*, 337–341. [CrossRef] [PubMed]
201. Obrdlik, A.; Louvet, E.; Kukalev, A.; Naschekin, D.; Kiseleva, E.; Fahrenkrog, B.; Percipalle, P. Nuclear myosin 1 is in complex with mature rRNA transcripts and associates with the nuclear pore basket. *FASEB J.* **2010**, *24*, 146–157. [CrossRef] [PubMed]

202. Cisterna, B.; Necchi, D.; Prosperi, E.; Biggiogera, M. Small ribosomal subunits associate with nuclear myosin and actin in transit to the nuclear pores. *FASEB J.* **2006**, *20*, 1901–1903. [CrossRef] [PubMed]
203. Mehta, I.S.; Elcock, L.S.; Amira, M.; Kill, I.R.; Bridger, J.M. Nuclear motors and nuclear structures containing A-type lamins and emerin: Is there a functional link? *Biochem. Soc. Trans.* **2008**, *36*, 1384–1388. [CrossRef] [PubMed]
204. Chuang, C.H.; Carpenter, A.E.; Fuchsova, B.; Johnson, T.; de Lanerolle, P.; Belmont, A.S. Long-range directional movement of an interphase chromosome site. *Curr. Biol.* **2006**, *16*, 825–831. [CrossRef] [PubMed]
205. Venit, T.; Dzijak, R.; Kalendova, A.; Kahle, M.; Rohozkova, J.; Schmidt, V.; Rulicke, T.; Rathkolb, B.; Hans, W.; Bohla, A.; et al. Mouse nuclear myosin I knock-out shows interchangeability and redundancy of myosin isoforms in the cell nucleus. *PLoS ONE* **2013**, *8*, e61406. [CrossRef] [PubMed]
206. Lu, Q.; Li, J.; Ye, F.; Zhang, M. Structure of myosin-1c tail bound to calmodulin provides insights into calcium-mediated conformational coupling. *Nat. Struct. Mol. Biol.* **2015**, *22*, 81–88. [CrossRef] [PubMed]
207. Pruschy, M.; Ju, Y.; Spitz, L.; Carafoli, E.; Goldfarb, D.S. Facilitated nuclear transport of calmodulin in tissue culture cells. *J. Cell Biol.* **1994**, *127*, 1527–1536. [CrossRef] [PubMed]
208. Visuttijai, K.; Pettersson, J.; Mehrbani Azar, Y.; van den Bout, I.; Orndal, C.; Marcickiewicz, J.; Nilsson, S.; Hornquist, M.; Olsson, B.; Ejeskar, K.; et al. Lowered Expression of Tumor Suppressor Candidate MYO1C Stimulates Cell Proliferation, Suppresses Cell Adhesion and Activates AKT. *PLoS ONE* **2016**, *11*, e0164063. [CrossRef] [PubMed]
209. Letellier, E.; Schmitz, M.; Ginolhac, A.; Rodriguez, F.; Ullmann, P.; Qureshi-Baig, K.; Frasquilho, S.; Antunes, L.; Haan, S. Loss of Myosin Vb in colorectal cancer is a strong prognostic factor for disease recurrence. *Br. J. Cancer* **2017**, *117*, 1689–1701. [CrossRef] [PubMed]
210. Buschman, M.D.; Field, S.J. MYO18A: An unusual myosin. *Adv. Biol. Regul.* **2018**, *67*, 84–92. [CrossRef] [PubMed]

Correcting Calcium Dysregulation in Chronic Heart Failure Using *SERCA2a* Gene Therapy

T. Jake Samuel [1], Ryan P. Rosenberry [1], Seungyong Lee [1] and Zui Pan [2,*]

1 Department of Kinesiology, College of Nursing and Health Innovation, The University of Texas at Arlington, Arlington, TX 76019, USA; Thomas.Samuel@uta.edu (T.J.S.); ryan.rosenberry@uta.edu (R.P.R.); seungyong.lee@uta.edu (S.L.)

2 Department of Graduate Nursing, College of Nursing and Health Innovation, The University of Texas at Arlington, Arlington, TX 76019, USA

* Correspondence: zui.pan@uta.edu

Abstract: Chronic heart failure (CHF) is a major contributor to cardiovascular disease and is the leading cause of hospitalization for those over the age of 65, which is estimated to account for close to seventy billion dollars in healthcare costs by 2030 in the US alone. The successful therapies for preventing and reversing CHF progression are urgently required. One strategy under active investigation is to restore dysregulated myocardial calcium (Ca^{2+}), a hallmark of CHF. It is well established that intracellular Ca^{2+} concentrations are tightly regulated to control efficient myocardial systolic contraction and diastolic relaxation. Among the many cell surface proteins and intracellular organelles that act as the warp and woof of the regulatory network controlling intracellular Ca^{2+} signals in cardiomyocytes, sarco/endoplasmic reticulum Ca^{2+} ATPase type 2a (SERCA2a) undoubtedly plays a central role. SERCA2a is responsible for sequestrating cytosolic Ca^{2+} back into the sarcoplasmic reticulum during diastole, allowing for efficient uncoupling of actin-myosin and subsequent ventricular relaxation. Accumulating evidence has demonstrated that the expression of SERCA2a is downregulated in CHF, which subsequently contributes to severe systolic and diastolic dysfunction. Therefore, restoring SERCA2a expression and improving cardiomyocyte Ca^{2+} handling provides an excellent alternative to currently used transplantation and mechanical assist devices in the treatment of CHF. Indeed, advancements in safe and effective gene delivery techniques have led to the emergence of *SERCA2a* gene therapy as a potential therapeutic choice for CHF patients. This mini-review will succinctly detail the progression of *SERCA2a* gene therapy from its inception in plasmid and animal models, to its clinical trials in CHF patients, highlighting potential avenues for future work along the way.

Keywords: calcium handling; SERCA2a; chronic heart failure; clinical trial; adeno-associated virus (AAV); gene delivery

---

1. Introduction

Chronic heart failure (CHF) is a major form of cardiovascular disease and is the leading cause of hospitalization for those over the age of 65 [1]. It is expected to account for almost seventy billion dollars in healthcare costs by 2030 in the US alone [2]. The current pharmacological therapies for CHF merely target symptom management, with limited success in treating the underlying etiology of the disease. The late stage CHF patients eventually need expensive and hard-to-obtain heart transplantation or mechanical assist devices. Therefore, successful therapies for preventing and reversing CHF progression are urgently required.

Understanding the basic mechanisms involved in the development of CHF has been an active field of research in the quest to identify abnormalities that could potentially be targeted by gene transfer.

It has been well established that myocardial calcium (Ca^{2+}) dysregulation is a hallmark of CHF. Tight control of cardiac intracellular Ca^{2+} handling plays an integral role in synchronous actin-myosin cross-bridge cycling and the resulting systolic contraction and diastolic relaxation. Many proteins either at the cell surface or intracellular organelles form the warp and woof of the regulatory network for intracellular Ca^{2+} signals in cardiomyocytes (Figure 1). During systole, Ca^{2+} enters the sarcolemma through L-type Ca^{2+} channels, i.e., dihydropyrodine receptor (DHPR) located on the transverse-tubules (T-tubules), before diffusing across the short distance to the ryanodine receptors (RyRs) located to the sarcoplasmic reticulum (SR). This small amount of Ca^{2+} influx triggers RyRs to generate further Ca^{2+}-induced-Ca^{2+}-release (CICR) from the SR to form magnified Ca^{2+} transient [3]. Ca^{2+} then binds to troponin-C in the sarcomere and stimulates actin-myosin cross-bridge linking, i.e., systolic contraction [4]. During diastole, in order to reduce the likelihood of Ca^{2+} binding to troponin-C and causing prolonged myocardial systolic contraction, Ca^{2+} is actively sequestrated back into the SR via a sarco/endoplasmic reticulum Ca^{2+} ATPase, SERCA2a in cardiomyocyte [5,6], or expelled from the sarcolemma by sodium-Ca^{2+} exchanger pumps (NCX) [3]. As such, SERCA2a is undoubtedly essential for complete myocyte Ca^{2+} homeostasis.

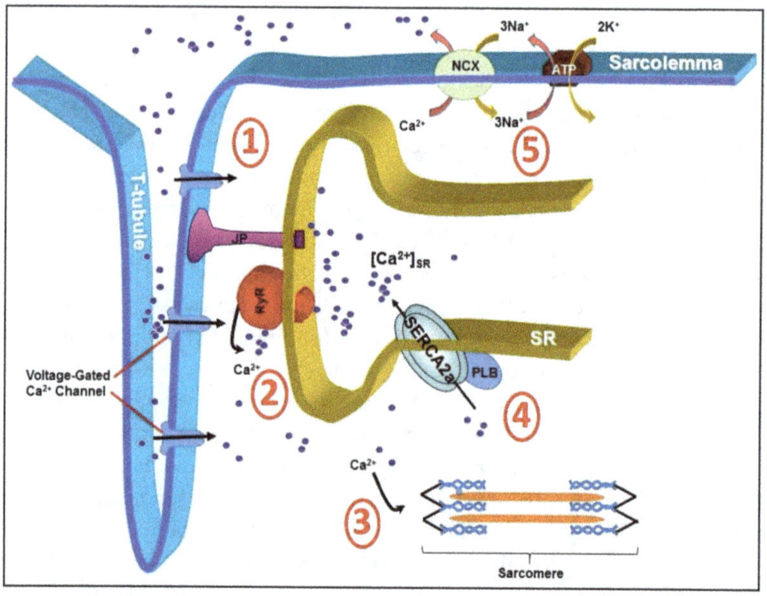

Figure 1. Cardiac intracellular Ca^{2+} is tightly regulated by several proteins. Efficient systolic contraction and diastolic relaxation is reliant on efficient Ca^{2+} handling through 5 main processes. (1) Diffusion of Ca^{2+} in to the cytosol via voltage-gated Ca^{2+} channels (DHPRs) located on the surface of the transverse-tubule (T-tubule); (2) Ca^{2+}-induced-Ca^{2+}-release from the ryanodine receptors (RyRs); (3) Binding of Ca^{2+} to troponin-C in the sarcomere, stimulating actin-myosin cross-linking; (4) Sequestration of Ca^{2+} back in to the sarcoplasmic reticulum via the important Ca^{2+} pump sarco/endoplasmic reticulum Ca^{2+} ATPase (SERCA2a); and (5) Expulsion of Ca^{2+} from the cell via sodium-calcium exchanger pumps (NCX). JP, junctophilins; PLB, phopholamban; ATP, ATP pump; $3Na^+$, sodium; K^+, potassium.

Dysregulation of any of the above Ca^{2+} handling processes will result in impaired ventricular contractility and impaired myocyte relaxation, leading to cardiac dysfunction [4,7,8]. While efforts have been devoted to investigating the therapeutic value of many Ca^{2+} channels and proteins in regulating cytosolic Ca^{2+} concentrations, SERCA2a has received the most interest in recent years.

Reduction in SERCA2a levels in the SR has been reported in failing heart tissues [9,10]. Impaired Ca^{2+}-ATPase pump function and decreased *SERCA2a* gene transcription has recently been attributed to delayed left ventricular relaxation and increased left ventricular filling pressures [11] which may lead to pulmonary edema, a common symptom in CHF (Figure 2). Biotechnological advances over the past two decades have led to gene therapy becoming a viable option for treating many pathologic conditions—including CHF [12–14]. Matter of fact, SERCA2a had been identified as a possible gene therapy target as early as 1978 [15]. This review will focus on *SERCA2a* gene therapy in CHF and attempt to highlight the seminal investigations using early gene delivery techniques in animals that contributed to the early promise for its use in humans. It will also provide a summary of the major human clinical trials which identified several limitations to these methods for treating human CHF. Finally, recent novel developments in gene delivery and targeting will be discussed and future directions for this field of research will be proposed.

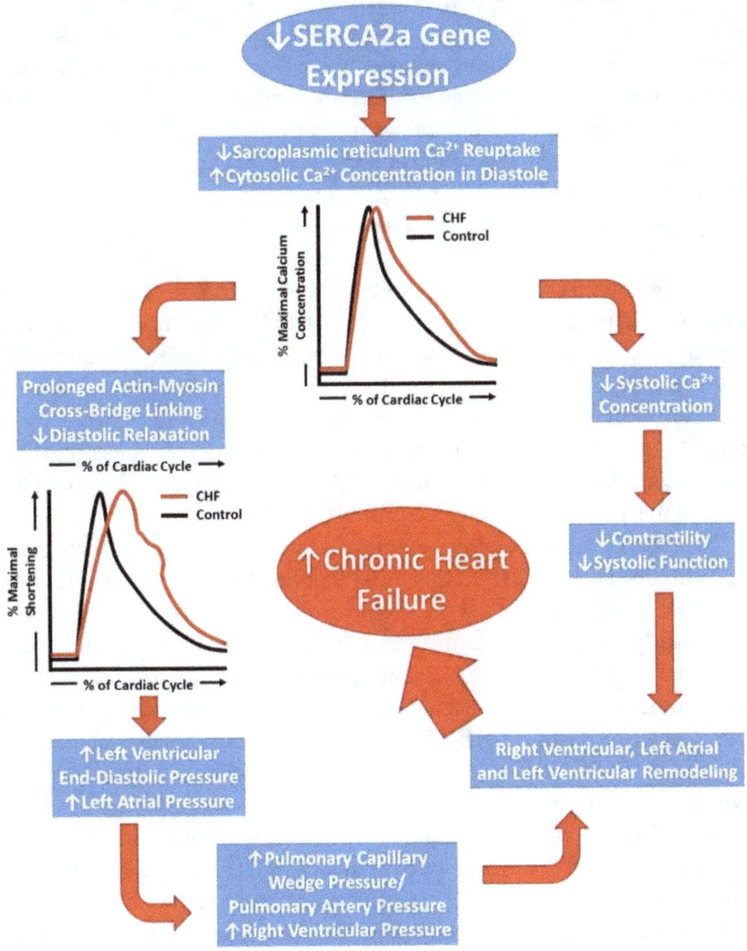

Figure 2. Schematic representation of the progression of chronic heart failure. Initial downregulation of SERCA2a function leads to increased cytosolic Ca^{2+}, which ultimately compromises left ventricular contractility (systolic function) and leads to prolonged left ventricular relaxation. Prolonged relaxation leads to increased filling pressures and a backlog of pressure in to the pulmonary circulation and right heart, the result of which leads to severe right and left heart remodeling and chronic heart failure development.

2. Early Techniques of *SERCA2a* Gene Manipulation in Animals

The role of cytosolic Ca^{2+} in cardiac myocyte contraction has been intensely studied for decades. As far back as 40 years ago, investigators were purifying SERCA proteins to better understand the molecular mechanisms through which Ca^{2+} is transported from the cytosol into the sarcoplasmic reticulum (SR) [15]. Shortly following the identification of Ca^{2+} channels as key mediators of cardiac function, they became therapeutic targets in the treatment of various cardiac ailments [16–18]. With the rise of interest in genomics and gene editing techniques, the direct manipulation of Ca^{2+} channels or pumps, and thus manipulation of Ca^{2+} handling in cardiac myocytes, became possible.

2.1. Plasmid and Direct DNA Injection

The earliest successful approaches of manipulating *SERCA2a* gene expression in animal models were performed using direct insertion of plasmids into developing mouse oocytes. The rat SERCA2a cDNA was cloned into plasmids containing mouse cardiac α-MHC promoter. While certain repetitive and unnecessary exons were removed, an additional human growth hormone polyadenylation site was included to promote both polyadenylation and termination [9]. Once these sequences are successfully cloned into the plasmid, restriction endonucleases are applied to cleave the gene from the plasmid. The linear DNA fragment can then be purified for direct injection into oocytes. The resulting transgenic mice possessed extra copies of the *SERCA2a* gene and exhibited increased expression of the protein [9,19,20].

Initial studies applying this transgenic technique sought to characterize and measure the effects of the increased SERCA2a expression in otherwise unaltered mice. Early investigations found that overexpression of the *SERCA2a* gene in a transgenic mouse model resulted in improved Ca^{2+} handling which translated to augmented myocyte contraction and relaxation [9,10,20]. Of note, the transgenic mice had a significantly shorter relaxation time indicating improved relaxation kinetics, consistent with improved Ca^{2+} sequestration via SERCA2a [20]. With these positive proof-of-concept results, investigators next sought to determine whether SERCA2a could improve cardiac functionality in disease stages. To test this potential treatment, an aortic stenosis-induced CHF model was employed in transgenic mice. The increased afterload from the artificial stenosis caused left ventricular hypertrophy, mimicking the conditions of early CHF. In this investigation, the transgenic expression of SERCA2a provided chronic protective effects against the progression of CHF [19].

Although these early studies demonstrated positive and profound effects from direct injection of DNA fragment containing *SERCA2a* gene, there are substantial limitations inherent to this technique. First, this laboratory procedure requires a significant time commitment and many uncontrollable steps before the success of the transfection can be determined. In the case of He et al. super-ovulated eggs were treated with purified SERCA2a DNA, gestated for 20 days, and the newborn mice were allowed to develop for three weeks before testing tail samples for SERCA2a expression by Southern blotting [20]. This process relies on integration of the gene into the embryonic genome, normal gestation and delivery, and inerrant expression in the developed animal. Even if each step is performed to perfection, the imprecision introduces excessive error and contributes to study failure. In fact, only three mice out of nine integrated the SERCA2a plasmid. Of these three mice, the single male was sterile, leaving only two female transgenic mice that successfully passed the gene on to their offspring [20].

A second limitation of this technique is that for additional copies of SERCA2a to be expressed, all biological machinery in the transcription/translation/post-translational modification pathway must also function in a normal and healthy manner. Direct injection of the DNA into oocytes carries the potential risk for improper integration of the gene into the genome. As evidenced by the low number of successfully reared transgenic mice, the possibility for lethality or sterility is considerable. Furthermore, this approach cannot be applied to developed organisms, hindering its expansion and/or adaptation into other animal models. To further investigate SERCA2a as an intervention, it was necessary to explore more advanced and versatile gene modifying techniques.

Last and foremost, direct manipulation of human oocytes raises serious ethical issues, which disqualify this approach from developing into a useful human gene therapy for CHF. Rather, these transgenic animal studies laid the theoretical understanding that prompted improvements in biotechnological techniques to tackle this problem. Alternative ethical and safer approaches have been called upon.

2.2. *Adenoviral (Ad)-Based Vectors*

Ad are double-strand DNA vectors that bind to and enter the cell membrane before being transported to the nucleus, allowing transfection of genes into a variety of cells such as cardiomyocytes, skeletal muscle and smooth muscle myocytes [21]. Delivery of Ad-based vectors to cardiomyocytes through intracoronary infusion and direct myocardial injection has showed some success in transduction of the target gene [22]. Although Ad-based vectors attain a high level of cellular transduction, the various Ad serotypes result in varied efficiency [21], leading to mixed success in modulating target gene expression.

Ad mediated *SERCA2a* gene delivery (Ad/SERCA2a) was reported to present improvements in cardiac function in CHF due to the upregulation of *SERCA2a* genes. Ad/*SERCA2a* gene transfer in mice with CHF enhances contractile function [23] and restores phosphocreatine and ATP level in the heart, which represents the recovery of heart energetic state and removal of myocardial ischemia [24]. Furthermore, there is a dose response to Ad/*SERCA2a* transfer as overexpression of the *SERCA2a* gene promotes an additive increase in SERCA2a activity which further influences improved ventricular contractility [25] but has no effect on the atrium [26]. In addition to improvements in left ventricular contractility, Ad/*SERCA2a* gene transfer has also been shown to promote reverse myocardial remodeling by reducing left ventricular anterior wall thickening in Wistar rats with CHF, which reduced arrhythmic events and caused the restoration of left ventricular function similar to the observation in control animals [27].

The benefits of *SERCA2a* gene transfer via Ad-based vectors were not only limited to the myocardial tissue, however, as it also showed benefits on the vascular smooth muscle cells by preventing vascular remodeling and inhibition of neointimal thickening in a human ex vivo model of the coronary artery [28] and augmented coronary blood flow [29]. Although the Ad-based vector technique was successful overall, there were some inherent limitations to this gene delivery method, such as the presence of neutralizing antibodies and the severe immune response against Ad vector and/or the gene-modified cells [21]. Therefore, the development of more sophisticated viral vectors techniques was explored to overcome these limitations.

2.3. *Adeno-Associated Virus (AAV)-Based Vectors*

The AAV-based vector has significant advantage over Ad-based vector in gene delivery. To date, several generations of AAV systems have been adapted and adopted in the *SERCA2a* gene delivery with each generation improving delivery efficiency and safety over the previous one. AAV-based vector has been shown to safely and successfully alter gene expression in cardiac tissue [21], with much less inflammation than that associated with the Ad-based vector technique [21]. AAVs comprise more than 100 serotypes and each serotype presents distinct transduction efficiency in different tissues [21]. Among these AAVs, the AAV-1 was the best studied and AAV1-based vector is most commonly used. Indeed, several earlier trials were conducted using the AAV1/SERCA2a intra-coronary injection method in animal CHF models with good success. Hadri and colleagues demonstrated that long-term SERCA2a overexpression via in vivo AAV1-mediated gene transfer improves left ventricular ejection fraction, coronary artery blood flow, and expression of endothelial nitric oxide synthase in a swine model of CHF [30]. Similarly, AAV1/*SERCA2a* gene delivery improves echocardiography derived myocardial function and decreases myocardial apoptosis in pigs [31], as well as restores SERCA2a activity and protein expression following atrial fibrillation-related declines in SERCA2a activity and expression [32]. Finally, *SERCA2a* gene transfer not only effects cardiac tissue but also directly influences vascular endothelial and smooth muscle cell function by improving Ca^{2+} signaling [33] which highlights SERCA2a as an important therapeutic target for treating cardiovascular disease—including CHF.

Overall, AAV1 and AAV6 mediated-*SERCA2a* gene delivery methods have provided beneficial improvements in myocardial and coronary artery Ca^{2+} handling, contributing to improved cardiac function and reverse remodeling in animal models of CHF. This initial success in improving cardiac function in animal models led to large-scale clinical trials in humans which will be discussed in the following section.

3. Human Trials

Over the past decade several large-scale clinical trials have been performed using AAV1/SERCA2a in the treatment of human CHF (Table 1) and are discussed below.

3.1. The Calcium Upregulation by Percutaneous Administration of Gene Therapy in Cardiac Disease (CUPID) Trial

In response to the promising preclinical models in animals that demonstrated AAV1/SERCA2a were well tolerated and improved cardiac function, CUPID was the first-in-human clinical trial to evaluate the efficacy of recombinant AAV1/SERCA2a in human CHF [12,13]. The initial CUPID multicenter trial included two phases, wherein phase 1 was an open-label dose escalation protocol and phase 2, a randomized, double-blind, placebo-controlled trial. The overall primary outcome of the CUPID trials was to monitor safety, while secondary outcomes included improvements in physical activity and efficacy.

In CUPID phase 1, 12 patients were administered single intracoronary infusions, comparing 3 dose levels of AAV1/SERCA2a (AAV1 capsid with human SERCA2a cDNA flanked with inverted terminal repeats) with placebo, and primary outcomes were measured at 6 and 12 months post-infusion. Patients were enrolled independent of CHF etiology, were NYHA class III/IV, and had exhausted all current pharmacologic therapy. The results of phase 1 suggested that intracoronary infusion of AAV1/SERCA2a demonstrated an acceptable safety profile given the advanced CHF population. Several of the patients exhibited improvements in CHF symptoms and left ventricular structure and function at 6 months follow-up [12]. Of note, two of the patients who did not respond to therapy already had pre-existing neutralizing antibodies (NAb) for the viral capsid proteins. Pre-existing natural exposure to NAb, which are known to inhibit vector uptake, limit the efficacy of AAV1 treatment in CHF [34–36]. Discussion of the impact of NAb on vector uptake are beyond the scope of the current review and have been comprehensively described elsewhere [37].

The relative success of phase 1 led to initiation of phase 2, including 39 advanced CHF patients randomized to 3 doses of viral vector; low (6×10^{11} DNase resistant particles; DRP), medium (3×10^{12} DRP), and high doses (1×10^{13} DRP). Endpoints included patients' symptomatic (NYHA functional class, Minnesota Living with Heart Failure Questionnaire; MLWHFQ) and functional (maximal oxygen uptake, 6-minute Walk Test) status, blood biomarker levels (N-Terminal-pro Brain Natriuretic Peptide, NT-proBNP), and left ventricular function and remodeling (Ejection fraction and End-systolic volume). One-year follow-up suggested the patients randomized to the "high" treatment dose of AAV1/SERCA2a improved in functional class (Mean reduction in MLWHFQ: -10.3 ± 12.21), blood biomarker (NT-proBNP mean reduction: $12.4 \pm 47.83\%$) and left ventricular function (end-systolic volume mean reduction: -9.6 ± 27.55 mL; $-4 \pm 13.76\%$). Importantly, there was a significant increase in the time to cardiac event in all AAV1/SERCA2a groups in the CUPID trial, suggesting that perhaps even the lower doses of viral vector had beneficial effects on patient outcome, independent of physiological changes that are not appreciable at smaller sample sizes [13].

Table 1. Current and previous clinical trials to test the efficacy of AAV1/SERCA2a on the Heart Failure.

National Clinical Trial Code	Status	Study Title	Conditions	Interventions	Study Results	References
NCT02772068	Recruiting	Hemodynamic Response to Exercise in HFpEF Patients After Upregulation of SERCA2a	Congestive Heart Failure	Drug: Istaroxime Other: Exercise	No Published Results	-
NCT01966887	Terminated	AAV1-CMV-Serca2a GENe Therapy Trial in Heart Failure (AGENT-HF)	Congestive Heart Failure Ischemic and non-ischemic Cardiomyopathies	Genetic: AAV1/SERCA2a (MYDICAR)-single intracoronary infusion Genetic: Placebo; single intracoronary infusion	No Positive or Negative Effects	[38]
NCT00534703	Terminated	Investigation of the Safety and Feasibility of AAV1/SERCA2a Gene Transfer in Patients with Chronic Heart Failure (SERCA-LVAD)	Chronic Heart Failure Left Ventricular Assist Device	Genetic: AAV1/SERCA2a Drug: Placebo	Terminated Early—No Results	-
NCT01643330	Completed	A Study of Genetically Targeted Enzyme Replacement Therapy for Advanced Heart Failure (CUPID-2b)	Ischemic and non-ischemic Cardiomyopathies Heart Failure	Genetic: AAV1/SERCA2a (MYDICAR) Genetic: Placebo	No Positive or Negative Effects	[12–14,37,39–41]
NCT00454818	Completed	Efficacy and Safety Study of Genetically Targeted Enzyme Replacement Therapy for Advanced Heart Failure (CUPID)	Heart Failure, Congestive Dilated Cardiomyopathy	Genetic: MYDICAR Phase 1 (Open-label, Serial Dose-Escalation Study) Procedure: Placebo Infusion Genetic: MYDICAR Phase 2 (Placebo-controlled, Randomized Study)	Positive Results	[12,13,40–42]
NCT02346422	Terminated	A Phase 1/2 Study of High-Dose Genetically Targeted Enzyme Replacement Therapy for Advanced Heart Failure	Heart Failure Cardiomyopathy	Genetic: MYICAR Phase 1/2 (Dose 2.5 × 1013 DRP)	Terminated Early—No Results	[12,13,40]

3.2. Post-CUPID

Since the conclusion of CUPID, three clinical trials utilizing AAV1 have targeted SERCA2a in human CHF [14,38,39,43]. CUPID2, a continuation of the CUPID clinical trials, was the first to recruit patients from outside the United States, enrolling a total of 250 patients with CHF. Patients were randomized to 10-minute intracoronary infusion of either placebo or "high" dose of AAV1/SERCA2a. In contrast to the results of the original CUPID trial, CUPID2 did not reduce either recurrent heart failure events, or terminal events in the study population [14,39].

The other two post-CUPID trails began in the United Kingdom, known as SERCA-LVAD and AGENT-HF. SERCA-LVAD aimed to take tissue biopsies at the time of left ventricular assist device implantation (pre-infusion of AAV1/SERCA2a) and at the time of transplant or LVAD removal (post-infusion) in order to correlate changes in clinical outcome with SERCA2a expression levels [43]. AGENT-HF utilized AAV1/SERCA2a with left ventricular structure and function as the primary outcome variables assessed at 6 months post-treatment [38]. Unfortunately, due to the negative results of the CUPID2 trial and lack of preliminary success, both studies were terminated prematurely. The studies showed no improvement in ventricular remodeling but were underpowered to demonstrate any effect of AAV1/*SERCA2a* gene therapy [38]. Taken together, the human trials utilizing AAV1/SERCA2a to treat HF have been widely unsuccessful to date.

The early clinical trial CUPID provided the basis for several more aggressive clinical trials (CUPID2, Agent-HF, SERCA-LVAD), all of which failed to demonstrate improvement in survival and/or were terminated prior to sufficient data collection. The reason for the failure of CUPID2 remains largely unknown. One possible reason is attributable to the insufficient delivery of SERCA2a DNA using intracoronary infusion of AAV1 [44], as intracoronary infusion may result in inadequate uptake of the viral vector into cardiomyocytes. In addition, work by Mingozzi and colleagues suggests that viral loads which include empty viral capsid particles enhance the gene delivery as they act as "decoys" and may block the inhibitory activity of antibodies [45]. Thus, *SERCA2a* gene therapy may not be at a dead end yet and there is some light through the tunnel to show it still as a valuable approach for HF patients. The challenge is rather how to effectively deliver the gene to cardiomyocytes. Are there better, more potent viral vectors and more advanced gene editing techniques for gene delivery and expression in human heart?

4. The Promise of Novel Gene Delivery Techniques in Animals

Among more than 100 wild-type AAV serotypes, some serotypes have been demonstrated to have distinct features on cardiomyocyte transduction [21]. Compared to AAV1, AAV8 showed 20-fold higher gene expression in the myocardium in mice. Systemic venous infusion of AAV9 triggers robust level (>200-fold) of gene expression in the cardiomyocytes when compared to AAV1 [46]. To date, AAV9-based vectors are by far the most efficient methods for delivering genes to cardiac tissue in mice [46–48]. However, AAV6 was also reported to mediate the most efficient transduction in mouse cardiac tissue [49].

Based on these findings, many recent studies have adopted the AAV9 method to transfer *SERCA2a* genes for treating CHF in animal models. Lyon et al. showed that *SERCA2a* gene transfer with AAV9 restored SERCA2a proteins in the rat heart and improved left ventricular function by reducing Ca^{2+} leakage from sarcoplasmic reticulum [50]. In addition, AAV9/SERCA2a therapy recovers the electrical signal constancy by improving ventricular arrhythmias [50,51] and the treatment rectifies ECG abnormalities such as tachycardia, shortened P-R interval, and prolonged Q-T intervals [52]. Similarly, *SERCA2a* gene therapy overexpressed SERCA2a protein and repressed atrial fibrillation induced by rapid pacing of the atrium in rabbit model [53]. Toward clinical application, this line of study has been moved forward to large animals as well. AAV9/*SERCA2a* gene therapies were performed in German pigs and ovine with ischemic CHF models [54]. AAV9 targeting the *S100A1* gene, which augments the activity of SERCA2a proteins so that Ca^{2+} signaling is improved, prevents left ventricular remodeling, and restores SERCA2a protein expression [54]. Likewise, injection of AAV9/SERCA2a into the coronary arteries significantly increased *SERCA2a* gene expression in the walls of the left ventricle in sheep [55]. This successful overexpression of SERCA2a triggered improvements in left ventricular systolic and diastolic function and subsequent left ventricular remodeling [55]. These positive outcomes provide the basis for future clinical trials in humans utilizing AAV9-based vectors. Current investigations in humans using AAV9-based vectors include Pompe disease (NCT02240407) and Batten disease (NCT02725580). Taken together, these data suggest that the AAV9/*SERCA2a* gene delivery is an appealing therapeutic method for targeting CHF due to its effects on the ventricular contractile function and restoration of electrical complications associated with the failing heart.

Besides AAVs, gene delivery method using other viral vectors is emerging. For example, lentiviral vectors have been demonstrated to be effective for SERCA2 gene transfer in rat ischemic HF model [56]. Since lentivirus can integrate the gene of interest into the host chromosome, this approach can achieve long-term effect of gene expression [57–59]. Many research groups have begun to examine safety and efficacy of this gene therapy approach in CHF animal models [56,60–62].

5. Concluding Remarks and Future Directions

The beneficial left ventricular functional and structural adaptations associated with AAV9/*SERCA2a* gene therapy provide solid preclinical evidence for future work to focus on translating these findings from animal work to human clinical applications. Compared to AAV1 which used in failed CUPID2 and other earlier trials, AAV9 presents superior efficiency and safety. Therefore, the improved Ca^{2+} handling in cardiac tissues in response to AAV9/SERCA2a would have significant benefits for patients suffering from CHF—particularly CHF of ischemic origin.

In addition to the focus on *SERCA2a* gene expression, there are a host of other molecules that both indirectly and directly modify SERCA2a function that are worthwhile investigating using modern gene therapy techniques. These molecules include, but are not limited to, small ubiquitin-related modifier-1 (SUMO-1) [63,64]. SUMO-1 is a post-transcriptional protein, which plays an important role in modifying other proteins involved in the preservation and stabilization of SERCA2a proteins. In porcine and human cardiac tissue, AAV1/SUMO-1 improved cardiac function. Concurrent transfection of both SUMO-1 and SERCA2a increased both expression and function of SERCA2a [63,64]. As such, this approach may represent a possible dual therapy in humans. Indeed, combined gene therapy has recently shown significant promise in the treatment of animal models of heart failure, with combined transfer of apelin, fibroblast growth factor-2, and SERCA2a demonstrating increased gene expression in ischemic heart failure model [61]. Therefore, future work should seek to translate the use of combined gene therapy in humans.

Moreover, novel pharmaceutical therapies targeting activation of SERCA2a have been developed. Istaroxime, a drug developed by Italian pharmaceutical company Sigma-Tau (Pomezia, Italy), is a lusitropic-inotropic drug recently approved for use in CHF [65]. In fact, currently a Phase 1 clinical trial at the University of Texas Southwestern is utilizing Istaroxime during exercise (NCT02772068), with primary outcome measure changes in cardiac filling pressures, and secondary outcome measure change in cardiac relaxation time. The study hopes to investigate the effect of SERCA2a activation on exercise-dependent changes in cardiac function (NCT02772068). At the same time, many other altered regulatory proteins involving in cardiomyocyte Ca^{2+} homeostasis could be potential targets for gene therapy in CHF as well, such as phospholamban, junctophilin-2 (JP2), RyRs and NCXs. For example, JP2 downregulation has been reported in human failing heart and restoring JP2 by AAV9-mediated gene therapy could rescue heart failure in mice [66–69].

In conclusion, the development of gene therapy and gene delivery methods have received significant progress over the past few decades, contributing to a novel therapeutic approach for treating CHF. Although the early promising AAV1/SERCA2a method in animal models failed to replicate the same results in humans, recent advancements in gene delivery techniques, especially AAV9/SERCA2a have provided compelling results for beneficial outcome of the upregulation of *SERCA2a* gene expression. Targeting SERCA2a for treatment of CHF in human patients should be still an option deserving further investigation. Future work should focus on translating the recent findings using AAV9/SERCA2a techniques into large-scale clinical trials in humans.

Acknowledgments: The authors gratefully acknowledge the contribution of Benjamin Young in helping prepare this manuscript.

References

1. Azad, N.; Lemay, G. Management of chronic heart failure in the older population. *J. Geriatr. Cardiol.* **2014**, *11*, 329–337. [PubMed]
2. Heidenreich, P.A.; Albert, N.M.; Allen, L.A.; Bluemke, D.A.; Butler, J.; Fonarow, G.C.; Ikonomidis, J.S.; Khavjou, O.; Konstam, M.A.; Maddox, T.M.; et al. Forecasting the impact of heart failure in the United States: A policy statement from the american heart association. *Circ. Heart Fail.* **2013**, *6*, 606–619. [CrossRef] [PubMed]
3. Berridge, M.J.; Bootman, M.D.; Roderick, H.L. Calcium signalling: Dynamics, homeostasis and remodelling. *Nat. Rev. Mol. Cell Biol.* **2003**, *4*, 517–529. [CrossRef] [PubMed]
4. Hasenfuss, G.; Pieske, B.; Holubarsch, C.; Alpert, N.R.; Just, H. Excitation-contraction coupling and contractile protein function in failing and nonfailing human myocardium. *Adv. Exp. Med. Biol.* **1993**, *346*, 91–100. [PubMed]
5. Balderas-Villalobos, J.; Molina-Munoz, T.; Mailloux-Salinas, P.; Bravo, G.; Carvajal, K.; Gomez-Viquez, N.L. Oxidative stress in cardiomyocytes contributes to decreased serca2a activity in rats with metabolic syndrome. *Am. J. Physiol. Heart Circ. Physiol.* **2013**, *305*, H1344–H1353. [CrossRef] [PubMed]
6. Kawase, Y.; Ly, H.Q.; Prunier, F.; Lebeche, D.; Shi, Y.; Jin, H.; Hadri, L.; Yoneyama, R.; Hoshino, K.; Takewa, Y.; et al. Reversal of cardiac dysfunction after long-term expression of SERCA2a by gene transfer in a pre-clinical model of heart failure. *J. Am. Coll. Cardiol.* **2008**, *51*, 1112–1119. [CrossRef] [PubMed]
7. Morgan, J.P. Abnormal intracellular modulation of calcium as a major cause of cardiac contractile dysfunction. *N. Engl. J. Med.* **1991**, *325*, 625–632. [CrossRef]
8. Schwarzl, M.; Ojeda, F.; Zeller, T.; Seiffert, M.; Becher, P.M.; Munzel, T.; Wild, P.S.; Blettner, M.; Lackner, K.J.; Pfeiffer, N.; et al. Risk factors for heart failure are associated with alterations of the LV end-diastolic pressure-volume relationship in non-heart failure individuals: Data from a large-scale, population-based cohort. *Eur. Heart J.* **2016**, *37*, 1807–1814. [CrossRef] [PubMed]
9. Baker, D.L.; Hashimoto, K.; Grupp, I.L.; Ji, Y.; Reed, T.; Loukianov, E.; Grupp, G.; Bhagwhat, A.; Hoit, B.; Walsh, R.; et al. Targeted overexpression of the sarcoplasmic reticulum Ca^{2+}-atpase increases cardiac contractility in transgenic mouse hearts. *Circ. Res.* **1998**, *83*, 1205–1214. [CrossRef] [PubMed]
10. Del Monte, F.; Harding, S.E.; Schmidt, U.; Matsui, T.; Kang, Z.B.; Dec, G.W.; Gwathmey, J.K.; Rosenzweig, A.; Hajjar, R.J. Restoration of contractile function in isolated cardiomyocytes from failing human hearts by gene transfer of SERCA2a. *Circulation* **1999**, *100*, 2308–2311. [CrossRef]
11. Studeli, R.; Jung, S.; Mohacsi, P.; Perruchoud, S.; Castiglioni, P.; Wenaweser, P.; Heimbeck, G.; Feller, M.; Hullin, R. Diastolic dysfunction in human cardiac allografts is related with reduced SERCA2a gene expression. *Am. J. Transplant.* **2006**, *6*, 775–782. [CrossRef] [PubMed]
12. Jaski, B.E.; Jessup, M.L.; Mancini, D.M.; Cappola, T.P.; Pauly, D.F.; Greenberg, B.; Borow, K.; Dittrich, H.; Zsebo, K.M.; Hajjar, R.J.; et al. Calcium upregulation by percutaneous administration of gene therapy in cardiac disease (CUPID Trial), a first-in-human phase 1/2 clinical trial. *J. Card. Fail.* **2009**, *15*, 171–181. [CrossRef] [PubMed]
13. Jessup, M.; Greenberg, B.; Mancini, D.; Cappola, T.; Pauly, D.F.; Jaski, B.; Yaroshinsky, A.; Zsebo, K.M.; Dittrich, H.; Hajjar, R.J.; et al. Calcium upregulation by percutaneous administration of gene therapy in cardiac disease (CUPID): A phase 2 trial of intracoronary gene therapy of sarcoplasmic reticulum Ca^{2+}-atpase in patients with advanced heart failure. *Circulation* **2011**, *124*, 304–313. [CrossRef] [PubMed]
14. Greenberg, B.; Butler, J.; Felker, G.M.; Ponikowski, P.; Voors, A.A.; Desai, A.S.; Barnard, D.; Bouchard, A.; Jaski, B.; Lyon, A.R.; et al. Calcium upregulation by percutaneous administration of gene therapy in patients with cardiac disease (CUPID 2): A randomised, multinational, double-blind, placebo-controlled, phase 2b trial. *Lancet* **2016**, *387*, 1178–1186. [CrossRef]

15. Tada, M.; Yamamoto, T.; Tonomura, Y. Molecular mechanism of active calcium transport by sarcoplasmic reticulum. *Physiol. Rev.* **1978**, *58*, 1–79. [CrossRef] [PubMed]
16. Leon, M.B.; Rosing, D.R.; Bonow, R.O.; Epstein, S.E. Combination therapy with calcium-channel blockers and β-blockers for chronic stable angina-pectoris. *Am. J. Cardiol.* **1985**, *55*, B69–B80. [CrossRef]
17. Cannon, R.O., 3rd; Watson, R.M.; Rosing, D.R.; Epstein, S.E. Efficacy of calcium channel blocker therapy for angina pectoris resulting from small-vessel coronary artery disease and abnormal vasodilator reserve. *Am. J. Cardiol.* **1985**, *56*, 242–246. [CrossRef]
18. Antman, E.M.; Stone, P.H.; Muller, J.E.; Braunwald, E. Calcium channel blocking agents in the treatment of cardiovascular disorders. Part I: Basic and clinical electrophysiologic effects. *Ann. Intern. Med.* **1980**, *93*, 875–885. [CrossRef] [PubMed]
19. Ito, K.; Yan, X.H.; Feng, X.; Manning, W.J.; Dillmann, W.H.; Lorell, B.H. Transgenic expression of sarcoplasmic reticulum Ca^{2+} atpase modifies the transition from hypertrophy to early heart failure. *Circ. Res.* **2001**, *89*, 422–429. [CrossRef] [PubMed]
20. He, H.P.; Giordano, F.J.; HilalDandan, R.; Choi, D.J.; Rockman, H.A.; McDonough, P.M.; Bluhm, W.F.; Meyer, M.; Sayen, M.R.; Swanson, E.; et al. Overexpression of the rat sarcoplasmic reticulum Ca^{2+} atpase gene in the heart of transgenic mice accelerates calcium transients and cardiac relaxation. *J. Clin. Investig.* **1997**, *100*, 380–389. [CrossRef] [PubMed]
21. Rincon, M.Y.; VandenDriessche, T.; Chuah, M.K. Gene therapy for cardiovascular disease: Advances in vector development, targeting, and delivery for clinical translation. *Cardiovasc. Res.* **2015**, *108*, 4–20. [CrossRef] [PubMed]
22. Williams, P.D.; Ranjzad, P.; Kakar, S.J.; Kingston, P.A. Development of viral vectors for use in cardiovascular gene therapy. *Viruses* **2010**, *2*, 334–371. [CrossRef] [PubMed]
23. Del Monte, F.; Lebeche, D.; Guerrero, J.L.; Tsuji, T.; Doye, A.A.; Gwathmey, J.K.; Hajjar, R.J. Abrogation of ventricular arrhythmias in a model of ischemia and reperfusion by targeting myocardial calcium cycling. *Proc. Natl. Acad. Sci. USA* **2004**, *101*, 5622–5627. [CrossRef] [PubMed]
24. Del Monte, F.; Williams, E.; Lebeche, D.; Schmidt, U.; Rosenzweig, A.; Gwathmey, J.K.; Lewandowski, E.D.; Hajjar, R.J. Improvement in survival and cardiac metabolism after gene transfer of sarcoplasmic reticulum Ca^{2+}-atpase in a rat model of heart failure. *Circulation* **2001**, *104*, 1424–1429. [CrossRef] [PubMed]
25. Chaudhri, B.; del Monte, F.; Hajjar, R.J.; Harding, S.E. Contractile effects of adenovirally-mediated increases in SERCA2a activity: A comparison between adult rat and rabbit ventricular myocytes. *Mol. Cell. Biochem.* **2003**, *251*, 103–109. [CrossRef] [PubMed]
26. Nassal, M.M.; Wan, X.; Laurita, K.R.; Cutler, M.J. Atrial SERCA2a overexpression has no affect on cardiac alternans but promotes arrhythmogenic SR Ca^{2+} triggers. *PLoS ONE* **2015**, *10*, e0137359. [CrossRef] [PubMed]
27. Miyamoto, M.I.; del Monte, F.; Schmidt, U.; DiSalvo, T.S.; Kang, Z.B.; Matsui, T.; Guerrero, J.L.; Gwathmey, J.K.; Rosenzweig, A.; Hajjar, R.J. Adenoviral gene transfer of SERCA2a improves left-ventricular function in aortic-banded rats in transition to heart failure. *Proc. Natl. Acad. Sci. USA* **2000**, *97*, 793–798. [CrossRef] [PubMed]
28. Lipskaia, L.; Hadri, L.; Le Prince, P.; Esposito, B.; Atassi, F.; Liang, L.; Glorian, M.; Limon, I.; Lompre, A.M.; Lehoux, S.; et al. SERCA2a gene transfer prevents intimal proliferation in an organ culture of human internal mammary artery. *Gene Ther.* **2013**, *20*, 396–406. [CrossRef] [PubMed]
29. Sakata, S.; Lebeche, D.; Sakata, Y.; Sakata, N.; Chemaly, E.R.; Liang, L.; Nakajima-Takenaka, C.; Tsuji, T.; Konishi, N.; del Monte, F.; et al. Transcoronary gene transfer of SERCA2a increases coronary blood flow and decreases cardiomyocyte size in a type 2 diabetic rat model. *Am. J. Physiol. Heart Circ. Physiol.* **2007**, *292*, H1204–H1207. [CrossRef] [PubMed]
30. Hadri, L.; Bobe, R.; Kawase, Y.; Ladage, D.; Ishikawa, K.; Atassi, F.; Lebeche, D.; Kranias, E.G.; Leopold, J.A.; Lompre, A.M.; et al. SERCA2a gene transfer enhances enos expression and activity in endothelial cells. *Mol. Ther.* **2010**, *18*, 1284–1292. [CrossRef] [PubMed]
31. Xin, W.; Lu, X.; Li, X.; Niu, K.; Cai, J. Attenuation of endoplasmic reticulum stress-related myocardial apoptosis by serca2a gene delivery in ischemic heart disease. *Mol. Med.* **2011**, *17*, 201–210. [CrossRef] [PubMed]
32. Kuken, B.N.; Aikemu, A.N.; Xiang, S.Y.; Wulasihan, M.H. Effect of SERCA2a overexpression in the pericardium mediated by the AAV1 gene transfer on rapid atrial pacing in rabbits. *Genet. Mol. Res.* **2015**, *14*, 13625–13632. [CrossRef] [PubMed]

33. Lipskaia, L.; Hadri, L.; Lopez, J.J.; Hajjar, R.J.; Bobe, R. Benefit of SERCA2a gene transfer to vascular endothelial and smooth muscle cells: A new aspect in therapy of cardiovascular diseases. *Curr. Vasc. Pharmacol.* **2013**, *11*, 465–479. [CrossRef] [PubMed]
34. Wobus, C.E.; Hugle-Dorr, B.; Girod, A.; Petersen, G.; Hallek, M.; Kleinschmidt, J.A. Monoclonal antibodies against the adeno-associated virus type 2 (AAV-2) capsid: Epitope mapping and identification of capsid domains involved in AAV-2-cell interaction and neutralization of AAV-2 infection. *J. Virol.* **2000**, *74*, 9281–9293. [CrossRef] [PubMed]
35. Scallan, C.D.; Jiang, H.; Liu, T.; Patarroyo-White, S.; Sommer, J.M.; Zhou, S.; Couto, L.B.; Pierce, G.F. Human immunoglobulin inhibits liver transduction by AAV vectors at low AAV2 neutralizing titers in scid mice. *Blood* **2006**, *107*, 1810–1817. [CrossRef] [PubMed]
36. Moskalenko, M.; Chen, L.; van Roey, M.; Donahue, B.A.; Snyder, R.O.; McArthur, J.G.; Patel, S.D. Epitope mapping of human anti-adeno-associated virus type 2 neutralizing antibodies: Implications for gene therapy and virus structure. *J. Virol.* **2000**, *74*, 1761–1766. [CrossRef] [PubMed]
37. Greenberg, B.; Butler, J.; Felker, G.M.; Ponikowski, P.; Voors, A.A.; Pogoda, J.M.; Provost, R.; Guerrero, J.; Hajjar, R.J.; Zsebo, K.M. Prevalence of AAV1 neutralizing antibodies and consequences for a clinical trial of gene transfer for advanced heart failure. *Gene Ther.* **2016**, *23*, 313–319. [CrossRef] [PubMed]
38. Hulot, J.S.; Salem, J.E.; Redheuil, A.; Collet, J.P.; Varnous, S.; Jourdain, P.; Logeart, D.; Gandjbakhch, E.; Bernard, C.; Hatem, S.N.; et al. Effect of intracoronary administration of AAV1/SERCA2a on ventricular remodelling in patients with advanced systolic heart failure: Results from the agent-hf randomized phase 2 trial. *Eur. J. Heart Fail.* **2017**, *19*, 1534–1541. [CrossRef] [PubMed]
39. Greenberg, B.; Yaroshinsky, A.; Zsebo, K.M.; Butler, J.; Felker, G.M.; Voors, A.A.; Rudy, J.J.; Wagner, K.; Hajjar, R.J. Design of a phase 2b trial of intracoronary administration of AAV1/SERCA2a in patients with advanced heart failure: The CUPID 2 trial (calcium up-regulation by percutaneous administration of gene therapy in cardiac disease phase 2b). *JACC Heart Fail.* **2014**, *2*, 84–92. [CrossRef] [PubMed]
40. Zsebo, K.; Yaroshinsky, A.; Rudy, J.J.; Wagner, K.; Greenberg, B.; Jessup, M.; Hajjar, R.J. Long-term effects of AAV1/SERCA2a gene transfer in patients with severe heart failure: Analysis of recurrent cardiovascular events and mortality. *Circ. Res.* **2014**, *114*, 101–108. [CrossRef] [PubMed]
41. Horowitz, J.D.; Rosenson, R.S.; McMurray, J.J.; Marx, N.; Remme, W.J. Clinical trials update aha congress 2010. *Cardiovasc. Drugs Ther.* **2011**, *25*, 69–76. [CrossRef] [PubMed]
42. Hajjar, R.J.; Zsebo, K.; Deckelbaum, L.; Thompson, C.; Rudy, J.; Yaroshinsky, A.; Ly, H.; Kawase, Y.; Wagner, K.; Borow, K.; et al. Design of a phase 1/2 trial of intracoronary administration of AAV1/SERCA2a in patients with heart failure. *J. Card. Fail.* **2008**, *14*, 355–367. [CrossRef] [PubMed]
43. Hulot, J.S.; Ishikawa, K.; Hajjar, R.J. Gene therapy for the treatment of heart failure: Promise postponed. *Eur. Heart J.* **2016**, *37*, 1651–1658. [CrossRef] [PubMed]
44. Yla-Herttuala, S. Gene therapy for heart failure: Back to the bench. *J. Am. Soc. Gene Ther.* **2015**, *23*, 1551–1552. [CrossRef] [PubMed]
45. Mingozzi, F.; Anguela, X.M.; Pavani, G.; Chen, Y.; Davidson, R.J.; Hui, D.J.; Yazicioglu, M.; Elkouby, L.; Hinderer, C.J.; Faella, A.; et al. Overcoming preexisting humoral immunity to AAV using capsid decoys. *Sci. Transl. Med.* **2013**, *5*, 194ra192. [CrossRef] [PubMed]
46. Pacak, C.A.; Mah, C.S.; Thattaliyath, B.D.; Conlon, T.J.; Lewis, M.A.; Cloutier, D.E.; Zolotukhin, I.; Tarantal, A.F.; Byrne, B.J. Recombinant adeno-associated virus serotype 9 leads to preferential cardiac transduction in vivo. *Circ. Res.* **2006**, *99*, e3–e9. [CrossRef] [PubMed]
47. Vandendriessche, T.; Thorrez, L.; Acosta-Sanchez, A.; Petrus, I.; Wang, L.; Ma, L.; Waele, D.E.; Iwasaki, Y.; Gillijns, V.; Wilson, J.M.; et al. Efficacy and safety of adeno-associated viral vectors based on serotype 8 and 9 vs. Lentiviral vectors for hemophilia B gene therapy. *J. Thromb. Haemost.* **2007**, *5*, 16–24. [CrossRef] [PubMed]
48. Inagaki, K.; Fuess, S.; Storm, T.A.; Gibson, G.A.; McTiernan, C.F.; Kay, M.A.; Nakai, H. Robust systemic transduction with AAV9 vectors in mice: Efficient global cardiac gene transfer superior to that of AAV8. *Mol. Ther.* **2006**, *14*, 45–53. [CrossRef] [PubMed]
49. Zincarelli, C.; Soltys, S.; Rengo, G.; Koch, W.J.; Rabinowitz, J.E. Comparative cardiac gene delivery of adeno-associated virus serotypes 1–9 reveals that AAV6 mediates the most efficient transduction in mouse heart. *Clin. Transl. Sci.* **2010**, *3*, 81–89. [CrossRef] [PubMed]

50. Lyon, A.R.; Bannister, M.L.; Collins, T.; Pearce, E.; Sepehripour, A.H.; Dubb, S.S.; Garcia, E.; O'Gara, P.; Liang, L.; Kohlbrenner, E.; et al. SERCA2a gene transfer decreases sarcoplasmic reticulum calcium leak and reduces ventricular arrhythmias in a model of chronic heart failure. *Circ. Arrhythm. Electrophysiol.* **2011**, *4*, 362–372. [CrossRef] [PubMed]
51. Cutler, M.J.; Wan, X.; Plummer, B.N.; Liu, H.; Deschenes, I.; Laurita, K.R.; Hajjar, R.J.; Rosenbaum, D.S. Targeted sarcoplasmic reticulum Ca^{2+} ATPase 2a gene delivery to restore electrical stability in the failing heart. *Circulation* **2012**, *126*, 2095–2104. [CrossRef] [PubMed]
52. Shin, J.H.; Bostick, B.; Yue, Y.; Hajjar, R.; Duan, D. SERCA2a gene transfer improves electrocardiographic performance in aged mdx mice. *J. Transl. Med.* **2011**, *9*, 132. [CrossRef] [PubMed]
53. Wang, H.L.; Zhou, X.H.; Li, Z.Q.; Fan, P.; Zhou, Q.N.; Li, Y.D.; Hou, Y.M.; Tang, B.P. Prevention of atrial fibrillation by using sarcoplasmic reticulum calcium atpase pump overexpression in a rabbit model of rapid atrial pacing. *Med. Sci. Monit.* **2017**, *23*, 3952–3960. [CrossRef] [PubMed]
54. Pleger, S.T.; Shan, C.; Ksienzyk, J.; Bekeredjian, R.; Boekstegers, P.; Hinkel, R.; Schinkel, S.; Leuchs, B.; Ludwig, J.; Qiu, G.; et al. Cardiac AAV9-S100A1 gene therapy rescues post-ischemic heart failure in a preclinical large animal model. *Sci. Transl. Med.* **2011**, *3*, 92ra64. [CrossRef] [PubMed]
55. Katz, M.G.; Fargnoli, A.S.; Williams, R.D.; Steuerwald, N.M.; Isidro, A.; Ivanina, A.V.; Sokolova, I.M.; Bridges, C.R. Safety and efficacy of high-dose adeno-associated virus 9 encoding sarcoplasmic reticulum Ca^{2+} adenosine triphosphatase delivered by molecular cardiac surgery with recirculating delivery in ovine ischemic cardiomyopathy. *J. Thorac. Cardiovasc. Surg.* **2014**, *148*, 1065–1072, 1073e1–1073e2; discussion 1072–1073. [CrossRef] [PubMed]
56. Niwano, K.; Arai, M.; Koitabashi, N.; Watanabe, A.; Ikeda, Y.; Miyoshi, H.; Kurabayashi, M. Lentiviral vector-mediated SERCA2 gene transfer protects against heart failure and left ventricular remodeling after myocardial infarction in rats. *J. Am. Soc. Gene Ther.* **2008**, *16*, 1026–1032. [CrossRef] [PubMed]
57. Fleury, S.; Simeoni, E.; Zuppinger, C.; Deglon, N.; Von Segesser, L.K.; Kappenberger, L.; Vassalli, G. Multiply attenuated, self-inactivating lentiviral vectors efficiently deliver and express genes for extended periods of time in adult rat cardiomyocytes in vivo. *Circulation* **2003**, *107*, 2375–2382. [CrossRef] [PubMed]
58. Miyoshi, H.; Takahashi, M.; Gage, F.H.; Verma, I.M. Stable and efficient gene transfer into the retina using an hiv-based lentiviral vector. *Proc. Natl. Acad. Sci. USA* **1997**, *94*, 10319–10323. [CrossRef] [PubMed]
59. Naldini, L.; Blomer, U.; Gallay, P.; Ory, D.; Mulligan, R.; Gage, F.H.; Verma, I.M.; Trono, D. In vivo gene delivery and stable transduction of nondividing cells by a lentiviral vector. *Science* **1996**, *272*, 263–267. [CrossRef] [PubMed]
60. Merentie, M.; Lottonen-Raikaslehto, L.; Parviainen, V.; Huusko, J.; Pikkarainen, S.; Mendel, M.; Laham-Karam, N.; Karja, V.; Rissanen, R.; Hedman, M.; et al. Efficacy and safety of myocardial gene transfer of adenovirus, adeno-associated virus and lentivirus vectors in the mouse heart. *Gene Ther.* **2016**, *23*, 296–305. [CrossRef] [PubMed]
61. Renaud-Gabardos, E.; Tatin, F.; Hantelys, F.; Lebas, B.; Calise, D.; Kunduzova, O.; Masri, B.; Pujol, F.; Sicard, P.; Valet, P.; et al. Therapeutic benefit and gene network regulation by combined gene transfer of apelin, FGF2, and SERCA2a into ischemic heart. *Mol. Ther.* **2018**, *26*, 902–916. [CrossRef] [PubMed]
62. Merentie, M.; Rissanen, R.; Lottonen-Raikaslehto, L.; Huusko, J.; Gurzeler, E.; Turunen, M.P.; Holappa, L.; Makinen, P.; Yla-Herttuala, S. Doxycycline modulates VEGF-A expression: Failure of doxycycline-inducible lentivirus shrna vector to knockdown VEGF-A expression in transgenic mice. *PLoS ONE* **2018**, *13*, e0190981. [CrossRef] [PubMed]
63. Tilemann, L.; Lee, A.; Ishikawa, K.; Aguero, J.; Rapti, K.; Santos-Gallego, C.; Kohlbrenner, E.; Fish, K.M.; Kho, C.; Hajjar, R.J. SUMO-1 gene transfer improves cardiac function in a large-animal model of heart failure. *Sci. Transl. Med.* **2013**, *5*, 211ra159. [CrossRef] [PubMed]
64. Kho, C.; Lee, A.; Jeong, D.; Oh, J.G.; Chaanine, A.H.; Kizana, E.; Park, W.J.; Hajjar, R.J. SUMO1-dependent modulation of SERCA2a in heart failure. *Nature* **2011**, *477*, 601–605. [CrossRef] [PubMed]
65. Ferrandi, M.; Barassi, P.; Tadini-Buoninsegni, F.; Bartolommei, G.; Molinari, I.; Tripodi, M.G.; Reina, C.; Moncelli, M.R.; Bianchi, G.; Ferrari, P. Istaroxime stimulates SERCA2a and accelerates calcium cycling in heart failure by relieving phospholamban inhibition. *Br. J. Pharmacol.* **2013**, *169*, 1849–1861. [CrossRef] [PubMed]

66. Reynolds, J.O.; Quick, A.P.; Wang, Q.; Beavers, D.L.; Philippen, L.E.; Showell, J.; Barreto-Torres, G.; Thuerauf, D.J.; Doroudgar, S.; Glembotski, C.C.; et al. Junctophilin-2 gene therapy rescues heart failure by normalizing RyR2-mediated Ca^{2+} release. *Int. J. Cardiol.* **2016**, *225*, 371–380. [CrossRef] [PubMed]
67. Beavers, D.L.; Landstrom, A.P.; Chiang, D.Y.; Wehrens, X.H. Emerging roles of Junctophilin-2 in the heart and implications for cardiac diseases. *Cardiovasc. Res.* **2014**, *103*, 198–205. [CrossRef] [PubMed]
68. Guo, A.; Zhang, X.; Iyer, V.R.; Chen, B.; Zhang, C.; Kutschke, W.J.; Weiss, R.M.; Franzini-Armstrong, C.; Song, L.S. Overexpression of Junctophilin-2 does not enhance baseline function but attenuates heart failure development after cardiac stress. *Proc. Natl. Acad. Sci. USA* **2014**, *111*, 12240–12245. [CrossRef] [PubMed]
69. Zhang, C.; Chen, B.; Guo, A.; Zhu, Y.; Miller, J.D.; Gao, S.; Yuan, C.; Kutschke, W.; Zimmerman, K.; Weiss, R.M.; et al. Microtubule-mediated defects in Junctophilin-2 trafficking contribute to myocyte transverse-tubule remodeling and Ca^{2+} handling dysfunction in heart failure. *Circulation* **2014**, *129*, 1742–1750. [CrossRef] [PubMed]

Permissions

All chapters in this book were first published in CSHHD, by MDPI; hereby published with permission under the Creative Commons Attribution License or equivalent. Every chapter published in this book has been scrutinized by our experts. Their significance has been extensively debated. The topics covered herein carry significant findings which will fuel the growth of the discipline. They may even be implemented as practical applications or may be referred to as a beginning point for another development.

The contributors of this book come from diverse backgrounds, making this book a truly international effort. This book will bring forth new frontiers with its revolutionizing research information and detailed analysis of the nascent developments around the world.

We would like to thank all the contributing authors for lending their expertise to make the book truly unique.

They have played a crucial role in the development of this book. Without their invaluable contributions this book wouldn't have been possible. They have made vital efforts to compile up to date information on the varied aspects of this subject to make this book a valuable addition to the collection of many professionals and students.

This book was conceptualized with the vision of imparting up-to-date information and advanced data in this field. To ensure the same, a matchless editorial board was set up. Every individual on the board went through rigorous rounds of assessment to prove their worth. After which they invested a large part of their time researching and compiling the most relevant data for our readers.

The editorial board has been involved in producing this book since its inception. They have spent rigorous hours researching and exploring the diverse topics which have resulted in the successful publishing of this book. They have passed on their knowledge of decades through this book. To expedite this challenging task, the publisher supported the team at every step. A small team of assistant editors was also appointed to further simplify the editing procedure and attain best results for the readers.

Apart from the editorial board, the designing team has also invested a significant amount of their time in understanding the subject and creating the most relevant covers. They scrutinized every image to scout for the most suitable representation of the subject and create an appropriate cover for the book.

The publishing team has been an ardent support to the editorial, designing and production team. Their endless efforts to recruit the best for this project, has resulted in the accomplishment of this book. They are a veteran in the field of academics and their pool of knowledge is as vast as their experience in printing. Their expertise and guidance has proved useful at every step. Their uncompromising quality standards have made this book an exceptional effort. Their encouragement from time to time has been an inspiration for everyone.

The publisher and the editorial board hope that this book will prove to be a valuable piece of knowledge for researchers, students, practitioners and scholars across the globe.

List of Contributors

Francesco Moccia
Laboratory of General Physiology, Department of Biology and Biotechnology "L. Spallanzani", University of Pavia

John A. D'Elia and Larry A. Weinrauch
E P Joslin Research Laboratory, Kidney and Hypertension Section, Joslin Diabetes Center, Department of Medicine, Mount Auburn Hospital, Harvard Medical School, Boston and Cambridge, 521 Mount Auburn Street Watertown, MA 02472, USA

Germano Guerra and Angelica Perna
Department of Medicine and Health Sciences "Vincenzo Tiberio", University of Molise, via F. De Santis, 86100 Campobasso, Italy

Angela Lucariello and Antonio De Luca
Department of Mental Health and Preventive Medicine, Section of Human Anatomy, University of Campania "L. Vanvitelli", 81100 Naples, Italy

Laura Botta and Francesco Moccia
Laboratory of General Physiology, Department of Biology and Biotechnology "L. Spallanzani", University of Pavia, via Forlanini 6, 27100 Pavia ltaly

Muhammad Hayder Bin Khalid, Jing Tao Qu, Feng Ling Fu, andWan Chen Li, Fu Ai Sun and Hao Qiang Yu
Maize Research Institute, Sichuan Agricultural University, Chengdu 611130, China

Muhammad Ali Raza and Ling Yang Feng
College of Agronomy, Sichuan Agricultural University, Chengdu 611130, China

Imran Khan
Department of Grassland Science, Sichuan Agricultural University, Chengdu 611130, China

Sudip Paudel, Regan Sindelar and Margaret Saha
College of William and Mary, Williamsburg, VA 23187, USA

Luigi Catacuzzeno and Fabio Franciolini
Department of Chemistry, Biology and Biotechnology, University of Perugia, 06134 Perugia, Italy

Andrea Gerbino and Matilde Colella
Department of Biosciences, Biotechnology and Biopharmaceutics, University of Bari, 70121 Bari, Italy

Claudia M. Tellez Freitas, Deborah K. Johnson and K. Scott Weber
Department of Microbiology and Molecular Biology, Brigham Young University, Provo, UT 84604, USA

Ivan V. Maly and Wilma A. Hofmann
Department of Physiology and Biophysics, Jacobs School of Medicine and Biomedical Sciences, University at Buffalo, 955 Main Street, Buffalo, NY 14203, USA

T. Jake Samuel Ryan P. Rosenberry and Seungyong Lee
1 Department of Kinesiology, College of Nursing and Health Innovation, The University of Texas at Arlington, Arlington, TX 76019, USA

Zui Pan
2 Department of Graduate Nursing, College of Nursing and Health Innovation, The University of Texas at Arlington, Arlington, TX 76019, USA

Index

L

M

N

O

P

S

T

V

W

X

Z

www.ingramcontent.com/pod-product-compliance
Lightning Source LLC
LaVergne TN
LVHW080612210926
848485LV00003B/144
* 9 7 8 1 6 3 9 8 7 0 9 1 2 *